Y0-BCV-203

Introduction to
POLYMER CHEMISTRY

Second Edition

Introduction to
POLYMER CHEMISTRY

Second Edition

Charles E. Carraher, Jr.

CRC Press
Taylor & Francis Group
Boca Raton London New York

CRC Press is an imprint of the
Taylor & Francis Group, an **informa** business

CRC Press
Taylor & Francis Group
6000 Broken Sound Parkway NW, Suite 300
Boca Raton, FL 33487-2742

© 2010 by Taylor and Francis Group, LLC
CRC Press is an imprint of Taylor & Francis Group, an Informa business

No claim to original U.S. Government works

Printed in the United States of America on acid-free paper
10 9 8 7 6 5 4 3 2

International Standard Book Number: 978-1-4398-0953-2 (Hardback)

Visit the Taylor & Francis Web site at
http://www.taylorandfrancis.com

and the CRC Press Web site at
http://www.crcpress.com

Contents

Preface

There is an appropriate and necessary move toward green materials and green chemistry. This trend is captured in this book as it is both integrated within the appropriate sections and treated separately in a section on green materials. There also exists a greater awareness of health concerns within our society, and this awareness is also mirrored in this book.

Polymers are all about us and are the basis of life itself; they are used for communication (both natural and synthetic), for our nutrition and clothing, for recording history, and for constructing buildings and highways. In fact it is difficult to imagine society without synthetic polymers, and life without natural polymers. Part of being an educated and responsible citizen involves knowing the correct questions to ask and knowing (possible) the correct answers. In our ever-increasingly technological world, science plays a crucial role in providing solutions to critical problems of food; clean and abundant water, energy, and air; and health. This book provides both the information and the insights that allow a better understanding of these large molecules that are all about us.

Most books on polymers are aimed at either graduate students or are hybrids aimed at both undergraduate, but mostly graduate students. This book is aimed mainly at undergraduate students but also serve graduate students. Thus, a strong bond is forged between science, history, and the crucible that is today's society. Information gained from the basic core science courses are brought to bear on understanding giant molecules. This information includes factual, theoretical, and practical concepts presented in science. It is of use to those who want to simply be well educated, as well as to those who want to pursue a career in medicine, engineering, physics, chemistry, biomedical sciences, law, and business.

This book provides a detailed coverage of polymers including natural and synthetic giant molecules; inorganic and organic polymers; and elastomers, adhesives, coatings, fibers, plastics, blends, caulks, composites, and ceramics. The basic principles that apply to one polymer group apply to all of the other groups when used with some simple fundamentals. These fundamentals are integrated into the fabric of this book.

We have kept in mind the saying that we should be students of history so we do not repeat the same mistakes; at the same time we should also be students of history so that we might repeat the successes. Brief case studies are woven within the fabric of this book as historical accounts illustrating the purposes in back of change as well as the societal and scientific contexts within which these changes occurred.

Acknowledgments

I would like to gratefully acknowledge the contributions and assistance of the following people in the preparation of this book: John Droske, Charles Pittman, Edward Kresge, Gerry Kirshenbaum, Sukumar Maiti, Alan MacDiarmid, Les Sperling, Otto Vogel, Thomas Miranda, Murry Morello, and Graham Allan; and a number of our children who gave valuable suggestions—Charles Carraher III, Shawn Carraher, Colleen Carraher-Schwarz, and Cara Carraher. Special thanks to Gerry Kirshenbaum for his kind permission to use portions of articles that I had written that appeared in *Polymer News*. This book could not have been written without the guidance of those who have gone before us, especially Raymond Seymour, Herman Mark, Charles Gebelein, Paul Flory, and Linus Pauling; all of these friends shepherded and helped me. My thanks to them.

I also thank Girish Barot, Amitabh Battin, James D. Johnson, Gauri Nayak, and Theodore Sabir for their assistance in proofing.

Author

Charles E. Carraher, Jr. is a professor of chemistry and biochemistry, Florida Atlantic University, Boca Raton, FL and an associate director of the Florida Center for Environmental Studies, Palm Beach Gardens, FL. He has been recognized as an outstanding chemist in the southeast United States (1992) by the American Chemical Society (ACS) and was the recipient of a distinguished service award for his efforts in science education (1995) from the ACS's Divisions of Polymer Chemistry and Polymeric Materials: Science and Engineering. He is also a fellow of the American Institute of Chemists and the American Chemical Society-PMSE. Currently, he serves as a cochair of the ACS's joint Polymer Education Committee and on the board of the Intersocietal Polymer Education Committee; he has also been a member of the ACS's Committee on Professional Training. He is an associate editor of the *Journal of Polymeric Materials* and serves on the board of the *Journal of Inorganic and Organometallic Polymers and Materials*. He has authored/coauthored nearly 60 books and over 1000 articles and has chaired/cochaired numerous national and international symposia. His research has led to the synthesis of over 75 new families of polymers. In 1984, he received the Outstanding Scientist and Engineering Award from the Engineers and Scientists Affiliate Societies Council for his work in science education and research, and in 1992 he received the Saltarilli Sigma Xi Award for his research efforts. Dr. Carraher was the recipient of the 2002 Distinguished Researcher Award from Allied Technologies.

How to Study Polymers

Studying polymers is similar to studying any other science. The following text provides some ideas that may prove to be helpful in your study.

Much of science is abstract. While much of the study of polymers is abstract, it is easier to conceptualize and create mental pictures of what a polymer is and how it should behave than it is in many other areas of science. In the case of linear polymers, think of a string or a rope. Long ropes often get entangled with themselves or with other ropes. In the same way, polymer chains get entangled with themselves and with chains of other polymers that are brought into contact with them. **Thus, create mental pictures of the polymer molecules as you study them.**

Polymers are real and they are all about us. We can look at giant molecules at a microscopic or an atomic level, or at a macroscopic level. The PET bottles we have may be composed of long chains of poly(ethylene terephthate) (PET) chains; the aramid tire cord is composed of aromatic polyamide chains; even our hair is made up of complex bundles of fibrous proteins, again polyamides. **Polymers are related to the real world in which we live. We experience these "large molecules" at the macroscopic level every day of our lives and this macroscopic behavior is a direct consequence of the atomic-level structure and behavior.** Make pictures in your mind that allow you to relate the atomic world to the macroscopic world.

At the introductory level, we often examine only the primary factors that may cause particular giant molecule behaviors. Other factors may assume importance only under particular conditions. **The polymer molecules you study at times examine only the primary factors that impact polymer behavior and structure. Even so, these primary factors form the basis for both complex and simple structure–property behavior.**

The structure–property relationships you will be studying are based on well-known basic chemical and physical relationships. **Such relationships build upon one another and need to be studied in an ongoing manner. Understand as you go along. Read the material BEFORE you go to class.**

This course is an introductory-level course. Each chapter or topic emphasizes knowledge about one or more areas. **The science and excitement of polymers has its own language. It is a language that requires you to understand and memorize certain key concepts.** Our memory can be short term or long term. Short-term memory may be considered as that used by an actor or actress for a TV drama. It really does not need to be totally understood, nor retained after the final "take." **Long-term memory is required when studying giant molecules since it will be used repeatedly to understand other concepts (i.e., it is built upon).**

When memorizing, learn how you do this best, i.e., time of day, setting, etc. Use as many senses as necessary—**be active**, read your assignment, write out what is needed to be known, say it, and listen to yourself say it. Also, look for patterns, create mnemonic devices, avoid cramming too much into too small a time, practice associations in all directions, and test yourself. **Memorization is hard work.**

While knowledge involves recalling memorized material, to really "know" something involves more than simple recall—it involves **comprehension, application, evaluation**, and **integration** of the **knowledge.** Comprehension is the interpretation of this knowledge—making predictions, applying it to different situations. Analysis involves the evaluation of the information and comparing it with other information, and synthesis (integration) has to do with integration of the information with other information.

When studying giant molecules please consider doing the following:

- Skim the text BEFORE the lecture
- Attend the lecture and take notes
- Organize your notes and relate information
- Read and study the assigned material
- Study your notes and the assigned material
- Review and self-test

Learning takes time and effort. Study by daily skimming the text and other material, think about it, visualize key points and concepts, write down important material, make outlines, take notes, and study sample problems. All of these help, but some may help you more than others; hence, focus on these modes of learning, but not at the exclusion of the other aspects.

When preparing for an exam, consider the following:

- Accomplish the above—DO NOT wait until the day before the exam to begin studying; create good study habits.
- Study wisely—study how YOU study best, i.e., time of day, surroundings, etc.
- Take care of yourself—get plenty of sleep the night before the exam.
- Attend to last-minute details—Is your calculator working, is it the right kind, do I have the needed pencils? Review the material once again.
- Know what kind of test it will be, if possible.
- Get copies of old exams, if possible; talk to others who might have already undergone the course.

During the test

- Stay cool, do NOT PANIC
- Read the directions, try to understand what is being asked for
- In an essay or similar exam, work for at least partial credit; plan your answers
- In a multiple choice or T/F exam, eliminate obviously wrong choices
- Look over the entire exam; work questions that you are sure of then go to less sure questions; check answers if time permits

The study of polymer molecules contains several types of content:

- **Facts**—the term *polymer* means "many" (poly) "units" (mers)
- **Concepts**—linear polymers are long molecules such as strings
- **Rules**—solutions that contain polymer chains are more viscous and slower flowing than solutions that do not contain polymers
- **Problems**—what is the approximate molecular weight of a single polystyrene chain that has 1000 styrene units in it?

These varied types of contents are often integrated within any topic. In this introduction to polymer molecules, the emphasis is often on concepts but all the aspects are equally important.

Polymer Nomenclature

As with most areas, the language of the area is important. Here we will focus on naming polymers with an emphasis on synthetic polymers. Short presentations on how to name proteins and nucleic acids are provided in Chapter 4 and for nylons in Chapter 5.

As synthetic polymer science developed in many venues before standard nomenclature groups were present to assist in the standardization of the naming approach, many popular polymers took on several names including common names. Many polymer scientists have not yet accepted the guidelines laid out by the official naming committee of the International Union of Pure and Applied Chemistry (IUPAC), because the common names have gained such widespread acceptance. Although there is a wide diversity in the practice of naming polymers, we will concentrate on the most used ones.

COMMON NAMES

Little rhyme or reason is associated with many of the common names of polymers. Some names are derived from the place of origin of the material, such as *Hevea brasilliensis*—literally "rubber from Brazil"—for natural rubber. Other polymers were named after their discoverer, e.g., Bakelite, the three-dimensional polymer produced by condensation of phenol and formaldehyde, which was commercialized by Leo Baekeland in 1905.

For some important groups of polymers, special names and systems of nomenclature were developed. For instance, nylons were named according to the number of carbons in the diamine and dicarboxylic acid reactants used in their syntheses. The nylon produced by the condensation of 1,6-hexamethylenediamine (six carbons) and adipic acid (six carbons) is called nylon-66. Even here, there is not a set standard as to how nylon-66 is to be written with alternatives including nylon-66 and nylon-6,6.

Nylon-66

SOURCE-BASED NAMES

Most common names are source based, i.e., they are based on the common name of the reactant monomer, preceded by the prefix "poly." For example, polystyrene is the most frequently used name for the polymer derived from the monomer 1-phenylethene, which has the common name styrene.

Styrene   Polystyrene

The vast majority of polymers based on the vinyl group ($H_2C = CHX$) or the vinylidene group ($H_2C = CX_2$) as the repeat unit are known by their source-based names. Thus, polyethylene is the name of the polymer synthesized from the monomer ethylene, poly(vinyl chloride) from the monomer vinyl chloride, and poly(methyl methacrylate) from the monomer methyl methacrylate.

Many condensation polymers are also named in this manner. In the case of poly(ethylene terephthalate), the glycol portion of the name of the monomer, ethylene glycol, is used in constructing the polymer name, so that the name is actually a hybrid of a source- and a structure-based name.

Poly(ethylene terephthalate)

This polymer is well known by a number of trade names, such as Dacron; its common grouping, polyester; and by an abbreviation, PET.

Although it is often suggested that parentheses be used in naming polymers of more than one word (e.g., poly(vinyl chloride)), but not for single-word polymers (e.g., polyethylene), some authors entirely omit the use of parentheses for either case (e.g., polyvinyl chloride); thus even here there exists a variety of practices. We will employ the use of parentheses for naming polymers of more than one word.

Copolymers are composed of two or more monomers. Source-based names are conveniently used to describe copolymers using an appropriate term between the names of the monomers. Any of a half dozen or so connecting terms may be used, depending on what is known about the structure of the copolymer. When no information is known or intended to be conveyed, the connective term "co" is employed in the general format poly(A-*co*-B), where A and B are the names of the two monomers. An unspecified copolymer of styrene and methyl methacrylate would be called poly[styrene-*co*-(methyl methacrylate)].

Kraton, the yellow rubberlike material often found on the bottom of running shoes, is a copolymer where structural information is known. It is formed from a group of styrene units, i.e., a "block" of polystyrene, attached to a group of butadiene units, or a block of polybutadiene, which is attached to another block of polystyrene, forming a triblock copolymer. The general representation of such a block might be –AAAAAAAABBBBBBBAAAAAAAA–, where each A and B represents an individual monomer unit. The proper source-based name for Kraton is polystyrene-*block*-polybutadiene-*block*-polystyrene, or poly-*block*-styrene-*block*-polybutadiene-*block*-polystyrene with the prefix "poly" being retained for each block. Again, some authors will omit the use of "poly," giving polystyrene-*block*-butadiene-*block*-styrene.

STRUCTURE-BASED NAMES

Although source-based names are generally employed for simple polymers, IUPAC has published a number of reports for naming polymers. These reports are widely accepted for the naming of complex polymers. A listing of such names is given in the references section. A listing of source- and structure-based names for some common polymers is given in Table 1.

LINKAGE-BASED NAMES

Many polymer "families" are referred to by the name of the particular linkage that connects the polymers (Table 2). The family name is "poly" followed by the linkage name. Thus, those polymers that contain an ester linkage are known as polyesters; those with an ether linkage are called polyethers.

TABLE 1
Source-Based and Structure-Based Names

Source-Based Names	Structure-Based Names
Polyacrylonitrile	Poly(1-cyanoethylene)
Poly(ethylene oxide)	Polyoxyethylene
Poly(ethylene terephthalate)	Polyoxyethyleneoxyterephthaloyl
Polyisobutylene	Poly(1,1-dimethylethylene)
Poly(methyl methacrylate)	Poly[(1-methoxycarbonyl)-1-methylethylene]
Polypropylene	Poly(1-methylethylene)
Polystyrene	Poly(1-phenylethylene)
Polytetrafluoroethylene	Polydifluoromethylene
Poly(vinylacetate)	Poly(1-acetoxyethylene)
Poly(vinyl alcohol)	Poly(1-hydroxyethylene)
Poly(vinyl chloride)	Poly(1-chloroethylene)
Poly(vinyl butyral)	Poly[(2-propyl-1,3-dioxane-4,6-diyl)methylene]

TABLE 2
Linkage-Based Names

Family Name	Linkage	Family Name	Linkage
Polyamide	$-N-\overset{\overset{O}{\|\|}}{C}-$	Polyvinyl	$-C-C-$
Polyester	$-O-\overset{\overset{O}{\|\|}}{C}-$	Polyanhydride	$-\overset{\overset{O}{\|\|}}{C}-O-\overset{\overset{O}{\|\|}}{C}-$
Polyurethane	$-O-\overset{\overset{O}{\|\|}}{C}-\overset{\overset{H}{\|}}{N}-$	Polyurea	$-\overset{\overset{H}{\|}}{N}-\overset{\overset{O}{\|\|}}{C}-\overset{\overset{H}{\|}}{N}-$
Polyether	$-O-$	Polycarbonate	$-O-\overset{\overset{O}{\|\|}}{C}-O-$
Polysiloxane	$-O-Si-$	Polysulfide	$-S-$

TABLE 3
Abbreviations for Selected Polymeric Materials

Abbreviation	Polymer	Abbreviation	Polymer
ABS	Acrylonitrile–butadiene– styrene terpolymer	CA	Cellulose acetate
EP	Epoxy	HIPS	High-impact polystyrene
MF	Melamine–formaldehyde	PAA	Poly(acrylic acid)
PAN	Polyacrylonitrile		
PBT	Poly(butylene terephthalate)	PC	Polycarbonate
PE	Polyethylene	PET	Poly(ethylene terephthalate)
PF	Phenyl-formaldehyde	PMMA	Poly(methyl methacrylate)
PP	Polypropylene	PPO	Poly(phenylene oxide)
PS	Polystyrene	PTFE	Polytetrafluoroethylene
PU	Polyurethane	PVA, PVAc	Poly(vinyl acetate)
PVA, PVAl	Poly(vinyl alcohol)	PVB	Poly(vinyl butyral)
PVC	Poly(vinyl chloride)	SAN	Styrene–acrylonitrile
UF	Urea–formaldehyde	SBR	Butadiene–styrene copolymer

TRADE NAMES, BRAND NAMES, AND ABBREVIATIONS

Trade (and/or brand) names and abbreviations are often used to describe material. They may be used to identify the product of a manufacturer, processor, or fabricator, and may be associated with a particular product or with a material or modified material, or a material grouping. Trade names (or tradenames) are also used to describe specific groups of materials that are produced by a specific company or under license of that company. Bakelite is the tradename given for the phenol–formaldehyde condensation developed by Baekeland. A sweater whose contents are described as containing Orlon contains polyacrylonitrile fibers that are "protected" under the Orlon trademark, and produced or licensed to be produced by the holder of the Orlon trademark. Carina, Cobex, Dacovin, Darvic, Elvic, Geon, Koroseal, Marvinol, Mipolam, Opalon, Pliofex, Rucon, Solvic, Trulon, Velon, Vinoflex, Vygen, and Vyram are all tradenames for poly(vinyl chloride) manufactured by different companies. Some polymers are better known by their trade name than their generic name. For instance, polytetrafluoroethylene is better known as Teflon, the trade name held by Dupont.

Abbreviations, generally initials in capital letters, are also employed to describe polymers. Table 3 contains a listing of some of the more widely used abbreviations and the polymer associated with the abbreviation.

SUMMARY

While there are several important approaches to the naming of polymers, we will use common and source-based names because these are the names that are most commonly utilized by polymer scientists and the general public, and these names, in particular the source-based names, allow a good vehicle to convey structure–property relationships.

SELECTED READINGS

Bikales, N. M. (1987): Nomenclature, in *Encyclopedia of Polymer Science and Engineering*, 2nd ed., Vol. 10, Wiley, New York, p. 191.

Carraher, C. (2003): *Polymer Chemistry*, 6th ed., Dekker, New York.

IUPAC (1991): *Compendium of Macromolecular Nomenclature*, CRC Press, Boca Raton, FL.

Jenkins, A. D. and Loening, K. L. (1989): Nomenclature, in *Comprehensive Polymer Science*, Vol. 1, Pergamon Press, Oxford, U.K., pp. 13–54.

1 Introduction to Polymers

1.1 HISTORY OF POLYMERS

Since most chemists, biochemists, and chemical engineers are now involved in some phase of polymer science or technology, some have called this the polymer age. Actually, we have always lived in a polymer age. The ancient Greeks classified all matter as animal, vegetable, and mineral. Minerals were emphasized by the alchemists, but medieval artisans emphasized animal and vegetable matter. All are largely polymeric and are important to life as we know it.

The word **polymer** is derived from the Greek *poly* and *meros*, meaning many and parts, respectively. Some scientists prefer to use the word *macromolecule*, or large molecule, instead of polymer. Others maintain that naturally occurring polymers, or *biopolymers*, and synthetic polymer should be studied in different courses. Others name these large molecules simply "giant molecules." However, the same principles apply to all polymers. If one discounts the end uses, the differences between all polymers, including plastics, fibers, and elastomers or rubbers, are determined primarily by the intermolecular and intramolecular forces between the molecules and within the individual molecule, respectively, and by the functional groups present, and most of all, by their size allowing an accumulation of these forces.

In addition to being the basis of life itself, protein is used as a source of amino acids and energy. The ancients degraded or depolymerized the protein in meat by aging and cooking, and they denatured egg albumin by heating or adding vinegar to the eggs. Early humans learned how to process, dye, and weave the natural proteinaceous fibers of wool and silk and the carbohydrate fibers from flax and cotton. Early South American civilizations, such as the Aztecs, used natural rubber (*Hevea brasiliensis*) for making elastic articles and for waterproofing fabrics.

There has always been an abundance of natural fibers and elastomers but few plastics. Of course, early humans employed a crude plastic art in tanning the protein in animal skins to make leather and in heat-formed tortoise shells. They also used naturally occurring tars as caulking materials and extracted shellac from the excrement of small coccid insects (*Coccus lacca*).

Until Wohler synthesized urea from inorganic compounds in 1828, there had been little progress in organic chemistry since the alchemists emphasized the transmutation of base metals to gold and believed in a vital force theory. Despite this essential breakthrough, little progress was made in understanding organic compounds until the 1850s when Kekule developed the presently accepted technique for writing structural formulas. However, polymer scientists displayed a talent for making empirical discoveries before this science was developed.

Charles Goodyear grew up in poverty. He was a Connecticut Yankee born in 1800. He began work in his father's farm implement business. Later, he moved to Philadelphia where he opened a retail hardware store that soon went bankrupt. Charles then turned to being an inventor. As a child he had noticed the magic material that formed a rubber bottle he had found. He visited the Roxbury India Rubber Company to try and interest them in his efforts to improve the properties of rubber. They assured him that there was no need to do so.

He started his experiments with a malodorous gum from South America in debtor's prison. In a small cottage on the grounds of the prison, he blended the gum, the raw rubber called hevea rubber, with anything he could find—ink, soup, caster oil, and so on. While rubber-based products were available, they were either sticky or became sticky in the summer's heat. He found that the treatment of the raw rubber with nitric acid allowed the material to resist heat and not to adhere to itself. This success attracted backers who helped to form a rubber company. After some effort, he obtained a contract to supply the U.S. Post Office with 150 rubber mailbags. He made the bags and stored them in a hot room while he and his family were away. When they returned they found the bags in a corner of the room, joined together as a single mass. The nitric acid treatment was sufficient to prevent surface stickiness, but the internal rubber remained tacky and susceptible to heat.

While doing experiments in 1839 at a Massachusetts rubber factory, Charles accidentally dropped a lump of rubber mixed with sulfur on the hot stove. The rubber did not melt, but rather charred. He had discovered vulcanization, the secret that was to make rubber a commercial success. While he had discovered vulcanization, it would take several years of ongoing experimentation before the process was commercially useful. During this time he and his family were near penniless. While he patented the process, the process was too easily copied and pirated so that he was not able to fully profit from his invention and years of hard work. Even so, he was able to develop a number of items.

Charles Goodyear, and his brother Nelson, transformed natural rubber, hevea rubber, from a heat-"softenable" thermoplastic to a less heat-sensitive product through the creation of cross-links between the individual polyisoprene chain-like molecules using sulfur as the cross-linking agent. *Thermoplastics* are two-dimensional molecules that may be softened by heat. *Thermosets* are materials that are three-dimensional networks that cannot be reshaped by heating. Rather than melting, thermosets degrade. As the amount of sulfur was increased, the rubber became harder becoming a hard rubberlike (ebonite) material.

The spring of 1851 found the construction of a remarkable building on the lawns of London's Hyde Park. The building was designed by a maker of greenhouses, so it was not unexpected that it had a "greenhouse look." This Crystal Palace was to house almost 14,000 exhibitors from all over the world. It was the chance for exhibitors to show their wares. Charles Goodyear, then 50 years old, used this opportunity to show off his over two decades worth of rubber-related products. He decorated his Vulcanite Court with rubber walls, roof, furniture, buttons, toys, carpet, combs, etc. Above it hung a giant 6 ft rubber raft and assorted balloons. The European public was introduced to the world of new man-made materials.

Within a little more than a decade Charles Goodyear was dead. Within a year of his death, the American Civil War broke out. The Union military used about $27 million worth of rubber products by 1865 helping launch the American rubber industry.

In 1862, Queen Victoria, while in mourning for her recently departed husband Albert, opened the world's fair in London. One of the exhibitors was Alexander Parks. He was displeased with the limited colors available for rubber products—generally dull and dark. In his workshop in Birmingham, England he was working with nitrocellulose, a material made from the treatment of cotton and nitric and sulfuric acids. Nitrocellulose solutions were made from dissolving the nitrocellulose in organic liquids, such as ethanol and ether. Thin films and coatings were made by simply pouring the nitrocellulose solutions onto the desired item or surface and allowing the solvent to evaporate. He wanted to make solid objects from nitrocellulose. After years of work, he developed a material he called Parkensine from which he made buttons, combs, and in fact many of the items that were made of rubber—except his materials could be brightly colored, clear, or made to shine like mother-of-pearl. At the London world's fair, he advertised "PATENT PARKESINE of various colours: hard elastic, transparent, opaque, and waterproof." Even with his work he had not developed a material that could be "worked" or was stable and even with his hype, the material never caught on except within exhibition halls.

About this time, John Wesley Hyatt, a printer from Albany, New York, in pursuit of a $10,000 prize for anyone who could come up with a material that was a substitute for ivory billiard balls, developed a material that was stable and could be "worked" from shellac and wood pulp. He then turned to nitrocellulose, discovering that shredded nitrocellulose could be mixed with camphor, heated under pressure, to produce a tough white mass that retained its shape. This material, dubbed celluloid, could be made into the usual rubberlike products, but also solid pieces like boxes, wipe-clean linen, collars, cuffs, and ping-pong balls. Celluloid could also, like the shellac-wood pulp mixture, be worked—cut, drilled, and sawed. But celluloid was flammable, and did not stand up well in hot water. The wearers of celluloid dentures truly could have their "teeth curled" when drinking a hot cup of coffee. One of its best qualities was that it could be made to "look like" other materials—it could be dyed to look like marble, swirled to mimic tortoiseshell and mother-of-pearl, and even look and feel like ivory. It did not make good billiard balls. One account has it that billiard balls hitting and exploding like a shot caused cowboys to draw their guns.

Both cellulose and cellulose nitrate are linear, or two-dimensional, polymers, but the former cannot be softened because of the presence of multitudinous hydrogen bonds between the chain-like molecules. When used as an explosive the cellulose nitrate is essentially completely nitrated, but the material used by Parks and Hyatt was a dinitrate, still potentially explosive, but less so. Parks added caster oil and Hyatt added camphor to plasticize—reduce the effect of the hydrogen bonding—the cellulose nitrate allowing it some flexibility.

With the invention of the air-filled or pneumatic tires by a Scotsman, John Dunlop in 1888, rubber gained a worldwide importance. He had a successful veterinarian practice in Belfast. In his off time he worked to improve the ride of his son's tricycle. His invention happened at the right time. The automobile was emerging and the air-filled tires offered a gentler ride. Thus began the tire industry.

All of these inventions utilized natural materials as at least one ingredient. In 1907, after years of work in his chemistry labs in Yonkers NY, Leo Baekeland announced in an American Chemical Society meeting the synthesis of the first truly synthetic polymeric material latter dubbed Bakelite.

Baekeland was born in Belgium in 1863, the son of an illiterate cobbler and a maidservant. He was bright and received, with highest honors, his doctorate at the age of 20. He could have spent the remaining part of his life in academics in Europe, but heading the words of Benjamin Franklin, he sailed to America. In the 1890s, he developed the first photographic paper, called Velox that could be developed in synthetic light rather than sunlight. George Eastman saw the importance of this discovery and paid Baekeland $750,000 for the rights to use this invention.

It was generally recognized by the leading organic chemists of the nineteenth century that phenol would condense with formaldehyde. Since they did not recognize the concept of functionality, Baeyer, Michael, and Kleeberg produced useless cross-linked goos, gunks, and messes and then returned to their research on reactions of monofunctional reactants. However, by the use of a large excess of phenol, Smith, Luft, and Blumer were able to obtain a hard, but meltable thermoplastic material.

With his $750,000, Baekeland set up a lab next to his home. He then sought to solve the problem of making the hard material made from phenol and formaldehyde soluble. After many failures, he thought about circumventing the problem by placing the reactants in a mold of the desired shape and allowing them to form the intractable solid material. After much effort, he found the conditions under which a hard, clear solid could be made—Bakelite was discovered. Bakelite could be worked, it was resistant to acids and organic liquids, stood up well to heat and electrical charge, and could be dyed to give colorful products. It was used to make bowling balls, phonograph records, telephone housings, gears, and cookware. His materials also made excellent billiard balls. Bakelite also acted as a binder for sawdust, textiles, and paper forming a wide range of composites including Formica laminates, many of which are still used. It was also used as an adhesive giving us plywood.

While there is no evidence that Baekeland recognized what polymers were, he appeared to have a grasp on functionality and how to "use" it to produce thermoplastic materials that could latter be converted to thermosets. Through the control of the ratio of phenol to formaldehyde, he was able to form a material that was a thermoplastic. He coined the term *A-stage resole resin* to describe this thermoplastic. This A-stage resole resin was converted to a thermoset cross-link, *C-stage Bakelite*, by additional heating. Baekeland also prepared thermoplastic resins called *novolacs* by the conden-sation of phenol with a lesser amount of formaldehyde under acidic conditions. The thermoplastic novolacs were converted to thermosets by the addition of more formaldehyde. While other polymers had been synthesized in the laboratory, Bakelite was the first truly synthetic plastic. The "recipes" used today differ little from the ones developed by Baekeland, showing his ingenuity and knowledge of the chemistry of the condensation of the trifunctional phenol and difunctional formaldehyde.

While poly(vinyl chloride) was initially formed by Baumann in 1872, it awaited interest until 1926, when B. F. Goodrich discovered how to make sheets and adhesives from poly(vinyl chloride), and the "vinyl age" began. While polystyrene was probably first formed by Simon in 1839, it was almost 100 years later, that is, in 1930, that the German giant company I. G. Farben placed polystyrene on the market. Polystyrene molded parts became commonplace. Rohm and Haas bought out Plexiglass from a British firm in 1935 and began the production of clear plastic parts and goods including replacements for glass as camera lenses, aircraft windows, clock faces, and car taillights.

During to this time, polymer science was largely empirical, instinctive, and intuitive. Prior to World War I, celluloid, shellac, Galalith (casein), Bakelite, and cellulose acetate plastics; hevea rubber, cotton, wool, silk rayon fibers; Glyptal polyester coatings; bitumen or asphalt; and coumarone-indene and petroleum resins were all commercially available. However, as evidenced by the chronological data shown in Table 1.1, there was little additional development in polymers prior to World War II, because of a general lack of fundamental knowledge of polymers. But the theoretical basis was being built. Only a few of many giants will be mentioned.

Over a century ago, Graham coined the term *colloid* for aggregates with dimensions in the range of 10^{-9}–10^{-7} m. Unfortunately, the size of many macromolecules is in this range, but it is important to remember that unlike colloids, whose connective forces are ionic and/or secondary forces, polymers are individual molecules whose size cannot be reduced without breaking the covalent bonds,

TABLE 1.1
Chronological Developments of Commercial Polymers (to 1991)

Before 1800	Cotton, flax, wool, and silk fibers; bitumens caulking materials; glass and hydraulic cements; leather and cellulose sheet (paper); natural rubber (*Hevea brasiliensis*), gutta percha, balata, and shellac
1839	Vulcanization of rubber (Charles Goodyear)
1845	Cellulose esters (Schonbein)
1846	Nitration of cellulose (Schonbein)
1851	Ebonite (hard rubber; Nelson Goodyear)
1860	Molding of shellac and gutta percha
1868	Celluloid (plasticized cellulose nitrate; Hyatt)
1888	Pneumatic tires (Dunlop)
1889	Cellulose nitrate photographic films (Reinchenbach)
1890	Cuprammonia rayon fibers (Despeisses)
1892	Viscose rayon fibers (Cross, Bevan, and Beadle)
1903	First tubeless tire (Litchfield of Goodyear Tire Co.)
1897	Poly(phenylene sulfide)
1901	Glyptal polyesters
1907	Phenol–formaldehyde resins (Bakelite; Baekeland)
1908	Cellulose acetate photographic fibers
1912	Regenerated cellulose sheet (cellophane)
1913	Poly(vinyl acetate)
1914	Simultaneous interpenetrating network (SIN)
1920	Urea–formaldehyde resins
1923	Cellulose nitrate automobile lacquers
1924	Cellulose acetate fibers
1926	Alkyd polyester (Kienle)
1927	Poly(vinyl chloride) wall covering
1927	Cellulose acetate sheet and rods

TABLE 1.1 (continued)
Chronological Developments of Commercial Polymers (to 1991)

1927	Graft copolymers
1928	Nylon (Carothers, DuPont)
1929	Polysulfide synthetic elastomer (Thiokol; Patrick)
1929	Urea–formaldehyde resins
1930	Polyethylene (Friedrich/Marvel)
1931	Poly(methyl methacrylate) (PMMA) plastics
1931	Polychloroprene elastomer (Neoprene; Carothers)
1934	Epoxy resins (Schlack)
1935	Ethylcellulose
1936	Poly(vinyl acetate)
1936	Poly(vinyl butyral) (safety glass)
1937	Polystyrene
1937	Sytrene–butadiene (Buna-S) and styrene–acrylonitrile (Buna-N) copolymer elastomers
1939	Melamine–formaldehyde resins
1939	Nylon 6 (Schlack)
1939	Nitrile rubber (NR)
1940	Isobutylene–isoprene elastomer (butyl rubber; Sparks and Thomas)
1941	Low-density polyethylene (LDPE)
1941	Poly(ethylene terephthalate) (PET)
1942	Butyl rubber
1942	Unsaturated polyesters (Ellis and Rust)
1943	Fluorocarbon resins (Teflon; Plunket)
1943	Silicones
1945	SBR
1946	Polysulfide rubber (Thiokol)
1948	Copolymers of acrylonitrile, butadiene, and styrene (ABS)
1949	Cyanoacrylate (Goodrich)
1950	Polyester fibers (Winfield and Dickson)
1950	Polyacrylonitrile fibers
1952	Block copolymers
1953	High-impact polystyrene (HIPS)
1953	Polycarbonates (Whinfield and Dickson)
1956	Poly(phenylene ether); poly(phenylene oxide) (GE)
1957	High-density polyethylene (HDPE)
1957	Polypropylene
1957	Polycarbonate
1958	Poly(dihydroxymethylcyclohexyl terephthate) (Kodel, Eastman Kodak)
1960	Ethylene–propylene copolymer elastomers
1961	Aromatic nylons (Aramids, Nomex, Dupont)
1962	Polyimide resins
1964	Poly(phenylene oxide)
1964	Ionomers
1965	Polysulfone
1965	Styrene–butadiene block copolymers
1966	Liquid crystals
1970	Poly(butylene terephthate)
1974	Polyacetylene
1982	Polyetherimide (GE)
1991	Carbon nanotubes (Iijima; NEC Lab)

which hold the atoms together. In 1860, an oligomer, a small polymer, was prepared from ethylene glycol and its structure correctly given as $HO-(-OCH_2CH_2-)_n-OH$. But when poly(methacrylic acid) was made by Fittig and Engelhorn in 1880, it was incorrectly assigned a cyclic structure. Polymers were thought of as being colloids, or cyclic compounds such as cyclohexane. By using the Raoult and van't Hoff concepts, several scientists obtained high molecular weight values for these materials and for a number of other polymeric materials. However, since the idea of large molecules was not yet accepted, they concluded that these techniques were not applicable to these molecules rather than accepting the presence of giant molecules.

The initial "tire-track in the sand" with respect to tires was the discovery of vulcanization of rubber by Charles Goodyear in 1844. The first rubber tires appeared in the mid-1880s. These tires were solid rubber with the rubber itself absorbing the bumps and potholes. John Dunlop invented the first practical pneumatic or inflatable tire with his patent granted in 1888. Andre Michelin was the first person to use the pneumatic tires for automobiles. The Michelin brothers, Andre and Edouard, equipped a racing car with pneumatic tires and drove it in the 1895 Paris–Bordeaux road race. They did not win but it was sufficient advertising to begin interest in pneumatic tires for automobiles. Further, because they did not cause as much damage to the roads, pneumatic tires were favored by legislation. It is interesting to see that the names of these three pioneers still figure prominently in the tire industry. Even so, another inventor had actually been given the first patent for a vulcanized rubber pneumatic tire in 1845 but it did not take off. Thompson was a prolific inventor also having patented a fountain pen in 1849 and a steam traction engine in 1867.

A number of the giant tire companies started at the turn of the century. In the United States, many of these companies centered about Akron, the capital of the rubber tire. In 1898, the Goodyear Tire and Rubber Company was started. The Firestone Tire and Rubber Company was started by Harvey Firestone in 1900. Other tire companies shortly followed.

Hermann Staudinger studied the polymerization of isoprene as early as 1910. Intrigued by the difference between this synthetic material and natural rubber, he began to focus more of his studies on such materials. His turn toward these questionable materials, of interest to industry but surely not academically important, was viewed unkindly by his fellow academics. He was told by one of his fellow scientists "Dear Colleague, Leave the concept of large molecules well alone… There can be no such thing as a macromolecule."

Staudinger, systematically synthesized a variety of polymers. In the paper "Uber Polymerization" in 1920, he summarized his findings and correctly proposed linear structures for such important polymers as polyoxymethylene and polystyrene. X-ray studies of many natural and synthetic materials were used as structural proof that polymers existed. Foremost in these efforts were Herman Mark and Linus Pauling. Both these giants contributed to other important areas of science. Pauling contributed to the fundamental understanding of bonding and the importance of vitamins. Mark helped to find the academic and communication (journals, short courses, workshops) basis that would allow polymers to grow from its very diverse roots.

Probably the first effort aimed at basic or fundamental research in the chemical sciences was by DuPont. Their initial venture in artificial fibers was made in 1920, when they purchased a 60% interest in Comptoir des Testiles Artificels, a French rayon company. The combined company was named the DuPont Fiber Company. It spent considerable effort and money on expanding the properties of rayon. In 1926, Charles M. A. Stine, director of the chemical department, circulated a memo to DuPont's executive committee suggesting that the company move from investing in already existing materials to investigating in new materials. This was a radical idea that a company supposedly focused on profit spends some of its effort on basic research. The executive approved much of Stine's proposal giving him $25,000 a month for the venture allowing him to hire 25 chemists for the task. The initial hiring was difficult because academic chemists did not trust DuPont to allow them to do basic research. A year later, he was able to make his central hiring, Wallace Hume Carothers.

Wallace Hume Carothers is the father of synthetic polymer science. History is often measured by the change in the flow of grains of sand in the hourglass of existence. Carothers is a granite boulder in this hourglass. Carothers was born, raised, and educated in the midwest of the United States. In 1920, he left Tarkio College with his BS degree and entered the University of Illinois where he received his MA in 1921. He then taught at the University of South Dakota where he published his first paper. He returned to receive his PhD under Roger Adams in 1924. In 1926, he became an instructor in organic chemistry at Harvard.

In 1927, the DuPont Company reached a decision to begin a program of fundamental research "without any regard or reference to commercial objectives." This was a radical departure since the bottom line was previously products marketed and not papers published. Charles Stine, the director of DuPont's chemical department was interested in pursuing fundamental research in the areas of colloid chemistry, catalysis, organic synthesis, and polymer formation and convinced the Board to hire the best chemists in each field to lead this research. Stine visited with many in the academic community including the then president of Harvard, one of my distant uncles, J. B. Conant, an outstanding chemist himself, who told him about Carothers. Carothers was persuaded to join the DuPont group attracted with a generous research budget and an approximate doubling of his academic salary to $6000. This was the birth of the Experimental Station at Wilmington, Delaware.

Up to this point, it was considered that universities were where discoveries were made and industry was where they were put to some practical use. This separation between basic and applied work was quite prominent at this juncture and continues in many areas even today in some fields of work though the difference has decreased. However, in polymers, most of the basic research was done in industry having as its inception the decision by DuPont to bridge this "unnatural" gap between fundamental knowledge and application. In truth, they can be considered as the two hands of an individual and in order to do manual work both hands are important.

Staudinger believed that large molecules were based on the jointing, through covalent bonding, of large numbers of atoms. Essentially he and fellow scientists like Karl Freudenberg, Herman Mark, Michael Polanyi, and Kurt Myer looked at already existing natural polymers. Carothers, however, looked at the construction of these giant molecules from small molecules forming synthetic polymers. His intention was to prepare molecules of known structure using known organic chemistry and to "investigate how the properties of these substances depended on constitution." Early work included the study of polyester formation through the reaction of diacids with diols forming polyesters. However, he could not achieve molecular weights greater than about 4000 below the size where many of the interesting so-called polymeric properties appear.

DuPont was looking for a synthetic rubber. Carothers assigned Arnold Collins to this task. Collin's initial task was to produce pure divinylacetylene. While performing the distillation of an acetylene reaction, in 1930, he obtained a small amount of an unknown liquid that he set aside in a test tube. After several days, the liquid turned to a solid. The solid bounced and eventually was shown to be a synthetic rubber polychloroprene whose properties were similar to those of vulcanized rubber but it was superior in its resistance to ozone, ordinary oxidation, and to most organic liquids. It was sold under its generic name "neoprene" and the trade name "Duprene."

Also in 1930, Carothers and Julian Hill designed a process to remove water that was formed during the esterification reaction. Essentially they simply froze the water as it was removed using another recent invention called a molecular still (basically, a heating plate coupled to vacuum) allowing the formation longer chains. In April, Hill synthesized a polyester using this approach and touched a glass-stirring rod to the hot mass and then pulled the rod away effectively forming strong fibers, the pulling helps to reorient the mobile polyester chains. The polyester had a molecular weight of about 12,000. By again pulling the cooled fibers, additional strength was achieved with further reorienting occurring. This process of "drawing" or pulling to produce stronger fibers is now known as "cold drawing" and is widely used in the formation of fibers today. Carothers' group discovered the process of "cold drawing." While this was an interesting discovery, the fibers were not considered to be of commercial use. Carothers and his group then moved to look at the reaction of diacids with diamines instead of diols. Again, fibers were formed but these initial materials were deemed not to be particularly interesting.

In 1934, Paul Flory was hired to work with Carothers to help gain a mathematical understanding of the polymerization process and relationships. Thus, there was an early association between theory and practice or structure–property relationships.

In 1934, Donald Coffman, a member of the Carothers team, pulled a fiber from an aminoethylester (polyamide) polymer. The fiber retained the elastic properties of the polyesters previously investigated, but had a higher melting point allowing it to be laundered and ironed. The field of candidates for further investigation was narrowed to two-polyamide 5,10 made from pentamethylene diamine and sebacic acid, and polyamide 6,6 synthesized from hexamethylenediamine and adipic acid. Polyamide 6,6 won because the monomers could be made from benzene, a readily available feedstock from coal tar.

The polyamide fiber project was begun again. The promising candidate was formed from the reaction of adipic acid with hexamethylenediamine, called fiber 66 because each carbon-containing unit had six carbons. It formed a strong, elastic, largely insoluble fiber with a relatively high melt temperature. DuPont chose this material for production. These polyamides were given the name "nylons." Thus was born nylon-6,6. It was the first synthetic material whose properties equaled or exceeded the natural analog, namely silk. (In reality, this may not be the truth, but at that time, it was believed to be true.)

The researchers had several names for polyamide 6,6 including rayon 66, fiber 66, and Duparon derived from "Dupont pulls a rabbit out [of] the hat nitrogen/nature/nature/nozzle/naphtha." The original "official" name was "Nuron" which implied newness and spelled "on run" somewhat backward. This name was too close to other trademarked names and was renamed "Nirton" and eventually to what we know today as "Nylon."

As women's hemlines rose in the 1930s, silk stockings were in great demand but were very expensive. Nylon changed this perspective. Nylon could be woven into sheer hosiery. The initial presentation of nylon hose to the public was by Stine at a forum of women's club members in New York City on October 24, 1938. Nearly 800,000 pairs were sold on May 15, 1940 alone—the first day they were on the market. By 1941, nylon hosiery held 30% of the market but by December 1941 nylon was diverted to make parachutes, etc.

From these studies, Carothers established several concepts. First, polymers could be formed by employing already known organic reactions but with reactants that had more than one reactive group per molecule. Second, the forces that bring together the individual polymer units are the same as those that hold together the starting materials, namely, primary covalent bonds. Much of the polymer chemistry names and ideas that permeate polymer science today were standardized through his efforts.

Representing the true multidisciplinary nature of polymers, early important contributions were also made by physicists, engineers, and those from biology, medicine, and mathematics including W. H. Bragg, Peter Debye, Albert Einstein, and R. Simha.

World War II helped shape the future of polymers. Wartime demands and shortages encouraged scientists to seek substitutes and materials that even excelled currently available materials. Polycarbonate (Kevlar), that could stop a "speeding bullet," was developed as was polytetrafluoroethylene (Teflon) that was super slick. New materials were developed spurred on by the needs of the military, electronics industry, food industry, etc. The creation of new materials continues at an even accelerated pace brought on by the need for materials with specific properties and the growing ability to tailor-make giant molecules–macromolecules–polymers.

Unlike other areas of chemistry, most of the basic research has been done in industry so that there is often a close tie between discoveries and their commercialization. Table 1.2 lists some of the dates of commercialization for some important polymer discoveries.

A number of Nobel Prizes have been given for polymer work. Table 1.3 contains winners whose efforts are related to synthetic polymers. In truth, there are many more since most of the prizes given out in medicine and biology deal with giant molecules.

Throughout this chapter, advances are placed in some historical setting. This setting adds some texture to the topics as well as acting as case histories that are widely used in some subject areas such as business and medicine.

TABLE 1.2
Commercialization of Selected Polymers

Polymer	Year	Company
Bakelite	1909	General Bakelite Corp.
Rayon	1910	American Viscose Company
Poly(vinyl chloride)	1927	Goodrich
Styrene–butadiene copolymer	1929	I. G. Farben
Polystyrene	1929/1930	I. G. Faben and Dow
Neoprene	1931	Dupont
Poly(methyl methacrylate)	1936	Rhom and Haas
Nylon 66	1939/1940	Dupont
Polyethylene (LDPE)	1939	ICI
Poly(dimethyl siloxane)	1943	Dow Corning
Acrylic fiber	1950	Dupont
Poly(ethylene terephthalate), PET	1953/1954	Dupont/ICI
Polyurethane block copolymers (Spandex)	1959	Dupont
Poly(phenylene terephthalamide)	1960	Dupont

TABLE 1.3
Nobel Prize Winners for Their Work with Synthetic Polymers

Scientist(s)	Year	Area
Herman Staudinger	1953	Polymer hypothesis
Karl Ziegler and Giulio Natta	1963	Stereoregulation of polymer structure
Paul Flory	1974	Organization of polymer chains
Bruce Merrifield	1984	Synthesis on a solid matrix
Pierre de Gennes	1991	Polymer structure and control at interfaces
A. J. Heeger, Alan Mac Diarmid, and H. Shirakawa	2000	Conductive polymers

1.2 WHY POLYMERS?

Polymers are all about us. They serve as the very basis of both plant and animal life as proteins, nucleic acids, and polysaccharides. In construction, they serve as the concrete, insulation, and wooden and composite beams. At home, they are found as the materials for our rugs, curtains, coatings, waste paper baskets, water pipes, window glass, ice cube trays, and pillows. In transportation, they are present in ever-increasing amounts in our aircraft, automobiles, ships, and trucks. In communication, they form critical components in our telephones, TVs, computers, CDs, newspaper, optical fibers, and cell phones. Plastics act as favorite materials for our toys such as toy soldiers, plastic models, toy cars, dolls, skip ropes, hula hoops, and corvettes. Our food, namely meats, vegetables, breads, and cookies, is polymer intense. In history, polymers have been the vehicle for the Magna Carter, Torah, Bible, Koran, and our Declaration of Independence. Outside our homes, they are present in our flowers, trees, soil, spider webs, and beaches. In fact, it is improbable that a polymer is not involved in your present activity—reading a paper book, holding a plastic-intense writing device, sitting on a cloth-covered chair or bed, and if your eyes need corrective vision, glasses of one variety or another.

Polymers gain their importance because of their size. Many polymers are made from inexpensive and readily available materials allowing vast quantities of products to be made for a high increase in value, but they are typically inexpensive compared to nonpolymer alternatives. They also often have desirable physical and chemical properties. Some polymers are stronger on a weight basis than steel. Most are resistant to rapid degradation and rusting. You will learn more about these essential materials for life and living in this book.

Polymers are often divided according to whether they can be melted and reshaped through the application of heat and pressure. These materials are called *thermoplastics*. The second general classification belongs to compounds that decompose before they can be melted or reshaped. These polymers are called *thermosets*. While both thermosets and thermoplastic polymers can be recycled, because thermoplastics can be reshaped simply through the application of heat and pressure, the recycling of thermoplastics is easier and more widespread.

In general, synthetic polymers are often described by their "use" and "appearance" as fibers, elastomers, plastics, adhesives, and coatings. A common toothbrush illustrates the three major physical forms of synthetic polymers—the rubbery (elastomeric) grips, plastic shaft, and fibrous brussels. The rubbery grips have a relatively soft touch; the plastic shaft is somewhat flexible and hard; and the brussels are highly flexible. Another illustration of the breath of polymers about us is given in Table 1.4 where polymers are divided according to source.

To get an idea of the pervasiveness of polymers in our everyday life, we can look at containers. Most containers are polymeric—glass, paper, and synthetic polymer. It is relatively easy to identify each of these general categories. Even, within the synthetic polymer grouping, it has become

TABLE 1.4
Common Polymers

Material/Name	Typical Polymer	Chapters
Styrofoam	Polystyrene	7
PVC pipe	Poly(vinyl chloride)	7
Nylon stockings	Polyamide, nylon 6,6	5
Concrete	Cement	10
Meat	Protein	4
Plexiglass	Poly(methyl methacrylate)	7
Automotive bumpers and side panels	Polyethylene and polyethylene/ polypropylene blends	6, 8
Potatoes	Starch	4
Compact disks (case)	Polycarbonate (polystyrene)	5
Hula hoop	Polypropylene, polyethylene	6
Diamond	Carbon	10
Silicon sealants	Polydimethylsiloxane	9
Bakelite	Phenol–formaldehyde cross-linked	5
Super glue	Poly(ethyl cyanoacrylate)	16
Cotton T-shirt	Cellulose	4
Fiberglass	Composite	14
Saran wrap	PVC copolymer	8
Velcro	Polyamide	5
Rubber band	Natural rubber	4
Soda bottle	Poly(ethylene terephthalate), PET	5
Teflon	Polytetrafluoroethylene	7
Orlon sweater	Polyacrylonitrile	7
Sand	Silicon dioxide	10
Pillow stuffing	Polyurethane	5
Wood, paper	Cellulose	4
Human genome	Nucleic acids	4

relatively easy to identify the particular polymer used in some applications such as with disposable containers. Most of these synthetic polymers are identified by an "identification code" that is imprinted somewhere on the plastic container, generally at their bottom. The numbers and letters are described in Figure 1.1. The recycling code was developed by the Society of Plastics Industry for

PETE

Poly(ethylene terephthalate)—**PET** or **PETE**
PET is the plastic used to package the majority of soft drinks. It is also used for some liquor bottles, peanut butter jars, and edible-oil bottles. About 25% of plastic bottles is PET. PET bottles can be clear; they are tough and hold carbon dioxide well.

HDPE

High-density polyethylene—**HDPE**
HDPE is a largely linear form of polyethylene. It accounts for over 50% of the plastic bottle market and is used to contain milk, juices, margarine, and some grocery snacks. It is easily formed through the application of heat and pressure and is relatively rigid and low cost.

V

Poly(vinyl chloride)—**PVC** or **V**
PVC is used "pure" or as a blend to make a wide variety of products including PVC pipes, food-packaging films, and containers for window cleaners, edible oils, and solid detergents. It accounts for only 5% of the container market.

LDPE

Low-density polyethylene—**LDPE**
LDPE has branching and is less crystalline, more flexible, and not as strong as HDPE. The greater amount of amorphous character makes it more porous than HDPE, but it offers a good inert barrier to moisture. It is a major material for films from which trash bags and bread bags are made.

PP

Polypropylene—**PP**
PP has good chemicals and fatigue resistance. Films and fibers are made from it. Few containers are made of PP. It is used to make some screw-on caps, lids, yogurt tubs, margarine cups, straws, and syrup bottles.

PS

Polystyrene—**PS**
PS is used to make a wide variety of containers, including those known as "Styrofoam" plates, dishes, cups, etc. Cups, yogurt containers, egg cartons, meat trays, and plates are made from PS.

OTHER

Other plastics
A wide variety of other plastics are coming to the marketplace including copolymers, blends, and multilayered combinations.

FIGURE 1.1 The Society of Plastics Industry recycling codes utilizing the numbers 1–7 and bold, capital letters to designate the material utilized to construct the container.

use with containers. Today, the "chasing-arrows" triangle is being used more widely for recycling by the public. A colorless somewhat hazy water container has a "2" within the "chasing" arrows and underneath it "HDPE" both indicating the bottle is made of high-density polyethylene. The clear, less flexible soda bottle has a "1" and "PETE" both signifying that the container is made out of poly(ethylene terephthalate), a polyester. A brownish clear medicine container has a "5" and the letters "PP" on its bottom conveying the information that the bottle is made of polypropylene. Thus, the ready identification of some common items is easy.

However, because of the use of many more complex combinations of polymers for many other items, such identification and identification schemes are not as straightforward. For some items, such as clothing and rugs, labels are present that tell us the major materials in the product. Thus, a T-shirt might have "cotton" on its label signifying that the T-shirt is largely made of cotton. A dress shirt's label may say 55% cotton and 45% polyester meaning it is made from two polymers. Some items are identified by trade names. Thus, a dress advertised as being made from Fortrel (where "Fortrel" is a trade name) means it is made largely of a polyester material, probably the same polyester, PET or PETE that made our soda bottle. Some everyday items are a complex of many materials but only some or none being noted. This is true for many running shoes and tires. Tires will often be described as being polyester (again, probably the same PETE) or nylon (or aramid). This describes only the composition of the tire cord but does not tell us what other materials are included in the tire's composition. Yet, those that deal with tires generally know what is used in the manufacture of the tire in addition to the "stated ingredients." You will be introduced, gently, to the identification of the main polymers that are present in many everyday items, either through looking at labels, researching on the Web, simply knowing what certain items are generally composed of, through the feel and gross physical properties (such as flexibility and stiffness) of the material, etc.

Further, the properties of essentially the same polymer can be varied through small structural changes giving materials with differing properties and uses. There is a match between desired properties and the particular material used. For instance, for plastic bags, strength and flexibility are needed. The bag material should be somewhat strong, inexpensive (as most bags are "throw-away" items), and readily available in large quantities. Increased strength is easily gained from increasing thickness. However, with increased thickness comes decreased flexibility, increased cost since more material is needed to make the thicker bags, and increased transportation (because of the additional weight) and storage costs. Thus, there is a balance between many competing factors. Plastic bags are typically made from three polymers, high-density polyethylene (HDPE), low-density polyethylene (LDPE), and linear low-density polyethylene (LLDPE; actually a copolymer with largely ethylene units). These different polyethylene polymers are similar differing only in the amount of branching that results in differing tendencies to form ordered (crystalline) and less ordered (amorphous) chain arrangements. You will learn more about them in Chapter 5. Grocery bags are generally made from HDPE, which is a largely linear polymer that has a high degree of crystallinity. Here, in comparison to LDPE film with the same strength, the bags are thinner allowing a decrease in the cost of materials, transportation cost, and storage space. The thinness allows good flexibility. LDPE is used for dry cleaning garment bags where simply covering the garments is the main objective rather than strength. The LDPE is less crystalline and weaker, but more flexible, because of the presence of more branching in comparison to HDPE. The thicker glossy shopping bags from malls are LLDPE, which, like HDPE, is largely linear. This increased thickness results in the bags being less flexible. These bags can be used many times.

Thus, most of the common items about us are polymeric. Table 1.4 gives a brief listing of some of these materials along with the locations where they will be dealt with in the book.

With the advent of the electronic age, we can access the Web to gather lots of general information about almost any topic including polymers. This book allows you to have a greater appreciation and understanding of such information and the products about us, including our own bodies.

TABLE 1.5
U.S. Production of Plastics (Millions of Pounds; 2007)

Thermosetting resins	
Epoxies	600
Ureas and melamines	3,200
Phenolics	4,000
Thermoplastics	
Polyethylenes	
Low density	7,200
High density	16,600
Linear low density	12,200
Polypropylene	17,600
Polystyrene	
Polystyrene	5,400
Acrylonitrile–butadiene–styrene and styrene–acrylonitrile	2,600
Polyamides, nylons	1,200
Poly(vinyl chloride) and copolymers	13,200
Polyesters	2,400

Source: American Plastics Council, Arlington, VA.

1.3 TODAY'S MARKETPLACE

As noted earlier, polymers are all around us. Over 100 billions pounds (50 million ton) of synthetic polymers are produced annually in the United States (Tables 1.5 through 1.8), and the growth of the industry is continuing at a faster rate. There is every reason to believe that this polymer age will continue as long as petroleum and other feedstocks are available and as long as consumers continue to enjoy the comfort, protection, and health benefits provided by elastomers, fibers, plastics, adhesives, and coatings. The 100 billion pounds of synthetic polymers consumed each year in the United States translates to over 300 lb for every man, woman, and child in the United States. This does not include paper and wood-related products, natural polymers such as cotton and wool, or inorganic polymers (Table 1.9).

Polymers are all about us. The soils we grow our foods from are largely polymeric as are the foods we eat. The plants about us are largely polymeric. We are walking exhibits as to the wide-spread nature of polymers—from our hair and finger nails, our skin, bones, tendons, and muscles; our clothing-socks, shoes, glasses, undergarments; the morning newspaper; major amounts of our automobiles, airplanes, trucks, boats, space craft; our chairs, waste paper baskets, pencils, tables, pictures, coaches, curtains, glass windows; the roads we drive on, the houses we live in, and the buildings we work in; the tapes and CDs we listen to music on;

TABLE 1.6
U.S. Production of Man-Made Fibers (Millions of Pounds; 2007)

Noncellulosic	
Acrylics	120
Nylons	1800
Olefins	2600
Polyesters	2400
Cellulosic	
Acetate and rayon	54

Source: Fiber Economics Bureau, Arlington, VA.

TABLE 1.7
U.S. Production of Paints and Coatings (Millions of Gallons)

Year →	1995	2000	2005
Architectural	620	650	860
Product	380	450	410
Special	195	180	175

Source: Department of Commerce, Washington, DC.

packaging—all are either totally polymeric or contain a large amount of polymeric materials. Table 1.9 lists some general groupings of important polymers. Welcome to the wonderful world of polymer science.

The number of professional chemists directly employed with polymers as part of their interest and assignment is estimated to be 40%–60% of all the chemists. As the diversity of chemistry increases, the dispersion of those dealing with polymers increases. Polymer chemistry is a major tool applied in biomedical research, synthesis, manufacturing, chemical engineering, pharmaceutical efforts, the environment, communications, etc. "As it was in the beginning" polymers continue to draw strength from those with diverse training allowing polymers to directly contribute to solutions for most of our technological problems including fuel and transportation, building and construction, communication, medicine and dentistry, etc. Analytical chemistry is applied to the analysis of materials; inorganic to the catalysts employed in the synthesis of natural and synthetic polymers; organic to the synthesis of diverse materials; physical to describe the kinetics, thermodynamics, and properties of macromolecules; and biological chemistry that deals intently with biopolymers.

Polymeric materials, along with several other chemical industrial products, contribute positively to the balance of trade (Table 1.10). In fact, plastics and resins show the greatest value increase of exports minus imports with about $20 billion net favoring exports. The polymer-intense materials numbers are higher than noted in Table 1.10 since fiber and rubber materials are absent as a separate entry. Even so, the figures demonstrate the positive nature polymers play in our balance of trade situation.

TABLE 1.8
U.S. Production of Synthetic Rubber (Millions of Pounds; 2008)

Styrene–butadiene	1750
Polybutadiene	1210
Nitrile	180
Ethylene–propylene	540
Other	1100

Source: International Institute of Synthetic Rubber Producers, Houston, TX.

TABLE 1.9
Polymer Classes—Natural and Synthetic

Polymeric Materials				
Inorganic		Organic/Inorganic	**Organic**	
Natural	**Synthetic**		**Natural**	**Synthetic**
Clays	Fibrous glass siloxanes		Proteins	Polyethylene
Cement	Poly(sulfur nitride)	Polyphosphazenes	Nucleic acids	Polystyrene
Pottery	Poly(boron nitride)	Polyphosphate esters	Lignins	Nylons
Bricks	Silicon carbide	Polysilanes	Polysaccharides	Polyesters
Sands		Sol–gel networks	Melanins	Polyurethanes
Glasses			Polyisoprenes	Poly(methyl methacrylate)
Rocklike				Polytetrafluoroethylene
Agate				Polyurethane
Talc				Poly(vinyl chloride)
Zirconia				Polycarbonate
Mica				Polypropylene
Asbestos				Poly(vinyl alcohol)
Quartz				
Ceramics				
Graphite/diamond				
Silicas				

TABLE 1.10
U.S. Chemical Trade—Imports and Exports (Millions of Dollars)

Year →	1985		1994		2007	
	Exports	**Imports**	**Exports**	**Imports**	**Exports**	**Imports**
Organic chemicals	6,000	4,600	12,800	10,800	34,400	42,200
Inorganic chemicals	2,000	2,000	4,100	4,100	11,000	13,400
Oils and perfumes	—	—	3,500	2,000	10,600	8,900
Dyes, colorants	—	—	2,300	1,900	5,900	3,100
Medicinals/pharmaceuticals	2,700	1,080	6,100	4,700	33,500	53,700
Fertilizers	2,160	1,000	2,700	1,300	3,700	5,000
Plastics and resins	3,800	1,600	8,500	3,300	37,700[b]	18,200[b]
Others	5,300	4,220	7,600	2,700	21,300	10,900
Total chemicals[a]	22,000	14,500	51,600	33,400	158,200	155,400
Total	213,000	345,000	502,800	669,100	1,162,700	1,855,100

Source: Department of Commerce, Washington, DC.

[a] Includes nonlisted chemicals.

[b] Includes both plastics in primary and nonprimary forms but does not include rubber and rubber products and fibers and coatings.

Essentially, all of the industrially advanced countries of the world have major chemical producers. These producers are involved directly and/or indirectly with some form of synthetic polymers. Table 1.11 contains a partial listing of these companies.

Thus, polymers play a critical role in our everyday lives, actually forming the basis for both plant and animal life, and represent an area where chemists continue to make important contributions.

1.4 ENVIRONMENTAL ASSESSMENT

Daily, we are becoming increasingly aware of the importance of environmental planning and good environmental actions (practices) and their effects on us. This emphasis is being driven by a number of pressures including federal and state laws and initiatives, industrial consciousness, reality, international efforts, individual actions, etc. Chemistry, chemists, and the chemical industry have been directing much effort in this direction for over several decades with this effort magnified over the past several years. This book highlights some of these efforts in polymers. Here a number of terms will be introduced related to these efforts. While these terms are described individually, the environmental activity is a matrix of actions and activities each one dependent on others to be successful. Many of these studies and assessments are governed somewhat by procedures described by the International Organization for Standardization, ISO. (More about ISO can be found on the Web and in Appendix E.)

1.4.1 ENVIRONMENTAL IMPACT ASSESSMENT

An environmental impact assessment, EIA, is simply an assessment of the possible impact that a project or material may have on the natural environment. This possible impact may be positive or negative, and often is a combination of positive and negative impacts. The intention of such EIAs is to identify where changes can and should be made and to make us aware of these instances. The International Association for Impact Assessment, IAIA, describes such impact assessments as a process for "identifying, predicting, evaluating and mitigating the biophysical, social, and other relevant effects of development proposals prior to major decisions being taken and commitments made."

Table 1.11
Major Chemical Producers Based on (Net) Sales
(>2500 Million Dollars; in Millions of U.S. Dollars; 2007)

United States

Dow	53,500	ExxonMobile	36,800
Dupont	29,200	Huntsman	19,700
Lyondell	16,200	Chevron Phillips	12,500
PPG	10,000	Praxair	9,400
Air products	8,800	Rhom and Haas	7,800
Eastman chemical	6,800	Celanese	6,400
Hexion specialty	5,800	Mosaic	5,700
Dow corning	4,900	Honeywell	4,900
Occidental petroleum	4,700	Lubrizol	4,500
Nalco	3,900	Chemtura	3,700
Monsanto	3,600	Cytec Ind.	3,500
Solutia	3,500	Westlake	3,200
W. R. Grace	3,100	Sunoco	2,800
CF	2,800	Rockwood Specialties	2,700
FMC	2,600	Cabot	2,600
Momentive	2,500		

Europe

BASF (Germany)	79,500	Bayer (Germany)	44,400
Linde (Germany)	16,900	Air Liquide (France)	16,300
Akzo Nobel (the Netherlands)	14,000	Solvay (Belgium)	13,100
DSM (the Netherlands)	12,200	ICI (United Kingdom)	10,600
Merck (Germany)	9,700	Lanxess (Germany)	9,100
BOC (United Kingdom)	8,400	Arkema (France)	7,800
Syngenta (Switzerland)	7,700	Clariant (Switzerland)	7,100
Rhodia (France)	7,000	Ciba (Switzerland)	5,400
Wacker (Germany)	5,200	Kemira (Finland)	3,900
Givaudan (Switzerland)	3,400		

Japan

Mitsubishi Chem. Inds.	24,900	Sumitomo Chemical	16,100
Mitsui Chem.	15,200	Asahi Kasei	14,400
Toray Inds.	14,000	Shin-Etsu	11,700
DIC	9,200	Teijin	8,800
Showa Denko	8,700	Taiyo Nippon Sanso	4,300
Kaneka	4,300	JSR Corp.	3,500

Canada

Nova Corp.	6,700	Agrium	5,500
Potash Corp.	5,200		

Source: C & EN, Washington, DC.

(Petts, J. (Ed.) (1999): *Handbook of Environmental Impact Assessment*, Vols. 1 and 2, Blackwell, Oxford, U.K.). The need and specifications for EIAs depends on the particular country. For the United States, it originated as part of the National Environmental Policy Act of 1970. States may also have other requirements.

1.4.2 ECOLOGICAL FOOTPRINT

The ecological footprint is a measure of our demand on the Earth's ecosystems. It includes our demand on both natural resources and on the ability for these resources to be regenerated. In the past, the methods and items to be measured varied widely, often based on factors suiting the particular sector making the footprint assessment. Today, more homogeneous standards are emerging. It is intended to reflect a measure of the land, fresh water, and ocean area required to produce, for instance, a product. Such footprints are often calculated to reflect other average measures. One such measure is a per capita ecological footprint that compares consumption and lifestyles with the natural ability to provide this consumption.

1.4.3 LIFE CYCLE ASSESSMENT

A life cycle assessment, LCA, investigates and evaluates the environmental impact of a product or service. It is also referred to as an ecobalance analysis, cradle-to-grave analysis, and life cycle analysis. Such assessments are intended to measure the effects of the cascade of technologies related to products and services. Here we will restrict our comments to products. "Life cycle" can refer to a holistic assessment of raw material used in the production of a product including energy consumption for the procurement and transport of the material, manufacture, distribution, use, and finally disposal, including recycling if possible, environmental costs. Often guesses and averages are included in these studies. Thus, the cost of transportation could include some proportion of the truck or train construction, road or rail construction, and the deterioration and repair of the same.

According to ISO 14040 and 14044 standards, a LCA contains four distinct phases. These phases are goal and scope, inventory analysis, impact assessment, and the interpretation of the first three phases. In the "goal and scope" phase, the functional unit is described with effort focused on defining the boundaries of the system or product to be studied. It includes describing the methods used for assessing possible environmental impacts and which impact categories are to be included in the study. In the "inventory" phase, data are collected including the description and verification of the data along with various modeling programs to be used. Generally, items considered include inputs such as quantities of materials, land usage, and energy and outputs such as air emission, solid waste, and water emissions. Software packages have been developed, and are being developed, to assist in such evaluations. The "impact" phase is intended to describe contributions to more global situations such as global warming and acidification. The final phase, "interpretation," brings together the other three phases and conclusions are made.

There are varying types of LCA studies that act to limit or define the type of LCA study being made. The *cradle-to-grave* assessment is a full LCA study from manufacture (cradle) to the disposal (grave) of a product. A *cradle-to-gate* study looks at the life cycle from manufacture (cradle) to factory gate (before it is sent to the consumer). These assessments are often used as a basis for environmental product declarations. A *cradle-to-cradle* assessment involves products where the product is recycled, so the study terminates when a new product is made from the recycled original product. A *life cycle energy analysis* looks at all energy inputs to a product and not solely direct energy inputs during manufacture including energy necessary to produce components and materials needed for the manufacturing process.

1.5   MANAGING SUSTAINABILITY

Earth, air, fire, and water, not the music group but rather the Greek classical elements, represent the critical elements of global sustainability. Our ability to manage these elements will determine the sustainability of human life, as we know it. The need for sustainability is not a recent phenomenon but one on the "front burner." The most cited definition of sustainability is from the UN World Commission on Environment and Development (Brundtland Commission) that says "Sustainable development is development that meets the needs of the present without compromising the ability of future generations to meet their own needs."

Earth represents food, EPA sites, etc.; *air* our atmosphere and need for breathable air; *water* the necessity for safe drinking water; and *fire* our energy. Currently, we are sinking further toward and past the "tipping point" for each of these. A tipping point is generally defined as when some feature reaches a point where various "feedback" loops are not able to compensate for unwanted movement toward (here unwanted) consequences and the movement toward these unwanted consequences becomes more rapid and/or permanent.

The magnitude of the problem or opportunity is great. Over 80 countries and 1 billion people do not have a good source of safe water. About 87 million barrels of nonrenewable crude oil are pumped daily resulting in over 3.5 billion gallons each day with most of this employed as fuel with the remaining used for the production of plastics, drugs, rubbers, paints, fibers, etc.

The polymer industry is part of this challenge and polymers play an essential role in solutions as the materials that allow us to perform the needed tasks, so that we can have safe water and air, the efficient use of fuel and an earth that allows us to thrive. Polymers are integral parts of filtering systems for water and air; lighter components to transportation systems, and serve as the basis for more fuel-efficient engines; polymeric growth hormones, and gene-engineered seeds can help to feed us.

Our reply to issues of sustainability is critical and involves many stakeholders. Managing this reply is critical. The following text illustrates some of the inner relationships and complexities related to the "elements" and their management.

The Florida peninsula receives unwanted mercury from three major sources. The three sources are natural, past agricultural practices, and atmospheric. The atmospheric mercury comes from two main sources. The first is the result of gold retrieval when gold ore is heated with mercury in open containers in South America allowing mercury to escape into air. The second source is Europe and is connected to ore production and coal burning. The atmospheric mercury finds its way over Florida and is deposed from the sky during violent summer storms. This example illustrates several key factors in addressing sustainability. First, the sources are identified through research highlighting the need for science to contribute to sustainability solutions. Second, there are no governing and enforcement agencies that successfully control such emissions. Third, globalization is a feature since the mercury is moved from country and continent to country and continent. Fourth, the problem is complex and not easily solved. Fifth, environmental problems are often comprised of past and present practices.

Another feature that should be recognized is unintended consequences. One of the initial efforts at producing genetically engineered corn was dropped because the plant killed butterflies that were a chief means of pollinating the corn. For several decades, no new petroleum-refining plants have been constructed mainly because of the seemingly excessive cost associated with meeting the "new" environmental regulations.

We will take a moment to describe the situation with respect to biofuels. While ethanol production helps to decrease our dependence on "foreign" oil, it is not our answer to energy independence. Ethanol can be made from corn, but this approach has many drawbacks. While corn is a renewable source of ethanol fuel it is also a source of food, so as corn production for food is diverted to fuel production the price of corn products increases. Only about a net 20% energy is realized from the production of ethanol from corn when including the energy needed for its production. The production of ethanol is water intensive. Further, calculations show ethanol production from corn is acre-limited with only a reasonable 10% of our overall fuel consumption possible through this venue. The problem of supplying ethanol as a fuel is partially solved with the blending of ethanol with regular petroleum-based fuels. The long-range fuel reduction and automotive performance and care has not yet been established.

Research shows some alternatives to the present ethanol production from corn. One alternative is the production of plants that give greater fuel production for acre. Another alternative, and one where the energy tab is low, employs especially designed microbes, mainly bacteria that convert carbohydrate biomasses into alcohols, mainly ethanol and methanol. These alternatives illustrate other possible solutions. "Total" solutions are often complex with multiple "solutions" needed. Energy solutions must depend on many sources including coal, biofuels, nuclear, hydrogeological, biosolar, etc.

Global warming illustrates the political nature of most sustainability issues. While the press is fully convinced that global warming is a fact and that we are responsible, many scientists are not on board with this conclusion. There are abundant scientific studies to support both sides. Even so, with a sustainability issue that can have such dramatic consequence affecting all of humanity, it is better to treat it as real until it is overwhelmingly shown to be false. Thus, there is sufficient scientific information that supports global warming, whether it is a majority or not, as a fact it must be dealt with in a reasonable manner. Note the "reasonable" clause. Today there are emerging countries, as well as first world nations, that must act as responsible citizens of Earth, but this means that the emerging countries be allowed to "emerge" and the current first world countries be allowed to exist without great damage to them. This is a balancing act and points out another factor that is important as we address global issues. This factor is to allow some independence of nations while recognizing their interdependence.

All of the major sustainability issues influence our personal health and our business and industrial economies.

We now move toward managing sustainability. The stakeholders are many. Here we will assume the most important stakeholder is you and me. We contribute by influencing legislation, voting, membership in groups, etc. We also contribute through recycling, the use of reduced energy consuming transportation, lights, appliances, etc. Sociological and psychological components are also important.

The important stakeholders include

1. Research
2. Business and industry
3. Governmental and legislative including enforcement
4. Environmental and related groups
5. Us

Thus far, all of these stakeholders have shown an inability to effect needed positive change on a global basis. On the research side, while lots of research is being done, it is not globally focused and it has not been shown to appropriately interact with some of the other major stakeholders. Most scientific (including engineering) organizations are small, national, and poorly funded. The largest single disciple organization is the American Chemical Society with 160,000 members, mostly in the United States. It is mainly funded by membership dues and information-related activities. While it attempts to highlight sustainability issues and has a governmental action arm, its lacks sufficient funding and legislative connections. Multidiscipline organizations such as Sigma Xi also suffer from inadequate funding, industrial and business, and legislative governmental connections.

Business and industry is already global and must play a critical role in managing sustainability. They have the multinational ability to influence change on a global way. They lack the primary focus, legislative and enforcement power as well as "free" funds to assist in the research necessary to understand and influence sustainability.

Governmental and legislative stakeholders have shown an inability to manage sustainability. While some good has come from such entities as the United Nations and World Bank, they lack the power and funds to adequately manage sustainability. Environmental and related groups are without sufficient funds and research capability. Even so, they are a critical stakeholder often influencing needed legislation that in turn acts as an incentive to obey directives.

Thus, each stakeholder is important and all lack some crucial aspects to lead in the management of sustainability.

Hopefully, this chapter will allow you to both see where polymers are part of the solution and to allow you to become informed in one of the most scientifically important disciplines, polymer science, which will supply many of the materials and knowledge needed to allow a sustainable world.

1.6   SUMMARY

After reading this chapter, you should understand the following concepts.

1. Polymers or macromolecules are giant molecules with large structures and high molecular weights. In spite of their varieties they are governed by the same laws that apply to small molecules.
2. If we disregard metals and some inorganic compounds, practically everything else in this world is polymeric. Polymers form the basis for life itself and for our communications, transportation, buildings, food, etc. Polymers include protein and nucleic acids in our bodies, the fibers (natural and synthetic) we use for clothing, the protein and starch we eat, the elastomers in our automotive tires, the paint, plastic wall and floor coverings, foam insulation, dishes, furniture, pipes, etc.
3. There are some systems in place that allow us to readily identify the nature of many polymeric materials including clothing and containers.
4. Early developments in polymers were largely empirical because of a lack of knowledge of polymer science. Advancements in polymers were rapid in the 1930s and 1940s because of the theories developed by Staudinger, Carothers, Mark, and many other scientists.
5. This is truly the age of the macromolecule. Essentially every important problem and advance includes polymers including synthetic (such as carbon nanotubes) and biological (such as the human genome and proteins). There are more chemists working with synthetic polymers than in all of the other areas of chemistry combined.
6. The so-called green chemistry is becoming increasingly important in today's chemical industry. The use of naturally derived monomers and polymers is increasing. We are part of the overall solution as informed citizens and scientists.

GLOSSARY

ABS: A polymer produced by the copolymerization of acrylonitrile, butadiene, and styrene.
Bakelite: A polymer produced by the condensation of phenol and formaldehyde.
Cellulose: A naturally occurring carbohydrate polymer.
Ecological footprint: A measure of our demand on the Earth's ecosystems.
Elastomer: A rubber.
Environmental impact assessment (EIA): An assessment of the possible impact that a project or material may have on the natural environment.
Filament: The individual extrudate emerging form the holes in a spinneret; forms fibers.
Functionality: The number of reactive groups.
Intermolecular forces: Secondary forces between macromolecules,
Intramolecular forces: Secondary forces within the same macromolecular chain.
Life cycle assessment (LCA): The investigation and evaluation of the environmental impact of a product or service. It is also referred to as an ecobalance analysis, cradle-to-grave analysis, and life cycle analysis.
Linear: A continuous chain.
Macromolecule: A polymer. Large chained molecular structure.
Natural rubber (NR): Polyisoprene obtained form rubber plants. *Hevea brasiliensis.*
Nylon-66: A polyamide produced from the condensation of adipic acid and 1,6-hexanediame.
Oligomer: Low molecular weight polymer with generally 2–10 repeat units.
Plasticizer: An additive that reduces intermolecular forces in polymers making it more flexible.
Polymer: A giant molecule, macromolecule, made up of multiple repeating units where the backbone is connected by covalent bonds.
Protein: A natural polyamide composed of many amino acid-derived repeat units.

Rayon: Regenerated cellulose in the form of filaments.
Sustainability: Development that meets the needs of the present without compromising the ability of future generations to meet their own needs.
Thermoplastic: A linear polymer that softens when heated.
Thermoset: A network polymer, containing chemical cross-linking that does not soften when heated.
Vital force concept: A hypothesis that stated that organic compounds can be produced only by natural processes and not in the laboratory.
Vulcanization: Process where elastomers such as natural rubber are cross-linked by heating with sulfur.

EXERCISES

1. Name six polymers that you encounter daily.
2. Why are there more chemists who work with polymers than with other areas?
3. Why are there so many outstanding polymer chemists alive today?
4. Which of the following are polymeric or contain polymers as major components?
   (a) Water, (b) Wood, (c) Meat, (d) Cotton, (e) Tires, and (f) Paint
5. Name three inorganic polymers.
6. Name three synthetic polymers.
7. Look at the bottom of several containers. Identify what the bottles are made of from the following recycling codes found on their bottoms: (a) 4 with the letters LDPE, (b) 6 with the letters PS, and (c) 5 and the letters PP.
8. Why is there a time delay between discovering a new polymer and commercializing it?
9. What are some advantages of polymers over metals?
10. It has been said that we are walking exhibitions of the importance of polymers. Explain.
11. Why are polymer-intense industries often located in the same geographical area?
12. Why might simple identification codes such as those employed for containers fail for objects such as sneakers and tires?
13. How can we become more environmentally responsible?

FURTHER READINGS

Allcock, H. (2008): *Introduction to Materials Chemistry*, Wiley, Hoboken, NJ.
Allcock, H., Lampe, F., and Mark, J. E. (2003): *Contemporary Polymer Chemistry*, 3rd ed., Prentice Hall, Upper Saddle River, NJ.
Bahadur, P. and Sastry, N., (2006): *Principles of Polymer Science*, 2nd ed., CRC Press, Boca Raton, FL.
Billmeyer, F. W. (1984): *Textbook of Polymer Science*, 3rd ed., Wiley, New York.
Bower, D. (2002): *Introduction to Polymer Physics*, Cambridge University Press, Cambridge, U.K.
Callister, W. D. (2000): *Materials Science and Engineering*, 5th ed., Wiley, New York.
Campbell, I. M. (2000): *Introduction to Synthetic Polymers*, Oxford University Press, Oxford, U.K.
Campo, E. A. (2007): *Industrial Polymers*, Hanser-Gardner, Cincinnati, OH.
Cardarelli, F. (2008): *Materials Handbook: A Concise Desktop Reference*, Springer, New York.
Carraher, C. (2004): *Giant Molecules*, Wiley, Hoboken, NJ.
Carraher, C. (2009): *Polymer Chemistry*, 7th ed., Taylor & Francis, Boca Raton, FL.
Carraher, C. and Gebelein, C. (1982): *Biological Activities of Polymers*, ACS, Washington, DC.
Challa, G. (2006): *Introduction to Polymer Science and Chemistry*, Taylor & Francis, Boca Raton, FL.
Cherdron, H. (2001): *Tailormade Polymers*, Wiley, New York.
Clark, H., Deswarte, F., and Clarke, J. ((2008): *The Introduction to Chemicals from Biomass*, Wiley, Hoboken, NJ.
Cowie, J. and Arrighi, V. (2007): *Polymers: Chemistry and Physics of Modern Material*, CRC Press, Boca Raton, FL.
Craver, C. and Carraher, C. (2000): *Applied Polymer Science*, Elsevier, New York.
Davis, F. J. (2004): *Polymer Chemistry*, Oxford University Press, New York.
Ebewele, R. O. (2000): *Polymer Science and Technology*, CRC Press, Boca Raton, FL.

Elias, H. G. (1997): *An Introduction to Polymers*, Wiley, New York.
Elias, H. G. (2008): *Macromolecules: Physical Structures and Properties*, Wiley, Hoboken, NJ.
Fischer, T. (2008): *Materials Science for Engineering Students*, Wiley, Hoboken, NJ.
Fried, J. R. (2003): *Polymer Science and Technology*, 2nd ed., Prentice-Hall, Upper Saddle River, NJ.
Gnanou, Y. and Fontanille, M. (2008): *Organic and Physical Chemistry of Polymers*, Wiley, Hoboken, NJ.
Gooch, J. (2007): *Encyclopedia Dictionary of Polymers*, Springer, New York.
Grosberg, A. and Khokhlov, A. R. (1997): *Giant Molecules*, Academic Press, Orlando, FL.
Hiemenz, P. and Lodge, T. (2007): *Polymer Chemistry*, 2nd ed., CRC Press, Boca Raton, FL.
Kricheldorf, H., Nuyken, O., and Swift, G. (2005): *Handbook of Polymer Synthesis*, Taylor & Francis,
    Boca Raton, FL.
Mark, H. (2004): *Encyclopedia of Polymer Science and Technology*, Wiley, Hoboken, NJ.
Mishra, A. (2008): *Polymer Science: A Test Book*, CRC Press, Boca Raton, FL.
Morawetz, H. (2003): *Polymers: The Origins and Growth of a Science*, Dover Publications, New York.
Munk, P. and Aminabhavi, T. M. (2002) *Introduction to Macromolecular Science*, 2nd ed., Wiley, Hoboken, NJ.
Newell, J. (2008): *Modern Materials Science and Engineering*, Wiley, Hoboken, NJ.
Nicholson, J. W. (2006): *The Chemistry of Polymers*, Royal Society of Chemistry, London, U.K.
Osswald, T. (2003): *Materials Science of Polymers for Engineering*, Hanser-Gardner, Cincinnati, OH.
Ravve, A. (2000), *Principles of Polymer Chemistry*, 2nd ed., Kluwer, New York.
Rodriguez, F. (1996): *Principles of Polymer Systems*, 4th ed., Taylor & Francis, Philadelphia, PA.
Sandler, S., Karo, W., Bonesteel, J., and Pearce, E. M. (1998): *Polymer Synthesis and Characterization:
    A Laboratory Manual*, Academic Press, Orlando, FL.
Shashoua, Y. (2008): *Conservation of Plastics*, Elsevier, New York.
Sorsenson, W., Sweeny, F., and Campbell, T. (2001): *Preparative Methods in Polymer Chemistry*, Wiley, New York.
Sperling, L. H. (2006): *Introduction to Physical Polymer Science*, 4th ed., Wiley, New York.
Stevens, M. P. (1998): *Polymer Chemistry*, 2nd ed., Oxford University Press, Oxford, England.
Strobl, G. (2007): *The Physics of Polymers*, Springer, New York.
Tanaka, T. (1999): *Experimental Methods in Polymer Science*, Academic Press, New York.
Tonelli, A. (2001): *Polymers from the Inside Out*, Wiley, New York.
Walton, D. (2001): *Polymers*, Oxford, New York.

General Encyclopedias and Dictionaries

Alger, M. (1997): *Polymer Science Dictionary*, 2nd ed., Chapman & Hall, London, U.K.
Brandrup, J., Immergut, E. H., and Grulke, E. A. (1999): *Polymer Handbook*, 4th ed., Wiley, New York.
Compendium of Macromolecular Nomenclature (1991): *IUPAC*, CRC Press, Boca Raton, FL.
Gooch, J. W. (2007): *Encyclopedic Dictionary of Polymers*, Springer, New York.
Harper, C. A. (2002): *Handbook of Plastics, Elastomers, and Composites*, McGraw-Hill, New York.
Kaplan, W. A. (Published yearly): *Modern Plastic World Encyclopedia*, McGraw-Hill, New York.
Kroschwitz, J. I. (2004): *Encyclopedia of Polymer Science and Engineering*, 3rd ed., Wiley, New York.
Mark, J. E. (1996): *Physical Properties of Polymers Handbook*, Springer, New York.
Mark, J. E. (1999): *Polymer Data Handbook*, Oxford University Press, New York.
Olabisi, O. (1997): *Handbook of Thermoplastics*, Dekker, New York.
Salamone, J. C. (1996): *Polymer Materials Encyclopedia*, CRC Press, Boca Raton, FL.
Wilkes, E. S. (2001): *Industrial Polymers Handbook*, Wiley-VCH, Weinheim, Germany.

2 Polymer Structure (Morphology)

The size and shape of polymers are intimately connected to their properties. The shape of polymers is also intimately connected to the size of the various units that comprise the macromolecule and the various primary and secondary bonding forces that are present within and between chains. This chapter covers the basic components that influence polymer shape or morphology.

We generally describe the structure of both synthetic and natural polymers in terms of four levels of structures. The *primary structure* describes the precise sequence of the individual atoms that compose the polymer chain. For polymers where there is only an average structure, such as proteins, polysaccharides, and nucleic acids, a representative chain structure is often given.

The structure can be given as a single repeat unit such that the full polymer structure can be obtained by simply repeating that unit 100, 500, 1000 times depending on the precise number of repeat units in the polymer chain. For poly(vinyl chloride), PVC, this is

$$R-(-CH_2-CH-)-R \quad \text{or} \quad R-(-CH_2-CH(Cl)-)-R \quad \text{or} \quad -(CH_2-CH-) \tag{2.1}$$
$$\qquad\qquad | \qquad\qquad\qquad\qquad\qquad\qquad\qquad\qquad\qquad\qquad\qquad | $$
$$\qquad\qquad Cl \qquad\qquad\qquad\qquad\qquad\qquad\qquad\qquad\qquad\qquad\qquad Cl$$

Or some fuller description of the primary structure may be given such as that below for three repeat units of PVC where the particular geometry about each chiral carbon is given:

$$\tag{2.2}$$

The ends may or may not be given depending on whether they are important to the particular point being made. Thus, for the single PVC repeat unit given above the end groups may be given as follows:

$$CH_3-CH-(-CH_2-CH-)_n-CH_2-CH_2Cl \tag{2.3}$$
$$\qquad\quad | \qquad\qquad\quad | $$
$$\qquad\quad Cl \qquad\qquad Cl$$

The repeat unit for cellulose is

$$\tag{2.4}$$

The *secondary structure* describes the molecular shape or the conformation of the polymer chain. For most linear polymers this shape approaches a helical or "pleated skirt" (or sheet) arrangement depending on the nature of the polymer, treatment, and function. Examples of secondary structures appear in Figure 2.13.

The *tertiary structure* describes the shaping or folding of the polymer. Examples of tertiary structures are given in Figures 2.16b and 2.17a and c.

Finally, the *quaternary structure* represents the overall shape of groups of the tertiary structures where the tertiary structures may be similar or different. Examples of quaternary structures are found in Figures 2.15, 2.16a, and 2.17b.

2.1 STEREOCHEMISTRY OF POLYMERS

The terms "memory" and "to remember" are similar and used by polymer chemists in similar, but different ways. The first use of the terms "memory" and "to remember" involves reversible changes in the polymer structure usually associated with stress–strain deformation of a rubber material where the dislodged, moved polymer segments are connected to one another through chemical and physical cross-links so that once the particular stress–strain is removed, the polymer returns to its original, prestress/strain arrangement of the particular polymer segments. Thus, the polymer "remembers" its initial segmental arrangement and returns to it through the guiding of the cross links.

The second use involves nonreversible changes of polymer segments and whole chain movements also brought about through stress–strain actions. These changes include any synthetic chain and segmental orientations as well as post-synthesis changes including fabrication effects. These changes involve "permanent" differences in chain and segmental orientations, and in some ways, these changes represent the total history of the polymer materials from inception (synthesis) through the moment when a particular property or behavior is measured. These irreversible or nonreversible changes occur with both cross-linked and non-cross-linked materials and are largely responsible for the change in polymer property as the material moves from being synthesized, processed, fabricated, and used in whatever capacity it finds itself. Thus, the polymeric material "remembers" its history with respect to changes and forces that influence chain and segmental changes. The ability of polymers to "remember" and have a "memory" is a direct consequence of their size.

We can get an idea of the influence of size in looking at the series of methylene hydrocarbons as the number of carbon atoms increases. For low numbers of carbons, methane, ethane, propane, butane, the materials are gases at room temperature. For the next groupings (Table 2.1), the materials

TABLE 2.1
Typical Properties of Straight Chain Hydrocarbons

Average Number of Carbon Atoms	Boiling Range (°C)	Name	Physical State at Room Temp.	Typical Uses
1–4	<30	Gas	Gas	Heating
5–10	30–180	Gasoline	Liquid	Automotive fuel
11–12	180–230	Kerosene	Liquid	Jet fuel, heating
13–17	230–300	Light gas oil	Liquid	Diesel fuel, heating
18–25	305–400	Heavy gas oil	Viscous liquid	Heating
26–50	Decomposes	Wax	Waxy	Wax candles
50–1000	Decomposes		Tough waxy to solid	Wax coatings of food containers
1000–5000	Decomposes	Polyethylene	Solid	Bottles, containers, films
>5000	Decomposes	Polyethylene	Solid	Waste bags, ballistic wear, fibers, automotive parts, truck liners

are liquids. Dispersion forces that are a sum of the individual methylene and end group forces hold the individual hydrocarbon chains together. There is a gradual increase in boiling point and total dispersion forces for the individual chains until the materials become waxy solids such as found in bees waxes, and finally where the total dispersion forces are sufficient to be greater than individual carbon–carbon bond strengths so that the chains decompose prior to their evaporation. Finally, the chain lengths are sufficient to give tough solids, which we call linear polyethylene. It is interesting to note that these linear long-chain hydrocarbons become very strong but they are brittle. They are crystalline and as with most other crystalline materials, such as many rocks and diamonds, they are strong but brittle. Fortunately, synthetic polyethylene contains both *crystalline regions* where the polymer chains are arranged in ordered lines and they contain regions where the chains are not arranged in as orderly lines. These latter arrangements are imposed on the polyethylene because of the presence of some branching in the linear polymer backbone and the need for long chains to be present in several crystalline structures with less-ordered regions present between the crystalline portions. These noncrystalline regions are referred to as *amorphous regions* and are responsible for allowing the polyethylene to have some flexibility. Thus, many polymers contain both amorphous and crystalline regions that provide both flexibility and strength.

The result of cumulative secondary forces can be readily seen by looking at the properties of linear alkanes as the number of methylene units increases. As noted before, methane, ethane, and propane are gases. As the number of methylene units increases to hexane, heptane, octane, the alkanes become liquids. Further increases give a variety of waxes including ones that make birthday candles. These materials are solids but not strong. As the number of methylene units increases to several hundreds, the material is referred to as low molecular weight polyethylene and it has some strength and now feels and acts more like a plastic material. Increasing the chain size gives eventually ultrahigh molecular weight polyethylene that is very strong and used in applications where strength is important such as in the construction of ballistic resistant body armor (i.e., bulletproof vests.) As a side note, low molecular weight polyethylene with appreciable side branching has a melting range generally below 100°C whereas high molecular weight polyethylene with few branches has a melting range approaching the theoretical value of about 145°C. It is interesting to note that candle wax is largely crystalline but it is brittle and lacks strength since the various crystalline groupings are not attached together. By comparison, linear crystalline polyethylene is strong because the chains are sufficiently long to allow them to connect the various crystalline groupings (domains) together. This connecting also allows applied forces to be distributed throughout the surrounding areas on a micro-level.

The polyethylene chains described in Table 2.1 exhibit irreversible and, when appropriate cross-linking is present, reversible memory because of their size.

High-density polyethylene (HDPE), formerly called low-pressure polyethylene, $[H(CH_2CH_2)_n)H]$, like other alkanes $[H(CH_2)_nH]$, may be used to illustrate a lot of polymer structures. As in introductory organic chemistry, we can comprehend almost all of the complex organic compounds if we understand their basic chemistry and geometry. HDPE, like decane $[H(CH_2)_8H]$ or paraffin $[H(CH_2)_{about\ 50}H]$, is a largely linear chain-like molecule consisting of catenated carbon atoms bonded covalently. The carbon atoms in all alkanes, including HDPE, are joined at characteristic tetrahedral bond angles of approximately 109.5°. While decane consists of 10-methylene groups, HDPE may contain more than 1000 of these methylene units (Figure 2.1). While we use the term normal or linear to describe nonbranched chains, we know that because of the tetrahedral bond angles and ability for twisting, the chains are zigzag shaped with many possible structural variations.

The distance between the carbon atoms is 1.54 Å or 0.154 nm. The apparent zigzag distance between carbon atoms in a chain of many carbon atoms is 0.126 nm. Thus, the length of an extended nonane chain is 8 units times 0.126 nm/units = 1.008 nm. For polyethylene, the repeat unit has two methylenes so that the apparent zigzag is 0.252 nm for each "ethylene" unit. The zigzag or contour length of a HDPE chain 1000 units long, $[H(CH_2CH_2)_{1000}H]$ is 1000 units times 0.252 nm/units or 252 nm.

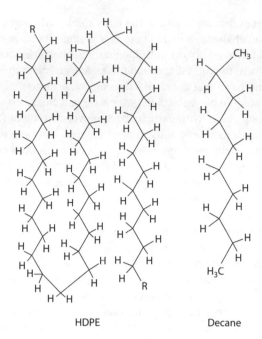

HDPE   Decane

FIGURE 2.1 Simulated structure of HDPE contrasted with the structural formula of linear or normal decane.

However, because of rotations about the carbon atoms, chains seldom extend to their full extended *contour* length but are present in many different shapes or **conformations**.

The *full contour length* of a polymer is obtained by multiplying the apparent repeat unit length (l), that is the length of each mer or unit, by the number of units (n); contour length = nl. Even so, it is important to know the average end-to-end distance of polymer chains. The statistical method for this determination, called the random flight technique, was developed by Lord Raleigh in 1919. The classical statistical approach may be used to show the distance traveled by a blindfolded person taking n number of steps of length l in a random walk or the distance flown by a confused moth, bird, or bee.

The distance traveled from start to finish is not the straight-line path measured as nl but the root-mean-square distance ($(r^2)^{1/2}$), which is equal to $ln^{1/2}$. Thus, the root-mean-square length of a flexible PE chain with 1000 units is 0.252 nm times $(1000)^{1/2}$ = 7.96 nm or about 3% of the contour length. Nobel laureate Paul Flory and others have introduced several corrections so that this random flight technique could be applied to polymer chains approaching a full contour length of nl; that is, rigid rod structures.

Each specific protein molecule has a specific chain length, like classical small molecules, and are said to be monodisperse with respect to chain length or molecular weight. However, most synthetic commercial polymers such as HDPE are composed of molecules of different lengths. Thus, the numerical value for n, or the *degree of polymerization* ($\overline{DP}$) should be considered an average $\overline{DP}$ or average molecular weight. This average notion is often noted by a bar over the top of the $\overline{DP}$ or M. Accordingly, the average molecular weight of a *polydisperse* polymer will equal the product of the average $\overline{DP}$ times the molecular weight of the repeating unit or *mer*.

In organic chemistry, it is customary to call a nonlinear molecule, such as isobutane, a branched chain. However, polymer scientists used the term *pendant group* to label any group present on the repeat unit. Thus, polypropylene

$$-(-CH_2-CH-)- \overset{\displaystyle CH_3}{\underset{\displaystyle |}{}}$$

(2.5)

FIGURE 2.2 Simulated structural formula for branched LDPE; compare with Figure 2.1 for HDPE.

has a methyl group as a pendant unit, but polypropylene (PP) is designated as a linear polymer. In contrast, low-density polyethylene (LDPE), formally called high-pressure polyethylene, is a branched polymer because chain extensions or branches of methylene units are present coming off of branch point along the typically linear backbone chain (Figure 2.2). For LDPE the frequency of this branching is about 1.5 per 20 methylene units to 1 per 2000 methylene units. This branching, like branching in simple alkanes, increases the specific volume and thus reduces the density of the polymer. The linearity provides strength while the branching provides flexibility and toughness. Recently, low-pressure processes have been developed that produce linear low-density polyethylene (LLDPE) that is largely linear but with some branching (Table 2.2).

Both linear and branched polymers are typically *thermoplastics* that melt when heated. However, cross-linked three-dimensional, or network, polymers are *thermoset polymers* that do not melt when heated but rather decompose prior to melting. The cross-link density can vary from low, as found in a rubber band, to high, as found in ebonite (Figure 2.3).

While there is only one possible segment arrangement for the repeat units in HDPE, there are two possible repeat units in PP and many other polymers. The units can be connected using

TABLE 2.2
Types of Commercial Polyethylene

	General Structure	Crystallinity (%)	Density (g/cc)
LDPE	Linear with branching	50	0.92–0.94
LLDPE	Linear with less branching	50	0.92–0.94
HDPE	Linear with little branching	90	0.95

(a)   (b)   (c)

FIGURE 2.3 Skeletal structural formulas of a linear polymer (a), and a network (cross-linked) polymer with low cross-linking density (b) and high density cross-linking (c).

Head-to-tail (2.6)   Head-to-head (2.7)

FIGURE 2.4 Simulated structural formulas showing the usual head-to-tail and unusual head-to-head configurations of polypropylene.

a *head-to-tail arrangement* or through a *head-to-head arrangement* (Figure 2.4). The usual arrangement by far is the head-to-tail so that pendant methyl groups appear on alternate backbone carbons.

The polymerization of monosubstituted vinyl compounds to give polymers such as polystyrene and polypropylene give polymer chains that possess chiral sites on every other carbon in the polymer backbone. Thus, the number of possible arrangements within a polymer chain is staggering since the number of possible isomers is 2^n where n is the number of chiral sites. For a relatively short chain containing 50 propylene units, the number of isomers is about 1 times 10^{15}. While the presence of such sites in smaller molecules can be the cause of optical activity, these polymers are not optically active since the combined interactions with light are negated by similar, but not identical, other sites contained on the particular and other polymer chains. Further, because of their chiral differences, it is quite possible that no two polymer chains made during a polymerization will be exactly identical.

The particular combinations of like and mirror image units within a polymer chain influence the polymer properties on a molecular level. On the bulk level, the average individual chain structure influences properties.

In the early 1950s, Nobel laureate Giulio Natta used stereospecific coordination catalysts to produce stereospecific isomers of PP. Natta used the term tacticity to describe the different possible structures. As shown in Figure 2.5 the isomer corresponding to the arrangement DDDDDD or

FIGURE 2.5 Skeletal formulas of isotactic, syndiotactic, and atactic of poly(vinyl chloride), PVC.

FIGURE 2.6 Simulated formulas of ditactic isomers where R_2 are chain extensions and R and R_1 are not hydrogen.

FIGURE 2.7 Newman projections of designated conformers of n-butane.

LLLLL is called *isotactic* (same). The isomer corresponding to the DLDLDLDL structural arrangement about carbon is called *syndiotactic* (alternating). The isomer arrangement that corresponds to some mix of stereo arrangements about the chiral carbons is called *atactic* (having nothing to do with). The differences in stereoregularity about the chiral carbon influence the physical properties of the polymers. Thus, those with isotactic or syndiotactic arrangements are more apt to form compact crystalline arrangements and those with atactic stereoregularity are more apt to form amorphous arrangements. Isotactic PP (iPP) has a MP of about 160°C and it is highly crystalline whereas atactic PP (aPP) melts at about 75°C and is amorphous. The term *eutactic* is used to describe either an isotactic or syndiotactic polymer or a mixture of both.

While most polymers contain only one chiral or asymmetrical center in each repeat unit, it is possible to have diisotacticity where two different substituents are present at chiral centers. These isomers are labeled erythro- and threodiisotactic and erythro- and threosyndiotactic isomers (Figure 2.6).

Many different conformers resulting from rotation about the carbon–carbon bonds in a simple molecule like *n*-butane may be shown by Newman projections (Figure 2.7). The most stable is the anti or trans projection where the steric hindrance is minimized. There are a number of eclipsed and gauche arrangements of which only one of each is shown in Figure 2.7. The energy difference between the anti and eclipsed, the least stable form, is about 12 kJ/mol.

The ease in going from one conformer to the other conformer decreases as the pendant groups increase in size and increase in secondary bonding. Thus, poly(methyl methacrylate) (PMMA) is hard at room temperature because of the polar groups and steric hindrance. By comparison, polyisobutylene, offering only a small amount of steric hindrance, is flexible at room temperature.

2.2 MOLECULAR INTERACTIONS

Forces in nature are often divided into primary forces (typically greater than 200 kJ/mol) and secondary forces (typically less than 40 kJ/mol). *Primary bonding* forces can further be subdivided into ionic (characterized by a lack of directional bonding; occurs between atoms of large differences in electronegative; normally not found in polymer backbones), metallic (the number of outer, valence electrons is too small to provide complete outer shells; often considered as charged atoms surrounded by a potentially fluid sea of electrons; lack of bonding direction), and covalent (including dative and coordinate) bonding (the major bonding in organic compounds and polymers; directional). The bonding length of primary bonds is generally about 0.1–0.22 nm. The carbon–carbon bond length is about 0.15–0.16 nm. Atoms in individual polymer chains are joined to one another by covalent bonds with bonding energies about 320–370 kJ/mol.

Polymer molecules are also attracted to one another through secondary forces. Secondary forces, often called *van der Waals forces* because they are the forces responsible for the van der Waals corrections to the ideal gas relationships, are of longer distance in interaction, in comparison to primary forces. Secondary bonding distances are generally about 0.25–0.5 nm. The force of these interactions is inversely proportional to some power of the distance, generally 2 or greater (i.e., force is proportional to 1/(distance)^r). Thus, many physical properties are sensitive to the polymer *conformation* (arrangements related to rotation about single bonds) and *configuration* (arrangements related to the actual chemical bonding about a given atom), since both affect the proximity one chain can be relative to another. Thus, amorphous PP is more flexible than crystalline PP because the crystalline PP has the units closer to one another allowing the secondary bonding to be stronger (Figure 2.8).

These intermolecular forces are also responsible for the increase in boiling points within a homologous series such as the alkanes, for the higher-than-expected boiling points of polar molecules such as alkyl chlorides, and for the abnormally high boiling points of water, alcohols, amines, and amides. While the forces responsible for these increases in boiling points are all van der Waals forces, the forces can be further subdivided in accordance to their source and intensity. Secondary forces include London dispersion forces, induced permanent forces, and dipolar forces including hydrogen bonding.

All molecules including nonpolar molecules such as heptane and polyethylene are attracted to each other by weak *London* or *dispersion forces* that result from induced dipole–dipole interactions.

FIGURE 2.8 Representation of a crystalline portion from (a) isotactic polypropylene and (b) an amorphous portion from atactic polypropylene.

The temporary or transient dipoles are due to instantaneous fluctuations in the electron cloud density. The energy range of these forces is fairly constant and about 8 kJ/mol. This force is independent of temperature and is the major force between chains in largely nonpolar polymers such as those in classical elastomers and soft plastics such as PE.

It is of interest to note that methane, ethane, and ethylene are all gases; hexane, octane, and nonane are all liquids (at room conditions); while low molecular weight PE is a waxy solid. This trend is primarily due to an increase in the mass per molecule and to an increase in the London forces per polymer chain. The London force interaction between methylene units is about 8 kcal/mol. Thus, for methane molecules the attractive forces is 8 kJ/mol; for octane it is 64 kJ/mol; and for polyethylene with 1000 ethylene (or 2000 methylenes) it is 2,000 methylene units × 8 kJ/mol per methylene unit = 16,000 kJ/mol well sufficient to make PE a solid and to break backbone bonds before it boils. (Polymers do not boil because the energy necessary to make a chain volatile is greater than the primary backbone bond energy.)

Polar molecules such as ethyl chloride and PVC are attracted to each other by both the London forces, but also *dipole–dipole interactions* resulting from the electrostatic attraction of a chlorine atom in one molecule to a hydrogen atom in another molecule. These dipole–dipole forces are of the order of 8–25 kJ/mol, generally greater than the London forces and they are temperature dependent. Hard plastics, such as PVC, have these dipole–dipole forces present between chains.

Strongly polar molecules such as ethanol, poly(vinyl alcohol) (PVA), cellulose, and proteins are attracted to each other by a special type of dipole–dipole force called *hydrogen bonding*. Hydrogen bonding occurs when hydrogen present on a highly negative element, such as nitrogen or oxygen, comes close to another highly electronegative element. These forces are in the order of 40 kJ/mol and for something like hydrogen fluoride, they are almost as strong as primary bonding. Intermolecular hydrogen bonding is usually present in classical fibers such as cotton, wool, silk, nylon (Figure 2.9), polyacrylonitrile, polyesters, and polyurethanes. Intramolecular hydrogen bonds are responsible for the helices observed in starch and globular proteins.

It is important to note that the high melting point of nylon-66 (Figure 2.9; 265°C) is a result of a combination of dipole–dipole, London, and hydrogen bonding forces. The amount of hydrogen bonding decreases as the number of methylene groups increases, such as for nylon-610, and a corresponding decrease is seen in the melting point.

In addition to the contribution of intermolecular forces, chain entanglement is also an important contributory factor to the physical properties of polymers. While paraffin wax and HDPE are homologs with relatively high molecular weights, the chain length of paraffin is too short to permit chain entanglement and hence it lacks the strength and other characteristic properties of HDPE.

Chain entanglement allows long chains to act as though they were even longer because entanglement causes the entangled chains to act together. Many of the physical properties, such as tensile strength, of polymers increase dramatically when chain entanglement occurs. The *critical chain*

FIGURE 2.9 Typical hydrogen bonding (shown as "–") between hydrogen on nitrogen and oxygen for nylon-66.

length (*z*) required for the onset of entanglement is dependent on the polarity and shape of the polymer. The number of atoms in the critical chain lengths of PMMA, polystyrene (PS), and poly-isobutylene are 208, 730, and 610 respectively.

Viscosity is a measure of the resistance to flow. The latter is a result of cooperative movement of the polymer segments from vacate location, hole, to another vacate location in a melted state. Chain entanglement, high intermolecular forces, the presence of reinforcing agents, and cross-links impede this movement. The *melt viscosity* (η) is often found to be proportional to the 3.4 power of the critical chain length, as shown in Equation 2.8, regardless of the polymer. The constant *K* is temperature dependent.

$$\log \eta = 3.4 \log z + \log K \tag{2.8}$$

The flexibility of amorphous polymers above the glassy state is governed by the same forces as melt viscosity and is dependent on a wriggling type of segment motion in the polymer chains. This flexibility is increased when many methylene groups ($-CH_2-$) or oxygen atoms ($-O-$) are present. Thus, the flexibility of aliphatic polyesters usually increases as *m* is increased (2.9).

$$-(-(CH_2)_m-O-C-(CH_2)_m-C-O-)-_n \tag{2.9}$$

Aliphatic polyester

Flexibilizing groups include methylene and ethylene oxides, dimethylsiloxanes, and methylene groups.

In contrast, the flexibility of amorphous polymer above the glass state is decreased with stiffening groups such as

1,4-Phenylene (2.10)   Amide (2.11)   Sulfone (2.12)   Carboxyl (2.13)

are present in the polymer backbone. Thus, poly(ethylene terephthate) (PET; 2.14) is stiffer and higher melting than poly(ethylene adipate; 2.15), and the former is stiffer than poly(butylene terephthalate) because of the presence of fewer methylene groups between the stiffening groups.

Poly(ethylene terephthalate) (2.14)   Poly(ethylene adipate) (2.15)

Small molecules such as water can exist in three phases—liquid, solid, and gas. Polymers do not boil but rather degrade before boiling, so they do not exist in the gas state. Polymers generally undergo several major thermal transitions. At low temperatures, polymers are brittle, glassy since there is not sufficient energy present to encourage even local or segmental chain movement. As the temperature is increased at some temperature, there is sufficient energy available to allow some chain mobility. For a polymer containing both amorphous and crystalline portions or is only amorphous, the onset of this segmental chain mobility for the amorphous segments is called the glass transition temperature, T_g. Because there is unoccupied volume in the amorphous polymer structure, some segmental chain movement occurs. This segmental chain movement is sometimes likened to a snake slithering "in place" within the grass. The localized chain movement causes a further increase in unoccupied volume and larger segments are able to move eventually allowing the snake further movement in the grass. As the temperature is increased, there is sufficient temperature to overcome the forces present in the crystalline portion of the polymer allowing a breaking up of the crystalline portion. This temperature is often referred to as the ***crystalline transition temperature***, T_c. This temperature is directly related to the melt transition temperature, T_m. Finally, as the temperature continues to rise there is sufficient energy available to overcome the primary bonds holding the polymer together and the polymer decomposes at the ***decomposition temperature***, T_d. The T_g and T_m are described further in the following text.

The flexibility of amorphous polymers is reduced drastically when they are cooled below a characteristic transition temperature called the ***glass transition temperature*** (T_g). At temperatures below T_g there is no segmental motion and any dimensional changes in the polymer chain are the result of temporary distortions of the primary covalent bonds. Amorphous plastics perform best below T_g but elastomers must be used above the brittle point, T_g, or they will act as a glass and be brittle and break when bent. The importance of a material being above its T_g to offer some flexibility was demonstrated by the space shuttle Challenger disaster where the cool temperature at the launch pad resulted in the "O" ring not being flexible so that fuel escaped resulting in the subsequent explosion.

The ***melting point***, T_m, is the temperature range where total or whole polymer chain mobility occurs. The T_m is called a first-order transition temperature, and T_g is sometimes referred to as a second-order transition. The values for T_m are usually 33%–100% greater than the T_g. Symmetrical polymers such as HDPE exhibit the greatest difference between T_m and T_g. The T_g values are low for elastomers and flexible polymers such as PE and dimethylsiloxane, and relatively high for hard amorphous plastics such as polyacrylonitrile and PET (Table 2.3).

TABLE 2.3
Approximate Glass Transition Temperatures (T_g)
for Selected Polymers

Polymer	T_g (K)	Polymer	T_g (K)
Cellulose acetate butyrate	323	Cellulose triacetate	430
Polyethylene (LDPE)	148	Polytetrafluoroethylene	160, 400[a]
a-Polypropylene	253	Poly(ethyl acrylate)	249
i-Polypropylene	373	Poly(methyl acrylate)	279
Polyacrylonitrile	378	a-Poly(butyl methacrylate)	339
Poly(vinyl acetate)	301	a-Poly(methyl acrylate)	378
Poly(vinyl alcohol)	358	Poly(vinyl chloride)	354
cis-Poly-1,3-butadiene	165	Nylon-66	330
trans-Poly-1,3-butadiene	255	Poly(ethylene adipate)	223
Polydimethylsiloxane	150	Poly(ethylene terephthalate)	342
Polystyrene	373		

[a] Two major transitions observed.

The T_g for isotactic PP is 373 K or 100°C yet, because of its high degree of crystallinity, it does not flow to any great extent below its melting point of 438 K (165°C). In contrast, the highly amorphous polyisobutylene, which has a T_g value of 203 K (−70°C), flows at room temperature (Table 2.3). T_g decreases as the size of the ester groups increases in polyacrylates and polymethylacrylates. The effect of the phenylene stiffening groups is also demonstrated with the T_g of PET being about 120°C higher than that of poly(ethylene adipate).

The main reasons why amorphous polymers go from a solid glassy state to a more flexible plastic state are the presence of sufficient energy and unoccupied volume. The energy is supplied by heating the sample, this provides the polymer sufficient energy for the chain segments to become separated and begin movement which in turn creates free or unoccupied volume that allows the chain segments to slip past one another resulting in the material being more flexible. In order for the chains to begin moving, the secondary forces that hold the chains together must be overcome. As movement begins, additional unoccupied volume is created and this expansion within a complex maze of intertwining chains creates additional free volume. A measure of this expansion is the ***thermal coefficient of expansion***. The temperature range where the available free volume and energy necessary to overcome segmental chain interactions is available is called the glass transition temperature, T_g. Since the specific volume of polymers increases at T_g in order to accommodate the increased segmental chain motion, T_g values can be estimated from plots of the change in specific volume with temperature (Figure 2.10).

Below the T_g, the chains are "frozen" into place and the material acts as a brittle solid or glass, hence the name glassy state.

Other properties such as stiffness (modulus), refractive index, dielectric properties, gas permeability, x-ray adsorption, and heat capacity (Figure 2.11) all change at T_g and have been used to determine the T_g. As seen in Figure 2.11 both T_g and T_m are endothermic because energy is absorbed when segmental mobility or melting occur. Moreover, causing whole sale mobility takes more energy than causing segmental mobility so that the area under the curve, a direct measure of the heat being taken on to effect the mobility, for the T_m is greater than for T_g where the amount of amorphous and crystalline regions are nearly equal. While T_g values are the most often reported to be related to the onset of segmental motion in the principle polymer backbone, other secondary values may be observed for the onset of motion of large pendant groups, branches, etc.

The number of possible conformers increases with chain length and can be shown statistically to equal 2^{2n} where n is the number of units. Thus, when $n = 1000$, the number of possible conformers of HDPE is 2^{2000} or 10^{600}, more than the grains of sand at all of our beaches. Four of these possible conformers are shown in Figure 2.12.

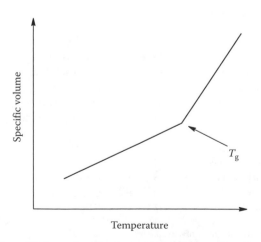

FIGURE 2.10 Determination of T_g by noting the abrupt change in specific volume.

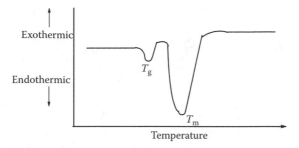

FIGURE 2.11 Typical DSC thermogram of a polymer. (From Gaylord, N., in *Linear and Stereoregular Addition Polymers*, Gaylord, N. and Mark, H., eds., Wiley, New York, 1959. With permission.)

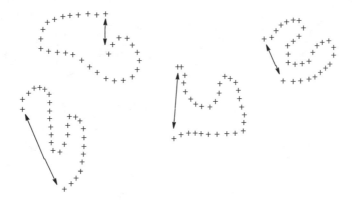

FIGURE 2.12 End-to-end distances for four 30 unit chains.

However, because there are so many possible ends in branched polymers, it is customary to use the radius of gyration (S) instead of r for such polymers. The radius of gyration is actually the root-mean-square distance of a chain end from the polymer's center of gravity. S is less than the end-to-end distance (r), and for linear polymers $r^2 = 6S^2$.

In general, polymers (both natural and synthetic) "emphasize" two general shapes—helical and pleated (Figures 2.9 and 2.13). The intermolecular bonds in many polyamides, including natural polyamides such as β-keratin, produce strong pleated sheets. Hair, fingernails and toenails, feather, and horns have a β-keratin structure. Helical structures are often found because they can minimize steric factors by "radiating" them outward from a core distributing the steric groups about a helical circle. Polyurethanes, polyacrylonitrile, and polyesters are characterized by the presence of strong hydrogen and polar bonding. In contrast, iPP, which has no hydrogen bonding, is also a strong fiber because of the ability of the similar chains to fit closely together.

α-Keratin (composed of parallel polypeptide α-helices) and most globular proteins are characterized by intramolecular bonds. These, and many other polymers, including nucleic acids, may form helices. Ribonucleic acid (RNA) exists as a single stranded helix, while deoxyribonucleic acid (DNA) exists as a double stranded helix.

2.3 POLYMER CRYSTALS

Prior to 1920, leading scientists not only stated that macromolecules were nonexistent, but they also believed, if they did exist, they could not exist as crystals. However, in the early 1920s, Haworth used x-ray diffraction techniques to show that elongated cellulose was a crystalline polymer consisting of repeat units of cellobiose. In 1925, Katz in jest placed a stretched natural rubber band in

FIGURE 2.13 Helical conformation of isotactic vinyl polymers. (From Gaylord, N., in *Linear and Stereoregular Addition Polymers*, Gaylord, N. and Mark, H., eds., Wiley, New York, 1959.)

an x-ray spectrometer and to his surprise observed an interference pattern typical of a crystalline substance. This phenomenon may be shown qualitatively by the development of opacity when a rubber band is stretched (try it yourself) and by the abnormal stiffening and whitening of unvulcanized rubber when it is stored for several days at 0°C. The opacity noted in stretched rubber and cold rubber is the result of the formation of crystallites, or regions of crystallinity. The latter were first explained by a fringed micelle model, which is now found not consistent with much of the current experimental findings (Figure 2.14).

Amorphous polymers with irregular bulky groups are seldom crystallizable, and unless special techniques are used, even ordered polymers are seldom 100% crystalline. The combination of amorphous and crystalline structures varies with the structure of the polymer and the precise conditions that have been imposed on the material. For instance, rapid cooling often decreases the amount of crystallinity because there is not sufficient time to allow the long chains to organize themselves into more ordered structures. The reason linear ordered polymers fail to be almost totally crystalline is largely kinetic, resulting from an inability of the long chains to totally disentangle and perfectly align themselves during the time the polymer chain is cooling and mobile.

FIGURE 2.14 Schematic two-dimensional representation of a modified micelle model of the crystalline-amorphous structure of polymers.

Mixtures of amorphous and mini-crystalline structures or regions may consist of somewhat random chains containing some chains that are parallel to one another forming short-range mini-crystalline regions. Crystalline regions may be formed from large range ordered platelet-like structures, including polymer single crystals, or they may form even larger organizations such as *spherulites* as shown in Figures 2.15 and 2.16. Short and longer range ordered structures can act as physical cross-links.

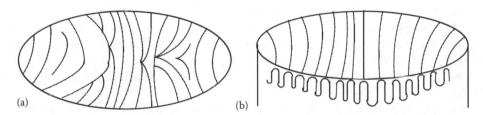

FIGURE 2.15 Structure of a spherulite from the bulk. Bottom shows a slice of a simple spherulite (b). As further growth occurs, filling in, branch points, etc. occur as shown on top (a). The contour lines are simply the hairpin turning points for the folded chains.

Lamellar chain-folded crystallite

Tie-chain segments/ amorphous region

(a)   (b)

FIGURE 2.16 (a) Spherulite structure showing the molecular-level lamellar chain-folded platelets and tie and frayed chain arrangements, and (b) a more complete model of two sets of three lamellar chain-folded platelets formed from polyethylene. Each platelet contains about 850 ethylene units as shown here.

In general, linear polymers form a variety of single crystals when crystallized from dilute solutions. For instance, highly linear PE can form diamond-shaped single crystals with a thickness in the order of 11–14 nm when crystallized from dilute solution. The surface consists of "hairpin turned" methylene units as pictured in Figures 2.15 and 2.16. The polymer chain axes are perpendicular to the large flat crystal faces. A single polymer chain with 1000 ethylene (2000 methylene) units might undergo about 50 of these hairpin turns on the top surface and another 50 turns on the bottom face with about 20 ethylene units between the two surfaces.

Many polymers form more complex single crystals when crystallized from dilute solution including hollow pyramids that often collapse on drying. As the polymer concentration increases, other structures occur, including twins, spirals, and multilayer dendritic structures with the main structure being spherulites.

When polymer solids are produced from their melts, the most common structures are these spherulites that can be seen by the naked eye and viewed as Maltese cross-like structures with polarized light and crossed Nicol prisms under a microscope.

For linear PE, the initial structure formed is a single crystal with folded chain *lamellae*. These quickly lead to the formation of sheaflike structures called axialites or hedrites. As growth proceeds, the lamellae develop on either side of a central reference point. The lamellae continue to fan out, occupying increasing volume sections through the formation of additional lamellae at appropriate branch points. The result is the formation of spherulites as pictured in Figures 2.15 and 2.16.

While the lamellar structures present in spherulites are similar to those present in polymer single crystals, the folding of chains in spherulites is less organized. Further, the structures that exist between these lamellar structures are generally occupied by amorphous structures including atactic chain segments, low molecular weight chains, and impurities.

The individual spherulite lamellae are bound together by "tie" molecules that are present in more than one spherulite. Sometimes these tie segments form intercrystalline links, which are threadlike structures, that are important in developing the characteristic good toughness found in semicrystalline polymers. They act to tie together the entire assembly of spherulites into a more or less coherent "package."

Depending upon the particular conditions of crystallization, a number of secondary and tertiary structures can be formed. In most cases, crystalline polymers attempt to form crystalline platelets. Under little or no externally applied stress, these platelets organize themselves in spherulites as pictured in Figures 2.15 and 2.16. They start by a nucleating process and begin to radiate outward from the central nucleating site. Amorphous chain segments get trapped between the forming crystalline platelet combinations resulting in a type fuzzy or frayed exterior. These platelets are generally either planar, as shown in Figure 2.17 or they can be helical or twisted. The platelets contain to grow until they butt up against other spherulites.

Under externally applied stress, including simple melt flow, the tertiary structure can approach a shish kebab arrangement where there are planes of platelets separated by areas where there exists both crystalline and amorphous regions as pictured in Figure 2.17a. These shish kebab structures often organize into quaternary structures consisting of bundles of shish kebab single strand filaments forming fibrils. Under more rapid flow conditions, the shish kebab itself become distorted, Figure 2.17c.

Interestingly, the amorphous regions within the spherulite confer onto the material some flexibility while to crystalline platelets give the material strength, just as in the case with largely amorphous materials. This theme of amorphous flexibility and crystalline strength (and brittleness) is a central idea in polymer structure–property relationships.

It must be remembered that the secondary structure of both the amorphous and crystalline regions typically tend toward a helical arrangement of the backbone for most polymers but not polyethylene that forms a crank-shaft structure because of the lack of steric restraints (i.e., lack of pendant groups off the backbone).

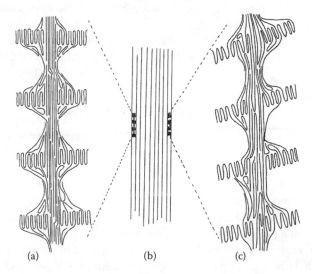

FIGURE 2.17 (a) Crystalline polymer structures formed under applied tension including flow conditions. (b) The tertiary mono-fibrillar structure including platelets and at the right shows these mono-fibrillar structures bundled together forming a quaternary structure fibril. (c) The distorted shish kebab formed with more rapid flow.

The kind, amount, and distribution of polymer chain order/disorder (crystalline/amorphous) is driven by the processing (including pre- and post-) conditions and thus it is possible to vary the polymer properties through a knowledge of and ability to control the molecular-level structures. The crystalline regions may be disrupted by processing techniques, such as thermoforming and extrusion of plastics and drawing of fibers. In the last process, which is descriptive of the others, the crystallites are ordered in the direction of the stress, the filament shrinks in diameter, and heat is evolved and reabsorbed as a result of additional orientation and crystallization.

In addition to crystallization of the backbone of polymers, crystallization may also occur in regularly spaced bulky groups even when an amorphous structure is maintained in the backbone. In general, the pendant group must contain at least 10 carbon atoms in order for this side-chain crystallization to occur. Ordered polymers with small pendant groups crystallize more readily than those with bulky groups do.

While polymeric hydrocarbons have been used as illustrations for simplicity, it is important to note that the principles discussed apply to all polymers, organic as well as inorganic, and natural as well as synthetic, and to elastomers, plastics, and fibers.

2.4 AMORPHOUS BULK STATES

An amorphous bulk polymer contains chains that are arranged in something less than a well-ordered, crystalline manner. Physically, amorphous polymers exhibit a T_g but not a T_m, and do not give a clear x-ray diffraction pattern. Amorphous polymer chains have been likened to spaghetti strands in a pot of spaghetti, but in actuality, the true extent of disorder that results in an amorphous polymer is still not fully understood.

Section 12.3 contains a discussion on a number of techniques employed in the search for the real structure of the amorphous bulk state. Briefly, evidence suggests that little order exists in the amorphous state, the order being similar to that observed with low molecular weight hydrocarbons. There is evidence that there is some short-range order and for long-range interactions and that the chains approximate a random coil with some portions paralleling one another. In 1953,

Flory and Mark suggested a random coil model whereby the chains had conformations similar to those present if the polymer were in a theta solvent. In 1957, Kargin suggested that amorphous polymer chains exist as aggregates in parallel alignment. Models continue to be developed, but all contain the elements of disorder/order suggested by Flory and Mark and the elements of order suggested by Kargin.

2.5 POLYMER STRUCTURE–PROPERTY RELATIONSHIPS

Throughout the text, we will relate polymer structure to the properties of the polymer. Polymer properties are related not only to the chemical nature of the polymer, but also to factors such as extent and distribution of crystallinity, distribution of polymer chain lengths, and nature and amount of additives, such as fillers, reinforcing agents, and plasticizers, to mention a few. These factors influence essentially all the polymeric properties to some extent including hardness, flammability, weatherability, chemical stability, biological response, comfort, flex life, moisture retention, appearance, dyeability, softening point, and electrical properties.

Materials must be varied to perform the many tasks required of them in today's society. Often they must perform them repeatedly and in a "special" manner. We get an idea of what materials can do by looking at some of the behavior of giant molecules in our body. While a plastic hinge must be able to work thousands of times, the human heart, a complex muscle largely composed of protein polymers (Section 4.7), provides about 2.5 billion beats within a lifetime moving oxygen (Section 11.8) throughout the approximately 144,000 km of the circulatory system with (some) blood vessels the thickness of hair and delivering about 8000 L of blood every day with little deterioration of the cell walls. The master design allows nerve impulses to travel within the body at the rate of about 300 m/min; again, polymers are the "enabling" material that allows this rapid and precise transfer of nerve impulses. Human bones, again largely polymeric, have a strength about five times that of steel (on a weight basis). Genes, again polymeric (Section 4.8), appear to be about 99.9% the same between humans, with the 0.1% functioning to give individuals the variety of size, abilities, etc., that confer uniqueness. In the synthetic realm, we are beginning to understand and mimic the complexities, strength, preciseness, and flexibility that are already present in natural polymers.

Here, we will briefly deal with the chemical and physical natures of polymeric materials that permit their division into three broad divisions—elastomers or rubbers, fibers, and plastics. *Elastomers* are high polymers possessing chemical and/or physical cross-links. For industrial application the "use" temperatures must be above T_g (to allow for ready "chain" mobility), and its normal state (unextended) must be amorphous. The restoring force, after elongation, is largely due to entropy. On release of the applied force, the chains tend to return to a more random state. Gross, actual mobility of chains must be low. The cohesive energy forces between chains should be low, permitting rapid, easy expansion. In its extended state, a chain should exhibit a high tensile strength, whereas at low extensions it should have a low tensile strength. Cross-linked vinyl polymers often meet the desired property requirements. The material, after deformation, should return to its original shape because of the cross-linking. This property is often referred to as an elastic "memory." Figure 2.18 illustrates force versus elongation for a typical elastomer. As the elastomer is pulled, the largely random chain segments become "stretched out" forming microcrystalline domains. Eventually, most of the chains

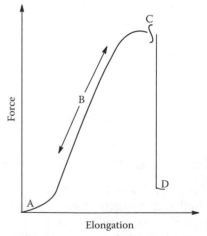

FIGURE 2.18 Elongation of an elastomer as a function of applied force, stress, where A is the original "relaxed" state, B represents movement to full extension, C is the point at which the elastomer "breaks," and D represents force necessary to pull two separate pieces of elastomer apart.

are part of these microcrystalline domains resulting in further elongation requiring much increased force (stress). This microcrystallinization also confers to the elastomer a greater brittleness, eventually resulting in the rubber breaking as additional stress is applied.

Fiber properties include high tensile strength and high modulus (high stress for small strains). These properties can be obtained from high molecular symmetry and high cohesive energies between chains, both requiring a fairly high degree of polymer crystallinity. Fibers are normally linear and drawn (oriented) in one direction, producing higher mechanical properties in that direction. Typical condensation polymers, such as polyesters and nylons, often exhibit these properties. If the fiber is to be ironed, its T_g should be above 200°C, and if it is to be drawn from the melt, its T_g should be below 300°C. Branching and cross-linking are undesirable since they disrupt crystalline formation, even though a small amount of cross-linking may increase some physical properties, if effected after the material is drawn and processed. Permanent press garments often have some cross-linking ensuring a "remembrance" of the "permanent press."

Products with properties intermediate between elastomers and fibers are grouped together under the heading "*plastics*."

Selected property–structure relationships are summarized in Tables 2.4 and 2.5. As noted before, some polymers can be classified in two categories, with properties being greatly varied by varying molecular weight, end group, processing, cross-linking, plasticizer, and so on. Nylon-66 in its more crystalline form behaves as a fiber, whereas less crystalline forms of nylon-66 are generally classified as plastics.

TABLE 2.4
Selected Property–Structure Relationships

Glass transition temperature
  Increases with the presence of
    Bulky pendant groups
    Stiffening groups as 1,4-phenylene
    Chain symmetry
    Polar groups
    Cross-linking
  Decreases with the presence of
    Additives like plasticizers
    Flexible main chain groups
    Nonpolar groups
    Dissymmetry
Solubility
  Favored by
    Longer chain lengths
    Low interchain forces
    Disorder and dissymmetry
    Increased temperature
    Compatible solvent
Crystallinity
  Favored by
    High interchain forces
    Regular structure; high symmetry
    Decrease in volume
    Increased stress
    Slow cooling from melt
    Homogeneous chain length

TABLE 2.5
General Property Performance–Structure Relationships

	Increased Crystallinity	Increased Cross-Linking	Increased Mol. Wt.	Increased Mol. Wt. Distribution	Addition of Polar Backbone Units	Addition of Backbone Stiffening Groups
Abrasion resistance	+	+	+	−	+	−
Brittleness	−	M	+	+	+	+
Chemical resistance	+	V	+	−	−	+
Hardness	+	+	+	+	+	+
T_g	+	+	+	−	+	+
Solubility	−	−	−	0	−	−
Tensile strength	+	M	+	−	+	+
Toughness	−	−	+	−	+	−
Yield	+	+	+	+	+	+

Note: +, increase in property; 0, little or no effect; −, decrease in property; M, property passes through a maximum; V, variable results dependent on particular sample and temperature.

Many polymers can be treated to express more than one behavior. As noted above, nylon-66 provides good fibrous material when aligned and behaves as a plastic if it is not aligned. Polyesters also exhibit the same tendencies. Other materials, such as PVC and siloxanes, can be processed to act as plastics or elastomers.

2.6 CRYSTALLINE AND AMORPHOUS COMBINATIONS

Most polymers consist of a combination of crystalline and amorphous regions. Even within polymer crystals such as spherulites (Figures 2.15 through 2.17), the regions between the ordered folded crystalline lamellae are less ordered, approximating amorphous regions. This combination of crystalline and amorphous regions is important for the formation of materials that have both good strength (contributed largely by the crystalline regions) and some flexibility or "softness" (derived from the amorphous portions). Figure 2.16 contains a space-filled model for polyethylene chains (a total of about 400 units with five branches, one longer and four shorter).

This model of polyethylene contains a mixture of amorphous and crystalline regions. Note the cavities within the amorphous regions with materials containing a majority of amorphous regions having a greater porosity and consequently a greater diffusion and greater susceptibility to chemical and nature attack. As noted before, materials that contain high amounts of the crystalline regions are referred to as being crystalline and they are less flexible and stronger and offer better stability to natural and attack by acids and bases, oils, etc. Moreover, as noted before, amorphous regions give the material flexibility while the crystalline regions give the material strength. Thus, many materials contain both crystalline and amorphous regions giving them a balance between strength and flexibility. The final properties of a material are then dependent on the molecular structure of that material.

Using specific treatment(s), the crystalline/amorphous regions can vary from being largely random to being preferentially oriented in one direction with a greater degree of "crystalline-type" structures when unidirectional stress is applied (Figure 2.19). Here, the amount of free space or volume is less with the overall order greater and properties associated with these changes also being changed. The material will be stronger, have a greater ability to resist attack by acids, bases, oils and other external agents, and the diffusion of gases and other agents through it is less. Polymers can be oriented (such as the pulling of fibers or films) in one or two directions. This preferential orientation

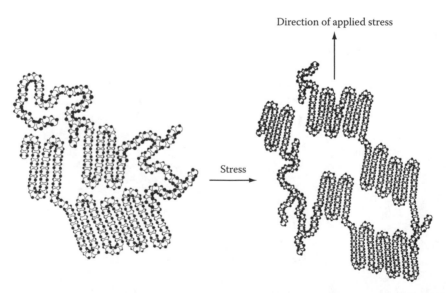

FIGURE 2.19 Idealized structure illustrating crystalline (ordered) and amorphous (unordered) regions of lightly branched polyethylene chains for a prestressed and stressed orientation.

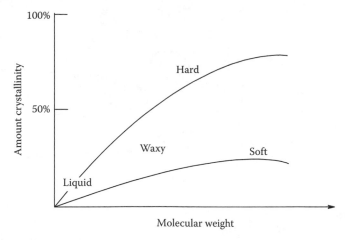

FIGURE 2.20   General physical states of materials as a function of crystallinity and molecular weight.

results in the fiber or film material with anisotropic properties, with the material showing greater strength along the axis of pull.

Figure 2.20 shows the general relationship between material "hardness/softness" and the proportion that is crystalline for largely linear polymers.

There are some general guidelines with respect to a materials T_g and T_m, its general amorphous/crystalline structure and the potential use area. *Elastomers* are cross-linked, amorphous polymers where the use temperature is above its T_g. An *adhesive* is a linear or branched amorphous polymer that is used above its T_g. *Coatings* are generally near their T_g when used so that some flexibility is present allowing the coatings material to withstand temperature changes without cracking and so coalescing occurs on drying. *Plastics* can be both amorphous or partially amorphous. Amorphous (or partially crystalline) plastics such as PP and PE should have a use temperature below the T_m but above the T_g. *Fibers* are composed of crystalline polymers where the use temperature is below the T_m.

2.7   CROSS-LINKING

Cross-linking is important since it can lock in structures, give structures "memory," and greatly influence the physical properties. Cross-linking can be divided into three general types. Chemical cross-linking consists of the formation of primary ionic or covalent bonds. Such cross-linking can be intentional such as formation of sulfur cross-links in many elastomers connecting chains giving the cross-linked material "memory" as well as other properties. Here, the cross-linking is referred to as "vulcanizing" and "curing" (Figure 2.21). For hair, the breakage and formation of thiol cross-linking is involved with permanents and the term "setting" is often employed to describe this process. Primary bond formation is also responsible for making some polymers brittle through exposure to UV sunlight. Primary chemical bond formation of cross-linking often leads to the formation of thermoset materials. Chemical cross-linking can result from heat, light, exposure to chemicals, etc.

The two main types of physical cross-linkings are illustrated in Figure 2.22. One method involves chain entanglement. Just as with string and yarn, the longer the strands the more apt they are to entangle. As one attempts to pull a single strand, other strands are pulled along. This is analogous to what occurs with polymer chains. For long chains, because of chain entanglement, as one chain is pulled other chains are also tugged on resulting in a great increase in the apparent chain length. Many properties such as tensile strength are dependent on the chain length and such properties have a marked increase as chain entanglement begins. For many polymers, such chain entanglement begins as the chain length approaches 100 units.

FIGURE 2.21   Chemical cross-linking of *cis*-1,4-butadiene through reaction with sulfur.

FIGURE 2.22   Illustration of two types of physical cross-linking-chain entanglements and crystalline regions.

The second type of physical cross-linking involves the formation of crystalline portions within an amorphous grouping. For vinyl polymers, the temperature that allows segmental chain mobility to occur is well below room temperature, but the temperature to disrupt crystalline formations is typically well above room temperature. Within mixtures of crystalline and amorphous structures below the T_m, the crystalline portions act to "tie-in" or connect the surrounding areas acting as cross-links. Formation of secondary bonding cross-linking is a primary method to retain desired three-dimensional structures (conformations) in proteins and nucleic acids. This secondary bonding generally entails a variety of forces including hydrogen bonding and the formation of such structures can also involve the formation of crystalline regions within nylons and polyesters.

The general name given to materials that do not melt when heated is ***thermoset***. Thermoset materials generally contain chemical cross-links. Since they do not melt when heated, it is more difficult to recycle such materials. By comparison, materials that do melt when heated are called ***thermoplastics***. Such materials do melt when heat is applied and so are more easily recycled. Examples of thermosets are some elastomers and phenolic and amino plastics. Polystyrene, polypropylene, polyethylene, nylon, and polyesters are examples of thermoplastic materials. By weight and value, we use many more synthetic thermoplastic materials. Even so, both thermoset and thermoplastic materials can be recycled.

2.8   SUMMARY

1. Polymers or macromolecules are high molecular weight compounds with chain lengths greater than the critical length required for the entanglement of these chains. There is an abrupt change in melt viscosity and other physical properties of high molecular weight polymers when the chain length exceeds the critical chain length.

2. While some naturally occurring polymers, such as proteins, are monodisperse, that is all have the same molecular weight, other natural and synthetic polymers, such as cellulose and PE, are polydisperse that is they consist of a mixture of chains of differing numbers of units. Thus, one uses the term average molecular weight when describing the molecular weight of these polydisperse materials.

3. Many polymers, such as cellulose and HDPE, are linear polymers containing long, continuous, covalently bonded atoms. Others may be branched or cross-linked. Both linear polymers and those with branching are generally thermoplastics that can be remolded by the application of heat and pressure. Others that are cross-linked are thermosets that cannot be remolded by the application of heat and pressure. Both groups of polymers can be recycled, but it is easier to recycle thermoplastic materials.

4. Functional groups in the polymer backbone, such as the methyl group in PP, are called pendant groups. Such polymers are formed giving a head-to-tail arrangement rather than a head-to-head arrangement.

5. The temperature at which local segmental mobility occurs is called the T_g and where whole sale polymer chain mobility occurs is called the T_m.

6. The first-order transition or melting point (T_m) is energy wise larger than the T_g. Entirely crystalline polymers will have only a T_m whereas a totally amorphous polymer will have only a T_g. Since most polymers are a combination of amorphous and crystalline regions, they have both a T_g and a T_m.

7. A polymer stretched out to its full contour length is only one of the myriad conformations possible. The chain length is expressed statistically as the root-mean-square distance, which is only a fraction of the contour length.

8. Since branched chains such as LDPE have many chain ends, it is customary to use the radius of gyration, S, as a measure of the distribution of polymer segments. The radius of gyration is the distance of a chain end from the polymer's center of gravity rather than a measure of the root-mean-square. Both measures are statistically related.

9. Fibers and stretched elastomers are translucent because of the presence of ordered crystallites or regions of crystallinity.

10. Crystalline regions of polymers can be represented as combinations of folded chains forming lamellar structures. Amorphous regions are less ordered than crystalline regions.

Additional orientation of polymer chains occurs when stress is applied resulting in increased strength in the direction of the applied stress.

11. The principle differences between elastomers, plastics, and fibers are the presence and absence of stiffening groups, molecular symmetry, and the strength of the intermolecular forces. Elastomers are typically characterized by the absence of stiffening groups, presence of molecular asymmetry, low amount of crystallinity, and overall absence of strong intermolecular forces. In contrast, fibers are characterized by the presence of stiffening groups, molecular symmetry, high amount of crystallinity, and presence of strong intermolecular forces. Fibers have a lack of branching and irregularly spaced pendant groups. Plastics have structures and properties that are between elastomers and fibers.

12. There are three types of cross-linkings. Cross-linked materials have good dimensional memory. The types of cross-linking are one chemical and two physical types of cross-linking. Physical cross-linking results from formation of crystalline regions within polymer structures and from chain entanglement. Chemically cross-linked materials do not dissolve and they do not melt.

GLOSSARY

Amorphous: Noncrystalline polymer or region in a polymer.

Atactic: Polymer in which there is a random arrangement of pendant groups on each side of the chain.

Backbone: Principle chain in a polymer.

Branched polymer: Polymer having extensions attached to the polymer backbone. Not pendant groups.

Bulky groups: Sterically large groups.

Chiral center: Asymmetric center such as a carbon atom with four different groups attached to it.

Cold drawing: Stretching a fiber under room temperature.

Configurations: Related chemical structures produced by the breaking and remaking of primary covalent bonds.

Conformations: Various shapes of polymer chains resulting from the rotation about single bonds in the polymer chain.

Conformer: Shape produced by a change in the conformation of a polymer.

Contour length: Fully extended length of a polymer chain; equal to the product of the length of a repeat unit times the number of units, or mers.

Critical chain length (z)**:** Minimum chain length required for entanglement of polymer chains.

Cross-linked density: Measure of the relative degree of cross-linking.

Crystalline polymer: Polymer with ordered structure.

Crystallites: Regions of crystallinity.

Degree of polymerization: Number of repeat units in a chain.

Dipole–dipole interactions: Moderate secondary forces between polar groups in different or the same polymer chain.

Dispersion forces: Low energy secondary forces due to the creation of momentary induced dipoles; also known as London forces.

Glass transition temperature (T_g)**:** Temperature range where a polymer gains local or segmental mobility.

Glassy state: Hard, brittle state; below T_g.

Gutta percha: Naturally occurring *trans* isomer of polyisoprene.

Heat-to-tail configuration: Normal sequence of mers in which the pendant groups are regularly spaced; for PVC, the chlorine atom appears on every other carbon.

Intermolecular forces: Secondary forces between different molecules.

Intramolecular forces: Secondary forces within the same molecule.

Isotactic: Polymer where the geometry of the pendant groups are all on the same side of the polymer backbone.

Lamellar: Platelike, or planar (flat), in shape.

Linear polymer: Polymer without chains extending off the backbone.

Low-density polyethylene (LDPE): A branched form of PE produced at high pressure by the free-radical initiated polymerization of ethylene.

Maltese cross: Cross with arms like arrowheads pointing inward.

Melting point (T_m): First-order transition when the solid and liquid phases are in equilibrium.

Mer: Repeat unit.

Modulus: Ratio of stress to strain, which is a measure of the stiffness of a polymer.

Monodisperse: Polymer mixture made up of molecules of one specific molecular weight.

Pendant groups: Groups attached to the main polymer backbone, like methyl groups in polypropylene.

Polydisperse: Polymer mixture containing chains of varying lengths.

Radius of gyration (S): Root-mean-square distance of a chain end to a polymer's center of gravity.

Root-mean-square distance [$(r^2)^{1/2}$]: Average end-to-end distance of polymer chains; $l(n^{1/2})$.

Side-chain crystallization: Crystallization related to that of regularly spaced long pendant groups.

Single polymer crystals: Lamellar structure consisting of folded chains of a linear polymer.

Spherulites: Three-dimensional aggregates of polymer crystallites.

Stiffening groups: Groups in a polymer backbone that decrease the segmental motion of the polymer chain.

Syndiotactic: Polymer in which the pendant groups are arranged alternately on each side of the carbon backbone.

Tacticity: Arrangement of the pendant groups in space; that is, isotactic, syndotactic, and atactic.

Thermoplastic: A linear polymer that softens when heated.

Thermoset: A network polymer containing chemical cross-linking that does not soften when heated.

van der Waals forces: Secondary forces based on the attraction between groups.

Viscosity: The measure of the resistance of a polymer or polymer solution to flow.

EXERCISES

(To answer some of these questions you may need to look at other parts in the book for structures and specific details.)

1. Make sketches or diagrams showing (a) a linear polymer, (b) a polymer with pendant groups, (c) a polymer with short branches, (d) a polymer with long branches, and cross-linked polymers with (e) low and (f) high cross-linked density.

2. Which has (a) the greater volume for the same weight of material and (b) the lower softening point: HDPEC or LDPE?

3. What is the approximate bond length of the carbon atoms in (a) a linear and (b) a cross-linked polymer.

4. What is the approximate contour length of an HDPE chain with an average degree of polymerization (chain length) of $n = 2000$ and of a PVC chain of the same number of repeating units?

5. Which of the following are monodisperse polymers with respect to chain length? (a) hevea rubber, (b) corn starch, (c) cellulose from cotton, (d) an enzyme, (e) HDPE, (f) PVC, (g) a specific DNA, (h) nylon 66, (i) a specific RNA?

6. What is the average degree of polymerization of LDPE having an average molecular weight of 28,000?

7. What is the structure of the repeating unit in (a) polypropylene, (b) poly(vinyl chloride), (c) hevea rubber?

8. Which of the following is a branched chain polymer: (a) HDPE, (b) Isotactic PP, (c) LDPE, (d) amylose starch?

9. Which of the following is a thermoplastic: (a) ebonite, (b) Bakelite, (c) vulcanized rubber, (d) HDPE, (e) celluloid, (f) PVC, (g) LDPE?

10. Which has the higher cross-linked density, (a) ebonite or (b) soft-vulcanized rubber?

11. Do HDPE and LDPE differ in (a) configuration or (b) conformation?

12. Which is a *trans* isomer: (a) gutta percha or (b) hevea rubber?

13. Which will have the higher softening point: (a) gutta percha or (b) hevea rubber?

14. Show (a) a heat-to-tail, and (b) a head-to-head configuration for PVC.

15. Show the structure of a typical portion of the chain of (a) s-PVC, (b) i-PVC.

16. Show Newman projections of the gauche forms of HDPE.

17. Name polymers whose intermolecular forces are principally (a) London forces, (b) dipole–dipole forces, (c) hydrogen bonding.

18. Which will be more flexible: (a) poly(ethylene terephthate), or (b) poly(butylene terephthalate)?

19. Which will have the higher glass transition temperature: (a) poly(methylene methacrylate) or (b) poly(butyl methacrylate)?

20. Which will have the higher T_g: (a) i-PP or (b) a-PP?

21. Which will be more permeable to a gas at room temperature: (a) i-PP or (b) a-PP?

22. Under what kind of physical conditions are you more apt to form spherulites?

23. What is the full contour length of a molecule of HDPE with a $\overline{DP}$ of 1500?

24. Which would be more flexible: (a) poly(methyl acrylate) or (b) poly(methyl methacrylate)?

25. Which would you expect to form "better" helical structures (a) i-polypropylene or (b) a-polypropylene?

26. Which would you expect to have a higher melting point (a) nylon-66 or (b) an aramide?

27. What type of hydrogen bonds are present in the internal structure of a globular protein?

28. Which would have the greater tendency to "cold flow" at room temperature: (a) poly(vinyl acetate) ($T_g = 301\ K$) or (b) polystyrene ($T_g = 375\ K$)?

29. Which would be least transparent: (a) a combination of amorphous and crystalline PS, (b) entirely crystalline PS, or (c) entirely amorphous PS?

30. Which would be more apt to produce crystallites: (a) HDPE or (b) poly(butyl methacrylate)?

31. Which of the following would you expect to provide strong fibers (a) nylon-66, (b) a-polypropylene, (c) wool.

32. Which would tend to be more crystalline when stretched: (a) unvulcanized rubber or (b) ebonite?

33. Which would be more apt to exhibit side-chain crystallization (a) poly(methyl methacrylate) or (b) poly(dodecyl methacrylate)?

34. What must be present in order for movement to occur within a polymer at T_g or T_m?

35. What are the three major forms of cross-linking?

FURTHER READINGS

Alfrey, T. and Gurnee, E. F. (1956): Dynamics of viscoelastic behavior, in *Rheology–Theory and Applications*, Eirich, F. R., ed., Academic Press, New York.

Bicerano, J. (2002): *Prediction of Polymer Properties*, Dekker, New York.

Bower, D. (2002): *An Introduction to Polymer Physics*, Cambridge University Press, Ithaca, NY.

Cheng, S. (2008): *Phase Transitions in Polymers: The Role of Metastable States*, Elsevier, New York.

Ellis, B. (2008): *Polymers: A Property Database*, CRC press, Boca Raton, FL.

Flory, P. J. (1953): *Principles of Polymer Chemistry*, Cornell University Press, Ithaca, NY.

Gaylord, N. (1959): *Linear and Stereoregular Addition Polymers*, Gaylord, N. and Mark, H., eds., Wiley, New York.

Geil, P. H. (1963): *Polymer Single Crystals*, Wiley, New York.

Henkel, M., Plemling, M., and Sanctuary, R. (2007): *Ageing and the Glass Transition*, Springer, New York.

Howell, B. and Zaikow, G. (2008): *Compounds and Materials with Specific Properties*, Nova, Hauppauge, NY.

Mandelkern, L. (2004): *Crystallization of Polymers—Kinetics and Mechanisms*, Cambridge University Press, Ithaca, NY.

Mark, J. (2009): *The Polymer Data Handbook*, Oxford University Press, Ithaca, NY.

Natta, G. (1955): Stereospecific macromolecules, *J. Polym. Sci.*, 16:143.

Pauling, L., Corey, R., and Branson, H. (1951): The structure of proteins, *Proc. Natl. Acad. Sci. USA*, 37:205.

Reiter, G. (2007): *Progress in Understanding of Polymer Crystallization*, Springer, New York.

Rotello, V. and Thayumanavan, S.(2008): *Molecular Recognition and Polymers: Control of Polymer Structure and Self-Assembly*, Wiley, Hoboken, NJ.

Schultz, J. (2001): *Polymer Crystallization*, Oxford University Press, Cary, NC.

Seymour, R. B. and Carraher, C. E. (1984): *Structure–Property Relationships in Polymers*, Plenum Press, New York.

Viney, C. (2003): *Techniques for Polymer Organization and Morphology Characterization*, Wiley, Hoboken, NJ.

3 Molecular Weight of Polymers

3.1 INTRODUCTION

It is the size of macromolecules that gives them their unique and useful properties. Size allows polymers to act more as a group, so that when one polymer chain moves, the surrounding chains are affected by that movement. Size also allows polymers to be nonvolatile since the secondary attractive forces are cumulative (for instance, the London dispersion forces are about 8 kJ/mol per repeat unit) and because of the shear size, the energy necessary to volatilize them is sufficient to degrade the polymer.

Generally, the larger the polymer, the higher the molecular weight. The average molecular weight (M) of a polymer is the product of the average number of repeat units for the polymer expressed as $\overline{DP}$ times the molecular weight for the repeating unit. Thus, for polyethylene (PE) with a $\overline{DP}$ of 100 the average molecular weight is simply 100 units times 28 Da/unit = 2800 Da. Note, that amu and daltons are the same units.

Polymerization reactions, both synthetic and natural (but not for all natural materials such as proteins and nucleic acids), lead to polymers with heterogeneous molecular weights, i.e., polymer chains with different number of mers. Molecular weight distributions (MWDs) may be rather broad (Figure 3.1), or relatively narrow, or may be mono-, bi-, tri-, or polymodal. A bimodal curve is often characteristic of a polymerization occurring under two different environments. Polymers consisting of chains of differing lengths are called polydisperse and polymers containing only one chain length, such as specific nucleic acids, are called monodisperse.

Many physical properties are related to molecular weight. Figure 3.2 lists three of these. The melt viscosity is typically proportional to the 3.4 power of the average chain length; so η is proportional to $M^{3.4}$. Thus, the melt viscosity increases rapidly as the chain length increases, and more energy is required for the processing and fabrication of large molecules. However, there is a trade-off between molecular weight related properties and chain size such that there is a range where acceptable physical properties are present but the energy required to cause the polymers to flow is acceptable. This range is called the ***commercial polymer range***. As seen in Figure 3.2, many physical properties tend to level off at some point, and an increased chain length provides little increase in that physical property. Thus, most commercial polymer ranges include the beginning of this leveling off threshold.

While a value above the ***threshold molecular weight value*** (TMWV; lowest molecular weight where the desired property value is achieved) is essential for most practical applications, the additional cost of energy required for processing extremely high polymer molecular weights is seldom justified. Accordingly, it is customary to establish a commercial polymer range above the TMWV but below the extremely high molecular weight range. However, it should be noted that since toughness increases with chain length, extremely high molecular weight polymers, such as ultrahigh molecular weight polyethylene (UHMWPE), are used for the production of tough articles such as waste barrels.

Some properties, such as heat capacity, refractive index, and density, are not as sensitive to molecular weight. Even so, many important properties are related to chain length.

Oligomers and other low molecular weight polymers are not useful for applications where high strength is required. The word oligomer is derived from the Greek word *oligos*, meaning "a few."

The value of TMWV is dependent on T_g, the cohesive energy density (CED) of amorphous polymers, the extent of crystallinity in crystalline polymers, and the effect of reinforcements in polymeric composites. Thus, while a low molecular weight amorphous polymer may be satisfactory for

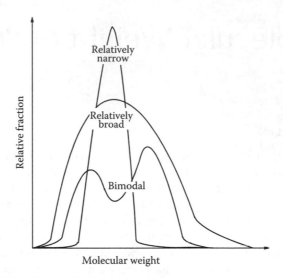

FIGURE 3.1 Relative differential weight distribution curves.

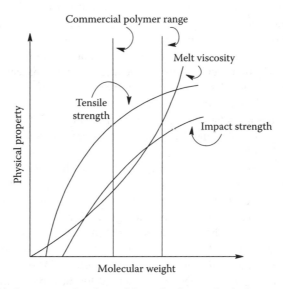

FIGURE 3.2 Relationship between molecular weight and polymer physical properties.

use as a coating or an adhesive, a chain length generally above 100 may be required if the polymer is to be used as an elastomer or a plastic.

3.2 SOLUBILITY

Polymer mobility is important in a number of aspects. Most processing procedures require a polymer to be fluid enough, either through the application of heat and/or pressure or with the polymer being present in a solution, to be processed. Here, we will focus on the general topic of polymer solubility and factors that influence polymer solubility.

Polymer mobility is an important aspect helping determine a polymer's physical, chemical, and biological behaviors. The lack of mobility, either because of interactions that are too swift to allow the segments within the polymer chain some movement or because there is not sufficient energy (such as a high enough temperature) available to create mobility, results in a brittle material. Many processing techniques require the polymer to have some mobility. This mobility can be achieved through the application of heat and/or pressure and through dissolving the polymer. Because of its size, the usual driving force for the mixing and dissolving of materials (entropy) is smaller for polymers in comparison to small molecules. Traditional molecular weight determinations require that the polymer be dissolved.

The physical properties of polymers, including solubility, are related to the strength of covalent bonds, the stiffness of the segments in the polymer backbone, the amount of crystallinity/amorphous, and the intermolecular forces between the polymer chains. The strength of the intermolecular forces is equal to the **CED**, which is the molar energy of vaporization per unit volume. Since intermolecular attractions of solvent and solute must be overcome when a solute (here the polymer) dissolves, CED values may be used to predict solubility.

When a polymer dissolves, the first step is often a slow swelling process called solvation in which the polymer molecules swell by a factor δ, which is related to CED. Linear and branched polymers dissolve in a second step, but network polymers remain in a swollen condition.

In order for a material to be dissolved it is essential that the free energy, G, which is the driving force in the solution process, decreases to below zero. ΔH and ΔS are the change in enthalpy and the change in entropy, and for constant temperature the relationship is the classical Gibbs equation.

$$\Delta G = \Delta H - T\Delta S \qquad (3.1)$$

"Like-likes-like best of all" is a description that is useful at appropriate times in science. It is true of solubility. Thus, water-likes-water best of all and is infinitely soluble in itself. Hexane-likes-hexane best of all and is infinitely soluble in itself. Hexane and water are not soluble in one another because hexane is nonpolar and water is polar; thus, they are not "like one another." In solubility, and in fact all of mixing, the ΔH term is always unfavorable when mixing or solubility occurs. Shortly, we will deal with attempts such as the CED and solubility parameter to minimize the unfavorable aspect of the ΔH term. Thus, it is the ΔS term that allows mixing and solubility to occur. As seen in Figure 3.3, the amount of randomness or disorder gain is great when pure materials, such as water and ethanol, are changed from the ordered pure materials to the disordered mixture.

By comparison, the increase in randomness, ΔS, is much smaller if one of the materials is a polymer since the arrangement of the polymer chains is much more limited when the polymer units are attached to one another and are not free to simply move about on their own. Figure 3.4 illustrates this with water and poly(ethylene glycol), PEG. We notice several aspects. First, as noted above, the number of arrangements of the PEG units is limited. Second, as in the case of an onion, each layer of PEG chains must be peeled back allowing water molecules to approach the inner layers before

FIGURE 3.3 Illustration of the mixing of small molecules.

FIGURE 3.4 Illustration of water-dissolving PEG.

complete solubility occurs and causes swelling. This results in polymer solubility often requiring a longer period of time, sometimes hours to week to months, in comparison to the solubility of smaller molecules where solubility can occur in seconds.

Polymer solubility, in comparison to small molecules, is

1. more limited with respect to the number of solvents as a result of the lower increase in randomness;
2. it is more limited with respect to extent of solubility; and
3. takes a longer time to occur.

The *theta* (θ) *temperature* is the lowest temperature at which a polymer of infinite molecular weight is completely miscible with a specific solvent. The coil expands above the theta temperature and contracts at lower temperatures.

As early as 1926, Hildebrand showed a relationship between the solubility and the internal pressure of the solvent, and in 1931, Scatchard incorporated the CED concept into Hildebrand's equation. This led to the concept of a *solubility parameter*, which is the square root of CED. Thus, as shown below, the solubility parameter, δ, for nonpolar solvents is equal to the square root of the heat of vaporization per unit volume:

$$\delta = \left(\frac{\Delta E}{V}\right)^{1/2} \tag{3.2}$$

According to Hildebrand, the heat of mixing a solute and a solvent is proportional to the square of the difference in solubility parameters, as shown below, where ϕ is the partial volume of each component, namely, solvent γ_1 and solute ϕ_2. Since typically the entropy term favors solution and the enthalpy term acts counter to solution, the objective is to match solvent and solute, so that the difference between their δ values is small allowing the entropy term to over-ride the enthalpy term.

$$\Delta H_m = \phi_1 \phi_2 (\delta_1 - \delta_2)^2 \tag{3.3}$$

The solubility parameter concept predicts the heat of mixing liquids and amorphous polymers. Hence, any nonpolar amorphous polymer will dissolve in a liquid or mixture of liquids having a solubility parameter that generally does not differ by more than ±1.8 (cal/cc)$^{1/2}$. The Hildebrand (H) with units of (cal/cc)$^{1/2}$ is preferred so ±1.8 (cal/cc)$^{1/2}$ becomes ±1.8 H.

The solubility parameter concept, like Flory's theta temperature, is based on Gibbs' free energy. Thus, as the term ΔH approaches zero, ΔG will have the negative value required for solution to occur. The entropy (S) increases in the solution process, hence the emphasis is on having low values for ΔH_m.

For nonpolar solvents, which were called regular solvents by Hildebrand, the solubility parameter is equal to the square root of the difference between the enthalpy of evaporation (H_v), and the product of the ideal gas constant (R) and the Kelvin (or absolute) temperature (T) divided by the molar volume (V), as shown in Equation 3.4.

$$\delta = \left(\frac{\Delta E}{V} \right)^{1/2} = \left(\frac{\Delta H_v - RT}{V} \right)^{1/2} \tag{3.4}$$

Since it is difficult to measure the molar volume, its equivalent the molecular weight (M) divided by density (D) is substituted for V as shown in Equation 3.5.

$$\delta = D \left(\frac{\Delta H_v - RT}{M} \right)^{1/2} \tag{3.5}$$

Because the law of mixtures applies to the solubility parameter, it is possible to blend nonsolvents forming mixtures that can serve as solvents. For example, an equimolar mixture of n-pentane ($\delta = 7.1$ H) and n-octant ($\delta = 7.6$ H) will have a solubility parameter value of 7.35 H.

The solubility parameter of a polymer is generally determined by noting the extent of swelling or the actual solution of small amounts of polymer in a series of solvents, and comparing the solubility values of the ones that swell or dissolve the polymer and assigning the polymer a solubility parameter value that is close to the solvents that dissolve/swell the polymer. The solubility parameter can also be determined by adding a nonsolvent to a polymer solution and noting the amount of nonsolvent needed to initiate the precipitation of the polymers.

Since polar forces are present in polar solvents and polar molecules, this must be considered when estimating solubilities with such "nonregular" solvents and polymers. Hydrogen bonding is also important for some solvents and polymers and again will influence the solubility parameters. Thus, special solubility values have been developed for such solvents (Tables 3.1 and 3.2).

Plasticizers help the flexibility of polymers and are chosen so that they do not dissolve the polymer but rather allow segmental mobility to occur. Through experience, it is found that the solubility parameter difference between the plasticizer and the polymer should be less than 1.8 H.

3.3 AVERAGE MOLECULAR WEIGHT VALUES

Small molecules, such as benzene and glucose, have precise structures such that each molecule of benzene has six carbon atoms and six hydrogen atoms. By comparison, each molecule of poly-1, 4-phenylene may have a differing number of benzene-derived units, while single molecules (single chains) of polyethylene may vary in the number of ethylene units, the extent and frequency of branching, the distribution of branching, and the length of branching. A few polymers, such as nucleic acids and many proteins, consist of molecules, individual polymer chains that must not vary, so they have a precise molecular weight. The concept of averaging in chemistry is not new. The atomic weights that appear in the periodic table are often averages dependent on the natural abundances of various isotopes. In a similar way, the molecular weights of polymeric materials that

TABLE 3.1

Solubility Parameters (δ) for Typical Solvents

Poorly Hydrogen Bonding		Moderately Hydrogen Bonding		Strongly Hydrogen Bonding	
Dimethylsiloxane	5.5	Diisopropyl ether	6.9	Diethylamine	8.0
Difluorodichloromethane	5.5	Diethylether	7.4	n-Amylamine	8.7
Neopentane	6.3	Isoamyl acetate	7.8	2-Ethylhexanol	9.5
Nitro-n-octane	7.0	Diisobutyl ketone	7.8	Isoamyl alcohol	10.0
n-Pentane	7.0	Di-n-propylether	7.8	Acetic acid	10.1
n-Octane	7.6	sec-Butyl acetate	8.2	m-Cresol	10.2
Turpentine	8.1	Isopropyl acetate	8.4	Analine	10.3
Cyclohexane	8.2	Methylamyl ketone	8.5	n-Octyl alcohol	10.3
Cymene	8.2	Ethyl acetate	9.0	t-Butyl alcohol	10.9
Carbon tetrachloride	8.6	Methyl ethyl ketone	9.3	n-Amyl alcohol	10.9
n-Propylbenzene	8.6	Butyl cellosolve	9.5	n-Butyl alcohol	11.4
p-Chlorotoluene	8.8	Methyl acetate	9.6	Isopropyl alcohol	11.5
Decalin	8.8	Dichloroethylether	9.8	Diethylene glycol	12.1
Xylene	8.8	Acetone	9.9	Furfuryl alcohol	12.5
Benzene	9.2	Dioxane	10.0	Ethanol	12.7
Styrene	9.3	Cyclopentanone	10.4	N-ethylformamide	13.9
Tetraline	9.4	Cellosolve	10.5	Methanol	14.5
Chlorobenzene	9.5	N,N-dimethylacetamide	10.8	Ethylene glycol	14.6
Ethylene dichloride	9.8	1,2-Propylene carbonate	13.3	Glycerol	16.5
p-Dichlorobenzene	10.0	Ethylene carbonate	14.7	Water	23.4
Nitroethane	11.1				
Acetonitrile	11.9				
Nitroethane	12.7				

contain chains of varying chain lengths are also an average dependent on the abundance of the various chain lengths present in the mixture.

While there are several statistically described averages dependent on how we calculate the final average molecular weight from the chain lengths of the individual molecules within the polymer sample, we will concentrate on the two that are most germane to polymers—number-average and weight-average molecular weights. These are averages based on statistical approaches that can be described mathematically and which correspond to measurements of specific values.

The number average value, corresponding to a measure of chain length average, is called the **number-average molecular weight, $\overline{M}_n$**. Physically, the number-average molecular weight can be measured using any technique that "counts" the molecules; i.e., is directly dependent on the number of chains. These techniques include vapor phase osmometry (VPO) and membrane osmometry, freezing point lowering, boiling point elevation, and end-group analysis.

We can describe the number average using a jar filled with plastic capsules such as those used to contain tiny prizes (Figure 3.5). Here, each capsule contains one polymer chain. All the capsules are of the same size, regardless of the size of the polymer chain. Capsules are then withdrawn, opened, and the individual chain length is measured and recorded (Figure 3.6). The probability of drawing a particular capsule is dependent on the **number** of each capsule and not on the size of the chain within the capsule.

The **weight-average molecular weight, $\overline{M}_w$**, is similarly described, except that the capsule size corresponds to the size of polymer chain contained within it (Figure 3.7). Thus, the probability of drawing out a particular chain length is dependent on the **size** of the capsule. Larger chains have a greater probability (at least in this exercise) of being drawn out because they are larger

TABLE 3.2
Approximate Solubility Parameter Values for Polymers

Polymer	Poorly Hydrogen Bonding	Moderately Hydrogen Bonding	Strongly Hydrogen Bonding
Polytetrafluoroethylene	5.8–6.4		
Poly(vinyl ethyl ether)	7.0–11.0	7.4–10.8	9.5–14.0
Poly(butyl acrylate)	7.0–12.5	7.4–11.5	
Poly(butyl methacrylate)	7.4–11.0	7.4–10.0	9.5–11.2
Polyisobutylene	7.5–8.0		
Polyethylene	7.7–8.2		
Poly(vinyl butyl ether)	7.8–10.6	7.5–10.0	9.5–11.2
Natural rubber	8.1–8.5		
Polystyrene	8.5–10.6	9.1–9.4	
Poly(vinyl acetate)	8.5–9.5		
Poly(vinyl chloride) (PVC)	8.5–11.0	7.8–10.5	
Buna N	8.7–9.3		
Poly(methyl methacrylate)	8.9–12.7	8.5–13.3	
Poly(ethylene oxide)	8.9–12.7	8.5–14.5	9.5–14.5
Poly(ethylene sulfide)	9.0–10.0		
Polycarbonate	9.5–10.6	9.5–10.0	
Poly(ethylene terephthate)	9.5–10.8	9.3–9.9	
Polyurethane	9.8–10.3		
Polymethacrylonitrile		10.6–11.0	
Cellulose acetate	11.1–12.5	10.0–14.5	
Nitrocellulose	11.1–12.5	8.0–14.5	12.5–14.5
Polyacrylonitrile		12.0–14.0	
Poly(vinyl alcohol)			12.0–13.0
Nylon-66			13.5–15.0
Cellulose			14.5–16.5

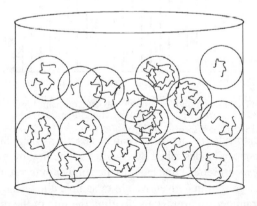

FIGURE 3.5 Jar with capsules, each capsule containing a single polymer chain where the capsule size is the same and independent of the chain size, illustrating the number-average dependence on the molecular weight.

and are contained within larger capsules. Again, a graph is constructed, and the maximum value is the weight-average molecular weight (Figure 3.8). The maximum in the graph is further along the molecular weight, M_i, axis in comparison to Figure 3.6 such that the weight-average molecular weight is greater than the number-average molecular weight.

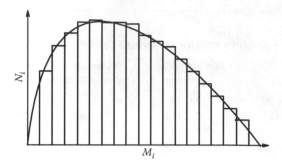

FIGURE 3.6 MWD for a polydisperse polymer sample constructed from Figure 3.5 "capsule-derived" data.

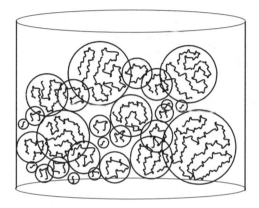

FIGURE 3.7 Jar with capsules, each containing a single polymer chain where the capsule size is directly related to the size of the polymer chain within the capsule.

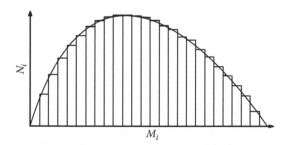

FIGURE 3.8 MWD for a polydisperse polymer sample constructed from Figure 3.7 "capsule-derived" data.

Several mathematical moments (about a mean) can be described using the differential or frequency distribution curve, and these can be described by equations. The first moment is the number-average molecular weight, $\bar{M}_n$. Any measurement that leads to the number of molecules, functional groups, end groups, or particles that are present in a given weight of sample allows the calculation of $\bar{M}_n$. The $\bar{M}_n$ is calculated like any other numerical average by dividing the sum of the individual molecular weight values by the number of molecules. Thus, $\bar{M}_n$ for three molecules having molecular weights of 1.00×10^5, 2.00×10^5, and 3.00×10^5 is $(6.00 \times 10^5)/3 = 2.00 \times 10^5$. Recalling that $W = \Sigma W_i = \Sigma M_i N_i$ the general solution is shown mathematically as

$$\bar{M}_n = \frac{\text{total weight of sample}}{\text{number of molecules of } N_i} = \frac{W}{\Sigma N_i} = \frac{\Sigma M_i N_i}{\Sigma N_i} \qquad (3.6)$$

Most thermodynamic properties are related to the number of particles present, and thus they are dependent on $\bar{M}_n$.

Colligative properties are dependent on the number of particles present and are thus related to $\bar{M}_n$. $\bar{M}_n$ values are independent of molecular size and are highly sensitive to small molecules present in the mixture. The values of $\bar{M}_n$ are determined by Raoult's techniques that are dependent on colligative properties such as ebulliometry (boiling point elevation), cryometry (freezing point depression), osmometry, and end-group analysis.

Weight-average molecular weight, $\bar{M}_w$, is determined from experiments in which each molecule or chain makes a contribution to the measured result relative to its size. This average is more dependent on the number of longer chains than is the number-average molecular weight, which is dependent simply on the total number of each chain.

The $\bar{M}_w$ is the second moment average and is shown mathematically as

$$\bar{M}_w = \frac{\sum W_i M_i}{\sum W_i} = \frac{\sum M_i^2 N_i}{\sum M_i N_i} \tag{3.7}$$

Thus, the $\bar{M}_w$ of the three chains cited before is

$$\frac{(1.00 \times 10^{10}) + (4.00 \times 10^{10}) + (9.00 \times 10^{10})}{6.00 \times 10^5} = 2.23 \times 10^5$$

Bulk properties associated with large deformations, such as viscosity and toughness, are most closely associated with $\bar{M}_w$. $\bar{M}_w$ values are most often determined by light-scattering photometry. However, melt elasticity is more closely related to the third moment known as the z-average molecular weight, $\bar{M}_z$. The $\bar{M}_z$ is most often determined using either light-scattering photometry or ultracentrifugation. It is shown mathematically as

$$\bar{M}_z = \frac{\sum M_i^3 N_i}{\sum M_i^2 N_i} \tag{3.8}$$

The $\bar{M}_z$ value for the three polymer chains cited above is 2.57×10^5:

$$\frac{(1 \times 10^{15}) + (8 \times 10^{15}) + (27 \times 10^{15})}{[(1 \times 10^{10}) + (4 \times 10^{10}) + (9 \times 10^{10})]} = 2.57 \times 10^5$$

While $Z + 1$ and higher average molecular weight values can be calculated, the major interests are in $\bar{M}_n$, $\bar{M}_v$, $\bar{M}_w$, and $\bar{M}_z$, which is the order of increasing size for a heterodisperse polymer sample as shown in Figure 3.6. Thus, for heterogeneous molecular weight systems $\bar{M}_z > \bar{M}_w > \bar{M}_n$. The ratio of $\bar{M}_w/\bar{M}_n$ is called the *polydispersity index*. The most probable polydispersity index for polymers produced by the condensation technique with respect to molecular weight is 2. As the heterogeneity decreases, the various molecular weight values converge until $\bar{M}_z = \bar{M}_w = \bar{M}_n$.

Typical techniques for determining molecular weight are given in Table 3.3. The most popular techniques will be considered briefly. All classical molecular weight techniques require dilute solutions, generally 0.01 g/mL or 1% (1 g/100 mL) solutions. To further minimize solute interactions, extrapolation of the measurements to infinite dilution is the normal practice.

For polydisperse polymer samples, measurements that lead directly to the determination of the molecular weight, such as light-scattering photometry and membrane osmometry, are referred to as "absolute molecular weight" methods. Techniques such as viscometry are not absolute molecular weight methods because they require calibration using an absolute molecular weight technique.

TABLE 3.3
Typical Molecular Weight Determination Methods[a]

Method	Type of Mol. Wt. Average	Applicable Wt. Range	Other Information
Light scattering	$\bar{M}_w$	To ∞	Can give other mol. wts. and shapes
Membrane osmometry	$\bar{M}_n$	10^4–10^6	
Vapor phase osmometry	$\bar{M}_n$	To 4×10^4	
Electron and x-ray microscopy	$\bar{M}_{n,w,z}$	10^2 to ∞	
Isopiestic method	$\bar{M}_n$	To 2×10^4	
Ebulliometry (BP elevation)	$\bar{M}_n$	To 4×10^4	
Cryoscopy (MP depression)	$\bar{M}_n$	To 5×10^4	
End-group analysis	$\bar{M}_n$	To 2×10^4	
Osmodialysis	$\bar{M}_n$	500–2.5×10^4	
Centrifugation			
Sedimentation equil.	$\bar{M}_z$	To ∞	
Archibald mod.	$\bar{M}_{z,w}$	To ∞	
Trautman's method	$\bar{M}_w$	To ∞	

Sedimentation velocity gives real M for only monodisperse systems

Method	Type of Mol. Wt. Average	Applicable Wt. Range	Other Information
Chromatography	Calibrated	To ∞	Gives mol. wt. distribution
Small-angle x-ray scattering (SAXS)	$\bar{M}_w$		
Mass spec. MALDI		To 10^7	Mol. wt. distribution
Viscometry	Calibrated		
Coupled chromatography-LS		To ∞	Mol. wt. distribution, shape, $\bar{M}_{n,w,z}$

[a] "To ∞" means that the molecular weight of the largest particles soluble in a suitable solvent can be, in theory, determined.

3.4 FRACTIONATION OF POLYDISPERSE SYSTEMS

The data plotted in Figure 3.9 were obtained by the fractionation of a polydisperse polymer sample. Polydisperse polymers can be fractionated by a number of techniques. The most widely used technique is chromatography. Other methods include the addition of a nonsolvent to a polymer solution, cooling a polymer solution, solvent evaporation, extraction, diffusion, or centrifugation. The molecular weight of the fractions may be determined using any of the classic techniques given in Table 3.3.

3.5 CHROMATOGRAPHY

As noted before, certain techniques such as colligative methods, light-scattering photometry, special mass spectral (MS) techniques, and ultracentrifugation allow the calculation of specific or absolute molecular weights. Under certain conditions, some of these also allow the calculation of the MWD.

There are a wide variety of chromatography techniques including paper and column techniques. Chromatographic techniques involve passing a solution containing the to-be-tested sample through a medium that shows selective absorption for the different components in the solution.

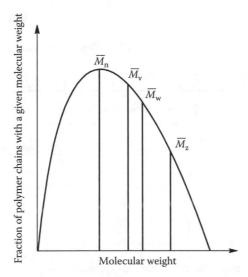

FIGURE 3.9   Molecular weight distributions.

Ion-exchange chromatography separates molecules on the basis of their electrical charge. Ion-exchange resins are either polyanions or polycations. For a polycation resin, those particles that are least attracted to the resin will flow more rapidly through the column and be emitted from the column first. This technique is most useful for polymers that contain charged moieties.

In *affinity chromatography*, the resin contains molecules that are especially selected and which will interact with the particular polymer(s) that is being studied. Thus, for a particular protein, the resin may be modified to contain a molecule that interacts with that protein type. The solution containing the mixture is passed through the column, and the modified resin preferentially associates with the desired protein allowing it to be preferentially removed from the solution. Later, the protein is washed through the column by the addition of a salt solution, and it is collected for further evaluation.

In *high-performance liquid chromatography, HPLC*, pressure is applied to the column that causes the solution to rapidly pass through the column allowing procedures to be completed in a fraction of the time in comparison to regular chromatography.

When an electric field is applied to a solution, polymers containing a charge will move toward either the cathode (positively charged species) or the anode (negatively charged species). This migration is called *electrophoresis*. The velocity at which molecules move is mainly dependent upon the electric field and the charge on the polymer driving the molecule toward one of the electrodes, and a frictional force dependent on the size and the structure of the macromolecule that opposes the movement. In general, the larger and more bulky the macromolecule, the greater the resistance to movement and the greater the applied field and charge on the molecule, and consequently, the more rapid the movement. While electrophoresis can be conducted on solution, it is customary to use a supporting medium of a paper or a gel. For a given system, it is possible to calibrate the rate of flow with the molecular weight and/or size of the molecule. Here, the flow characteristics of the calibration material must be similar to those of the unknown.

Generally though, electrophoresis is often employed in the separation of complex molecules, such as proteins, where the primary factor in the separation is the charge on the species. Some amino acids such as aspartic acid and glutamic acid contain an "additional" acid functional group, while amino acids such as lysine, arginine, and histidine contain "additional" basic groups. The presence of these units will confer to the protein tendencies to move toward the anode or the cathode. The rate of movement is dependent on a number of factors including the relative abundance and accessibility of these acidic and basic functional groups.

FIGURE 3.10   Basic components of an electrophoresis apparatus.

Figure 3.10 contains an illustration of the basic components of a typical electrophoresis apparatus. The troughs at either end contain an electrolyte buffer solution. The sample to be separated is placed in the approximate center of the electrophoresis strip.

Gel-permeation chromatography, GPC, is a form of chromatography that is based on separation by molecular size rather than chemical properties. GPC or *Size-exclusion chromatography, SEC*, is widely used for molecular weight and MWD, determination. In itself, SEC does not give an absolute molecular weight and must be calibrated against polymer samples whose molecular weights have been determined by a technique that does give an absolute molecular weight.

SEC is a HPLC technique whereby the polymer chains are separated according to differences in hydrodynamic volume. This separation is made possible by the use of special packing material in the column. Polymeric porous spheres, often composed of polystyrene cross-linked by addition of varying amounts of divinylbenzene, are generally used as the packing material. Retention in the column is mainly governed by the partitioning (or exchanging) of polymer chains between the mobile (or eluent) phase flowing through the column and the stagnate liquid phase that is present in the interior of the packing material.

Through control of the amount of cross-linking, nature of the packing material, and specific processing procedures, spheres of widely varying porosity are available. The motion in and out of the stationary phase is dependent on a number of factors including Brownian motion, chain size, and conformation. The latter two are related to the polymer chain's hydrodynamic volume—the real, excluded volume occupied by the polymer chain. Since smaller chains preferentially permeate the gel particles, the largest chains are eluted first. As noted above, the fractions are separated on the basis of size.

The resulting chromatogram is then a molecular size distribution. The relationship between molecular size and molecular weight is dependent on the conformation of the polymer in solution. As long as the polymer conformation remains constant, which is generally the case, molecular size increases with an increase in molecular weight. The precise relationship between molecular size and molecular weight is conformation dependent. For random coils, molecular size as measured by the polymer's radius of gyration, R, and molecular weight, M, R is proportional to M^b, where "b" is a constant dependent on the solvent, the polymer concentration, and the temperature. Such values are known and appear in the literature for many polymers allowing the ready conversion of molecular size data collected by SEC into molecular weight and MWD.

There is a wide variety of instrumentation ranging from simple manually operated devices to completely automated systems. Briefly, the polymer-containing solution and solvent alone are introduced into the system and pumped through separate columns at a specific rate. The differences in refractive index between the solvent itself and the polymer solution are determined using a differential refractometer. This allows the calculation of the amount of polymer present as the solution passes out of the column.

The unautomated procedure was first used to separate protein oligomers using Sephadex gels. Today, there is a wide variety of specialized and general gels used as column packing. The efficiency of these packed columns can be determined by calculating the height in feet equivalent to a theoretical plate (HETP), which is the reciprocal of the plate count per feet (P). P is directly proportional to the square of the elution volume (V_c) and inversely proportional to the height of the column in feet and the square of the baseline (D) as follows:

$$P = \left(\frac{16}{f}\right)\left[\left(\frac{V_c}{D}\right)^2\right] \tag{3.9}$$

The conversion of retention volume for a given column to molecular weight can be accomplished using several approaches including peak position, universal calibration, broad standard molecular weight determination, and actual molecular weight determination by coupling the SEC to an instrument that gives absolute molecular weight.

In the **peak-position approach**, well-characterized narrow fraction samples of known molecular weight are used to calibrate the column, and retention times are determined. A plot of log M vs. retention is made and used for the determination of samples of unknown molecular weights. Unless properly treated, such molecular weights are subject to error. The best results are obtained when the structures of the samples used in the calibration and those of the test polymers are the same.

The **universal calibration approach** is based on the product of the limiting viscosity number, LVN, and the molecular weight being proportional to the hydrodynamic volume. Benoit showed that for different polymers, the elution volume plotted against the log LVN times molecular weight gave a common line. In one approach, molecular weight is determined by constructing a "universal calibration line" through plotting the product of log LVN for polymer fractions with narrow MWDs as a function of the retention of these standard polymer samples for a given column. Molecular weight is then found from the retention time of the polymer sample using the calibration line.

Probably, the most accurate approach is to directly connect, couple, the SEC to a device, such as a light-scattering photometer, that directly measures the molecular weight for each elution fraction. Here both molecular weight and MWD are accurately determined.

3.6 COLLIGATIVE MOLECULAR WEIGHTS

3.6.1 Osmometry

A measure of any of the colligative properties involves counting solute (polymer) molecules in a given amount of solvent. The most common technique for polymers is **membrane osmometry**. The technique is based on the use of a semipermeable membrane through which solvent molecules freely pass, but through which the large polymer molecules are unable to pass. Existing membranes only approximate ideal semipermeability, the chief limitation being the passage of low molecular weight chains through the membrane.

There is a thermodynamic drive toward the dilution of the polymer-containing solution with a net flow of solvent toward the cell containing the polymer. This results in an increase in liquid in that cell causing a rise in the liquid level in the corresponding measuring tube. This rise in the liquid level is opposed and balanced by a hydrostatic pressure, resulting in a difference in the liquid levels of the two measuring tubes. The difference is directly related to the osmotic pressure of the polymer-containing solution. Thus, solvent molecules pass though the semipermeable membrane, reaching a "static" equilibrium.

Since osmotic pressure is dependent on the number of particles present, the measurement of this osmotic pressure can be used to determine the $\bar{M}_n$ of the dissolved polymer. The difference in height (Δh) of the liquids in the columns is converted to osmotic pressure (π) by multiplying the gravity (g) and the density of the solution (ρ), i.e., $\pi = \Delta h \rho g$. In the old static osmometers, it might take weeks to months for equilibrium to become established allowing the excessive passage of polymer chains through the membrane. Today, automated osmometers allow molecular weight measurements to occur in minutes with a minimal of the passage of polymer chains through the membrane. The relationship between molecular weight and osmotic pressure is given in the following van't Hoff equation:

$$\pi = \frac{RTC}{\bar{M}_n} + BC^2 \qquad (3.10)$$

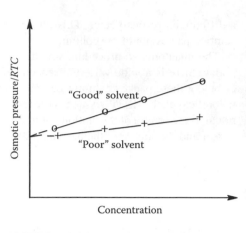

FIGURE 3.11 Plots of osmotic pressure, π, divided by RTC used to determine $1/\bar{M}_n$ in osmometry.

Thus, the reciprocal of $\bar{M}_n$ is the intercept when data for π/RTC vs. C are extrapolated to zero concentration (Figure 3.11).

The slop of the lines in Figure 3.11, i.e., the virial constant B, is related to the CED. The value for B would be zero at the theta temperature. Since this slope increases as the solvency increases, it is advantageous to use a dilute solution consisting of a polymer and a poor solvent to minimize extrapolation errors.

In the VPO technique, drops of solvent and solution are placed in an insulated chamber in proximity to thermistor probes. Since the solvent molecules evaporate more rapidly from the solvent than from the polymer solution, a temperature difference results that is related to the molarity of the polymer (M) that can be determined if the heat of vaporization per gram of solvent (λ) is known using the following relationship:

$$\Delta T = \frac{RT^2 M}{\lambda 100} \qquad (3.11)$$

3.6.2 END-GROUP ANALYSIS

In cases where the end groups are known and their concentration can be determined, knowledge of their abundance allows determination of $\bar{M}_n$. The sensitivity of this method decreases, and the chain length becomes greater. Some end groups can be determined using spectroscopic techniques and others through titration.

3.6.3 EUBLLIOMETRY AND CRYOMETRY

Raoult's law works for small polymers as well as small molecules. The determination of $\bar{M}_n$ is based for both **eublliometry** (boiling point elevation) and **cryometry** (freezing point lowering) on the Clausius–Clapeyron equation (Equation 3.12):

$$\bar{M}_n = \left(\frac{RT^2 V}{\Delta H} \right) \left(\frac{C}{\Delta T} \right)_{C \to 0} \qquad (3.12)$$

By use of sensitive thermocouples and care, molecular weights to about 50,000 Da can be determined.

3.7 LIGHT-SCATTERING PHOTOMETRY

Ever watch a dog or young child chase moonbeams? The illumination of dust particles is an illustration of light scattering, not of reflection. Reflection is the deviation of incident light through one particular angle such that the angle of incidence is equal to the angle of reflection. Scattering is the radiation of light in all direction. Thus, in observing the moonbeam, the dust particle directs a beam toward you regardless of your angle in relation to the scattering particle. The energy scattered per second (scattered flux) is related to the size and shape of the scattering particle and to the scattering angle.

Scattering of light is all about us—the fact that the sky above us appears blue, the clouds white, and the sunset is shades of reds and oranges is a consequence of the preferential scattering of light from air molecules, water droplets, and dust particles. This scattered light carries messages about the scattering objects.

The measurement of light scattering is the most widely used technique for the determination of $\bar{M}_w$. This technique is based on the optical heterogeneity of polymer solutions and was developed by Nobel Laureate Peter Debye in 1944.

Today, modern instruments use lasers as the radiation source because they provide a monochromatic, intense, and well-defined light source. Depending upon the size of the scattering object, the intensity of light can be essentially the same or vary greatly with respect to the direction of the oncoming radiation. For small particles, the light is scattered equally independent of the angle of the observer to the incoming light. For larger particles, the intensity of scattered light varies with respect to the angle of the observer to the incoming light. For small molecules, at low concentrations this scattering is described in terms of the Rayleigh ratio.

In 1871, Rayleigh showed that induced oscillatory dipoles were developed when light passed through gases and that the amount (intensity) of scattered light (τ) was inversely proportional to the fourth power of the wavelength of light. This investigation was extended to liquids by Einstein and Smoluchowski in 1908. These oscillations reradiate the light energy producing turbidity, i.e., the Tyndall effect. Other sources of energy, such as x-rays or laser beams, may be used in place of visible light sources.

For light-scattering measurements, the total amount of the scattered light is deduced from the decrease in the intensity of the incident beam, I_0, as it passes through a polymer sample. This can be described in terms of Beer's law for the absorption of light as follows:

$$\frac{I}{I_0} = e^{-tl} \tag{3.13}$$

where τ is the measure of the decrease in the incident beam intensity per unit length (l) of a given solution and is called the turbidity of the solution.

The intensity of scattered light or turbidity (τ) is proportional to the square of the difference between the index of refraction (n) of the polymer solution and of the solvent (n_0), to the molecular weight of the polymer ($\bar{M}_w$), and to the inverse fourth power of the wavelength of light used (λ). Thus

$$\frac{Hc}{\tau} = \frac{1}{\bar{M}_w P_0 (1 + 2Bc + Cc^2 + \cdots)} \tag{3.14}$$

where the expression for the constant H is as follows:

$$H = \left[\frac{32\pi^2}{3}\right]\left[\frac{n_0^2\left(\dfrac{dn}{dc}\right)^2}{\lambda^4 N}\right] \quad \text{and} \quad \tau = K'n^2\left(\frac{i_{90}}{i_0}\right) \tag{3.15}$$

where
  n_0 is the index of refraction of the solvent
  n is the index of refraction of the solution
  c is the polymer concentration
  B, C, etc., are virial constants related to the interaction of the solvent
  P is the particle scattering factor
  N is Avogadro's number

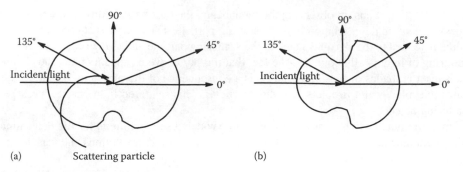

(a) Scattering particle (b)

FIGURE 3.12 Light scattering envelopes. Distance for the scattering particle to the boundaries of the envelope represents an equal magnitude of scattered light as a function of angle for a small scattering particle (a) and a large scattering particle (b).

The expression dn/dc is the specific refractive increment and is determined by taking the slope of the refractive index readings as a function of polymer concentration.

In the determination of $\bar{M}_w$, one measures the intensity of scattered light at different concentrations and at different angles (θ). The incident light sends out a scattering envelope that has four equal quadrants (Figure 3.12a) for small particles. The ratio of scattering at 45° compared with that at 135° is called the *dissymmetry factor* or *dissymmetry ratio* Z. The reduced dissymmetry factor Z_0 is the intercept of the plot of Z as a function of concentration extrapolated to zero concentration.

For polymer solutions containing polymers of moderate to low molecular weights P_0 is 1, and Equation 3.14 reduces to Equation 3.18, an equation for a straight line ($y = b + mx$):

$$\frac{Hc}{\tau} = \frac{1}{\bar{M}_w}(1 + 2Bc + Cc^2 + \cdots) \tag{3.16}$$

Several expressions are generally used in describing the relationship between values measured by light-scattering photometry and molecular weight. One is given in Equation 3.14 and others, such as Equation 3.17, are analogous except that constants have been rearranged.

$$\frac{Kc}{R} = \frac{1}{\bar{M}_w}(1 + 2Bc + Cc^2 + \cdots) \tag{3.17}$$

At low polymer concentrations, Equation 3.14 reduces to Equation 3.18 that is an equation of a straight line ($y = b + mx$):

$$\frac{Hc}{\tau} = \frac{1}{\bar{M}_w} \times \frac{2Bc}{\bar{M}_w} \tag{3.18}$$

When the ratio of the concentration c to the turbidity τ (tau; related to the intensity of scattering at 0° and 90°) multiplied by the constant H is plotted against concentration (Figure 3.13), the intercept of the extrapolated line is the reciprocal of $\bar{M}_w$ and the slope contains the virial constant B. Z_0 is directly related to the particle scattering factor, and both are related to both the size and shape of the scattering particle. As the size of the scattering particle, the individual polymer chain, approaches about one-twentieth the wavelength of the incident light, scattering interference occurs giving a scattering envelope

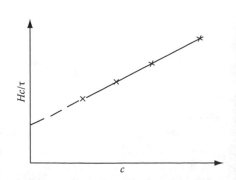

FIGURE 3.13 Typical simple plot used to determine $1/\bar{M}_w$ from scattering data.

that is no longer symmetrical (Figure 3.12b). Here the scattering dependency on molecular weight reverts back to the relationship given in Equation 3.14.

The molecular weight for dilute polymer solutions is typically found using one of the two techniques. The first technique is called the dissymmetrical method or approach because it uses the determination of Z_0 as a function of the particle scattering factor as a function of polymer shape. $\bar{M}_w$ is determined from the intercept through the substitution of the determined particle scattering factor. The weakness in this approach is the necessity of having to assume a shape for the polymer in a particular solution. For small Z_0 values, choosing an incorrect polymer shape results in a small error, but for larger Z_0 values, the error becomes significant.

The second approach uses multiple detectors (Figure 3.14), allowing a double extrapolation to zero concentration and zero angle with the data forming what is called a **_Zimm plot_** (Figure 3.15). The extrapolation to zero angle corrects for finite particle size effects. The radius of gyration, related to polymer shape and size, can also be determined from this plot. The second extrapolation to zero concentration corrects for concentration factors. The intercepts of both plots are equal to $1/\bar{M}_w$.

The Zimm plot approach does not require knowing or having to assume a particular shape for the polymer in solution.

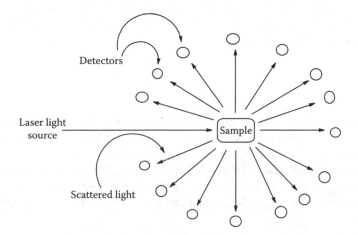

FIGURE 3.14 Multiple detector arrangement showing a sample surrounded by an array of detectors.

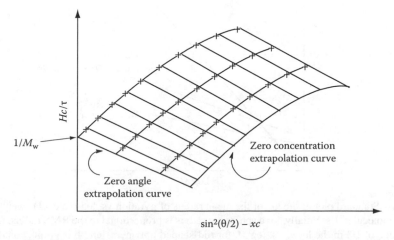

FIGURE 3.15 Zimm plot for a polymer scaled with a negative concentration coefficient (x) coefficient to improve data aesthetics and accessibility.

Related to the Zimm plot is the Debye plot. In the Zimm approach, different concentrations of the polymer solution are used. In the Debye approach, one low concentration sample is used with $1/\bar{M}_w$ plotted against $\sin^2(\theta/2)$, essentially one-half of the Zimm plot.

Low-angle laser light-scattering photometry, LALLS, and *multiangle low angle laser light-scattering photometry, MALS*, take advantage of the fact that at low or small angles, the scattering particle factor becomes one reducing Equation 3.14 to Equation 3.16 and at low concentrations to Equation 3.18.

A number of automated systems exist with varying capabilities. Some internally carry out dilutions and refractive index measurements allowing molecular weight to be directly determined without additional sample treatment. The correct determination of dn/dc is very important since any error in its determination is magnified because it appears as the squared value in the expression relating light scattering and molecular weight.

Low-angle and multiangle light-scattering photometers are available that allow not only the determination of $\bar{M}_w$, but also additional values under appropriate conditions. For instance, a Zimm plot as shown in Figure 3.16 allows both $\bar{M}_w$ and $\bar{M}_n$ to be determined as well as the mean radius independent of the molecular conformation and branching.

These systems may also allow the determination of molecular conformation matching the radius and molecular weight to graphs showing the change in the root mean square radius of gyration, RMS, and the molecular weight for different shaped molecules (Figure 3.15). The expression for the mean square radius of gyration is given as

$$r_g^2 = \frac{\sum r_i^2 m_i}{\sum m_i} \tag{3.19}$$

One of the most important advances in polymer molecular weight determination is the "coupling" of SEC and light-scattering photometry, specifically LALLS or MALS. As noted in Section 3.5, SEC allows the determination of the MWD. In its usual operational configuration, it does not by itself allow the calculation of an absolute molecular weight but relies on calibration with the polymers of known molecular weight. By coupling HPLC and light-scattering photometry, the molecular weight of each fraction can be determined giving an MWD and various molecular weight values ($\bar{M}_w$, $\bar{M}_z$, $\bar{M}_n$).

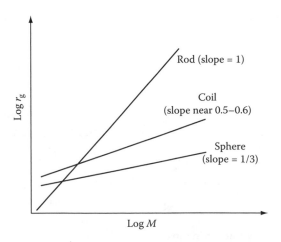

FIGURE 3.16 Standard plot of the log of the mean radius of gyration vs. log molecular weight for different shaped macromolecules. Essentially, for a sphere the radius is proportional to the RMS radius, rms radius, and $M^{1/3}$ with a slope of 1/3 in the log r_g vs. log M; for rod-shaped polymers, length is proportional to rms radius and M with a slope of 1; and for random coils the end-to-end distance is proportional to the rms radius and $M^{1/2}$ with a slope of about 0.5–0.6.

FIGURE 3.17 Typical SEC-MALS setup including refractive index refractometer. (Courtesy of Wyatt Technology Corporation, Santa Barbara, CA; wyatt@wyatt.com)

The LALLS or MALS detector measures τ-related values, a differential refractive index, DRI, detector is used to measure concentration, and the SEC supplies samples containing "fractionated" polymer solutions allowing both molecular weight and MWD to be determined. Further, polymer shape can be determined. This combination represents the most powerful one, based on the ease of operation, variety of samples readily used, cost, means to determine polymer size, shape, and MWD available today.

A general assembly for a SEC-MALS instrument is given in Figure 3.17. A typical three-dimensional plot obtained from such an assembly is shown in Figure 3.18.

Dynamic light scattering (DLS) is similar in principle to typical light scattering. When several particles are hit by oncoming laser light, a spotted pattern appears with the spots originating from the interference between the scattered light from each particle giving a collection of dark (from destructive interference) and light (from constructive interference) spots. This pattern of spots varies with time because of the Brownian motion of the individual scattering particles. The rate of change in the pattern of spots is dependent on a number of features including particle size. The larger the particle, the slower the Brownian motion, and consequently, the slower the change in the pattern. The measurement of these intensity fluctuations with time allows the calculation of the translational diffusion constant of the scattering particles. The technique for making these measurements is given several names including DLS with the name emphasizing the fact that it is the difference in the scattered light with time that is being measured; photon correlation spectroscopy, PCS, with the name emphasizing the particular mathematical technique employed to analyze the light-scattering data; and quasielastic light scattering, QELS, with the name emphasizing the fact that no energy is lost between the collision between the particle and the light photon.

FIGURE 3.18 Three-dimensional plot of scattering intensity as a function of scattering angle and elution volume for a broad MWD polystyrene (NITS standard reference 706). (Courtesy of Wyatt Technology Corporation, Santa Barbara, CA; wyatt@wyatt.com)

The effect of subtle particle changes as a function of temperature, sample preparation, time, solvent, and other changes can be measured using DLS. Such changes can then be related to performance variations, eventually interrelating structure shape and biological/physical property.

Another variation that is useful while employing coupled light scattering is referred to as *a triple detection set*. The set consists of three detectors—a light-scattering detector, a differential refractometer detector, and a capillary differential viscometer. It functions in concert with a GPC and a light-scattering source. The GPC separates the polymer mixture into molecular weight fractions. According to the Einstein equation, the intrinsic viscosity times the molecular weight is equal to the hydrodynamic volume or the size of polymers in solution. Thus, the molecular weight is determined by using light-scattering photometry, viscometry gives the intrinsic viscosity, and the equation is solved for size.

Generally, light-scattering photometry has a limit in determining molecular size with the lower limit being about 10 nm. The addition of the viscometer allows molecular sizes to be determined for oligomeric materials to about 1 nm. The assembly allows an independent measure of size and molecular weight as well as additional conformational and aggregation information, including small conformational changes. The assembly also allows good molecular determination to occur even with small dn/dc values, low-molecular weight fractions, absorbing and fluorescent polymers, copolymers with varying dn/dc values, and chiral polymers that depolarize the incident beam.

Figure 3.19 contains data on myoglobin obtained using a triple detection setup. A molecular weight of 21,100 is found with a viscosity of 0.0247 dL/g, and from this a hydrodynamic radius of 2.06 nm that is essentially the same as the Stokes value of 2.0 nm is reported for myoglobin.

3.8 OTHER TECHNIQUES

3.8.1 ULTRACENTRIFUGATION

Since the kinetic energy of solvent molecules is greater than the sedimentation force of gravity, polymer molecules remain suspended in solution. However, this gravitational field, which permits Brownian motion, may be overcome by increasing this force by the use of high centrifugal forces, such as the ultracentrifugal forces, developed by Nobel Laureates, The Svedberg in 1925.

FIGURE 3.19 Response to selected detectors as a function of retention volume for myoglobin (dissolved in PBS buffer at a pH of 6.9). The three detectors are the RI = refractive index signal, LS = light-scattering signal, and $\overline{DP}$ = differential pressure transducer (viscosity signal). (Courtesy of Viscotek, Houston, TX.)

Both $\overline{M}_w$ and $\overline{M}_z$ may be determined by subjecting dilute polymer solutions to high centrifugal forces. Solvents with densities and indices of refraction different from the polymers are chosen to ensure polymer motion and optical detection of this motion. In sedimentation velocity experiments, the ultracentrifuge is operated at extremely high rotational speeds up to over 70,000 rpm to transport the denser polymer molecules through the less dense solvent to the cell bottom or to the top if the density of the solvent is greater than the density of the polymer. The boundary movement during ultracentrifugation can be followed using optical measurement to monitor the sharp change in refractive index (n) between the solvent and the solution.

The rate of sedimentation is defined by the sedimentation constant s, which is directly proportional to the polymer mass m, the solution density ρ, and specific volume of the polymer V, and inversely proportional to the square of the angular velocity of rotation ω, the distance from the center of rotation to the point of observation in the cell r, and the fractional coefficient f, which is inversely related to the diffusion coefficient D extrapolated to infinite dilution. These relationships are shown in the following equations in which $(1 - V_p)$ is called the buoyancy factor since it determines the direction of macromolecular transport in the cell:

$$s = \left(\frac{1}{\omega^2 r}\right)\frac{dr}{dt} = \frac{m(1-V_p)}{f} \tag{3.20}$$

$$D = \frac{RT}{Nf} \quad \text{and} \quad mN = M \tag{3.21}$$

$$\frac{D}{s} = \frac{RT}{\overline{M}_w(1-V_p)} \tag{3.22}$$

The sedimentation velocity determination is dynamic and can be completed in a short period of time. The sedimentation equilibrium method gives quantitative results, but long periods of time are required for centrifugation at relatively low velocities to establish equilibrium between sedimentation and diffusion.

The $\overline{M}_w$ is directly proportional to the temperature T and the ln of the ratio of concentration c_2/c_1 at distances r_1 and r_2 from the center of rotation and the point of observation in the cell and inversely proportional to the buoyancy factor, the square of the angular velocity of rotation and the difference between the squares of the distances r_1 and r_2 as shown below:

$$\bar{M} = \frac{\left[2RT\left(\ln \frac{c_2}{c_1} \right) \right]}{(1 - V_p)\omega^2 (r_2^2 - r_1^2)} \tag{3.23}$$

3.8.2 MASS SPECTROMETRY

Certain MS procedures allow the determination of the molecular weight or the molecular mass of oligomeric to polymeric materials (Table 3.4).

In *matrix-assisted laser desorption/ionization mass spectroscopy, MALDI MS*, the polymer is dissolved, along with a "matrix chemical," and the solution deposited onto a sample probe. The solution is dried. MALDI MS depends on the sample having a strong UV absorption at the wavelength of the laser used. This helps minimize fragmentation, since it is the matrix UV-absorbing material that absorbs most of the laser energy. Often employed UV-matrix materials are 2,5-dihydroxybenzoic acid, sinnapinic acid, picplinic acids, and α-cyano-4-hydroxycinnamic acid. The high energy of the laser allows both the matrix material and the test sample to be volatilized. Such techniques are referred to as "soft" since the test sample is not subjected to (much) ionizing radiation, and hence little fragmentation occurs. Mass accuracy in the order of a few parts per million are obtained. Thus, chain content can be determined for copolymers and other chains with unlike repeat units. Polymer MWDs can also be determined using MALDI MS and related MS techniques.

MALDI MS was developed for the analysis of nonvolatile samples and was heralded as an exciting new MS technique for the identification of materials with a special use in the identification of polymers. It has fulfilled this promise to only a limited extent. While it has become a well-used and essential tool for biochemists in exploring mainly nucleic acids and proteins, it has been only sparsely employed by synthetic polymer chemists. This is because of the lack of congruency between the requirements of MALDI MS and most synthetic polymers. MALDI MS requires that the test material, polymer, be soluble in a relatively volatile solvent. Proteins, nucleic acids, and PEGs are water soluble, so allow the use of MALDI MS for the analysis of their chain lengths. Most polymers are not readily soluble in such solvents, so MALDI MS is of little use in the analysis of these materials. Carraher and coworkers have developed an approach that is applicable to most materials that focuses on analyzing the fragmentation of the polymers. While this technique is not able to give molecular weights and MWDs, it allows the identification of ion fragments and ion fragment clusters up to several thousand daltons.

TABLE 3.4
Mass Spectrometry Approaches Used in the Determination of Molecular Weights of Oligomeric and Polymeric Materials

MS-Type	(Typical) Upper Molecular Weight Range (Da)
(Usual) electron impact, EI	Up to 2000
Fast atom bombardment, FAB	Up to 2000
Direct laser desorption, direct LD	Up to 10^4
Matrix-assisted laser desorption/ionization, MALDI	Up to 10^7

Recently, MS combinations have been available including the TG–MS combination developed by Carraher that allows the continuous characterization of evolved materials as a polymer undergoes controlled thermal degradation.

3.9 VISCOMETRY

Viscosity is a measure of the resistance to flow of a material, mixture, or solution. Here, we will consider the viscosity of solutions containing small, generally 1 g/100 cc (called 1% solutions) and less, amounts of polymer. The study of such dilute polymer solutions allows a determination of a "relative" molecular weight. The molecular weight is referred to as "relative" since viscosity measurements have not been directly related, through rigorous mathematical relationships, to a specific molecular weight. By comparison, measurements made using light-scattering photometry and other methods covered before are relatable to specific molecular weight values, and these techniques are said to give us "absolute" molecular weights.

The relationship between the force f necessary to move a plane of area A relative to another plane at a distance d from the initial plane (Figure 3.20), is described as

$$f \propto \frac{A}{d} \tag{3.24}$$

To make this a direct relationship a proportionality factor is introduced. This factor is called the *coefficient of shear viscosity* or simply *viscosity*:

$$f = \eta \left(\frac{A}{d} \right) \tag{3.25}$$

Viscosity is then a measure of the resistance of a material to flow. In fact, the inverse of viscosity is given the name *fluidicity*. As a material's resistance to flow increases, its viscosity increases. Viscosities have been reported using a number of different names. The CGS unit of viscosity is called the poise, which is dyne seconds per square centimeter. Another widely used unit is the pascal (or Pa), which is Newton seconds per square centimeter. In fact, 1 Pa = 10 poise.

Table 3.5 contains the general magnitude of viscosity for some common materials. It is important to note the wide variety of viscosities of materials from gases, such as air, to viscoelastic solids, such as glass.

In polymer science, we typically do not measure viscosity directly, but rather we look at relative viscosity measures by measuring the flow rate of one material relative to that of a second material. Viscosity is one of the most widely used methods for the characterization of polymer molecular weight because it provides the easiest and most rapid means of obtaining molecular weight-related data that requires minimal instrumentation. A most obvious characteristic of polymer solutions is their high viscosity, even when the amount of added polymer is small. This is because polymers reside in several flow planes (Figure 3.20b) acting to resist the flow of one plane relative to another flow plane.

(a)   (b)

FIGURE 3.20 (a) Representation of Equation 3.24. (b) Illustrates a polymer chain between two flow planes.

TABLE 3.5
Viscosities of Selected Common Materials

Substances	General Viscosity (MPa)
Air	0.00001
Water	0.001
Polymer latexes/paints	0.01
PVC plastols	0.1
Glycerol	10
Polymer resins and "pancake" syrups	100
Liquid polyurethanes	1,000
Polymer "melts"	10,000
Pitch	100,000,000
Glass	1,000,000,000,000,000,000,000

The ratio of the viscosity of a polymer solution to that of the solvent is called the *relative viscosity* (η_r). This value minus 1 is called the **specific viscosity (η_{sp})**, and the **reduced viscosity (η_{red})** or **viscosity number** is obtained by dividing η_{sp} by the polymer concentration c; i.e., η_{sp}/c. The **intrinsic viscosity**, or **LVN**, is obtained by extrapolating η_{sp}/c to zero polymer concentration. These relationships are given in Table 3.6, and a typical plot of η_{sp}/c and $\ln \eta_r/c$ is given in Figure 3.21.

Table 3.6
Commonly Used Viscosity Terms

Common Name	Recommended Name (IUPAC)	Definition	Symbol
Relative viscosity	Viscosity ratio	η/η_0	$\eta_{rel} = \eta_r$
Specific viscosity	—	$(\eta/\eta_0) - 1$ or $\eta - \eta_0/\eta_0$	η_{sp}
Reduced viscosity	Viscosity number	η_{sp}/c	η_{red} or η_{sp}/c
Inherent viscosity	Logarithmic viscosity number	$\ln(\eta_r/c)$	η_{inh} or $\ln(\eta_r/c)$
Intrinsic viscosity	Limiting viscosity number	Limit $(\eta_{sp}/c)_{c \to 0}$ or limit $\ln(\eta_r/c)_{c \to 0}$	LVN

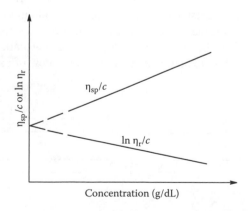

FIGURE 3.21   Reduced and inherent viscosity–concentration lines for a dilute polymer solution.

Staudinger showed that the intrinsic viscosity of a solution ([η]) or the LVN is related to the molecular weight of the polymer. The present form of this relationship was developed by Mark–Houwink (and is known as the **Mark–Houwink equation**) in which the proportionality constant K is the characteristic of the polymer and the solvent, and the exponential a is a function of the shape of the polymer in a solution. For theta solvents, the value of a is 0.5. This value, which is actually a measure of the interaction of the solvent and the polymer, increases as the coil expands, and the value is between 1.8 and 2.0 for rigid polymer chains extended to their full contour length and 0 for spheres. When a is 1.0, the Mark–Houwink equation (Equation 3.26) becomes the Staudinger viscosity equation.

$$\text{LVN} = KM^a \tag{3.26}$$

Values of "a" and "K" have been determined and compiled in several polymer handbooks and are dispersed throughout the literature. With known a and K values, molecular weight molecular weight can be calculated using Equation 3.26. As noted before, viscosity is unable to give absolute molecular weight values and must be calibrated, i.e., values of a and K must be determined using polymer samples where their molecular weights have been determined using some absolute molecular weight method, such as light-scattering photometry. It is customary in determining the a and K values to make a plot of log LVN vs. log M since the log of Equation 3.26, i.e., Equation 3.27, is a straight line relationship where the slope is a and the intercept is K. In reality, a is determined from the slope but K is determined by simply selecting a known LVN–M couple and using the determined a value to calculate the K value.

$$\log \text{LVN} = a \log M + \log K \tag{3.27}$$

The intrinsic viscosity or the LVN, like melt viscosity, is temperature dependent and decreases as temperature increases as shown in Equation 3.28.

$$\text{LVN} = A\, e^{E/RT} \tag{3.28}$$

However, if the original temperature is below the theta temperature, the viscosity will increase when the mixture of polymer and solvent is heated to a temperature slightly above the theta temperature.

Viscosity measurements of dilute polymer solutions are carried out using a viscometer, such as any of those pictured in Figure 3.22. The viscometer is placed in a constant temperature bath, and the time taken to flow through a space is measured.

Flory, Debye, and Kirkwood showed that [η] is directly proportional to the effective hydrodynamic volume of the polymer in solution and inversely proportional to the molecular weight, M. The effective hydrodynamic volume is the cube of the RMS end-to-end distance, $(r^2)^{3/2}$. This proportionality constant, N, in the Flory equation for hydrodynamic volume, Equation 3.29, has been considered to be a universal constant independent of solvent, polymer, temperature, and molecular weight:

$$[\eta] = \frac{N(r^2)^{3/2}}{M} \tag{3.29}$$

The actual average end-to-end distance, $\bar{r}$, is related to the nonsolvent expanded average end-to-end distance r_0 using the Flory expansion factor, α, as follows:

$$\bar{r} = \alpha \bar{r}_0 \tag{3.30}$$

Substitution of Equation 3.30 into Equation 3.29 and rearrangement gives

$$[\eta]M = N(\bar{r}_0^2)^{3/2}\alpha^2 \tag{3.31}$$

FIGURE 3.22   Common solution viscometers.

Values for α vary from 0.5 for Flory theta solvents to about 3 for polymers in good solvents.

In Equation 3.26, "a" values for random coils range from 0.5 for theta solvents to 0.8 for good solvents, 0 for hard spheres, about 1 for semicoils, and 2 for rigid rods.

The theta temperature corresponds to the Boyle point in an imperfect gas and is the range in which the virial coefficient B in the expanded gas law becomes zero. This same concept applies to the modification of the gas law ($PV = nRT$) used to determine the osmotic pressure of a polymer solution and is simply the van't Hoff equation that reduces to Equation 3.32 when $B = 0$:

$$\pi = \frac{RTC}{\bar{M}_n} \tag{3.32}$$

For linear polymers at their theta temperature, i.e., the temperature where the chain attains unperturbed dimensions, the Flory equation resembles the Mark–Houwink equation where α is equal to 1.0 as shown below:

$$[\eta] = KM^{1/2}\alpha^3 = K'M^{1/2} \tag{3.33}$$

The intrinsic viscosity of a solution, like the melt viscosity, is temperature dependent and decreases as the temperature increases:

$$[\eta] = A\, e^{E/RT} \tag{3.34}$$

While the description of viscosity is complex, the relative viscosity is directly related to the flow-through times using the same viscometer as shown in Equation 3.35 where t and t_0 are the flow times for the polymer solution and the solvent, respectively, and the density of the solution (ρ) and the solvent (ρ_0) are related as in Equation 3.35.

$$\frac{\eta}{\eta_0} = \frac{\rho t}{\rho_0 t_0} = \eta_r \tag{3.35}$$

Since the densities of the dilute solution and solvent are almost the same, they are normally cancelled giving Equation 3.36.

$$\frac{\eta}{\eta_0} = \frac{t}{t_0} = \eta_r \tag{3.36}$$

Thus, the relative viscosity is simply a ratio of flow times for the polymer solution and the solvent. The reduced viscosity is related to the LVN by a virial equation (Equation 3.37)

$$\frac{\eta_{sp}}{c} = [\eta] + k_1[\eta]^2 c + k'[\eta]^3 c^2 + \cdots \tag{3.37}$$

For most solutions, Equation 3.37 reduces to the **Huggins viscosity relationship**, Equation 3.38,

$$\frac{\eta_{sp}}{c} = [\eta] + k_1[\eta]^2 c \tag{3.38}$$

which allows $[\eta]$ to be determined from the intercept of the plot of η_{sp}/c vs. c and k_1 is the basis for the top plot given in Figure 3.21.

Another relationship often used in determining $[\eta]$ is called the inherent viscosity equation and is given in Equation 3.39.

$$\ln \frac{\eta_r}{c} = [\eta] - k_2[\eta]^2 c \tag{3.39}$$

Again, a plot of $\ln \eta_r/c$ vs. c gives a straight line with the intercept $[\eta]$ (or LVN) after extrapolation to zero polymer concentration. This is the basis of the lower plot in Figure 3.17. While k_1 and k_2 are mathematically such that

$$k_1 + k_2 = 0.5 \tag{3.40}$$

but many systems do not appear to follow this relationship.

We will now turn our attention from the viscosity of dilute solutions and look at the viscosity of melted polymers. The viscosity of melted polymers is important in transferring resins and in polymer processing such as determining the correct conditions to have a specific flow rate for injection processing and in determining the optimum conditions to get the necessary dimensions of extruded shapes. Fillers, plasticizers, temperature, solvents, and molecular weight are just some of the variables that influence the viscosity of polymer melts. Here, we will look at the dependence of melt viscosity on polymer molecular weight. Polymer melts have viscosities in the order of 10,000 MPa (1 centiposes is equal to 0.001 Pa/s).

For largely linear polymers, such as polystyrene, where particularly bulky side chains are not present the viscosity or the flow is mainly dependent on the chain length. In most polymers, the melt viscosity–chain length relationship has two distinct regions where the region division occurs when the chain length reaches some length called the **critical entanglement chain length, Z**, where the intermolecular entanglement occurs. This intermolecular entanglement causes the individual chains in the melt to act as being much more massive because of the entanglement. Thus, the resistance to flow is a combination of the friction and the entanglement between chains as they slide past one another. Below the critical entanglement length, where only the friction part is important, the melt viscosity, η, is related to the weight-average molecular weight by

$$\eta = K_1 \bar{M}_w^{1.0} \tag{3.41}$$

and above the critical chain length, where both the friction and the entanglement are important, the relationship is

$$\eta = K_h \bar{M}_w^{3.4} \tag{3.42}$$

where
  K_1 is a constant for the precritical entanglement chain length
  K_h is for the situation above Z and where both K values are temperature dependent

The first power dependence is due to the simple increase in molecular weight as chain length increases, but the 3.4 power relationship is due to a complex relationship between chain movement as related to entanglement and diffusion and chain length.

The critical chain length is often the onset of "strength"-related properties and is often considered the lower end for useful mechanical properties. The Z value for polymers varies but is generally between about 200 and 1000 units in length. For instance, the Z value for polystyrene is about 700, for polyisobutylene about 600, for poly(decamethylene sebacate) about 300, for poly(methyl methacrylate) about 200, and for poly(dimethyl siloxane) about 1000.

A number of techniques have been developed to measure melt viscosity. Some of these are listed in Table 3.7. Rotational viscometers are of varied structures. The Couette cup and bob viscometer consists of a stationary inner cylinder, the bob, and an outer cylinder, cup, which is rotated. Shear stress is measured in terms of the required torque needed to achieve a fixed rotation rate for a specific radius differential between the radius of the bob and the cup. The Brookfield viscometer is a bob and cup viscometer. The Mooney viscometer, often used in the rubber industry, measures the torque needed to revolve a rotor at a specified rate. In the cone and plate assemblies, the melt is sheared between a flat plat and a broad cone whose apex contacts the plate containing the melt.

A number of capillary viscometers or rheometers have been employed to measure melt viscosity. In some sense, these operate on a principle similar to the simple observation of a trapped bubble moving from the bottom of a shampoo bottle when it is turned upside down. The more viscous the shampoo, the longer it takes for the bubble to move through the shampoo.

TABLE 3.7
Viscosity Measuring Techniques
and Their Usual Ranges

Technique	Typical Range, Poises
Capillary pipette	0.01–1,000
Falling sphere	1–100,000
Parallel plate	10,000–10^9
Falling coaxial cylinder	100,000–10^{11}
Stress relaxation	1,000–10^{10}
Rotating cylinder	1–10^{12}
Tensile creep	100,000–greater than 10^{12}

3.10 SUMMARY

1. Some naturally occurring polymers, such as certain proteins and nucleic acids, consist of molecules with a specific molecular weight and are called monodisperse. However, many other natural polymers, such as cellulose and natural rubber, and most synthetic polymers consist of molecules with different molecular weights and are called polydisperse. Many properties of polymers are dependent on their chain length. Since the melt viscosity increases exponentially with chain length, the high-energy costs of processing high molecular weight polymers are not often justified.

2. The distribution of chain lengths in a polydisperse system may be represented on a typical probability-like curve. The $\bar{M}_n$ is the smallest in magnitude of the typically obtained molecular weights and is a simple arithmetic mean that can be determined using any technique based on colligative properties, such as osmotic pressure, boiling point elevation, freezing point depression, and end-group determination. $\bar{M}_w$ is larger than $\bar{M}_n$ and is referred to as the second power relationship for disperse polymer chains. This value is most often determined by light-scattering photometry. Light-scattering photometry and the colligative related values are referred to as absolute molecular weight values because there is a direct mathematical connection between molecular weight and the particular property used to determine molecular weight.

3. For monodisperse samples, $\bar{M}_n = \bar{M}_w$. For polydisperse samples, the ratio of $\bar{M}_w/\bar{M}_n$ is a measure of the polydisparity and is given the name polydispersity index. The viscosity molecular weight must be calibrated using samples whose molecular weight has been determined using an absolute molecular weight determination technique; thus, it is not an absolute molecular weight determining technique, but it requires simple equipment, and is easy to measure. The Mark–Houwink relationship, $LVN = KM^a$, is used to relate molecular weight and viscosity.

4. The number-average molecular weight is dependent on the number of polymer chains while the weight-average molecular weight is dependent on the size of the chains. Thus, there is a correlation between the way the molecular weight is obtained and the type of molecular weight obtained.

5. MWD is most often measured using some form of chromatography. In GPC, cross-linked polymers are used in a column and act as a sieve allowing the larger molecules to elute first. After calibration, the molecular weight of the various fractions of the polymer can be determined. Combinations, such as chromatography coupled with light-scattering photometry, are used to obtain the molecular weight of the various fractions in a continuous manner.

6. While some techniques such as membrane osmometry and light-scattering photometry give absolute molecular weight, other techniques such as viscometry give only relative molecular weights unless calibrated employing a technique that gives absolute molecular weight. After calibration between viscometry values and chain length through some absolute molecular weight method, viscometry is a fast, inexpensive, and simple method to monitor molecular weight.

7. In general, polymers are soluble in fewer solvents and to a lower concentration than similar smaller molecules. This is because entropy is the driving force for solubility, and smaller molecules have larger entropy values when solubility is achieved in comparison to polymers. Polymers also take longer to dissolve since it takes time for the solvent molecules to penetrate the polymer matrix.

8. Flory and Huggins developed an interaction parameter that may be used as a measure of the solvent power of solvents for amorphous polymers. Flory and Krigbaum introduced the idea of a theta temperature, which is the temperature at which an infinitely long polymer chain exists as a statistical coil in a solvent.

9. Hildebrand developed solubility parameters to predict the solubility of nonpolar polymers in nonpolar solvents. The solubility parameter is the square root of the CED. For polar solvents, special solvent–polymer interactions can be incorporated into the solubility parameter approach.

GLOSSARY

Affinity chromatography: Chromatography in which the resin is designed to contain moieties that interact with particular molecules and/or units within a polymer chain.

Bingham plastic: Plastic that does not flow until the external stress exceeds a critical threshold value.

Brownian motion: Movement of larger molecules that results from a bombardment of smaller molecules in a liquid.

Buoyancy factor: In ultracentrifugation experiments, it determines the direction of polymer transport under the effect of centrifugal forces in the cell.

Chromatography: Family of separation techniques based on the use of a medium that shows selective absorption.

Colligative properties: Properties of a solution that are dependent on the number of solute molecules present.

Cloud point: Temperature at which a polymer starts to precipitate when the temperature is lowered.

Cohesive energy density (CED): Heat of vaporization per unit volume.

Commercial polymer range: Molecular weight range high enough to have good physical properties but not too high for economical processing.

Cryometry: Measurement of number-average molecular weight from freezing point depression.

Ebulliometry: Measurement of number-average molecular weight from boiling point elevation.

Effective hydrodynamic volume: Cube of the RMS end-to-end distance of a polymer chain.

Electrophoresis: Form of chromatography that uses an electric field to separate molecules.

End group analysis: Determination of number-average molecular weight by the determination of end groups.

Flory–Huggins theory: Theory used to predict the equilibrium behavior between liquid phases containing polymer.

Fractional precipitation: Fractionation of polydisperse systems by the addition of small amounts of nonsolvent to a solution of polymer.

Fractionation of polymers: Separation of a polydisperse polymer into fractions of similar molecular weight.

Gel-permeation chromatography: Type of liquid–solid elution chromatography that separates solutions of polydisperse polymer solutions into fractions containing more homogeneous chain sizes by means of a sieving action of a swollen cross-linked polymeric gel. Also called SEC.

High-performance liquid chromatography (HPLC): Chromatography in which the pressure applied causes the solution to pass more rapidly through the column.

Hildebrand (H): Unit used for solubility parameter values.

Ion-exchange chromatography: Chromatography that separates molecules on the basis of their electrical charges employing polyanionic or polycationic resins.

Kauri–Butanol values: Measure of the aromaticity of a solvent.

Low-angle laser light-scattering photometry (LALLS): Light scattering that employs low-angle measurements, minimizing the effect of polymer shape on the scattering.

Mark–Houwink equation: Relates LVN to molecular weight; $LVN = KM^a$.

Matrix-assisted laser desorption/ionization mass spectrometry (MALDI MS): Mass spectrometry in which the sample is placed in a matrix that contains a strong UV absorber chosen to match the UV absorption of the laser, which allows the molecules to become volatilized with minimal fragmentation.

Melt index: Measure of the flow related inversely to melt viscosity.

Monodisperse: System containing molecules of only one chain length.

Multiangle low-angle laser light-scattering photometry (MALS): Similar to LALLS except where the necessary angle ratios are made together; employs low-angle measurements minimizing the effect of polymer shape on the scattered light.

Number-average molecular weight: Arithmetical mean value obtained by dividing the sum of the molecular weights by the number of molecules.

Oligomer: Polymers with 2–10 repeat units. *Oligos* means "few."

Osmometry: Gives number-average molecular weight from osmotic pressure measurements.

Polydisperse: Mixture of polymer chains of different lengths.

Raoult's law: States that the vapor pressure of a solvent in equilibrium with a solution is equal to the product of the mole fraction of the solvent and the vapor pressure of the pure solvent. This relationship is used to obtain number-average molecular weights.

SEC-MALS and SEC-LALLS: Coupled chromatography and light-scattering photometry that allows the determination of a number of important values along with chain length distribution.

Sedimentation equilibrium experiment: Ultracentrifugation technique that allows chain length information to be determined.

Semipermeable membrane: Membrane that permits the diffusion of solvent molecules but not large molecules.

Size-exclusion chromatography (SEC): Chromatography in which separation is based on molecular size or differences in hydrodynamic volume; also called GPC; can use the universal calibration approach to obtain molecular weight.

Solubility parameter: A numerical value equal to the square root of the CED, which is used to predict polymer solubility.

Theta solvent: Solvent in which the polymer chain exists as a statistical coil.

Theta temperature: Temperature at which a polymer of infinite molecular weight begins to precipitate.

Ultracentrifuge: Centrifuge that increases the force of gravity by as much as 100,000 times, causing a distribution of materials in a solution to separate in accordance with chain length.

Vapor pressure osmometry: Technique for determining number-average molecular weight by measuring the relative heats of evaporation of a solvent from a solution and a pure solvent.

Viscosity: Resistance to flow.

Intrinsic viscosity: The LVN obtained by the extrapolation of the reduced viscosity to zero concentration.

Reduced viscosity: Specific viscosity divided by the polymer concentration.

Relative viscosity: Ratio of the viscosities of a solution and its solvent.

Specific viscosity: Difference between the relative viscosity and 1.

Weight-average molecular weight: Second power average of molecular weight; dependent on the size of the particular chains.

Zimm plot: Type of double extrapolation used to determine the weight-average molecular weight in light-scattering photometry.

EXERCISES

1. Which of the following is polydisperse with respect to chain length: (a) casein, (b) commercial polystyrene, (c) paraffin wax, (d) cellulose, or (e) *Hevea brasiliensis*?
2. If the number-average molecular weight for LDPE is 1.4 million, what is the corresponding average chain length?
3. What are the number-average molecular weights and weight-average molecular weights for a mixture of five molecules each having the following molecular weights: 1.25×10^6; 1.35×10^6; 1.5×10^6; 1.75×10^6; and 2.00×10^6?

4. What is the most probably value for the polydispersity index for (a) a monodisperse polymer and (b) a polydisperse polymer synthesized by a condensation technique?
5. List in increasing values: $\bar{M}_z$, $\bar{M}_n$, $\bar{M}_w$, and $\bar{M}_v$.
6. Which of the following provides an absolute measure of the molecular weight of polymers: (a) viscometry, (b) cryometry, (c) osmometry, (d) light scattering photometry, (e) GCP?
7. What is the relationship between the intrinsic viscosity or the LVN and the average-molecular weight?
8. What molecular weight determination techniques can be used to fractionate polydisperse polymers?
9. Which of the following techniques yields a number-average molecular weight: (a) viscometry, (b) light-scattering photometry, (c) ultracentrifugation, (d) osmometry, (e) ebulliometry, (f) cryometry?
10. What kind of molecular weight do you generally get from light-scattering photometry?
11. What is the value of the exponent a in the Mark–Houwink equation for polymers in theta solvents?
12. How many amino groups are present in each molecule of nylon-66 made from an excess of hexamethylenediamine?
13. What is the value of the exponent in the Mark–Houwink equation for a rigid rod?
14. If the values of K and a in the Mark–Houwink equation are 1×10^{-2} cm^3/g and 0.5 respectively, what is the average molecular weight of a polymer whose solution has an intrinsic viscosity of 150 cc/g?
15. Which polymer of ethylene will have the highest molecular weight: (a) a trimer, (b) an oligomer, or (c) UHMWPE?
16. What is a Zimm plot?
17. What type of molecular weight average, $\bar{M}_n$ or $\bar{M}_w$, is based on colligative properties?
18. What principle is used in the determination of molecular weight by vapor pressure osmometry?
19. Why does the melt viscosity increase faster with molecular weight increase than other properties such as tensile strength?
20. In spite of the high cost of processing, UHMWPE is used for making trash cans and other durable goods. Why?
21. Under what conditions are the weight-average molecular weight and number-average molecular weight the same?
22. What is the driving force for polymer solubility?
23. What are colligative methods for measuring molecular weight, and what kind of molecular weight do you get?
24. What is the advantage of using viscometry to measure molecular weight?
25. Which will yield the higher apparent molecular weight values in the light-scattering method: (a) a dust-free system or (b) one in which dust particles are present?
26. Does HPLC need to be calibrated before it can give absolute molecular weights?
27. Which of the following does modern LC allow the calculation of (a) weight-average molecular weight, (b) radius of gyration, (c) number-average molecular weight, (d) MWD, and (e) polydispersity index?
28. What is the significance of the virial constant B in osmometry and light-scattering equations?
29. According to Hildebrand, what is a regular solvent?
30. Which of the two steps that occur in the solution process, (a) swelling and (b) dispersion of the polymer particles, can be accelerated by agitation?
31. Define CED.
32. For solution to occur the change in Gibbs free energy must be: (a) = (0), (b) < 0, or (c) > 0.
33. Will a polymer swollen by a solvent have higher or lower entropy than the solid polymer?
34. Define the change in entropy in the Gibbs free energy equation.

35. Is a liquid that has a value of 0.3 for its interaction parameter a good solvent or a poor solvent?
36. What is the value of the Gibbs free energy change at the theta temperature?
37. What term is used to describe the temperature at which a polymer of infinite molecular weight precipitates from a dilute solution?
38. At which temperature will the polymer coil be larger in a poor solvent: (a) at the theta temperature, (b) below the theta temperature, or (c) above the theta temperature?
39. If the solubility parameter for water is 23.4 H, what is the CED for water?
40. What is the heat of mixing of two solvents having identical solubility parameters?
41. If the density of a polymer is 0.85 g/cc and the molar volume is 1,176,470 cc, what is the molecular weight?
42. Name some steps that occur when a polymer is dissolved.
43. Why is it important to determine polymer chain length?
44. Why do δ values decrease as the molecular weight increases in a homologous series of aliphatic polar solvents?
45. Which would be a better solvent for polystyrene: (a) *n*-pentane, (b) benzene, or (c) acetonitrile.
46. Which will have the higher or greater slop when its reduced viscosity or viscosity number is plotted against concentration: a solution of polystyrene (a) in benzene, or (b) in nonane?
47. What are general typical values for a in the viscosity relationship to molecular weight?
48. When is the Flory equation similar to the Mark–Houwink equation?
49. What is the term used for the cube root of the hydrodynamic volume?
50. Explain why the viscosity of a polymer solution decreases as the temperature increases.
51. Is MALDI MS restricted for use for natural polymers such as proteins and nucleic acids?
52. Which of the following would you expect to be most soluble in water—ethanol, hexane, or benzene. Why?
53. If entropy is the driving force for mixing and solubility, why is there such a focus on enthalpy through the various approaches with solubility parameters and other similar values?
54. If MALDI MS is of such great use to biopolymer chemists, why is it not more widely used by synthetic polymer chemists?

FURTHER READINGS

Albertsson, A. (2008): *Chromatography for Sustainable Polymeric Materials: Renewable, Degradable, and Recyclable*, Springer, New York.

Brown, W. (1996): *Light Scattering Principles and Development*, Springer-Verlag, New York.

Carraher, C., Sabir, T., and Carraher, C. L. (2009): *Inorganic and Organometallic Macromolecules*, Springer, New York.

Chabra, R. and Richardson, J. (2008): *Non-Newtonian Flow and Applied Rheology: Engineering Applications*, Elsevier, New York.

Debye, P. J. (1944): Light scattering analysis, *J. Appl. Phys.*, 15:338.

Debye, P. J. and Bueche, A. M. (1948): Intrinsic viscosity, diffusion, and sedimentation rates of polymers in solution, *J. Chem. Phys.*, 16:573.

Einstein, A. (1910): Theory of the opalescence of homogeneous liquids and liquid mixtures in the neighborhood of the critical state, *Ann. Physik.*, 33:1275.

Hansen, C. (2000): *Hanson Solubility Parameters*, CRC Press, Boca Raton, FL.

Hansen, C. (2007): *Hansen Solubility Parameters: A User's Handbook*, CRC Press, Boca Raton, FL.

Huggins, M. L. (1942): The viscosity of dilute solutions of long-chain molecules. IV. Dependence on concentration, *J. Am. Chem. Soc.*, 64:2716.

Krigbaum, W. R. and Flory, P. J. (1952): Treatment of osmotic pressure data, *J. Polym. Sci.*, 9:503.

Kulichke, W. (2004): *Viscometry of Polymers and Polyelectrolytes*, Springer, New York.

Mark, H. and Whitby, G. S. (1940): *Collected Papers of Wallace Hume Carothers on High Polymeric Substances*, Interscience, New York.

Oliver, R. (1998): *HPLC of Macromolecules*, 2nd ed., Oxford University Press, Cary, NY.

Pasch, H. (2003) *MALDI-TOF Mass Spectrometry of Synthetic Polymers*, Springer, New York.

Pasch, H. and Trathnigg, B. (1999): *HPLC of Polymers*, Springer-Verlag, New York.

Rayleigh, J. W. S. (Lord) (1871): On the light from the sky, its polarization and color, *Phil. Mag.*, 41:107.

Schartl, W. (2007): *Light Scattering from Polymer Solutions and Nanoparticle Dispersions*, Springer, New York.

Staudinger, H. (1928): *Ber. Bunsenges Phys. Chem.*, 61:2427.

Svedberg, T. and Pederson, K. O. (1940): *The Ultracentrifuge*, Clarendon Press, Oxford, U.K.

Yau, W. (2009): *Size-Exclusion Chromatography*, Wiley, Hoboken, NJ.

Zimm, B. H. (1948): The scattering of light and the radical distribution function of high polymer solutions, *J. Chem. Phys.*, 16:1093 and Apparatus and methods for measurement and interpretation of the angular variation of light scattering: Preliminary results on polystyrene solutions, *J. Chem. Phys.*, 37:19.

4 Naturally Occurring Polymers

One of the strongest, most rapidly growing areas of science and polymers involves natural polymers. Our bodies are largely composed of polymers: DNA, RNA, proteins, and polycarbohydrates. The "health" of these polymers is related to aging, awareness, mobility, strength, and so on, i.e., all the characteristics of being alive. Many medical, health, and biological advances focus on polymers. There is an increasing emphasis on such natural polymers. The emphasis on the human genome and relationships between genes, proteins, and our health underlie much of this movement. Thus, an understanding of polymeric principles is advantageous to those desiring to pursue a career related to their natural environment, be it medicine, biomedical science, biological science, bioengineering, etc.

Physically, there is no difference in the behavior, study, or testing of natural and synthetic polymers. The techniques suitable for application to synthetic polymers are equally applicable to the study of the behavior of natural polymers.

While the specific chemistry and physics dealing with synthetic polymer is complicated, the chemistry and physics of natural polymers is even more complex because of a number of related factors, including (1) the fact that many natural polymers are composed of different, often similar but not identical, repeat units; (2) a greater dependency on the exact natural polymer environment; (3) the question of real structure of many natural polymers in their natural environment is still not well known; and (4) the fact that polymer shape and size are even more important and complex for natural polymers than for synthetic polymers.

Industrially, we are undergoing a reemergence of the use of natural polymers as feedstocks and materials in many old and new areas. Since natural polymers are typically regeneratable or renewable resources, nature continues to synthesize them as we harvest them. Many natural polymers are available in large quantities. For instance, cellulose makes up about one-third of the bulk of the entire vegetable kingdom, being present in corn stocks, tree leaves, grass, and so on. With the realization that we must conserve and regulate our chemical resources comes the awareness that we must find substitutes for resources that are not self-renewing, thus, the reason for the increased emphasis in polymer chemistry toward the use and modification of natural, renewable polymers by industry.

Natural feedstocks must serve many human purposes. Carbohydrates as raw materials are valuable due to their actual or potential value. For example, commercial plants are already utilizing rapidly reproducing reengineered bacteria that metabolize cellulose wastes converting it to more protein-rich bacteria that is harvested and then used as a protein source feed-meal for animals. Further, natural materials can be used in applications now reserved largely only for synthetic polymers. Sufficient natural materials are available to supply both food and polymer needs.

When plant or animal tissues are extracted with nonpolar solvents, a portion of the material dissolves. The components of this soluble fraction are called lipids, and they include fatty acids, triacylglycerols, waxes, terpenes, postagladins, and steroids. The insoluble portion contains the more polar plant components including carbohydrates, lignin, proteins, and nucleic acids.

Many renewable feedstocks are currently summarily destroyed (through rotting or burning) or utilized in a noneconomical manner, for example, leaves are "ritualistically" burned each fall. A number of these seemingly useless natural materials have already been utilized as feedstock sources for industrial products, with more becoming available.

Biological polymers represent successful strategies that are being studied by scientists as avenues to different and better polymers and polymer structure control. Sample "design rules" and approaches that are emerging include the following:

- Identification of mer sequences that give materials with particular properties
- Identification of mer sequences that cause certain structural changes
- The formation of a broad range of materials with a wide variety of general/specific properties and functions (such as proteins/enzymes) through a controlled sequence assembly from a fixed number of feedstock molecules (proteins—about 20 different amino acids; 5 bases for nucleic acids and 2 sugar units)
- Integrated, in situ (in cells) polymer productions with precise nanoscale control
- Repetitive use of proven strategies with seemingly minor structural differences but resulting in quite divergent results (protein for skin, hair, and muscle)
- Control of polymerizing conditions that allow steady state production far from equilibrium

There often occurs a difference in "mind-set" between the nucleic acid and protein biopolymers covered in this chapter and other biopolymers and synthetic polymers covered in other chapters. Nucleic acids and proteins are site specific with a single conformation. Generally, if it differs in structure or geometry from the specific macromolecule needed, it is discarded. Nucleic acids and proteins are not a statistical average, but rather specific materials with specific chain lengths and conformations. By comparison, synthetic and many other biopolymers are statistical averages of chain lengths and conformations. All of these distributions are often kinetic and/or thermodynamic driven.

This difference between the two divisions of biologically important polymers is also reflected in the likelihood that there are two molecules with the exact same structure. For molecules such as polysaccharides and those based on terpene-like structures, the precise structures of individual molecules vary, but for proteins and nucleic acids, the structures are identical from molecule to molecule. This can be considered a consequence of the general function of the macromolecule. For polysaccharides, the major functions, though not the sole function, are providing energy and structure. For proteins and nucleic acids, main functions include memory and replication, in addition, proteins sometimes also serve a structural function.

Another difference between proteins and nucleic acids and other biopolymers and synthetic polymers involves the influence of stress–strain activities on the materials properties. Thus, the application of stress on many synthetic polymers and some biopolymers encourages a realignment of polymer chains and regions often resulting in a material with greater order and strength. Counter, the application of stress to certain biopolymers, such as proteins and nucleic acids, causes a decrease in performance (through denaturation, etc.) and strength. For these biopolymers, this is a result of the biopolymer already existing in a compact and "energy-favored" form and already existing in the "appropriate" form for the desired performance. The performance requirements for the two classifications of polymers are different. For one set, including most synthetic and some biopolymers, performance behavior involves response to stress–strain application with respect to certain responses such as chemical resistance, absorption enhancement, and other physical properties. By comparison, the most cited performances for nucleic acids and proteins involve selected biological responses requiring specific interactions occurring within a highly structured environment with specific shape and electronic requirements.

Proteins and nucleic acids typically act as individual units, the nanoworld in action, while many other natural polymers and synthetic polymers act in concert with one another. This is not entirely true since proteins and nucleic acids, while acting as individual units, act with other essential biologically important units to carry out their tasks. Synthetic polymers generally act as groups of chains through chain entanglement giving the overall aggregate desired properties such as strength. Actually, the behavior of branched natural polymers such as amylopectin is similar to the branched low-density polyethylene while the behavior of linear amylose is similar to that of the largely linear high-density polyethylene.

Just to bring natural polymers a little closer to home, the following might be our breakfast from the viewpoint of natural polymers.

Milk	Coffee cake
Proteins	Gluten
Fruit	Starches
Starches	Dextrins
Cellulose	Scrambled eggs
Pectin	Ovalbumin
	Conalbumin
	Ovomucoid
	Mucins
	Globins

4.1 POLYSACCHARIDES

Carbohydrates are the most abundant organic compounds, constituting three-fourths of the dry weight of the plant world. They represent a great storehouse of energy as a food for humans and animals. About 400 billion ton of sugars are produced annually through photosynthesis, dwarfing the production of other natural polymers, with the exception of lignin. Much of this synthesis occurs in the oceans, pointing to the importance of harnessing this untapped food, energy, and renewable feedstocks storehouse.

The potential complexity of even simple aldohexose monosaccharides is indicated by the presence of five different chiral centers, giving rise to 2^5 or 32 possible stereoisomeric forms of the basic structure, two of which are glucose and mannose. While these sugars differ in specific biological activity, their gross chemical reactivities are almost identical, permitting one to often employ mixtures within chemical reactions without regard to actual structure. Their physical properties are also almost the same, again allowing for the mixing of these structures with little loss in their physical behaviors.

Carbohydrates are diverse with respect to occurrence and size. Familiar mono and disaccharides include glucose, fructose, sucrose (table sugar), cellobiose, and mannose. Familiar polysaccharides are listed in Table 4.1 along with their source, purity, and molecular weight range.

The most important polysaccharides are cellulose and starch. These may be hydrolyzed to lower molecular weight carbohydrates (oligosaccharides) and finally to D-glucose. Glucose is the building block for many carbohydrate polymers. It is called a monosaccharide since it cannot be further hydrolyzed while retaining the ring. Three major types of representations are used to reflect saccharide structures. These are given in Figure 4.1. Here, we will mainly use the Boeseken–Haworth planer hexagonal rings to represent polysaccharide structures.

4.2 CELLULOSE

Cellulose was originally "discovered" by Payen in 1838. For thousands of years, impure cellulose formed the basis of much of our fuel and construction systems in the form of wood, lumber (cut wood), and dried plant material; served as the vehicle for the retention and conveying of knowledge and information in the form of paper; and clothing in the form of cotton, ramie, and flax. Much of the earliest research was aimed at developing stronger materials with greater resistance to the natural elements (including cleaning) and to improve dyeability, so that the color chosen by people for their clothing material could be other than a drab white. In fact, the dyeing of textile materials, mainly cotton, was a major driving force in the expansion of the chemical industry in the latter part of the nineteenth century.

TABLE 4.1
Naturally Occurring Polysaccharides

Polysaccharide	Source	Monomeric Sugar Unit(s)	Structure	Molecular Weight
Amylopectin	Corn, potatoes	D-Glucose	Branched	10^6–10^7
Amylose	Plants	D-Glucose	Linear	10^4–10^6
Chitin	Animals	2-Acetamidoglucose		
Glucogen	Animals (muscles)	D-Glucose	Branched	$>10^8$
Inulin	Arichokes	D-Fructose	(Largely) Linear	10^3–10^6
Mannan	Yeast	D-Mannose	Linear	—
Cellulose	Plants	D-Glucose	Linear (2D)	10^6
Xylan	Plants	D-Xylose	(Largely) Linear	—
Lichenan	Iceland moss	D-Glucose	Linear	10^5
Galactan	Plants	D-Galactose	Branched	10^4
Arabinoxylan	Cereal grains	L-Arabinofuranose linked to xylose chain	Branched	$>10^4$
Galactomannans	Seed mucilages	D-Mannopyrnose chains with D-Galactose side chains	(Largely) Linear	10^5
Arabinogalactan	Lupin, soybean, coffee beans	D-Galactopyranose chain, side chain galactose, arabinose	Branched	10^5
Carrageenan	Seaweeds	Complex	Linear	10^5–10^6
Agar	Red seaweeds	Complex	Linear	—
Alginic	Brown seaweeds	β-D-Mannuronic acid and α-L-guluronic acid	Linear	—

FIGURE 4.1   β-D-glucose using the Boeseken–Haworth, chair, and linear (from left to right) representations.

Cellulose is a polydisperse polymer with an average $\overline{\mathrm{DP}}$ in the general range of 3,500–36,000. Native cellulose is widely distributed in nature, and it is the principal constituent of cotton, kapok, flax, hemp, jute, ramie, and wood. Cellulose comprises more than one-third of all vegetable matter and is the world's largest renewable resource. Approximately 50 billion ton is produced annually by land plants, which absorb 4×10^{20} cal of solar energy. Natural cotton fibers, which are the seed hairs from *Gossypium*, are about 1–2 cm in length and about 5–20 μm in diameter. The molecules in native cellulose are present in threadlike strands or bundles called fibrils. Cellulose is not found in a pure form, but rather it is associated with other materials such as lignin and hemicelluloses. Cotton contains the purest form of cellulose. Wood, in its dry state, contains 40%–55% cellulose; 15%–35% lignin; and 25%–40% hemicellulose. Plant pulp is the major source of commercial cellulose. The extraction of cellulose from plants is called pulping. Pulping is achieved using thermomechanical, chemical, or mechanical approaches. Plant pulp, from wood, is the major source of nontextile fibers while cotton is the major source of textile fibers.

Cellulose is used in the textile industry in cloths, cartons, carpets, blankets, and sheets. Paper is made from cellulose. Cellulosic fibers are also used as filter materials in artificial kidneys and reverse osmosis, though today most kidney dialysis units use cuprammonium tubular films derived from cellulose rather than cellulose itself.

While the celluloses are often largely linear, they are not soluble in water because of the presence of strong intermolecular hydrogen bonding and sometimes the presence of a small amount of cross-linking. Highly ordered crystalline cellulose has a density as high as 1.63 g/cc, while amorphous cellulose has a density as low as 1.47 g/cc. High molecular weight native cellulose, which is insoluble in 17.5% aqueous sodium hydroxide, is called α-cellulose. The fraction that is soluble in 17.5% sodium hydroxide but insoluble in 8% solution is called β-cellulose, and that which is soluble in 8% sodium hydroxide is called γ-cellulose.

Strong basic solutions, such as sodium hydroxide, penetrate the crystalline lattice of α-cellulose producing an alkoxide called alkali or soda cellulose. Mercerized cotton is produced by the aqueous extraction of the sodium hydroxide.

Most linear celluloses may be dissolved in solvents capable of breaking the strong hydrogen bonds. These solutions include aqueous solutions of inorganic acids, calcium thiocyanate, zinc chloride, lithium chloride, ammonium hydroxide, iron sodium tartrate, and cadmium or copper ammonium hydroxide (Schweitzer's reagent). The product precipitated by the addition of a nonsolvent to these solutions is highly amorphous regenerated cellulose.

Cellulose

(4.1)

Cellulose 3D structure

(4.2)

Structure (4.1) is most commonly employed as a description of the repeat unit of cellulose but the lower structure (4.2) more nearly represents the actual 3D structure with each D-glucosyl unit rotated through 180°. We will employ a combination of these two structural representations. Numbering is shown in (4.3) and the type of linkage is written as 1 → 4 since the units are connected through oxygen atoms contained on carbon atoms 1 and 4 as below.

$$(4.3)$$

We also call the linkage, by agreement with the anometic nature of the particular carbons involved in linking together the glucosyl units, a β or β-linkage. Thus, this linkage is a β 1 → 4 linkage. The other similar 1 → 4 linkage found in starch is called an α or α-linkage. The geometric consequence of this difference is great. The linear arrangement of cellulose with the β-linkage gives an arrangement where the OH groups reside somewhat uniformly on the outside of the chain allowing close contact and ready hydrogen-bond formation between chains. This arrangement results in a tough, insoluble, rigid, and fibrous material that is well suited as cell wall material for plants. By comparison, the α-linkage of starch (namely amylose) results in a helical structure where the hydrogen bonding is both interior and exterior to the chain allowing better wettability. This difference in bonding also results in one material being a "meal" for humans (the α-linkage) whereas the other is a meal for termites. The reason for this is the difference in the composition of enzymes present in the two species –humans with the enzyme capability to lyse or break α-linkages and for cows and termites and other species that contain symbiotic bacteria in their digestive systems that furnish the enzymes capable of digesting or breaking the β-glucoside linkages.

While the stomach acids in humans and most animals can degrade polysaccharides to the energy-giving monomeric units, this is not efficient unless there is a specific enzyme, normally present in the gut that allows the ready and rapid degradation of polysaccharide. Since these enzymes are somewhat specific, their ability to degrade is polysaccharide specific.

The various crystalline modifications have different physical properties and chemical reactivities. These variations are a consequence of the properties varying according to plant source, location in the plant, plant age, season, seasonal conditions, treatment, etc. Thus, in general, the bulk properties of polysaccharides are generally measured with average values and tendencies given. These variations are not important for most applications but possibly are important for specific biological applications where the polysaccharide is employed as a drug, within a drug delivery system, or as a biomaterial within the body.

4.2.1 PAPER

It is believed that paper was invented by Ts'ai in China around the second century AD. The original paper was a mixture of bark and hemp. Paper was first produced in the United States by William Rittenhouse in Germantown, PA in 1690 and this paper was made from rags. Paper was named after the papyrus plant, *Cyperus papyrus.*

Paper comes in many forms with many uses. The book you are reading is made from paper, we have paper plates, paper napkins, newspapers and magazines, cardboard boxes, in fact, the amount of paper items is probably over twice, by weight, that of all the synthetic polymers combined. About 30% of all paper is writing and printing paper. The rest is mainly used for tissues, toweling, and packaging. If you rip a piece of ordinary paper, not your book page please, you will see that it consists of small fibers. Most of these cellulosic fibers are randomly oriented, but a small percentage are preferentially oriented in one direction because the paper is made from a cellulose-derived watery slurry with the water largely removed through the use of heated rollers that somewhat orient the fibers.

Modern paper is made from wood pulp, largely cellulose, which is obtained by the removal of lignin from debarked wood chips by the use of chemical treatments with sodium hydroxide, sodium sulfite, or sodium sulfate. Newsprint and paperboard, which are thicker than paper, often contain a greater amount of residual lignin.

Wood is almost entirely composed of cellulose and lignin. In the simplest paper-making scheme, the wood is chopped, actually torn, into smaller fibrous particles as it is pressed against a rapidly moving pulpstone. A stream of water washes the fibers away, dissolving much of the water-soluble lignin. The insoluble cellulosic fibers are concentrated into a paste called the pulp. The pulp is layered into thin sheets, and rollers are used to both squeeze out much of the water and assist in achieving a paper of uniform thickness. This paper is not very white. It is also not very strong. The remaining lignin is somewhat acidic (lignin contains acidic phenolic groups that hydrolyze to give a weakly acidic aqueous solution) that causes the hydrolytic breakdown of the cellulose. Most of the newsprint is of this type or it is regenerated reused paper.

Pulping processes are designed to remove the nonsaccharide lignin portion of wood, which constitutes about 25% of the dry weight. The remaining is mostly cellulose with about 25% hemicellulose (noncellulose cell wall polysaccharides that are easily extracted by dilute aqueous base solutions). Pulping procedures can be generally classified as semichemical, chemical, and semimechanical procedures. In semimechanical pulping, the wood is treated with water or sulfate, bisulfite, or bicarbonate solution that softens the lignin. The pulp is then ground or shredded to remove much of the lignin giving purified or enriched cellulose content. The semichemical process is similar but digestion times are longer and digesting solutions more concentrated giving a product with less lignin, but the overall yield of cellulose-intense material is lowered by 70%–80%. Further, some degradation of the cellulose occurs.

Most paper is produced by the chemical process where chemicals are employed to solubilize and remove most of the lignin. While overall yields are lower than the other two main processes, the product gives good quality writing and printing paper. Three main chemical processes are used. In the soda process, extracting solutions containing about 25% sodium hydroxide and 2.4% sodium carbonate are used. In the sulfite process, the extracting solution contains a mixture of calcium dihydrogen sulfite and sulfur dioxide. The sulfide process utilizes sodium hydroxide, sodium monosulfide, and sodium carbonate in the extracting solution.

After the chemical treatment, the pulped wood is removed, washed, and screened. Unbleached, brown-colored paper is made directly for this material. Most whitened or bleached paper is made from the treatment of the pulp with chlorine, chlorine dioxide, hypochlorite, and/or alkaline extraction. In general, sulfate-pulped paper is darker and requires more bleaching and alkaline extraction to give a "white" pulp.

The sulfide process, also called the kraft process (the term "kraft" comes from the Swedish word for strong since stronger paper is produced), is more commonly used. The kraft process is favored over the sulfite treatment of the paper because of environmental considerations. The sulfite process employs more chemicals that must be disposed of, particularly, mercaptans, RSHs, that are quite odorous. Research continues on reclaiming and recycling pulping chemicals.

If pure cellulose were solely used to make paper, the fiber mat would be somewhat water soluble with only particle surface polar groups and internal hydrogen bonding acting to hold the fibers together. White pigments such as clay and titanium dioxide are added to help to "cement" the fibers together and to fill voids producing a firm, white writing surface. This often occurs as part of an overall coating process.

Most paper is coated to provide added strength and smoothness. The coating is basically an inexpensive paint that contains a pigment and a small amount of polymeric binder. Unlike most painted surfaces, most paper products are manufactured with a short lifetime in mind with moderate performance requirements. Typical pigments are inexpensive low–refractive index materials such as platelike clay and ground natural calcium carbonate. Titanium dioxide is used only when high opacity is required. The binder may be a starch or latex or a combination of these. The latexes are usually copolymers of styrene, butadiene, acrylic, and vinyl acetate. Other additives and coloring agents may also be added for special performance papers. Resins in the form of surface-coating agents and other special surface treatments (such as coating with polypropylene and polyethylene) are used for paper products intended for special uses such as milk cartons, ice cream cartons, light building materials, and drinking cups. The cellulose supplies the majority of the weight (typically about 90%) and strength with the special additives and coatings providing special properties needed for the intended use.

A better understanding of the nature of paper and films made from synthetic polymers such as polyethylene can be seen when considering why authorities worry about anthrax escaping from a paper envelope, yet confining anthrax in a plastic container with no fear of it escaping. When you hold good paper up to the light or tear it, you will observe the tiny fibers that compose paper. Even when these fibers appear tiny, they are very large in comparison to individual polymer chains, be they cellulose or polyethylene. The web of chains for cellulose eventually compose the cellulose fibers that are put together physically forming paper with lots of "unoccupied" spaces between the fibers with the spaces of sufficient size to allow the escape of the anthrax. By comparison, polyethylene film has little "unoccupied" spaces between the individual chains with the entire film composed of these individual chains with no large spaces so that the anthrax cannot escape.

The recycling of paper continues. Today, up to about one-half of our paper products are recycled, and this fraction is increasing as we do a better job of collecting and recycling paper products.

4.3   CELLULOSE—REGENERATING PROCESSES

Cellulose is sometimes used in its original or native form as fibers for textile and paper, but often it is modified through dissolving and reprecipitation or through chemical reaction. The *xanthate viscose process*, which is used for the production of rayon and cellophane, is the most widely used regeneration process. The cellulose obtained by the removal of lignin from wood pulp is converted to alkali cellulose. The addition of carbon disulfide to the latter produces cellulose xanthate.

While the terminal hydroxyl and aldehyde groups, such as present in cellobiose, are also present in cellulose, they are not significant because they are only present at the ends of very long chains.

The hydroxyl groups are not equivalent. For instance, the pK_a values of the two ring hydroxyl groups are about 10 and 12, which is about the same as the hydroxyl groups on hydroquinone and the first value about the same as the hydroxyl on phenol. The pK_a value of the nonring or methylene hydroxyl group is about 14 about the same as found for typical aliphatic hydroxyl groups.

Hydroquinone   Phenol
(4.4)   (4.5)

In the cellulose-regenerating process, sodium hydroxide is initially added such that approximately one hydrogen atom, believed to be predominately a mixture of the hydroxyl groups on carbon atoms 2 and 3, is replaced by the sodium ion. This is followed by treatment with carbon disulfide forming cellulose xanthate, which is eventually changed back again, regenerated, to cellulose. This sequence is depicted below.

$$(4.6)$$

Cellulose $\longrightarrow$ Sodium salt $\longrightarrow$ Cellulose xanthate $\longrightarrow$ Regenerated cellulose rayon or cellophane

The orange-colored xanthate solution, or viscose, is allowed to age and is then extruded as a filament through holes in a spinneret. The filament is converted to cellulose when it is immersed in a solution of sodium bisulfite, zinc II sulfate, and dilute sulfuric acid. The tenacity, or tensile strength, of this regenerated cellulose is increased by a stretching process that reorients the molecules so that the amorphous polymer becomes more crystalline. Cellophane is produced by passing the viscose solution through a slit die into an acid bath. Important noncellulosic textile fibers are given in Table 4.2 and a listing of important cellulosic textile fibers is given in Table 4.3.

TABLE 4.2
Noncellulosic Textile Fibers

Fiber Name	Definition	Properties	Typical Uses	Patent Names (Sample)
Acrylic	Acrylonitrile units, 85% or more	Warm, lightweight, shape retentive, resilient, quick drying, resistant to sunlight	Carpeting, sweaters, skirts, baby cloths, socks, slacks, blankets, draperies	Orlon
Modacrylic	Acrylonitrile units, 35%–85%	Resilient, softenable at low temps., easy to dye, abrasion resistant, quick drying, shape retentive, resistant to acids, bases	Simulated fur, scatter rugs, stuffed toys, paint rollers, carpets, hairpieces, wigs fleece fabrics	Verel, Dynel
Polyester	Dihydric acid–terephtalic acid ester, 85% or more	Strong, resistant to stretching and shrinking, easy to dye, quick drying, resistant to most chemicals, easily washed, abrasion resistant; retains heat-set (permanent press)	Permanent press ware, skirts, slacks, underwear, blouses, rope, fish nets, tire cord, sails, thread	Vycron, Kodel, Fortrel, Chemstrand Dacron
Spandex	Segmented polyurethane, 85% or more	Light, soft, smooth, resistant to body oils, can be stretched often, retain original form, abrasion resistant	Girdles, bras, slacks bathing suits, pillows	Lycra
Nylon	Reoccurring amide groups	Very strong, elastic, lustrous, easy to wash, abrasion resistant, smooth, resilient, low moisture absorbency, recovers quickly from extensions	Carpeting, upholstery, blouses, tents, sails, hosiery, suits, tire cord, fabrics, rope, nets	Caprolan, Astroturf, Celanese polyester

TABLE 4.3
Cellulosic Fibers

Fiber Name	Definition	Properties	Typical Uses	Patent Names (Sample)
Rayon	Regenerated cellulose with less than 15% OH substituted	Highly absorbent, soft, comfortable, easy to dye, good drapability	Dresses, suits, slacks, blouses, coats, tire cord, ties, curtains, blankets	Avril, Cuprel, Zantel
Acetate	Not less than 92% OH groups acetylated	Fast drying, supple, wide range of dyeability, shrink resistant	Dresses, shirts, slacks, draperies, upholstery, cigarette filters	Estron, Celanese acetate
Triacetate	Derived from cellulose by combining with acetic acid and/or acetic anhydride	Resistant to shrinking, wrinkling, and fading; easily washed	Skirts, dresses, sportswear	Arnel

Since an average of only one hydroxyl group in each repeating glucose unit in cellulose reacts with carbon disulfide, the xanthate product is said to have a degree of substitution (DS) of 1 out of a potential DS of 3.

Partially degraded cellulose is called hydrocellulose or oxycellulose, depending on the agent used for degradation. The term holocellulose is used to describe the residue after lignin has been removed from wood pulp.

The control of the regeneration conditions, together with a wide variety of modifications, allows the production of a wide variety of products including high wet modulus fibers, hollow fibers, crimped fibers, and flame-resistant fibers. While almost all rayon is produced using the viscose process, some fibers are still produced using the cuprammonium process whereby cellulose is dissolved in an ammonium–copper II alkaline solution.

4.4 ESTERS AND ETHERS OF CELLULOSE

It must be remembered that the three hydroxyl groups on the individual cellulose rings are not equivalent. The two ring hydroxyls are acidic with pK_as similar to hydroquinone while the third nonring hydroxyl is similar to an aliphatic hydroxyl in acidity. Thus, in an aqueous sodium hydroxide solution the two ring hydroxyls will be deprotonated at high pH values. In theory, all three hydroxyls can undergo reaction, but in actuality less than three undergo reaction either because of reactivity restrictions and/or because of steric limitations. With many of the electrophilic/nucleophilic reactions, it is the ring hydroxyls that are favored to react initially. The average number of hydroxyl groups that are reacted is often given as the DS.

4.4.1 INORGANIC ESTERS

The most widely used so-called inorganic ester of cellulose is ***cellulose nitrate (CN)***, also called nitrocellulose and gun cotton. Celluloid is produced from a mixture of CN and camphor. CN was first made about 1833 when cellulose-containing linen, paper, or sawdust was reacted with concentrated nitric acid. It was the first "synthetic" cellulose recognized product. Initially, CN was used as a military explosive and further improvements allowed the manufacture of smokeless powder. A representation of CN is given in structure (4.7).

$$(4.7)$$

Cellulose nitrate

The development of solvents and plastizing agents for CN led to the production of many new and useful nonexplosive products. Celluloid was produced in 1870 from a mixture of CN and camphor. Films were cast from solution and served as the basis for the original still and motion pictures. After World War I, the development of stable CN solutions allowed the production of fast-drying lacquer coatings.

While CN played an important role in the development of technology, its importance today is greatly diminished. It is still used as a protective and decorative lacquer coating, in gravure inks, in water-based emulsions as coatings, and to a lesser extent in plastics and films.

Cellulose phosphate esters are produced from reaction with phosphoric acid and urea. The products are used to treat hypercalciuria because of its ability to bind calcium. It has also been used for the treatment of kidney stones.

4.4.2 ORGANIC ESTERS

The most important cellulose ester is cellulose acetate because of its use in fibers and plastics. They were first made in 1865 by heating cotton with acetic anhydride. During World War I, a cellulose acetate replaced the highly flammable CN coating on airplane wings and fuselage fabrics.

Varying properties are achieved by varying the amount of substitution. The melting point generally decreases with decreasing acetylation. Lower acetylation gives products with greater solubility in polar solvents and corresponding decreased moisture resistance. Cellulose acetate is made using heterogeneous solutions containing the cellulose, sulfuric acid as the catalyst, and acetic anhydride in acetic acid. Reaction occurs; beginning with the surface or outermost layer and continues on layer by layer as new areas are exposed. When more homogeneous modification is desired, preswelling of the cellulose in water, aqueous acetic acid solutions, or in glacial acetic acid is carried out.

Reaction occurs differently since there are two "types" of hydroxyl groups (as noted before), the two ring hydroxyls and the methylene hydroxyl. In the typical formation of esters, such as cellulose acetate, the ring hydroxyl groups are acetylated initially (4.8) prior to the C-6 exocyclic hydroxyl. Under the appropriate reaction conditions, reaction continues to almost completion with all three of the hydroxyl groups esterified (4.9). In triacetate products only small amounts (on the order of 1%) of the hydroxyls remain free and of these generally about 80% are the C-6 hydroxyl.

$$(4.8)$$

$$(4.9)$$

The most common commercial products are the triacetate (DS approaching 3) and the secondary acetate (DS about 2.45).

While other organic esters are commercially available, namely cellulose butyrate and cellulose propionate, by far the most widely used one is cellulose acetate. Cellulose acetate is available as plastics, in films, sheets, fibers, and in lacquers. Cellulose acetate is used in the manufacture of display packaging and as extruded film for decorative signs, and to coat a variety of fibers. Injected molded products include toothbrush handles, combs, and brushes. It is also used in lacquers and protective coatings for metal, glass, and paper. Cellulose acetate films are used in reverse osmosis to purify blood, fruit juices, and brackish water. Some eyeglass frames are made of cellulose acetate. Biodegradable films, sponges, and the microencapsulation of drugs for control release also utilize cellulose acetate. Cellulose triacetate is used for photographic film bases. Numerous continuous filament yarns, tows, staples, and fibers are made from cellulose acetate. The precise form of filament produced is controlled by a number of factors including the shape of the die.

As in all large-scale industrial processes, the formation of the cellulose esters involves the recovery of materials. Acetic anhydride is generally employed. After reaction, acetic acid and solvent is recovered. The recovered acetic acid is employed in the production of additional acetic anhydride. The recovered solvent is also reintroduced after treatment.

Cellulose esters are used as plastics for the formation by extrusion of films and sheets and by injection molding of parts. They are thermoplastics and can be fabricated employing most of the usual techniques of (largely compression and injection) molding, extrusion and casting. Cellulose esters plastics are noted for their toughness, smoothness, clarity, and surface gloss.

Acetate fiber is the generic name of a fiber that is partially acetylated cellulose. They are also known as cellulose acetate and triacetate fibers. They are nontoxic and generally nonallergic, so are ideal from this aspect as clothing material.

While acetate and triacetate differ only moderately in the degree of acetylation, this small difference accounts for differences in the physical and chemical behaviors for these two fiber materials. Triacetate fiber is hydrophobic, and the application of heat can bring about a high degree of crystallinity that is employed to "lock in" desired shapes (such as permanent press). Cellulose acetate fibers have a low degree of crystallinity and orientation even after heat treatment. Both readily develop static charge and thus antistatic surfaces are typically employed to clothing made from them.

For clothing application, there are a number of important performance properties that depend on the form of the textile. These properties include wrinkle resistance, drape, strength, and flexibility. These properties are determined using ASTM tests that often involve stress–strain behavior. Thus, the ability of a textile to resist deformation under an applied tensile stress is measured in terms of its modulus of elasticity or Young's modulus. As with any area of materials, speciality tests are developed to measure certain properties. Some of these are more standard tests like the aforementioned Young's modulus while others are specific to the desired property measured for a specific application. For instance, resistance to slightly acidic and basic conditions is important for textiles that are to be laundered. These are tested employing standard test procedures. In general, triacetate materials are more resistant than acetate textiles to basic conditions. Both are resistant to mild acid solutions but degrade when exposed to strong mineral acids. Further, behavior to various dry cleaning agents is important. As the nature of dry cleaning agents changes, additional testing and modification in the fabric treatments are undertaken to offer a textile that stands up well to the currently employed cleaning procedures. Again, both are stable to perchloroethylene dry cleaning solvents but can soften when exposed to trichloroethylene for extended treatment. Their stability to light is dependent upon the wavelength, humidity present, etc. In general, they offer a comparable stability to light as that offered by cotton and rayon.

While cellulose acetates are the most important cellulose esters, they suffer by their relatively poor moisture sensitivity, limited compatibility with other synthetic resins, and a relatively high processing temperature.

4.4.3 ORGANIC ETHERS

Reaction with an epoxide, such as ethylene oxide, under alkaline conditions gives hydroxyethylcellulose, HEC:

$$\text{Cellulose–OH, NaOH} + \text{H}_2\text{C——CH}_2 \overset{O}{\longrightarrow} \text{Cellulose–(–O–CH}_2\text{–CH}_2\text{–)}_n\text{–OH} + \text{NaOH} \qquad (4.10)$$

Ethylene oxide   Hydroxyethylcellulose

This is an S_N2 reaction with the reaction proportional to the concentration of the epoxide and the alkali cellulose, but since the base is regenerated, it is first order in epoxide:

$$\text{Rate} = k[\text{Epoxide}] \qquad (4.11)$$

Industrially, HECs with DS values below 2 are used. Low DS materials (to about 0.5) are soluble only in basic solutions, while those with DS values of about 1.5 are water soluble. The concentrated solutions of HEC are pseudoplastic with their apparent viscosities decreasing with increased rates of shear. Dilute solutions approach being Newtonian in their flow properties, even under a wide range of shear rates.

HEC is used as a protective colloid in latex coatings, and pharmaceutical emulsions; as a film former for fabric finishes, fibrous glass, and in aerosol starches; thickener for adhesives, latex coatings, toothpaste, shampoos and hair dressings, cosmetic creams and lotions, inks and joint cements; lubricant for wallpaper adhesives and in pharmaceutical gels; and as a water-binding agent for cements, plastics, texture coatings, ceramic glazes, and in printing inks.

Sodium carboxymethylcellulose is formed by the reaction of sodium chloroacetate with basic cellulose solutions. The sodium form of carboxymethylcellulose is known as CMC and as a food grade product as cellulose gum. It is soluble in both hot and cold water.

$$\text{Cellulose–OH, NaOH} + \text{Cl–CH}_2\text{COONa} \longrightarrow \text{Cellulose–O–CH}_2\text{–COONa} \qquad (4.12)$$

Sodium carboxymethylcellulose

The most widely used cellulose gums have DS values about 0.65–1.0. CMCs are used as thickening, binding, stabilizing, and film-forming agents.

Carboxymethylhydroxyethylcellulose, CMHEC, is synthesized from the reaction of hydroxyethylcellulose with sodium chloroacetate. The product is a mixed ether. It has properties similar to both CMC and HEC. Like CMC, it exhibits a high water-binding ability and good flocculating action on suspended solids but it is more compatible than CMC with salts. It forms ionic cross-links in the presence of salt solutions containing multivalent cations, allowing its viscosity to be greatly increased by the presence of such cations. Solutions can be gelled by the addition of solutions of aluminum and iron. It is a water-soluble material used in oil recovery and in hydraulic fracturing fluids.

Methyl and hydroxyalkylmethylcelluloses are nonionic polymers soluble in cool water. Methylcellulose, MC, hydroxyethylmethylcellulose, HEMC, and hydroxypropylmethylcellulose, HPMC, do not interact with cations forming insoluble salts, but electrolytes that compete with MC for water can cause precipitation.

MC is formed from basic cellulose on its reaction with chloromethane.

$$\text{Cellulose–OH, NaOH} + CH_3Cl \longrightarrow \text{Cellulose–O–CH}_3 \qquad (4.13)$$
$$\text{Methylcellulose}$$

MC is used as an adhesive; in ceramics to provide water retention and lubricity; in cosmetics to control rheological properties and in the stabilization of foams; in foods as a binder, emulsifier, stabilizer, thickener, and suspending agent; in paints, paper products, plywood as a rheology control for the adhesive; in inks, and in textiles as a binder, and for coatings.

Ethylhydroxyethylcellulose, EHEC is a nonionic mixed ether available in a wide variety of substitutions with corresponding variations in the aqueous and organic liquid solubilities. It is compatible with many oils, resins, and plasticizers along with other polymers such as nitrocellulose. EHEC is synthesized through a two-step process beginning with the formation of the HEC-like product through reaction between the basic cellulose and ethylene oxide. The second step involves further reaction with ethyl chloride.

$$\text{Cellulose–(–O–CH}_2\text{CH}_2\text{–)–CH}_2\text{CH}_3$$
$$\text{Ethylhydroxyethylcellulose} \qquad (4.14)$$

Uses for the water-soluble EHEC include in waterborne paints, pastes, polymer dispersions, ceramics and cosmetics and pharmaceuticals. Uses for organic soluble EHEC include in inks, lacquers, and as coatings.

Cellulose undergoes reactions with activated ethylenic compounds, such as acrylonitrile, giving cyanoethylcellulose via a Michael addition.

$$\text{Cellulose–OH} + H_2C\!=\!CHCN \longrightarrow \text{Cellulose–O–CH}_2\text{–CH}_2\text{–CN} \qquad (4.15)$$

4.5 STARCH

While cellulose is the major structural polysaccharide, plant energy storage and regulation utilize a combination of similar polysaccharides that combined are referred to as starch. Starch can be divided into two general structures, branched *amylopectin* and largely linear *amylose*.

$$(4.16)$$

Linear amylose

Most starches contain about 10%–20% amylose and 80%–90% amylopectin though the ratio can vary greatly. While cellulose can be considered a highly regular polymer of D-glucose with units linked through a β-1,4 linkage, amylose is a linear polysaccharide with glucose units linked in an α-1,4-fashion while amylopectin contains glucose units with chains of α-1,4 glucopyranosyl units but with branching occurring every 20–30 units, with the chain-branch occurring from the 6 position. While this difference in orientation in how the glucose units are connected appears small, it causes great differences in the physical and biological properties of cellulose and starch. As noted before, humans contain enzymes that degrade the α-glucose units but are unable to digest the β units. Thus, starch is a food source for us, but cellulose is not. Also, the individual units of cellulose can exist in the chair conformation with all of the substituents equatorial, yet amylose must either have the glycosyl substituent at the 1 position in an axial orientation or exist in a nonchair conformation.

(4.17)

Branched amylopectin

Amylose typically consists of over 1000 D-glucopyranoside units. Amylopectin is a larger molecule containing about 6,000–1,000,000 hexose rings essentially connected with branching occurring at intervals of 20–30 glucose units. Branches also occur on these branches giving amylopectin a fan or treelike structure similar to that of glycogen. Thus, amylopectin is a highly structurally complex material. Unlike nucleic acids and proteins where specificity and being identical are trademarks, most complex polysaccharides can boast of having the "mold broken" once a particular chain is made so that the chances of finding two identical molecules is very low.

Commercially, starch is prepared from corn, white potatoes, wheat, rice, barely, millet, cassava, tapioca, and sorghum. The fraction of amylose and amylopectin varies between plant species and even within the same species varies depending on location, weather, age, and soil conditions. Amylose serves as a protective colloid. The mixtures of amylose and amylopectin, found combined in nature, form suspensions when placed in cold water. Starch granules are insoluble in cold water but swell in hot water, first reversibly until gelatinization occurs at which point the swelling is irreversible. At this point, the starch loses its birefringence, the granules burst, and some starch material is leached into solution. As the water temperature continues to increase close to 100°C, a starch dispersion is obtained. Oxygen must be avoided during heating or oxidative degradation occurs. Both amylose and amylopectin are then water soluble at elevated temperatures. Amylose chains tend to assume a helical arrangement giving it a compact structure. Each turn contains six glucose units.

The flexibility of amylose and its ability to take on different conformations are responsible for the "retrogradation" and gelation of dispersions of starch. Slow cooling allows the chains to align to take advantage of inter- and intrachain hydrogen bonding squeezing out the water molecules, leading to the precipitation of the starch. This process gives retrograded starch, either in the presence of amylose alone or combined in native starch, which is more difficult to redisperse. The rapid cooling of starch allows some inter- and intrachain hydrogen bonding, but also allows water molecules to be captured within the precipitating starch allowing it to be more easily redispersed (Figure 4.2).

Most uses of starch make use of the high viscosity of its solutions and its gelling characteristics. The modification of starch through reaction with the hydroxyl groups lowers the gelation tendencies, decreasing the tendency for retrogradation. Starch is the major source of corn syrup and corn sugar (dextrose or glucose). In addition to its use as a food, starch is used as an adhesive for paper and as a textile-sizing agent.

Oligomeric materials called *cyclodextrins* are formed when starch is treated with *Bacillus macerans*. These oligomeric derivatives generally consist of six, seven, eight, and greater numbers of glucose units

FIGURE 4.2 Behavior of amylose in a concentrated aqueous solution as a function of cooling rate.

joined through 1,4-α-linkages to form rings. These rings are doughnut-like with the hydroxyl groups pointing upward and downward along the rim. Like crown ethers used in phase transfer reactions, cyclodextrins can act as "host" to "guest" molecules. In contrast to most phase transfer agents, cyclodextrins have a polar exterior and a nonpolar interior. The polar exterior allows guest molecules to be water soluble. The nonpolar interior allows nonpolar molecules also to be guest molecules. Cyclodextrins are being used as enzyme models since they can first bind a substrate and through substituent groups can act on the guest molecule—similar to the sequence carried out by enzymes.

A major commercial effort is the free-radical grafting of various styrenic, vinylic, and acrylic monomers onto cellulose, starch, dextran, and chitosan. The grafting has been achieved using a wide variety of approaches including ionizing and ultraviolet/visible radiations, charge-transfer agents, and various redox systems. Much of this effort is aimed at modifying the native properties such as tensile (abrasion resistance and strength) and care (crease resistance and increased soil and stain release) related properties, increased flame resistance, and modified water absorption. One area of emphasis has been the modification of cotton and starch in the production of superabsorbent material through grafting. These materials are competing with all synthetic cross-linked acrylate materials that are finding use in diapers, feminine hygiene products, wound dressings, and sanitary undergarments.

4.6 OTHER POLYSACCHARIDES

The best know homopolysaccharides are derived from D-glucose and are known as glucans. Glucose has a number of reactive sites, and a wide variety of polymers formed utilizing the combinations of these reactive sites are found in nature. We have already visited the two most well-known members of this group—cellulose and starch containing amylose and amylopectin. Here, we will visit some other important members.

Glycogen (4.18) is a very highly branched glucan or polysaccharide formed from glucose. It is structurally similar to amylopectin though more highly branched. This greater branching gives glycogen greater water solubility. Glycogens are the principal carbohydrate food reserve materials in animals. They are found in both invertebrates and vertebrates, and they are likely to be found in all animal cells. The highest concentration of glycogen is found in muscle with the greatest amount found in our liver, the tissue from which it is most often isolated.

Starch and glycogen are produced when the amount of glucose is high and are readily degraded back to glucose when energy is needed. In plants, this degradation occurs mainly through the action of two enzymes known as α-and β-amylase. Interestingly, while the α-amylase can degrade starch and glycogen completely to glucose, the β-amylase is not able to degrade the branch points.

In animals, glycogen degradation to give the glucose that is needed as an energy source or to increase the blood sugar concentration begins with the action of phosphorylase. Phosphorylase occurs in both active, a, and inactive, b, forms. Phosphorylase b is converted into phosphorylase a by phosphorylation that occurs at the end of a series of events initiated by an increased intercellular concentration of cAMP and an activation of the protein kinase (see Figure 4.8). This is reversed by a phosphoprotein phosphatase whose activity is hormonally regulated. Thus, phosphorylation initiated by increased intracellular concentrations of cAMP inactivates glycogen synthetase and activates phosphorylase. This is an example of the complex steps that are carried out in our bodies on a moment-by-moment basis. Here, enzymes that are responsible for the glycogen metabolism do not act directly on glycogen but regulate the activity of other enzymes.

(4.18)

Representative unit in glycogen

Skeletal muscle glycogen delivers glucose primarily as a response to contractile stress. Regulation occurs through both the modification of the enzyme phosphorylase, primarily by the action of epinephrine–adrenaline and the allosteric regulation of phosphorylase related to a demand for ATP. Glycogen found in the liver is seldom utilized as a source of energy but rather is employed to regulate blood sugar levels. Some tissues, such as nerve and brain tissue, rely solely on glucose as their energy source so that a steady supply of sugar is essential to their well-being.

Dextrans are high molecular weight branched, extracellular polysaccharides synthesized by bacteria. These bacteria are found in many places including the human mouth where they flourish on sucrose-containing food, which becomes trapped between our teeth. The generated dextrans become part of dental plaque, and thus are involved in tooth decay. Dextran-causing bacteria can also infect sugar cane and sugar beets after harvest and act to not only decrease the yield of sucrose but also interfere with sugar refining by clogging filters and pipelines. These bacteria can also contaminate fruit juices and wines, in fact, any ready source of glucose or sucrose.

On the positive side, dextran itself has been refined and employed as a therapeutic agent in restoring blood volume for mass casualties. Natural dextrans are very high molecular weight compounds (on the order of 10^8–10^9 Da) and are found to be unsuitable as a blood-plasma substitute. Lower molecular weight (about 10^6 Da) dextran is suitable and is often referred to as clinical dextran.

Dextran gels are commercially used. The gel formed from reaction with epichlorohydrin gives a cross-linked material that is used as a molecular sieve. Commercial cross-linked dextran is know as Sephadex™. Sephadex is obtained in bead form by dissolving dextran in sodium hydroxide solution followed by dispersion in an immiscible organic liquid such as toluene containing poly(vinyl acetate) and finally added to epichlorohydrin. Different series of Sephadex are used in industries and in research. Ionic groups are often incorporated to give the anionic and cationic dextrans and ion-exchange molecular sieves. The sulfate esters of dextran are also used in separations.

Chitin (4.19) is generally a homopolymer of 2-acetamido-2-deoxy-D-glucose (*N*-acetylglucosamine) $1 \rightarrow 4$ linked in a β configuration; it is thus an amino sugar analog of cellulose. While it is widely distributed in bacteria and fungi, the major source is crustaceans. In fact, chitin is the most abundant organic skeletal component of invertebrates. It is believed to be the most widely distributed polysaccharide with the Copepoda alone synthesizing on the order of 10^9 ton each year. It is an important structural material often replacing cellulose in cell walls of lower plants. It is generally found covalently bonded to protein. Invertebrate exoskeletons often contain chitin that provides strength with some flexibility along with inorganic salts, such as calcium carbonate, that provide strength. In a real sense, this is a composite where the chitin holds together the calcium carbonate domains.

Chitosan (4.20) is produced from the deacetylation of chitin. Chitosan is employed in the food industry. It is a hemostatic material from which blood anticoagulants and antithrombogenic agents have been formed. It is often sold as a body fat–reducing agent or to be taken along with eating to encapsulate fat particles.

Chitin (4.19)   Chitosan (4.20)

Both chitosan and chitin are greatly underused, readily available abundant materials that deserve additional study as commercial materials and feedstocks. Chitin itself is not antigenic to human tissue, and it can be inserted under the skin or in contact with bodily fluids generally without harm. In the body, chitin is slowly hydrolyzed by lysozyme and absorbed. Chitin and chitosan can be safely ingested by us, and often we eat some since mushrooms, crabs, shrimp, many breads, and beer that contain some chitin. Chitin and chitosan are believed to accelerate wound healing. Chitosan is also reported to exhibit bactericidal and fungicidal properties. Chitosan solutions are reported to be effective against topical fungal infections such as athlete's foot.

A continuing problem related to the introduction of bioengineering materials into our bodies is their incompatibility with blood. Many materials cause blood to clot (thrombosis) on the surfaces of the introduced material. Heparin is an anticoagulant, nontoxic material that prevents clot formation when coated on vascular implants. While chitosan is a hemostatic material (stops bleeding by enhancing clotting), chitosan sulfate has the same anticoagulant behavior as heparin.

Cardiovascular disease is the second leading cause of death in America. A contribution factor to cardiovascular disease is serum cholesterol. When ingested, chitosan exhibits hypocholesterolemic activity. Chitosan dissolves in the low pH found in the stomach and reprecipitates in the more alkaline intestinal fluid entrapping cholic acid as an ionic salt, preventing its absorption by the liver. The cholic acid is then digested by bacteria in the large intestine. Chitosan may also act to increase the ratio of high-density lipoprotein to total cholesterol. Chitosan has been studied in the formation of films including membrane gels that immobilize enzymes and other materials because of the mild conditions under which they can be formed.

Chitosan has been used as a flocculate in wastewater treatment. The presence of the amine gives coacervation with negatively charged materials such as negatively charged proteins allowing the removal of unwanted protein waste. The amine groups also capture metal ions, in particular poly-valent and heavy metal ions such as iron, lead, mercury, and uranium. Carraher, Francis, and Louda have also used chitosan to chelate with platinum salts to form materials with structures similar to the anticancer drug *cis*-dichlrodiamineplatinum II chloride. The amine and hydroxyl groups can be modified using a wide range of reactions including the formation of amides and esters.

Glycosaminoglycans are polysaccharides that contain aminosugar units. Most are of animal origin. *Heparin* (4.21) is complex containing D-glucuronic acid, L-iduronic acid, and D-glucosamine units. It is found in the lung, the liver, and the arterial walls of mammals. It is also found in the intracellular granules of mast cells that line arterial walls and is released through injury. The glucuronic acid and iduronic acid units are not randomly present but occur in blocks. Heparin is found as a free polysaccharide and is bonded to protein. Heparin acts as an anticoagulant, an inhibitor of blood clotting, and is widely used for this in medicine. In nature, its purpose appears to be the prevention of uncontrolled clotting.

(4.21)

Representative units in heparin

Hyaluronic acid (4.22) is found in connective tissues, umbilical cords, skin, and it is the synovial fluid of joints. It can have very large molecular weights, up to 10^7 Da making solutions of hyaluronic acid quite viscous. They can form gels. As a synovial fluid in joints, it acts as a lubricant and in the cartilage, it may also act, along with chondroitin sulfates, as a shock absorber. In some diseases, such as osteoarthritis, the hyaluronic acid of the joints is partially degraded resulting in a loss of elasticity of the area. The molecules can adopt a helical structure.

(4.22)

Representative unit of hyaluronic acid

Chondroitin sulfate is found in bone, skin, and cartilage but not as a free polysaccharide. Rather it exists as proteoglycan complexes where the polysaccharide is covalently bonded to a protein. The proteoglycan of cartilage contains about 10% protein, keratan sulfate, and chondroitin sulfate, mainly the 4-sulfate in humans. The function of proteoglycan in cartilage is similar to that of noncellulosic polysaccharides and proteins in plant cell walls. In cartilage, collagen fibers provide the necessary strength that is provided in plants by cellulose fibers. Thus, cartilage proteoglycan is an important part of the matrix that surrounds the collagen fibers giving it rigidity and incompressibility. This network can also act as a shock absorber since on compression the water is squeezed out to a nearby uncompressed region acting to "share the load" by distributing a shock or stress–strain. Chondroitin sulfate is sold as a health aid to "maintain healthy mobile joints and cartilage."

Arabinogalactan contains unmodified galactose and is contained in many plant gums, carra-geenans, and *agar*. Seaweeds represent a source of many polysaccharides including alginic acid, agar, and carrageenan. Alginic acid is a polymer of D-mannuronic acid and L-guluronic acid that may be arranged in a somewhat random fashion or in blocks. It is used as a stabilizer for ice cream, in paper coating, in the manufacture of explosives, and in latex emulsions.

Carrageenans and *agar* are able to form double helices containing two parallel staggered chains creating a gel. There are three disaccharide units per helix turn. The sulfate units are located on the outside of the helix with the helical structure stabilized by internal hydrogen bonds. In nature, red seaweeds contain an enzyme that converts the galactose-6-sulfate of the κ-carrageenan to 3,6-anhydrogalactose that causes a stiffening of the helix. It has been found that red seaweeds found in areas where there is strong wave action contain a high proportion of anhydrogalactose. Thus, it appears that the seaweed is able to control its structure in response to external stimuli to minimize shredding by the increased wave action. Because of its gelling ability, carrageenan is widely used as a food thickener and emulsion stabilizer in the food industry and is present in many dairy products including less expensive ice cream and other desert products providing a smooth, creamy texture. It is used as a stabilizer in foods such as chocolate milk.

Agar is employed as the basis of many microbiological media and in canned food because it can be sterilized. The latter is an advantage over gelatin, which is not able to withstand sterilization. *Agarose* is the agar polysaccharide with the greatest gelling tendency. Agarose gels are employed in gel permeation chromatography, GPC, and gel electrophoresis.

4.7 PROTEINS

The many different monodisperse polymers of amino acids, which are essentially components of plants and animals, are called *proteins*. This word is derived from the Greek *porteios*, "of chief importance." The 20 different α-amino acids are joined together by peptide linkages

$$
\begin{array}{cc}
\text{O} & \text{R} \\
\| & | \\
\end{array}
$$
$$-(-C-NH-CH-)-$$

(4.23)

and are also called polyamides or polypeptides. The latter term is often used by biochemists to denote oligomers or relatively low molecular weight proteins. (Note the structural similarities and differences between proteins and polyamides–nylons (Section 5.6).)

All α-amino acids found in proteins are of the general structure

$$
\begin{array}{c}
\text{NH}_2 \\
| \\
\text{R-CH-COOH}
\end{array}
$$

(4.24)

except glycine

$$
\begin{array}{c}
\text{NH}_2 \\
| \\
\text{H}_2\text{C-COOH}
\end{array}
$$

(4.25)

and contain a chiral carbon atom, and they are L-amino acids. The net ionic charge of an amino acid varies with the charges in the solution pH. The pH at which an amino acid is electrically neutral is called the *isoelectric point*. For simple amino acids (containing only one acid and one amine), this occurs at a pH of about 6 with the formation of a dipolar or *zwitterion* as shown below.

$$
\begin{array}{c}
\text{R} \\
| \\
\text{H}_3\text{N}^+\text{-CH-COO}^-
\end{array}
$$

(4.26)

Hence, α-amino acids, like other salts, are water-soluble, high-melting, polar compounds, which migrate toward an electrode at pH values other than the isoelectric point in a process called electrophoresis.

In writing out sequences for polypeptides, it is usual to use a three-letter abbreviation or a one-letter abbreviation starting with the N-terminus to the left and going to the C–O terminus to the right. Thus the trimer

$$
\begin{array}{cccc}
H & O & O & O \\
| & \| & \| & \| \\
HN-CH_2-C-O-NH-CH-C-O-NH-CH-C-O-H \\
& | & | \\
& CH_3 & CH_2 \\
& & | \\
& & OH
\end{array}
\tag{4.27}
$$

becomes Gly-Ala-Ser or GlyAlaSer or EGAS where the E signals the N-terminus.

It is important to remember that all proteins are polypeptides. The size of proteins is generally noted in either the number of amino acid units or more likely the molecular weight in kilodaltons, kDa. Proteins are synthesized in nature from the N-terminus to the C-terminus.

The amino acids may be neutral, acidic, or basic, in accordance with the relative number of the amino and carboxylic acid groups present. Cations can be formed with amino acids like tryptophan, lysine, histidine, and arginine, which have amino groups while others that contain acidic groups can be hydrolyzed to form anions like aspartic acid and glutamic acid. The presence of varying amounts of these amino acid moieties within a protein are primary driving forces for the separation of proteins using electrophoresis, and this results in polypeptides having different isoelectric points. If there are a number of acidic and/or basic groups on the polypeptide, the molecule is said to be a *polyampholyte* or if they contain only positive or negative charges they are called *polyelectrolytes*. The behavior of these charged polypeptides is similar to the behavior of other charged polymers. Thus, a fully hydrolyzed poly(acrylic acid) acts as a rod because the negative sites repel one another while a polypeptide with a large number of negative sites will also be elongated. The spacing and number of these charged sites help to determine the tertiary structure of such polypeptides.

Even though the atoms within a peptide bond are coplanar, they can exist in two possible configurations, *cis* and *trans*.

$$
\tag{4.28}
$$

trans   *cis*

The *trans* form is usually favored whenever there is a bulky group on the adjacent α-carbon(s) because the groups will interfere more in the *cis* structure.

While humans synthesize about a dozen of the 20 amino acids needed for good health, the other 8 are obtained from outside our bodies, generally from eating foods that supply these *essential amino acids*. Different foods are good sources of different amino acids. Cereals are generally deficient in lysine. Thus, diets that emphasize cereals will also have other foods that can supply lysine. In the orient, the combination of soybean and rice supplies the essential amino acids, while in Central America bean and corn are used.

Almost all of the sulfur needed for healthy bodies is found in amino acids as cysteine and methionine. Sulfur serves several important roles including as a cross-linking agent similar to that served by sulfur in the cross-linking, vulcanization of rubber. This cross-linking allows the various chains, that are connected by these cross-links, to "remember" where they are relative to one another. This cross-linking allows natural macromolecules to retain critical shapes to perform necessary roles.

The most widely used technique for producing polypeptides with specific sequences is the solid-phase technique developed by Nobel Prize winner Bruce Merrifield in which all reactions take place on the surface of cross-linked polystyrene beads. The process begins with the attachment of the C-terminal amino acid to the chloromethylated polymer. Nucleophilic substitution by the carboxylate anion of an N-blocked protected C-terminal amino acid displaces chloride from the chloromethyl group forming an ester, protecting the C site while attaching it to the solid support. The blocking group is removed by the addition of acid, and the polymer containing the nonprotected N terminus is washed to remove unwanted by-products. A peptide bond is formed by condensation to an N-blocked protected amino acid. Again, the solid phase system is washed removing by-products. The blocking group is removed, and the site is ready for attachment by another amino acid. This cycle is repeated eventually producing the desired polypeptide without the isolation of intermediate products. These steps are outlined below.

$$PS{-}Ph{-}CH_2{-}Cl + HOOC{-}\overset{\overset{R}{|}}{CH}{-}NH{-}BG \overset{-HCl}{\longrightarrow} PS{-}Ph{-}CH_2{-}O{-}\overset{\overset{O}{||}}{C}{-}\overset{\overset{R}{|}}{CH}{-}NH{-}BG \overset{H^+}{\longrightarrow}$$

(4.29)

$$PS{-}Ph{-}CH_2{-}O{-}\overset{\overset{O}{||}}{C}{-}\overset{\overset{R}{|}}{CH}{-}NH \longrightarrow \longrightarrow \longrightarrow \longrightarrow$$

where PS, polystyrene bead; Ph, phenyl group; BG, blocking group.

The shapes of macromolecules, both synthetic and natural, can be described in terms of the primary, secondary, tertiary, and quaternary structures (Figure 4.3). Protein structure will be used to illustrate these structures.

4.7.1 PRIMARY STRUCTURE

The term primary structure is used to describe the sequence of amino acid units (configuration) in a polypeptide chain. Thus, Equation 4.27 describes a primary structure.

4.7.2 SECONDARY STRUCTURE

The term *secondary structure* is used to describe the molecular shape or the conformation of a molecule. The most important factor in determining the secondary structure of materials is its precise structure. For proteins, it is the amino acid sequence. Hydrogen bonding is also an important factor

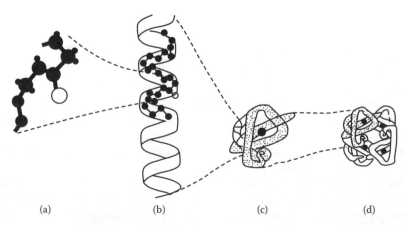

(a)   (b)   (c)   (d)

FIGURE 4.3 Four levels of structure elucidation. (a) Primary, (b) secondary, (c) tertiary, and (d) quaternary structures illustrated using a globular protein segment.

in determining the secondary structures of natural materials and those synthetic materials that can form hydrogen bonds. In fact, for proteins, secondary structures are generally those that allow a maximum amount of hydrogen bonding. This hydrogen bonding also acts to stabilize the secondary structure while cross-linking acts to lock in a structure.

In nature, the two most common secondary structures are helices and sheets. In nature, extended helical conformations appear to be utilized in two major ways: to provide linear systems for the storage, the duplication, and the transmission of information (DNA, RNA), and to provide inelastic fibers for the generation and transmission of forces (F-actin, myosin, and collagen). The examples of the various helical forms found in nature are single helix (messenger and ribosomal DNA), double helix (DNA), triple helix (collagen fibrils), and complex multiple helices (myosin). Generally, these helices are readily soluble in dilute aqueous solution. Often solubility is only achieved after the inter- and intrahydrogen bonding are broken.

There are varieties of examples in which linear or helical polypeptide chains are arranged in parallel rows. The two major forms that exist for proteins are illustrated in Figures 4.4 through 4.7. The chains can have the N $\rightarrow$ C directions running parallel making a parallel β sheet (Figure 4.6), or they can have the N $\rightarrow$ C directions running antiparallel giving antiparallel β sheets (Figure 4.7).

FIGURE 4.4 Commonly occurring repetitive helical patterns for polypeptides. (From Coates, J. and Carraher, C., *Polym. News*, 9, 77, 1983.)

FIGURE 4.5 A β arrangement or pleated sheet conformation of proteins.

FIGURE 4.6   α-Keratin helix for the copolymer derived from glycine and leucine.

FIGURE 4.7   β-Keratin sheet for the copolymer derived from glycine and alanine.

The structure of proteins generally falls into three groupings: fiber, membrane and globular. The structural proteins such as the keratins, collagen, and elastin are largely fibrous. A reoccurring theme with respect to conformation is that the preferential secondary structures of the fibrous synthetic and natural polymers approximate that of a pleated sheet (or skirt) or helix. The pleated sheet structures in proteins are referred to as β arrangements (Figure 4.7). In general, proteins with bulky groups take on a helical secondary structure while those with less bulky groups exist as β sheets.

Helices can be described by the number of amino acid residues in a complete "turn." In order to fit into a "good" helix, the amino acids must have the same configuration. For proteins, that configuration is described as an L-configuration, with the helix being a "right-handed" helix. This right-handed helix is referred to as an α helix.

4.7.2.1 Keratins

Keratins are structural proteins. As noted above, two basic "ordered" secondary structures predominate in synthetic and natural polymers. These are the helices and the pleated sheet structures. These two structures are illustrated by the group of proteins called the keratins. It is important to remember that hydrogen bonding is critical in both structures. For helices, the hydrogen bonding occurs within a single strand, whereas in the sheets, the hydrogen bonding occurs between adjacent chains.

Helices are often described in terms of a repeat distance—the distance parallel to the axis in which the structure repeats itself; pitch—the distance parallel to the helix axis in which the helix makes one turn; and rise—the distance parallel to the axis from the level of one repeat unit to the next.

Helices generally do not have an integral number of repeat units or residues per turn. The α helix repeats after 18 amino acid residues taking five turns to repeat. Thus, the number of residues per turn is 18/5 = 3.6 residues/turn. For polypeptides, each carbonyl oxygen is hydrogen bonded to the amino proton on the fourth residue giving a "hydrogen-bonded loop" with 13 atoms. Helices are often described in terms of this number, n. Thus, the α helix is described as a 3.6_{13} helix. Because hydrogen bonds tend to be linear, the hydrogen bonding in proteins approximates this with the $-N-H\cdots\cdots O=C$ in a straight line. The "rise," h, of α-keratin (Figure 4.6) is found by x-ray spectroscopy to be about 0.15 nm for each amino acid residue. The pitch, p, of a helix is given by $p = nh$. For α-keratin, p = 3.6 amino acid residues/turn × 0.15 nm/amino acid residue = 0.54 nm/turn.

Hair and wool are composed of α-keratin. A single hair on our head is composed of many strands of keratin. Coiled, α-helices, chains of α-keratin intertwine to form protofibrils that in turn are clustered with other protofibrils forming a microfibril. Hundreds of these microfibrils, in turn, are embedded in a protein matrix giving a macrofibril that in turn combines giving a human hair. While combing will align the various hairs in a desired shape, after a while, the hair will return to its "natural" shape through the action of the sulfur cross-links pulling the hair back to its original shape.

Secondary bonding is involved in forming the helical structures allowing the various bundles of α-keratin to be connected by weak secondary interactions that in turn allow them to readily slide past one another. This sliding or slippage along with the "unscrewing" of the helices allows our hair to be flexible. Some coloring agents and most permanent waving of our hair involve the breakage of the sulfur cross-links and a reforming of the sulfur cross-links at new sites to "lock in" the desired hair shape.

Our fingernails are also composed of α-keratin, but keratin with a greater amount of sulfur cross-links giving a more rigid material. In general, for both synthetic and natural polymers, increased cross-linking leads to increased rigidity.

The other major structural feature is pleated sheets. Two kinds of pleated sheets are found. When the chains have their $N \rightarrow C$ directions running parallel they are called parallel β sheets. The $N \rightarrow C$ directions can run opposite to one another giving what is called an antiparallel β sheet. The β keratin (Figure 4.7) that occurs in silk produced by insects and spiders is of the antiparallel variety. While α-keratin is especially rich in glycine and leucine, β-keratin is mostly composed of glycine and alanine with smaller amounts of other amino acids including serine and tyrosine. Size wise, leucine offers a much larger grouping attached to the α-carbon than does alanine. The larger size of the leucine causes the α-keratin to form a helical structure to minimize steric factors. By comparison, the smaller size of the alanine allows the β-keratin to form sheets. This sheet structure is partially responsible for the "softness" felt when we touch silk. While silk is not easily elongated because the protein chains are almost fully extended, β-keratin is flexible because of the low secondary bonding between sheets allowing the sheets to flow past one another.

In the silk fibroin structure, almost every other residue is glycine with either alanine or serine between them allowing the sheets to fit closely together. While most of the fibroin exists as β-sheets, regions that contain more bulky amino acid residues interrupt the ordered β structure. Such disordered regions allow some elongation of the silk. Thus, in the crystalline segments of silk fibroin there exists directional segregation using three types of bonding: covalent bonding in the first dimension, hydrogen bonding in the second dimension, and hydrophobic bonding in the third dimension. The polypeptide chains are virtually fully extended; there is a little puckering to allow for optimum hydrogen bonding. Thus, the structure is not extensible in the direction of the polypeptide chains. By comparison, the less specific hydrophobic (dispersive) forces between the sheets produce considerable flexibility. The crystalline regions in the polymers are interspersed with amorphous regions in which glycine and alanine are replaced by amino acids with bulkier pendant groups that prevent the ordered arrangements to occur. Furthermore, different silk worm species spin silks with differing amino acid compositions and thus, with differing degrees of crystallinity. The correlation between the extent of crystallinity and the extension at the break point is given in Table 4.4.

Another natural key that illustrates the tight relationship between structure and property is found in spider webs. The composition within a spider web is not all the same. We can look briefly at two of the general types of threads. One is known as the network or frame threads also called the dragline fabric. It is generally stiff and strong. The second variety is the catching or capture threads that are made of viscid silk that is strong, stretchy, and covered with the droplets of glue. The frame threads are about as stiff as nylon-6,6 thread and on a weight basis stronger than steel cable. Capture thread is not stiff but is more elastomeric-like and on a weight basis about one-third as strong as frame thread. While there are synthetic materials that can match the silks in both stiffness and strength there are few that come near the silk threads in toughness and their ability to withstand a sudden impact without breaking. Kevlar, which is used in bullet-resistant clothing, has less energy-absorbing capacity in comparison to either frame or capture threads. In fact, when weight is dropped onto frame silk, it adsorbs up to 10 times more energy than Kevlar. On impact with frame thread, most of the kinetic energy dissipates as heat which, according to a hungry spider, is better than transforming it into elastic energy which might simply act to "bounce" the prey out of the web.

The frame threads are composed of two major components—highly organized microcrystals compose about one-quarter of the mass and the other three-quarters are composed of amorphous spaghettilike tangles. The amorphous chains connect the stronger crystalline portions. The amorphous tangles are dry, glassy-like acting as a material below its T_g. The amorphous chains are largely oriented along the thread length as are the microcrystals giving the material good longitudinal strength. As the frame threads are stretched, the tangles straighten out allowing it to stretch without breaking. Because of the extent of the tangling, there is a lessening in the tendency to form

---

TABLE 4.4
Selected Properties as a Function of Silk Worm Species

Silk Worm Species	Approximate Crystallinity (%)	% Extension at Break Point
Anaphe moloneyi	95	12.5
Bombyx mori	60	24
Antherea mylitta	30	Flows then extends to 35%

Source: Coates, J. and Carraher, C., *Polym. News*, 9, 77, 1983.

---

micro-ordered domains as the material is stretched though that also occurs. Frame thread can be reversibly stretched to about 5%. Greater stretching causes permanent creep. Thread rupture does not occur until greater extension, such as 30%. By comparison, Kevlar fibers break when extended only by 3%.

The capture threads are also composed of the same kind of components but here the microcrystals compose less than 5% of the thread with both the amorphous and microcrystalline portions arranged in a more random fashion within the thread. Hydrated glue, that coats the thread, acts as a plasticizer imparting to the chains a greater mobility and flexibility. It stretches several times its length when pulled, and it is able to withstand numerous shocks and pulls that are appropriate to contain the prey as it attempts to escape. Further, most threads are spun as two lines so that the resulting thread has a kind of build in redundancy.

The spinning of each type of thread comes from a different emission site on the spider and the spider leaves little to waste, using the unwanted and used web parts as another source of protein.

The cloning of certain spider genes have been included in goats to specify the production of proteins that call for the production of silklike fibroin threads that allow the production and subsequent capture of spiderlike threads as part of the goats milk.

The β-keratin structure is also found in the feathers and scales of birds and reptiles.

Wool, while naturally existing in the helical form, forms a pleated skirt (or sheet) like structure when stretched. If subjected to tension in the direction of the helix axes, the hydrogen bonds parallel to the axes are broken and the structure can be irreversibly elongated to an extent of about 100%.

As an aside, wool is self-extinguishing so that flame retardants are not needed in wool rugs and suits. Further, we need not worry about the herds of flammable sheep running about the hillsides.

We are beginning to understand better how we can utilize the secondary structure of polymers as the tools of synthesis. One area where this is being applied is "folded oligomers." Here, the secondary structure of the oligomer can be controlled through its primary structure and the use of solvents. Once the preferred structure is achieved, the oligomers are incorporated into larger chains eventually forming synthetic polymers with several precise structures "embedded" within them. The secondary structure of the larger polymers can also be influenced by the nature of the solvent allowing further structural variety. Further, other species, such as metal ions, can be added to assist in locking-in certain desired structures and they can also be used to drive further structure modifications.

4.7.3 Tertiary Structure

The term *tertiary structure* is used to describe the shaping or folding of macromolecules. These larger structures generally contain the elements of the secondary structures. Often hydrogen bonding, salt bridges, posttranslational modifications, and disulfide cross-linking lock in such structures. As noted above, proteins can be divided into three broad groups—fibrous or fibrillar proteins, membrane proteins, and globular proteins.

4.7.3.1 Globular Proteins

There is a wide variety of so-called globular proteins. Almost all globular proteins are soluble in acidic, basic, or neutral aqueous solutions. Many globular proteins are enzymes. Many of these have varieties of α and β-structures imbedded within the overall globular structure. β-sheets are often twisted or wrapped into a "barrel-like" structure. They contain portions that are β-sheet structures and portions that are in an α-conformation. Further, some portions of the globular protein may not be conveniently classified as either an α- or β-structure.

These proteins are often globular in shape so as to offer a different "look" or polar nature to its outside. Hydrophobic residues are generally found in the interior while hydrophilic residues are found on the surface interacting with the hydrophilic water-intense external environment.

(This theme is often found for synthetic polymers that contain both the polar and nonpolar portions. Thus, when polymers are formed or reformed in a regular water-filled atmosphere, many polymers will favor the presence of polar moieties on their surfaces.)

Folding depends on a number of interrelated factors. The chain-folding process involves going from a system of random β- and α-conformations to a single folded structure. This change involves a decrease in the entropy or randomness. In order for folding to occur, this increase in order must be overcome by energy-related factors to allow the overall free energy to be favorable. These energy factors include charge–charge interactions, hydrogen bonding, van der Waals interactions, and hydrophilic/hydrophobic effects.

Within a particular globular polymer, there may be one or more polypeptide chains folded backward and forward forming quite distinct structural domains. Each domain is characterized by a particular style of coiling or "sheeting" which may be nonrepetitive with respect to its peptide chain geometry or may be repetitive, conforming to one of several now well-recognized patterns. The specific chain conformations are determined by the side-chain interactions of the amino acids superimposed on intrapeptide hydrogen bonding along the chain. The form of chain folding is thus ultimately determined by the amino acid sequence and the polymeric nature of the polypeptide chains, and is fundamental to the specific geometry of the given protein.

Protein units can be either negatively or positively charged. The attractions between unlike charges are important as are the repulsions between like changed units. As expected, these associations are pH dependent and are one route for conformation changes to occur. The ability to hydrogen bond is also an important factor with respect to the internal-folding scheme. Because the proteins are tightly packed, the weak van der Waals interactions can also play an important role in determining chain folding. The tendency for polarity-like segments to congregate can also be an important factor in chain folding. Thus, hydrophilic groupings generally are clustered to the outside of the globular protein allowing them to take an advantage of hydrogen bonding and other polar bonding opportunities while the hydrophobic clusters of amino acid units occupy the internal regions of the protein taking an advantage of hydrophobic interactions.

Globular proteins act in maintenance and regulatory roles—functions that often require mobility and thus some solubility. Included within the globular grouping are enzymes, most hormones, hemoglobin, and fibrinogen that is changed into an insoluble fibrous protein fibrin that causes blood clotting.

Denaturation is the irreversible precipitation of proteins caused by heating, such as the coagulation of egg white as an egg is cooked, or by the addition of strong acids, bases, or other chemicals. This denaturation causes permanent changes in the overall structure of the protein and because of the ease with which proteins are denatured, it makes it difficult to study "natural" protein structure. Nucleic acids also undergo denaturation.

Small changes in the primary structure of proteins can result in large changes in the secondary structure. For instance, researchers have interchanged the positions of two adjacent amino acid residues of a globular protein portion resulting in a β-strand becoming a right-handed helix.

Molecular recognition is one of the keys to life. Scientists are discovering ways to both modify molecular recognition sites and to "copy" such sites. One approach to modifying molecular recognition sites, namely enzymatic protein sites, is through what is referred to as **directed evolution**. Arnold and coworkers have employed the combinatorial approach by taking an enzyme with a desired catalytic activity and encouraging it to undergo mutation; selecting out those mutations that perform in the desired manner; and repeating this cycle until the new enzymes perform as desired. Ratner and coworkers have taken another approach whereby templates containing the desired catalytic sites are made. First, the protein is mounted on a mica support. The target protein is coated with a sugar monolayer that allows for specific recognition. A fluoropolymer plasma film is deposited over the sugar monolayer. The fluoropolymer reverse-image is attached to a support surface using an epoxy resin. Solvents are then added to etch away mica, sugar, and original protein leaving behind a "nano-pit" template that conforms to the shape of the original protein.

Enzymes generally act to lower the activation energy through a combination of holding the reactants in the correct geometry and making the number of "hits" or connections much less through holding the reactant in the proximity of one another. The increases in reaction rate are generally huge. In the case of the enzyme orotate decarboxylase, the increase is 10^{17} allowing a reaction that might take 80 million years without the enzyme to occur in about 20 ms in the presence of the enzyme. There are about 4000 reactions known to be catalyzed by enzymes within the human body. While the size of the enzyme is large, the active site is generally only 3–4 amino acid units long. Enzymes are involved in much of our metabolism, in the synthesis and repair of DNA, and in transcription.

4.7.3.2 Fibrous Proteins

Fibrous proteins are long macromolecules that are attached through either inter- or intrahydrogen bonding of the individual residues within the chain. Solubility, partial or total, occurs when these hydrogen bonds are broken. In general, they confer stiffness and rigidity to biological systems that are themselves fluid.

Fibrous proteins are found in animals. They appear as filaments and are often formed from a limited number of amino acid units. They are generally insoluble in water with hydrophobic portions protruding from the central core. Examples are collagen, elastin, and keratin. Actin and tubulin are globular soluble monomers that polymerize forming long stiff structures that makeup the cytoskeleton, which allows cells to maintain their shape. Structural proteins are used to construct tendons, bone matrices, connective tissues, and muscle fibers. They can also be used for storage. They are not as easily denatured as globular proteins.

Some structural proteins, such as kinesin, dynein, and myosin, serve as so-called motor proteins that can generate mechanical forces as muscles and are involved in allowing the movement of cells, such as the sperm cell, involved in sexual reproduction in multicellular organisms.

4.7.3.3 Collagen

Collagen is the most abundant single protein in vertebrates making up to one-third of the total protein mass. Collagen fibers form the matrix or the cement material in our bones where mineral materials precipitate. Collagen fibers constitute a major part of our tendons and act as a major part of our skin. Hence, it is collagen that is largely responsible for holding us together.

The basic building block of collagen is a triple helix of three polypeptide chains called the *tropocollagen unit*. Each chain is about 1000 residues long. The individual collagen chains form left-handed helices with about 3.3 residues per turn. In order to form this triple-stranded helix, every third residue must be glycine because glycine offers a minimum of bulk. Another interesting theme in collagen is the additional hydrogen bonding that occurs because of the presence of hydroxyproline derived from the conversion of proline to hydroxproline. The conversion of proline to hydroxproline involves vitamin C. Interestingly, scurvy, the consequence of a lack of vitamin C, is a weakening of collagen fibers giving way to lesions in the skin and gums and hence weakens blood vessels. Collagen fibers are strong. In tendons, the collagen fibers have strength similar to that of a hard-drawn copper wire.

Much of the toughness of collagen is the result of the cross-linking of the tropocollagen units to one another through a reaction involving lysine side chains. Lysine side chains are oxidized to aldehydes that react with either a lysine residue or with one another through an aldol condensation and dehydration resulting in a cross-link. This process continues throughout our life resulting in our bones and tendons becoming less elastic and more brittle. Again, a little cross-linking is essential, but more cross-linking leads to increased fracture and brittleness.

Collagen is a major ingredient in some "gelation" materials. Here, collagen forms a triple helix for some of its structures while other parts are more randomly flowing single collagen chain segments. The bundled triple helical structure acts as the rigid part of the polymer while the less ordered amorphous chain acts as the soft part of the chain. The triple helix also acts as a noncovalently bonded cross-link.

4.7.3.4 Elastin

Collagen is found where strength is needed, but some tissues, such as arterial blood vessels and ligaments, need materials that are elastic. Elastin is the protein of choice for such applications. Elastin is rich in glycine, alanine, and valine; it is easily extended and flexible. Its conformation approaches that of a random coil so that secondary forces are relatively weak, allowing elastin to be readily extended as tension is applied. The structure also contains some lysine side chains that are involved in cross-linking. The cross-linking is accomplished when four lysine side chains are combined to form a desmosine cross-link. This cross-link prevents the elastin chains from being fully extended and causes the extended fiber to return to its original dimensions when tension is removed. One of the areas of current research is the synthesis of polymers with desired properties based on natural analogs. Thus, elastin-like materials have been synthesized using glycine, alanine, and valine, and some cross-linking. These materials are similar to elastin in their elasticity.

4.7.3.5 Membrane Proteins

Membrane proteins are attached to or associated with the membrane of a cell. More than one-half of the proteins interact with these membranes. Membrane proteins are generally divided according to their attachment to a membrane. Transmembrane proteins span the entire membrane. Integral proteins are permanently attached to only one side of the membranes. Peripheral membrane proteins are temporarily attached to integral proteins or lipid bilayers through the combination of noncovalent bonding such as hydrophobic bonding and electrostatic bonding. These membranes often act as receptors or provide channels for charged or polar molecules to pass through them.

4.7.4 QUATERNARY STRUCTURE

The term quaternary structure is employed to describe the overall shape of the groups of chains of proteins, or other molecular arrangements. For instance, hemoglobin is composed of four distinctly different myoglobin units, each with its own tertiary structure, which come together giving the hemoglobin structure.

Both synthetic and natural polymers have superstructures that influence/dictate the properties of the materials. Many of these primary, secondary, tertiary, and quaternary structures are influenced in a similar manner. Thus, primary structure is a driving force for secondary structure. Allowed and preferred primary and secondary bondings influence structure. For most natural and synthetic polymers, the hydrophobic and hydrophilic domains tend to cluster. Thus, most helical structures will have a somewhat hydrophobic/hydrophilic inner core and the opposite outer core resulting from a balance between the secondary and primary bonding factors and the steric and bond angle constraints. Nature has used these differences in domain character to create the world around us.

As noted before, some proteins are linear with inner- and intrachain associations largely occurring because of hydrogen bonding. Influences on globular protein structures are more complex, but again, the same forces and features are at work. Globular proteins have irregular 3D structures that are compact but which, when brought together, form quaternary structures that approach being spherical. While the overall structure is spherical, the surface is irregular, with the irregularity allowing the proteins to perform specific biological functions.

The preferred folding conformation is again influenced by the same factors of bonding type, polarity, size, flexibility, and preferred bond angles. The folded conformations are possible because of the flexibility of the primary bonding that is present within proteins. Thus, polar portions, namely the amine and carbonyl moieties, are more fixed, but the carbon between them is more flexible. Again, the folding characteristic conformations are driven by the secondary bonding. Some folding is chemically "fixed" through the use of cross-links. In hair, these cross-links are often disulfides, –S–S–.

As noted before, the flexibility of proteins allows them to carry out a wide variety of tasks, and the building sites of proteins produce a variety of different proteins. Our cells often build about 60,000 different kinds of proteins. A bacterial cell will synthesize only a little over 1000 different kinds of proteins.

When a protein contains roughly more than about 200 amino acid groups, it often assumes two or more somewhat spherical tertiary structural units. These units are often referred to as domains. Thus, hemoglobin is a combination of four myoglobin units with each of the four units influenced by the other three, and where each unit contains a site to interact with oxygen.

The specificity of enzymatic catalytic activity is dependent on the tertiary structure. Phosphoglycerate kinase (Figure 4.8) is a protein composed of 415 amino acids. The protein chain is folded back and forward forming a clawlike structure that has two rigid domains divided by a more flexible hinge region. Phosphoglycerate kinase binds to phosphoglycerate, a fragment of glucose, transferring the phosphate forming ATP. If the enzyme binds to phosphoglycerate and water, the phosphate could react with the water rather than forming ATP. The protein is designed to circumvent this by having the essential parts present in both the halves of the claw. As the enzyme binds the phosphoglycerate and the ADP, the claw closes bringing together the two necessary active parts, excluding water and other nonessential compounds. In Figure 4.8, the active site phosphoglycerate is indicated by the "dotted" area about half way up one of the two sections of the claw.

Figure 4.8 also illustrates the usual arrangement whereby hydrophilic areas, noted as darkened areas, radiate outward from the enzyme surface while the less hydrophilic and hydrophobic areas tend to reside within the enzyme.

FIGURE 4.8   Protein enzyme phosphoglycerate kinase involved in the formation of adenosine-5′-triphosphate (ATP) from adenosine-5′-diphosphate (ADP).

TABLE 4.5
Shapes of Selected Biologically Important Proteins

Protein	Shape	Molecular Weight	Comments
Myoglobin	3D, oblate spheroid	1.7×10^4	Temporary oxygen storage in muscles
Hemoglobin	3D, more spherical than myoglobin	6.4×10^4	Oxygen transport through body
Cytochrome c	3D, prolate spheroid	$1.2–1.3 \times 10^4$	Heme-cont. protein, transports electrons rather than oxygen
Lysozyme	3D, short α-helical portions, region of antiparallel pleated sheets	1.46×10^4	Well studied, good illustration of structure activity
Chymotrypsin and trypsin	3D, extensive β-structure	—	Hydrolysis of peptide bonds on the carboxyl side of certain amino acids
Insulin	3D, two α-helical section in A chain, B chain has α-helix and remainder is extended linear central core	6×10^3	Regulation of fat, carbohydrate, and amino acid metabolism
Somatotropin (human)	3D, 50% α-helix	2.2×10^4	Pituitary hormone
Keratin	Varies with source, 3D or 2D; most contain α-helix sections	$10^4–10^5$	Most abundant protein; major part of skin, teeth, bones, cartilage, and tendon
Fibroin	Varies with source, fibrous-linear with cross-links; crystalline regions contain antiparallel, pleated sheets	3.65×10^5	Major constituent of silk
Elastin	Varies with source; cross-linked, mostly Random coil with some α-helix	$>7 \times 10^4$	Many properties similar to rubber; gives elasticity to arterial walls and ligaments

Table 4.5 contains a listing of some important proteins. Protein purification must be done under conditions where the conformational and configurational changes are minimal. Such purification is most often carried out using varieties of chromatography including affinity chromatography and electrophoresis. Somewhat common features of enzymes are the following:

1. α-Helix content is not as high as myoglobin, but areas of β-sheeting are not unusual.
2. Water-soluble enzymes have a large number of charged groups on the surface and those that are not on the surface are involved in the active site. Large parts of the interior are hydrophobic.
3. The active site is found either as a cleft in the macromolecule or as a shallow depression on its surface.

While enzymes are effective catalysts inside in the body, we have developed techniques for "capturing" some of this activity by immobilizing enzymes. The activity of many enzymes continues in this immobilized condition. In one approach, the enzyme is isolated and coupled to solvent-swellable gels using polyacrylamide copolymers that contain *N*-acryloxysuccinimide repeat units. The pendant groups react with the amino "ends" of the enzyme effectively coupling or immobilizing the enzyme. The modifications of this procedure have been used to immobilize a wide variety of enzymes. For instance, the particular reactive or anchoring group on the gel can be especially modified for a particular enzyme. Spacers and variations in the active coupling end are often employed. The amine groups on gels are easily modified to give other functional groups including alcohols, acids, nitriles, and acids. Recently, other entities, such as fungi, bacteria, and cells have been immobilized successfully. This technique allows the continuous synthesis of specific molecules.

4.8   NUCLEIC ACIDS

Nucleoproteins, which are conjugated proteins, may be separated into nucleic acids and proteins. The name "nuclein," which was coined by Miescher in 1869 to describe products isolated form nuclei in pus, was later changed to nucleic acid. Somewhat pure nucleic acid was isolated by Levene in the early 1900s. He showed that D-ribose and D-deoxyribose (Figure 4.9) were present in what are now known as ribonucleic acid (RNA) and deoxyribonucleic acid (DNA), respectively. They consist of two sugars that are identical except that the deoxyribose contains a hydrogen on carbon 2 (thus the name deoxy or without one "oxy" or hydroxyl (Figure 4.9)). These specific compounds were originally obtained from yeast (DNA) and the thymus gland (RNA).

In 1944, Avery showed that DNA was able to change one strain of bacteria to another. It is now known that nucleic acids direct the synthesis of proteins. Thus, our modern knowledge of heredity and molecular biology is based on the knowledge of nucleic acids. Recently, it was announced that the human genome was decoded. This is one of the most important events to date. The human genome is composed of nature's most complex, exacting, and important macromolecule. It is composed of nucleic acids that appear complex in comparison to simpler molecules such as methane and ethylene, but simple in comparison to their result on the human body. Each unit is essentially the same containing a phosphate, a deoxyribose sugar, and one of the four bases (Figure 4.9), with each base typically represented by the capital of the first letter of their names, G, C, A, and T. In fact, the complexity is less than having four separate and independent bases because the bases come in matched sets, they are paired. The mimetic Gee CAT allows an easy way to remember this pairing (G–C and A–T). The base, the sugar, and the phosphate combine forming nucleotides such as adenylic acid, adenosine-3′-phosphate (4.30) and are represented by the symbols A, dA, and dAMP.

(4.30)

Adenylic acid, adenosine-3'-phosphate

The backbone of nucleic acids is connected through the 3′ and 5′ sites on the sugar with the base attached at the 1′ site. Because the sugar molecule is not symmetrical, each unit can be connected differently, but there is an order (also called sense or directionality) in the sequence of this connection

FIGURE 4.9 Components of nucleic acids.

so that phosphodiester linkage between units is between the 3′ carbon of one unit and the 5′ carbon of the next unit. Thus, nucleic acids consist of units connected so that the repeat unit is a 3′–5′ (by agreement we consider the start at the 3′ and the end at the 5′ though we could just as easily describe this repeat as being 5′-3′) linkage. Thus, the two ends are not identical—one contains an unreacted 3′-hydroxyl and the other an unreacted 5′-hydroxyl.

A shorthand is used to describe nucleic acid sequences. Following is a trimer containing in order the bases cytosine, adenine, and thymine.

(4.31)

This sequence is described as

p-5′-C-3′-p-5′-A-3′-p-5′-T-3′ or pCpApT or usually as simply CAT.

Nobel Laureates Watson and Crick correctly deduced that DNA consisted of a double-stranded helix in which a pyrimidine base on one chain or strand was hydrogen bonded to a purine base on the other chain (Figure 4.10). The bonding distances are not the same with the GC pairing more compact. These uneven pairing distances result in a DNA with a characteristic twisting giving

FIGURE 4.10 Hydrogen bonding between preferred base pairs in DNA.

unique structures. It is this twisting, and the particular base sequence, that eventually results in the varying chemical and subsequently biological activities for various combinations.

The stability of the DNA is due to both the internal and external hydrogen bondings as well as the ionic and other bondings. First, the internal hydrogen bonding is between the complementary purine–pyrimidine base pairs. Second, the external hydrogen bonding occurs between the polar sites along the exterior sugar and the phosphate moieties and water molecules. Third, ionic bonding occurs between the negatively charged phosphate groups situated on the exterior surface of the DNA and electrolyte cations such as Mg^{+2}. Fourth, the core consists of the base pairs, which, along with being hydrogen bonded, stack together through hydrophobic interactions and van der Waals forces. In order to take good advantage of pi-electron cloud interactions, the bases stack with the flat "sides" over one another so that they are approximately perpendicular to the long axis.

The AT and CG base pairs are oriented in such a manner so that the sugar–phosphate backbones of the two twined chains are in opposite or antiparallel directions, with one strand starting at the 5′ and ending at the 3′, and the other strand starting at the 3′ end and ending at the 5′ end. Thus, the two chains "run" in opposite directions.

The glucose bonds holding the bases onto the backbone are not directly across the helix from one another. Thus, the sugar–phosphate repeat units are not the same. This dislocation creates structures referred to as the major and minor grooves as pictured in Figure 4.16. It is known that at least some proteins that bind to DNA recognize the specific nucleotide sequences by "reading" the hydrogen-bonding pattern presented by the edges of these grooves.

In solution, DNA is a dynamic, flexible molecule. It undergoes elastic motions on a nanosecond time scale most closely related to changes in the rotational angles of the bonds within the DNA backbone. The net result of these bendings and twistings is that DNA assumes a roughly globular or spherical tertiary shape. The overall structure of the DNA surface is not that of a reoccurring "barber pole" but rather because of the particular base sequence composition each sequence will have its own characteristic features of hills, valleys, bumps, etc.

As the two strands in a double helix separate, they act as a template for the construction of a complementary strand. This process occurs enzymatically with each nucleotide being introduced into the growing chain through matching with its complementary base on the existing chain. Thus, two identical strands are produced when one double-helix combination replicates.

DNA chains can contain 1 million subunits with an end-to-end contour length of about 1 mm. Even with the complexity of these large macromolecules, the synthesis of new chains generally occurs without any change in the molecule. Even when changes occur, these giant machines have built into them "correcting" mechanisms that re-correct when mistakes occur.

The transcription product of DNA is always a single-stranded RNA. The single strand generally assumes a right-handed helical conformation mainly caused by base-stacking interactions also present in the DNA. The order of interaction is purine–purine >> purine–pyrimidine >

pyrimidine–pyrimidine. The purine–purine interaction is so strong that a pyrimidine separating two purines is often displaced from the stacking order to allow the interaction between the two purines to occur. Base paring is similar to that of the DNA except that uracil generally replaces thymine. For coupled RNA, the two strands are antiparallel as in DNA. Where complementary sequences are present, the predominant double-stranded structure is an A form right-handed double helix. Many RNAs are combinations of complementary two-stranded helices, single-stranded segments, as well as other complex structures. Hairpin curves are the most common type of more complex structures in RNA. Specific sequences, such as UUCG, are generally found at the ends of RNA hairpin curves. Such sequences can act as starting points for the folding of an RNA into its precise 3D structure. The tertiary structures for RNAs are complex with combinations being present. For instance, the tertiary structure for the tRNA of yeast for phenylalanine consists of a cloverleaf including three loops formed by hairpin curves and double helix regions stabilized by hydrogen bonding. Hydrogen-bonding sites that are not significant in the DNA structures are important. Thus, the free hydroxyl on the ribose sugar moiety can hydrogen bond with other units.

There are four major kinds of RNA. Messenger RNA, mRNA, varies greatly in size from about 75 units to over 3,000 nucleotide units giving a molecular weight of 25,000 to 1 million. It is present at a percentage of about 2% of the total RNA in a cell. Transfer RNA, tRNA, has about 73–94 nucleotides with a corresponding molecular weight range of 23,000–30,000. It is present in the cell at a level of about 16%. The most abundant RNA, 82%, is the ribose RNA, rRNA, which has several groupings of molecular weight with the major ones being about 35,000 (about 120 nucleotide units), 550,000 (about 1550 units) and 1,100,000 (about 2900 units). Eukaryotic cells contain an additional type called small nuclear RNA, snRNA.

tRNA, generally contains 73–94 nucleotides in a single chain with a majority of the bases hydrogen bonded to one another. Hairpin curves promote complementary stretches of base bonding giving regions where helical double stranding occurs. The usual overall structure can be represented as a cloverleaf with each cloverleaf containing four of these helical double-stranded units. One of the loops acts as the acceptor stem that serves as the amino acid–donating moiety in protein synthesis.

rRNA, is a part of the protein-synthesizing machinery of cells, ribosomes. Ribosomes contain two subunits called "small" and "large" with ribosomal RNAs being part of both of these units. rRNAs contain a large amount of intrastrand complementary sequences and are generally highly folded. Interestingly, there is a similarity between the folded structures of rRNA from different sources even though the primary structure, base sequence, is quite varied. Thus, there appears to be preferred folding patterns for rRNAs.

mRNA, is the carrier of messages that are encoded in genes to the sites of protein synthesis in the cell where this message is translated into a polypeptide sequence. Because mRNAs are transcribed copies of the genetic unit, they are sometimes referred to as being the "DNA-like RNA." mRNA is made during transcription, an enzymatic sequence in which a specific RNA sequence is "copied" from a gene site. rRNA and tRNA are also made by the transcription of DNA sequences but unlike mRNA, they are not subsequently translated to form proteins.

Actual reproduction steps involving DNA and RNA often occur in concert with proteins where the protein can act as a clamp or vice holding the various important members involved in the particular reproduction step in place. Thus, the protein complex acts as an assembly line tunnel or doughnut with the reactants present within the interior.

4.8.1 FLOW OF BIOLOGICAL INFORMATION

Nucleic acids, proteins, some carbohydrates, and hormones are informational molecules. They carry directions for the control of biological processes. With the exception of hormones, these are macromolecules. In all these interactions, secondary forces such as hydrogen bonding and van der Waals forces, and ionic bonds and hydrophobic/hydrophilic character play critical roles. *Molecular recognition* is the term used to describe the ability of molecules to recognize and interact and bond-specifically with other molecules. This molecular recognition is based on a combination of these interactions and on structure.

Molecular recognition interactions have several common characteristics. First, the forces that are involved in these interactions are relatively weak and they are noncovalent. They are on the order of about 1–8 kcal/mol (4–30 kJ/mol) compared to covalent bonds of the order of about 80 kcal/mol (300 kJ/mol) for a C–C sigma bond. A single secondary bond is generally not capable of holding molecules together for any length of time. However, for macromolecules, there is a cumulative effect so that the forces are not singular but are multiplied by the number of such interactions that are occurring within the particular domain. Second, these interactions are reversible. Initial contact occurs as the molecules come into contact with one another often through simple diffusion or movement of the molecules or segments of the molecules. These initial contacts are often not sufficient to cause the needed binding through some transitory interactions to occur. Even so, in some cases the cumulative bonding is sufficient to allow a transient but significant interaction to occur. This complex can then begin a specific biological process. Eventually, thermal motions and geometrical changes cause the complex to dissociate. Ready reversibility is an important key that allows relatively few "signaling" molecules to carry out their mission. Third, bonding between the particular molecular sites is specific. There must exist a combination of complementary bonding, hydrophobic/hydrophilic sites, ionic charge, and geometry that allow effective long-term (generally no more than several seconds) interactions to occur.

In general, the flow of biological information can be mapped as follows:

$$\text{DNA} \rightarrow \text{RNA} \rightarrow \text{Protein} \rightarrow \text{Cell structure and function} \tag{4.32}$$

The total genetic information for each cell, called the *genome*, exists in the coded two-stranded DNA. This genetic information is expressed or processed either through the duplication of the DNA so it can be transferred during cell division to a daughter cell or it can be transferred to manufactured RNA that in turn transfers the information to proteins that carry out the activities of the cell.

The duplication of double-stranded DNA is self-directed. The DNA, along with accessory proteins, directs the **replication** or the construction of two complementary-strands forming a new, exact replicate of the original DNA template. As each base site on the DNA becomes available through the unraveling of the double-stranded helix, a new nucleotide is brought into the process held in place by hydrogen bonding and van der Waals forces, so that the bases are complementary. It is then covalently bonded through the action of an enzyme called DNA polymerase. After duplication, each DNA contains one DNA strand from the original double-stranded helix and one newly formed DNA strand. This is called **semiconservative replication** and increases the chance that if an error occurs, then the original base sequence will be retained.

How is DNA suitable as a carrier of genetic information? While we do not entirely understand this, several features are present in DNA. First, because of the double-stranded nature and mode of replication, retention is enhanced. Second, DNA is particularly stable within both the cellular and extracellular environments, including a good stability to hydrolysis within an aqueous environment. Plant and animal DNA have survived thousands of years. Using polymerase chain reactions, we can reconstruct DNA segments allowing comparisons to modern DNA.

Transcription is the term used to describe the transfer of information from DNA to RNA. The genome is quite large, on the order of a millimeter in length if unraveled, but within it exists coding regions called genes. Transcription is similar to DNA replication except that ribonucleotides are the building units instead of deoxyribonucleotides; the base thymine is replaced by uracil; the DNA:RNA duplex unravels releasing the DNA to again form its double-stranded helix and the single-stranded RNA; and the enzyme linking the ribonucleotides together is called RNA polymerase.

Many viruses and retroviruses have genomes that are single-stranded RNA instead of DNA. These include the AIDS virus and some retroviruses that cause cancer. Here, an enzyme called reverse transcriptase converts the RNA genome of the virus into the DNA of the host cell genome, thus infecting the host.

The transcription of the DNA gives three kinds of RNA—ribosomal, messenger, and transfer. The most abundant RNA is rRNA. Most rRNA is large and found in combination with proteins in the ribonucleoprotein complexes called ribosomes. Ribosomes are subcellular sites for protein synthesis.

tRNA, is the smallest of the RNAs being less than 100 nucleotides long. tRNA combines with an amino acid incorporating it into a growing protein. There is at least one tRNA for each of the 20 amino acids used in protein synthesis. mRNA, is varied in size but each carries the message found in a single gene or group of genes. The sequence of bases in mRNA is complementary to the sequence of DNA bases. mRNA is unstable and short lived, so its message for protein synthesis must be rapidly decoded. The message is decoded by the ribosomes that make several copies of the protein for each mRNA.

The ultimate purpose of DNA expression is protein synthesis. mRNA serves as the intermediate carrier of the genetic information for protein synthesis. The DNA message is carried in the form of base sequences that are transferred to RNA, also in terms of base sequences, and finally these are transferred into amino acid sequences through a translation process based on the genetic code. This process of information from the RNA to the protein is called *translation*.

A set of coding rules are in action as in the translation process. Briefly, these are as follows. First, a set of three adjacent nucleotides compose the code for each amino acid. A single amino acid can have several triplet codes or *codons*. Since there are four different nucleotides (or four different bases) in DNA and RNA, there exist 4^3, or 64 trinucleotide combinations. For instance, using U as a symbol for uracil, present in RNA, the triplet code or codon UUU is specific for phenylalanine.

Second, the code is nonoverlapping so that every three nucleotides code for an amino acid and the next three code for a second amino acid and the third set code for a third amino acid, etc. Third, the sets of nucleotides are read sequentially without punctuation. Fourth, the code is nearly universal. Fifth, there are codes for other than amino acids including stop or terminate, UAG, and start or initiate, AUG.

In essence, tRNA has two active sites—one that is specific for a given amino acid and the second that is specific for a given set of three bases. The tRNA "collects" an appropriate amino acid and brings it to the growing polypeptide chain inserting it as directed by the mRNA. There is then a collinear relationship between the nucleotide base sequence of a gene and the amino acid sequence of a protein.

The amount, presence, or absence of a particular protein is generally controlled by the DNA in the cell. Protein synthesis can be signaled external to the cell or within the cell. Growth factors and hormones form a part of this secondary messenger service.

The translation and transcription of DNA information is polymer synthesis and behavior; the particular governing factors and features that control these reactions are present in the synthesis and behavior of other macromolecules—synthetic and biological.

For the human genome, there exists so-called coding or active regions called *exons* and noncoding regions called *introns*. The average size of an exon is about 120–150 nucleotide units long or coding for about 40–50 amino acids. Introns vary widely in size from about 50 to over 20,000 units. About 5% of the genome is used for coding. It was thought that the other 95% was silent or junk DNA. We are finding that the intron regions play essential roles. Interestingly introns are absent in the most basic prokaryotes, only occasionally found in eukaryotes, but common in animals.

There are two types of cells. *Prokaryote cells* lack a cell nucleus or any other membrane bound organelles. Most are single celled. Bacteria and archaea are the examples of organisms with prokaryote cells. *Eukaryote cells* possess enclosed membranes and form the basis of animals, plants, and fungi. In prokaryote cells, the mRNA can be used immediately after it is produced or it may be bound to a ribosome. In comparison, in eukaryote cells mRNA is made in the cell nucleus and after it is synthesized it is moved across the nuclear membrane into the cytoplasm where protein synthesis occurs. Protein synthesis occurs on the average of two amino acid units every second.

4.9 NATURALLY OCCURRING POLYISOPRENES

Polyisoprenes occur in nature as hard plastics called gutta percha and balata, and as an elastomer or rubber known as *Hevea brasiliensis*, or natural rubber (NR). Approximately 50% of the 500 ton of gutta percha produced annually is produced annually from trees grown on plantations in Java and Malaya. Balata and about 50% of the gutta percha used are obtained from trees in the jungles of South America and the East Indies. The first gutta-insulated submarine cable was installed between England and France in 1859. Gutta percha (*Palaquium oblongifolium*) continues to be used for wire insulation, and polyisoprenes balata (*Mimusops globosa*) and gutta percha are used as covers for some golf balls.

$$\begin{array}{c} CH_3 \\ | \\ -(-CH_2-C=CH-CH_2-)- \end{array} \qquad (4.33)$$

The hardness of some of the polydisperse naturally occurring crystalline polymers is the result of a *trans* configuration in 1,4-polyisoprene (Figures 4.11 and 4.12). This geometry facilitates good fit between chains and results in inflexibility in the chains. The hevea or NR is, by comparison, soft because it is a *cis* configuration (Figures 4.11 and 4.12) that does not allow a close fit between chains, resulting in flexibility and a "rubbery" behavior. Both the *trans* and *cis* isomers of polyisoprene are present in chicle, obtained from the *Achras sapota* tree in Central America. Both the *trans* and *cis* isomers are also synthesized commercially.

Natural rubber was used by the Mayan civilization in Central and South America before the twentieth century. In addition to using the latex from the ule tree for waterproofing of clothing, they played a game called "tlachtli" with large hevea rubber balls. The object of the game was to insert the ball into a tight-fitting stone hole in a vertical wall using only the shoulder or the thigh. The game ended once a goal was scored, and the members of the losing team could be sacrificed to the gods.

FIGURE 4.11 Abbreviated structural formulas for polyisoprenes.

(a)   (b)

FIGURE 4.12 Ball-and-stick models of (a) "soft" hevea rubber (*cis*-1,4-polyisoprene) and (b) "hard" gutta percha (*trans*-1,4-polyisoprene).

While NR or caoutchouc was brought to Europe by Columbus, little use beyond erasing pencil marks was made of this important elastomer until the nineteenth century. Priestly coined the name India rubber to describe its major use at that time. Hevea latex is also present in many other plants in addition to *H. brasiliensis*. There has been interest in growing some of these plants and harvesting the NR but thus far, other than the guayule, none have become commercial-scale enterprises.

Charles Goodyear was born in 1800 in New Haven, Connecticut. He became driven to work with rubber to try to make it more temperature stable. This passion affected his health and took what little money he had. On more than one occasion, he lived in debtor's prison. One of his jobs was to supply the U.S. government with waterproof mailbags, but the mailbags he prepared were sticky and malformed. Another failure. After many unsuccessful attempts, one of which was to mix the rubber with sulfur, he accidentally allowed a mixture of sulfur and rubber to touch a hot stove. The rubber did not melt but only charred a little. As are many of the so-called discoveries by chance or accident, his mind was ready for the result and by 1844 had been given a patent for a process he called "vulcanization" after the Roman god of fire, Vulcan. Vulcanization is the cross-linking reaction between the rubber chains and the sulfur.

Goodyear had trouble defending his patent, piling up huge debts before he died in 1860. Daniel Webster defended him in one of his patent infringement cases. By 1858, the value of rubber goods was about $5 million. The major rubber-producing plants clustered about Akron, Ohio with the Goodyear Company founded in 1870.

Chemists learned about the structure of rubber by degrading it through heating and analyzing the evolved products. One of the evolved products was isoprene, a five-carbon hydrocarbon containing a double bond. Isoprene is a basic building block in nature serving as the "repeat" unit in rubber, and it is also the building block of steroids such as cholesterol.

$$H_2C{=}C{-}CH_2{-}CH_3$$
$$\underset{\displaystyle CH_3}{|}$$

Isoprene

(4.34)

With knowledge that natural rubber had isoprene units, chemists worked to duplicate the synthesis of rubber except using synthetic monomers. These attempts failed until two factors were realized. First, after much effort it was discovered that the methyl groups were present in a *"cis"* arrangement. Second, it was not until the discovery of stereoregular catalysts that the chemists had the ability to form natural rubberlike material from butadiene.

The synthesis of a purely synthetic rubber (SR), structurally similar to natural rubber, involved a number of scientists building upon one another's work—along with a little creativity. Nieuwland, a Catholic priest, President of Notre Dame University, and a chemist did extensive work on acetylene. He found that acetylene could be made to add to itself forming dimers and trimers.

Calcott, a Dupont chemist, attempted to make polymers from acetylene, reasoning that if acetylene formed dimers and trimers, conditions could be found to produce polymers. He failed, but went to Carothers who had one of his chemists, Arnold Collins, work on the project. Collins ran the reaction described by Nieuwland, purifying the reaction mixture. He found a small amount of material that was not vinylacetylene or divinylacetylene. He set the liquid aside. When he came back, the liquid had solidified giving a material that seemed rubbery and even bounced. They analyzed the rubbery material and found that it was not a hydrocarbon, but it had chlorine in it. The chlorine had come from HCl that was used in Nieuwland's procedure to make the dimers and trimers adding to the vinylacetylene forming chloroprene (4.36).

$$HC\equiv CH \longrightarrow H_2C=CH-C\equiv CH + H_2C=CH-C\equiv C-CH=CH_2 \tag{4.35}$$

Acetylene Vinylacetylene Divinylacetylene

This new rubber was given the name Neoprene (4.36). Neoprene had outstanding resistance to gasoline, ozone, and oil in contrast to natural rubber. Today, Neoprene is used in a variety of applications including as electrical cable jacketing, window gaskets, shoe soles, industrial hose, and heavy-duty drive belts.

$$H_2C=C-CH=CH_2 \longrightarrow -(-CH_2C=CH-CH_2-)- \tag{4.36}$$

$$\underset{Cl}{|} \qquad\qquad \underset{Cl}{|}$$

Chloroprene Polychloroprene (Neoprene)

Prior to the discovery of the vulcanization or cross-linking of hevea rubber with sulfur by Charles Goodyear in 1838, Faraday has shown that the empirical formula of this elastomer is C_5H_8, making it a member of the terpene family. The product obtained by the pyrolysis of rubber was named isoprene by Williams in 1860 and it was converted to a solid (polymerized) by Bouchardat in 1879.

Natural rubber crystallized when stretched in a reversible process. However, the sample remains in its stretched form (racked rubber) if it is cooled below its T_g. The racked rubber will snap back and approach its original form when it is heated above its T_g. The delay in returning to the original form is called *hysteresis*. These and other elastic properties of NR and other elastomers above the T_g are based on long-range elasticity. Stretching causes an uncoiling of the polymer chains, but these chains assume a more ordered microstructure creating crystalline domains acting to oppose further stretching. If the rubber is held in the stretched state for some time, slippage of chains occurs and precise return to its original shape does not occur when the stress is released.

The absence of strong intermolecular forces, the presence of pendant methyl groups discouraging close association, and crankshaft action associated with the *cis* isomer all contribute to the flexibility of NR. The introduction of a few cross-links by vulcanization with sulfur reduces the slippage of chains but still permits good flexibility.

When a strip of NR or SR is stretched at a constant rate, the tensile strength required for stretching (stress, s) increases slowly until an elongation (strain) of several hundred percent is observed. This initial process is associated with an uncoiling of the polymer chains in the regions that are not cross-linked. Considerably more stress is required for a greater elongation to about 800%. This rapid increase in modulus (G) is associated with better alignment of the polymer chains along the axis of elongation, crystallization, and decrease in entropy (ΔS). The work done in the stretching process (W_1) is equal to the product of the retractile force (F) and the change in length (dl). Therefore, the force is equal to the work per change in length:

$$W_1 = f\,dl \quad \text{or} \quad f = \frac{W_1}{dl} \tag{4.37}$$

W_1 is equal to the change in Gibbs free energy (dG), which under the conditions of constant pressure is equal to the change in internal energy (dE) minus the product of the change in the entropy and the Kelvin temperature as follows:

$$f = \frac{W_1}{dl} = \frac{dG_1}{dl} = \frac{dE}{dl} - T\left(\frac{dS}{dl}\right) \tag{4.38}$$

The first term (dE/dl) in Equation 4.38 is important in the initial low-modulus stretching process, and the second term [$T(dS/dl)$] predominates in the second high-modulus stretching process. For an ideal rubber, only the second term is involved.

As observed by Gough in 1805 and verified by Joule in 1859, the temperature of rubber increases as it is stretched, and the stretched sample cools as it snaps back to its original condition. (This can be easily confirmed by you by rapidly stretching a rubber band and placing it to your lips, noting that heating has occurred, and then rapidly releasing the tension and again placing the rubber band to your lips.) This effect was expressed mathematically by Kelvin and Clausius in the 1850s. The ratio of the rate of change in the retractive force (df) to the change in the Kelvin temperature (dT) in an adiabatic process is equal to the specific heat of the elastomer (C_p) per degree temperature (T) times the change in temperature (dT) times the change in length (dl):

$$\frac{df}{dT} = -\left(\frac{C_p}{T}\right)\left(\frac{dT}{dl}\right) \tag{4.39}$$

Equation 4.39 may be transformed to Equation 4.40.

$$\frac{dT}{df} = -\left(\frac{T}{C_p}\right)\left(\frac{dl}{dT}\right) \tag{4.40}$$

Unlike most solids, natural rubber and other elastomers contract when heated.

4.9.1 BALLOONS

Balloons have been part of our everyday lives from birthdays to graduations. As with many simple objects, their origin is unknown. Balloon-like objects have been part of ancient stories and probably were initially used in sports. These early balloons were made of animal bladders and intestines, both being protein based, and hence polymeric. European jesters inflated the animal entrails using them to entertain others. Galileo inflated a pig's bladder to help in measuring the weight of air.

While there are also new world mentions of similar balloon construction, it was not until rubber arrived on the scene that the manufacture of balloons, as balloons we know today, began.

In 1824, Michael Faraday used natural rubber to produce balloons for his hydrogen experiments at the Royal Institution in London. He writes "…The caoutchouc is exceedingly elastic….Bags made of it… have been expanded by having air forced into them, until the caoutchouc was quite transparent, and when expanded by hydrogen they were so light as to form balloons with considerable ascending power…" Faraday constructed his balloons by cutting two-layered sheets of rubber and then pressing the edges together. Flour was applied to the inside of the balloons because the native rubber was tacky and would adhere preventing it from being inflated.

Toy balloons were introduced by Thomas Hancock in 1825 as a do-it-yourself kit that consisted of a rubber solution and a syringe. Vulcanized toy balloons were initially manufactured by J. G. Ingram of London in 1847. The vulcanizing caused the balloons to be nontacky and not susceptible to becoming excessively tacky on hot days. Montgomery Ward had balloons in their catalog by 1889.

In the United States, the story of the balloon coincided with the story of the tyre, both requiring an understanding and use of rubber. The initial manufacture of balloons in the United States was in 1907 by the Anderson Rubber Company in Akron, Ohio. In 1912, Harry Rose Gill, founder of the National Latex Rubber Products of Ashland, Ohio made a nonspherical cigar-shaped balloon. Gill also began packaging balloons in packs, the initial sanitary balloon package.

Neil Tillotson, at 16, began work at the Hood Rubber Company in Boston, Massachusetts. After a 2 year period in the Seventh Cavalry during World War I, he returned to the Hood Rubber Company. Of about 25 Hood chemists, Neil was the only one without a college degree. But, he was inventive and with the first load of raw rubber latex in the 1920s, he started a lifelong romance with rubber. The companies' initial efforts with the raw latex were unsuccessful and the company largely dropped its efforts to produce balloons. Tillotson persevered, working out of his own home. Eventually, he managed to "tame" the latex sufficiently to make stylistic balloons using techniques similar to what is employed today. He cut a catlike face from cardboard and dipped the catlike mold in latex and allowed it to dry. Inflated, it did resembled a cat. His first sale was an order for 15 gross for use at the annual Patriots Day Parade on April 19, 1931. Next, he formed a family business later in 1931, using his family as an assembly line in the production of balloons. During the great depression, Neil traveled by bus around the country to sell his balloons. The Tillotson Rubber Company is still in operation with the balloon division named Dipco, a nice reminder of the way balloons are still made.

Rubber latex is sensitive to the influence of external forces or chemical agents. Thus, the so-called alligator balloon is formed when solid colored balloons are dipped in an acid bath causing the balloon to turn into two shades, supposedly like an alligator. The radium balloon is produced by taking a solid-colored balloon and dipping it a second time into another color.

Commercially inflated balloons held hydrogen and helium. Hydrogen was preferred because of its somewhat, about 10%, greater lifting ability. Hydrogen balloons were first produced by Faraday, as noted before. But, along with its greater lifting power, hydrogen offers a danger. On ignition, hydrogen burns rapidly forming water and energy, generally producing a flame. As early as 1914, firemen tried to curtail the use of hydrogen-filled balloons. In 1922, New York City banned the use of hydrogen-filled balloons because a city official was badly burned because of a fire caused by hydrogen-filled balloons.

There is an experiment that is often used in polymer demonstrations. Pointed objects, such as long pins and sticks, are gently pushed into a blow-up balloon without exploding the balloon. This results because the rubber polymer chains form about the impacting object to seal up the "hole" created by the object. Thus, even in its seemingly solid state, the rubber balloon's polymer chains are sufficiently mobile to allow it to "heal" itself.

Balloons today are still made from the rubber latex from the rubber tree using molds. Color is added as desired.

Foil balloons are derived from the effort of NASA Space Command with the concept and technology to metallize plastic sheeting. While the balloons are often referred to as Mylar or silver Mylars, they are not made from Mylar, a trade name for certain polyester film. They should be referred to as simply foil balloons. They are made from nylon film coated on one side with polyethylene and metallized on the other. The nylon film gives the balloon some strength, and the polyethylene gives it some flexibility and helps to retard the release of the held gas.

4.9.2 RESINS

Shellac, which was used by Edison for molding his first photograph records and is still used as an alcoholic solution (spirit varnish) for coating wood, is a cross-linked polymer consisting largely of derivatives of aleuritic acid (9,10,16-triphydroxyhexadecanoic acid; 4.41). Shellac is excreted by small coccid insects (*Coccus lacca*), which feed on the twigs of trees in Southeast Asia. Over 2 million insects must be dissolved in ethanol to produce 1 kg of shellac.

$$(4.41)$$

Aleuritic acid

Rosin, the residue left in the manufacture of turpentine by distillation, is a mixture of the diterpene, abietic acid (4.42), and its anhydride. It is a nonpolymeric compound but is used in the manufacture of synthetic resins and varnishes. Easter gum, a cross-linked ester, is obtained by the esterfication of glycerol or pentaerythritol with rosin. Many natural resins are fossil resins exuded from trees thousands of years ago. Recent exudates are called recent resins, and those obtained from dead trees are called semifossil resins. *Humic acid* is a fossil resin found with peat, coal, or lignite deposits throughout the world. It is used as a soil conditioner, as a component of oil drilling muds, and as a scavenger for heavy metals. Amber is a fossil resin found in the Baltic Sea regions, and sandarac and copals are found in Morocco and Oceania, respectively. *Casein*, a protein from milk, under the name of Galalith, has been used as a molding resin and as an adhesive.

$$(4.42)$$

Abietic acid

4.10 LIGNIN

Lignin is the second most widely produced organic material, after the saccharides. It is found in essentially all living plants and is the major noncellulosic constituent of wood. It is produced at an annual rate of about 2×10^{10} ton with the biosphere containing a total of about 3×10^{11} ton. It contains a variety of structural units including those pictured in Figure 4.13.

Lignin has a complex structure that varies with the source, the growing conditions, etc. This complex and varied structure is typical of many plant-derived macromolecules. Lignin is generally considered as being formed from three different phenylpropanoid alcohols—coniferyl, coumaryl, and sinapyl alcohols, which are synthesized from phenylalanine via various cinnamic acid derivatives. Commercially it is sometimes treated as being composed of a C_9 repeat unit where the superstructure contains aromatic and aliphatic alcohols and ethers, and aliphatic aldehydes and vinyl units (Figure 4.13).

Lignin is found in plant cell walls of supporting and conducting tissue, mostly the tracheids and vessel parts of the xylem. It is largely found in the thickened secondary wall but can occur elsewhere close to the celluloses and the hemicelluloses.

The presence of rigid aromatic groups and hydrogen bonding by the alcohol, aldehyde, and ether groups give a fairly rigid material that strengthens stems and vascular tissues in plants allowing

FIGURE 4.13   Representative structure of lignin.

upward growth. It also allows water and minerals to be conducted through the xylem under negative pressure without collapse of the plant. This structure can be made flexible through the introduction of a plasticizer, in nature mainly water. The presence of the hydrophilic aromatic groups helps to ward off excessive amounts of water, allowing the material to have a variable flexibility but to maintain some strength. This type of balance between flexibility and strength is also utilized by polymer chemists as they work to modify lignin as well as synthetic polymers to give strong, semiflexible materials. Without the presence of water, lignin is brittle but with the presence of water the tough lignin provides plants with some degree of protection from animals.

Its chemical durability also makes it indigestible to plant eaters; its bonding to cellulose and protein material in the plant also discourages plant eaters from eating it. The formation of lignin also helps to block the growth of pathogens and is often the response to partial plant destruction.

The role of transferring water by lignin makes the ability to produce lignin critical to any plants' survival and permits even primitive plants to colonize dry land.

During the synthesis of plant cell walls, polysaccharides are generally initially laid down. This is followed by the biosynthesis of lignin that fills the spaces between the polysaccharide fibers acting to cement them together. The filling of cell wall spaces results in at least some of the lignin having a somewhat 2D structure similar to a sheet of paper rather than the typical 3D structure for most polysaccharides and other natural macromolecules. The lignin sheets act as a barrier toward the outside elements including marauding pests as noted above.

Lignin is generally considered as being insoluble but it can be solubilized utilizing special systems, such as strong basic solutions, that disrupt the internal hydrogen bonding. It is not clear even now if the undegraded lignin is solubilized or if degradation accompanies lignin solubilization. Weight average molecular weights for some lignin samples vary depending on the type of solvent employed consistent with degradation occurring for at least some of these systems. Molecular weight values to about 10^7 have been reported for some alkali-extracted fractions.

Solubilized lignin solutions are easily oxidized and the presence of the aromatic units containing the electron-donating ether and alcohol moieties makes it available for electrophilic substitution reactions, such as nitration, halogenation, and hydroxylation.

Since its removal is the major step in the production of paper pulp, vast amounts of lignin are available as a by-product of paper manufacture. Its industrial use is just beginning, and it remains one of the major underused materials. Its sulfonic acid derivative is used as an extender for phenolic resins, as a wetting agent for oil drilling muds, and for the production of vanillin. Combined, this accounts for less than 1% of this important "by-product." In the enlightened age of green chemistry, greater emphasis on the use of lignin itself and in products derived from it must occur.

4.11   MELANINS

Light is continuous ranging from wavelengths smaller than 10^{-12} cm (gamma radiation) to greater than 10^8 cm. Radiation serves as the basis for the synthesis of many natural macromolecules via photosynthesis. Radiation is used commercially to increase the wood pulp yield through cross-linking and grafting of lignin and other wood components onto cellulosic chains. Radiation is also used in the synthesis and cross-linking of many synthetic polymers.

Radiation is also important in the synthesis and rearrangement of important "surface" macro-molecules. The tanning of human skin involves the activation of the polypeptide hormone β MSH that in turn eventually leads to the phenomena of tanning. An exposure to higher energy light from about 297–315 nm results in both tanning and burning whereas an exposure to light within the 315–330 nm region results in mainly only tanning.

UV radiation activates enzymes that modify the amino acid tyrosine in pigment-producing cells, the melanocytes. The enzyme tyrosinase, a copper-containing oxygenase, catalyzes the initial step which is the hydroxylation of tyrosine to 3,4-dihydroxyphenylalanine that is oxidized to dopaqui-none subsequently forming the melanins (Figure 4.14). The concentration of tyrosine is relatively

FIGURE 4.14   Representative structure of melanin.

high in skin protein. These modified tyrosine molecules undergo condensation forming macromolecules known as melanins. Melanins have an extended chain resonance where the pi electrons are associated with the growing melamine structure. As the melamine structure grows, it becomes more colored giving various shades of brown color to our skin. This brown coloration acts to help protect deeper skin elements from being damaged by the UV radiation. The absence of the enzyme tyrosinase, which converts tyrosine to melanin, can lead to albinism.

At least two colored melanins are formed—a series of black melanins and a series of so-called red melanins. Our skin pigmentation is determined by the relative amounts of these red and black melanins in our skin.

The concentration of melanin also contributes to the color of our hair (except for redheads where the iron-rich pigment trichosiderin dominates). The bleaching of hair, generally achieved through exposure to hydrogen peroxide, is a partial result of the oxidation of the melamine. A side reaction of bleaching is the formation of more sulfur cross-links leading to bleached hair being more brittle because of the increased cross-linking leading to a decrease in hair flexibility.

Melamin also provides a dark background in our eye's iris, is involved in animal color changes (such as the octopus and chameleon), is formed when fruit is bruised, and is partially responsible for the coloration of tea.

4.12 POLYMER STRUCTURE

Linus Pauling, in 1954, received the Nobel Prize for his insights into the structure of materials, mainly proteins. Pauling showed that only certain conformations are preferred because of inter- and intramolecular hydrogen bonding. While we know much about the structures of natural macromolecules, there is still much to become known.

Two major secondary structures are found in nature—the helix and the sheet. These two structures are also the major secondary structures found in synthetic polymers. The helix takes advantage of both the formation of intermolecular secondary bonding and the relief of steric constraints. Some materials utilize a combination of helix and sheet structures such as wool that consists of helical protein chains connected to give a "pleated" skect.

The Watson and Crick model for DNA as a double helix is only a generalized model to describe much more complex structures. Along with the typical double helix, there exist structural elements such as supercoils, kinks, cruciforms, bends, loops, and triple strands and the major and minor groves. Each of these structural elements can vary in length, shape, location, and frequency. Even the "simple" DNA double helix can vary in pitch (number of bases per helical turn), sugar pucker conformation, and helical sense (is the helix left- or right-handed).

Electron microscopy shows that DNA consists of either linear or circular structures. The chromosomal DNA in bacteria is a closed circle, a result of covalent joining of the two ends of the double helix. The chromosomal DNA in eukaryotic cells, like ours, is believed to be linear.

The most important of the secondary structures is supercoiling. *Supercoiling* simply is the coiling of a coil or in this case a coiling of the already helical DNA. The typical DNA structure is the thermally stable form. Two divergent mechanisms are believed to be responsible for supercoiling. The first, and less prevalent, is illustrated by a telephone cord. The telephone cord is typically coiled and represents the "at rest" or "unstressed" coupled DNA. As I answer the telephone, I have a tendency to twist it in one direction and after answering and hanging up the telephone for a while it begins forming additional coils. Thus, additional coiling tends to result in supercoiling. The second, and more common form, involves the presence of less than normal coiling. This can be illustrated by taking a rubber band, breaking one end, and then attaching it about a stationary object. Begin braiding the two ends until just prior to a bunching or the formation of supercoiling through over coiling. Then separate the two ends pulling them apart. The resulting strain produces supercoiling and illustrates that supercoiling results in under coiling or under winding. Thus, under winding occurs when there are fewer helical turns than would be expected. Purified DNA is rarely relaxed.

→

30 nm

→

Histone core ←

DNA-linker and "wrapper" →

FIGURE 4.15 Illustration of regularly spaced nucleosomes consisting of histone protein bound to super-coiled DNA with DNA links between the histone-bound units forming a 30-nm higher-order fiber.

Supercoiling with bacterial DNA gives a largely open, extended, and narrow, rather than compacted, multi-branched structure. By comparison, the DNA in eukaryotic cells is present in very compacted packages. Supercoiling forms the basis for the basic folding pattern in eukaryotic cells that eventually results in this very compacted structure. The subjection of chromosomes to treatments that partially unfold them show a structure where the DNA is tightly wound about "beads of proteins" forming a necklace-like arrangement where the protein beads represent precious stones imbedded within the necklace fabric (Figure 4.15). This combination forms the nucleosome, the fundamental unit of organization upon which higher-order packing occurs. The bead of each nucleosome contains eight histone proteins. Histone proteins are small basic proteins with molecular weights between 11,000 and 21,000 and specified by names such as H1, H2, and so on. H1 is especially important and its structure varies to a good degree from species to species whereas some of the other histones, such as H3 and H4, are very similar. Histones are rich in amino acid residues, arginine, and lysine.

The wrapping of DNA about a nucleosome core compacts the DNA length about sevenfold. The overall compacting though is about 10,000-fold. Additional compacting of about 100-fold is gained from the formation of the so-called 30-nm fibers. These fibers contain one H1 within the nucleosome core. This organization does not occur over the entire chromosome but rather is punctuated by areas containing sequence-specific (nonhistone containing) DNA-binding proteins. The name "30-nm fibers" occurs because the overall shape is of a fiber with a 30-nm thickness. The additional modes of compaction are just beginning to be understood but may involve scaffold assisting. Thus, certain DNA regions are separated by the loops of DNA with about 20,000–100,000 base pairs with each loop possibly containing sets of related genes.

The scaffold contains several proteins, especially H1 in the core and topoisomerase II. Both appear important to the compaction of the chromosome. In fact, the relationship between topoisomerase II and chromosome folding is so vital that the inhibitors of this enzyme can kill rapidly dividing cells and several drugs used in the treatment of cancer are topoisomerase II inhibitors.

The central theme concerning the major secondary structures found in nature is also illustrated with the two major polysaccharides derived from sucrose, i.e., cellulose and a major component of starch—amylose. Glucose exists in one of the two forms—an α form and a β form where the terms α and β refer to the geometry of the oxygen connecting the glucose ring to the fructose ring.

FIGURE 4.16 Helical structural arrangement of amylose derived from α-D-glucose units.

Cellulose is a largely linear homopolysaccharide of the β-D-glucose. Because of the geometry of the β linkage, individual cellulose chains are generally found to exist as sheets, the individual chains connected through hydrogen bonding. The sheet structure gives cellulose-containing materials good mechanical strength allowing them to act as structural units in plants. Amylose, by comparison, is a linear poly(α-D-glucose). Its usual conformation is a helix with six units per turn (Figure 4.16). Amylose is a major energy source occurring in plants as granules.

4.13 GENETIC ENGINEERING

Genetic engineering is the alteration of an organism's genetic material. The aim is to introduce into the organism's genetic material some desirable trait that is otherwise absent. The alternation of genetic material entails the use of polymer chemistry on a molecular (or nano) level making use of somewhat straightforward chemical reactions, many of the reactions employing biological entities, such as enzymes, to carry out these reactions.

Essentially, gene segments are replaced to inject, into the altered microorganism, genetic material that expresses the desired trait. Today, routine gene alteration is taught in undergraduate laboratories. Even so, specific gene alteration requires extensive planning and is conducted in major research laboratories.

In the broadest sense, genetic engineering refers to any artificial process that alters the genetic composition of an organism. Such alterations can be carried out indirectly through chemical methods, radiation, or selective breeding. Today, the term usually refers to the process whereby genes or portions of chromosomes are chemically altered.

After the alteration of a single, or a few genes, the altered genes reproduce giving much larger numbers of genes with the alternation incorporated into their genome. The term "clone" comes from the Greek work *klon*, meaning a cutting used to propagate a plant. ***Cell cloning*** is the production of identical cells from a single cell. In like manner, gene cloning is the production of identical genes form a single gene, introduced into a host cell. Today, the term cloning refers to one special type of genetic engineering.

Genes are a chromosomal portion that code for a single polypeptide or an RNA. Gene splicing is currently practiced as the enzymatic attachment of one gene or gene segment to another gene or gene segment. Genes are composed of DNA, which can be considered as a specialized polyphosphate polymer. The manipulation of DNA can occur for many reasons. One of these is the production of recombinant DNA. Here we will focus on the production of recombinant DNA. DNA cannot be directly transferred from one organism, the donor, to another recipient organism, the host. Instead, the donor DNA segment is cut and then recombined with DNA from a host. *Escherichia coli*, *E. coli*, is typically employed as the host cell since it is itself a harmless bacterium that reproduces rapidly. (However, under the wrong conditions *E. coli* is responsible for many types of food poisoning.) The *E. coli* then acts as a "factory" that reproduces bacteria that contain the desired modification.

Enzymes, specialized proteins, are used as designing tools for the genetic engineering. One of these enzyme tools consists of *restriction endonucleases* that recognize a specific series of base pairs. They split the DNA at these specific points. This splitting is called "lysing," which in reality is simply the hydrolysis of DNA units as shown below.

$$\text{(4.43)}$$

Organisms produce restriction endonucleases that are specific for that organism. Certain restriction endonucleases cut double-stranded DNA asymmetrically in regions called palindromes, that is regions that "read" (have identical sequences) the same way from left to right on one strand as right to left on the other strand. This produces what is referred to as "sticky ends" that form not only a "cleft" for attachment but also a single-stranded end that has the ability to pair with another complementary single-stranded stranded end. Both strands of the original donor twin strand have a tendency to recombine with the complementary strands of DNA from a host that has been treated to produce the complementary strands. The sticky ends, when mixed under the proper conditions in the presence of another enzyme, DNA ligase, combine. The hydrogen bonding between complementary sticky ends reinforce the recombination reaction. The resulting recombination reaction results in a variety of products including the desired recombination of the host DNA and the donor DNA as well as the combination of the original donor strands and uncombined DNA. The mixture is often treated in one of the two manners. The simplest case requires a chemical-resistant gene that is resistant to the employed chemical agent such as tetracycline. The desired recombinant genes survive and are then transferred into the host organism so that the new gene can express itself.

In some cases, such as the synthesis of insulin, the recombination mixture is added to a host organism, here *E. coli*. This infected mixture is then plated out and the individual colonies tested for insulin production. Those colonies that produce insulin are further plated out and grown for mass insulin production.

Cells that accept the recombinant DNA are called ***transformed***. More specialized sequences have been developed to increase the probability of gene incorporation and its successful reproduction.

A second tool employed by the genetic engineer is the enzyme terminal transferase that adds deoxyribonucleotide residues to the 3′ end of DNA strands creating 3′ tails of a single type of residue.

Special modified *plasmid* DNAs, called ***vectors*** or carriers, are used as host or targets for gene modification. These circularly shaped vectors reproduce autonomously in host cells. Plasmids have two other important properties. First, they can pass from one cell to another allowing a single "modified" bacterial cell to inject neighboring bacterial cells with this "modification." Second, gene material from other cells can be easily formed into plasmids, allowing ready construction of modified carriers.

The steps involved in gene splicing, emphasizing the chemical nature of the individual steps, are as follows:

1. Lysing (which is really simply the hydrolysis of DNA units as shown above)
2. Construction of staggered, sticky ends
3. Recombination or lysation, the reverse of lysing, chemically formation of a phosphate ester as below connecting the desired segment to the DNA of the host cell

$$
R_1\!-\!OH \;+\; \begin{array}{c} R \\ \backslash \\ O \end{array}\!-\!\overset{\displaystyle O}{\underset{\displaystyle O^-}{\overset{\|}{P}}}\!-\!OH \longrightarrow \begin{array}{c} R \\ \backslash \\ O \end{array}\!-\!\overset{\displaystyle O}{\underset{\displaystyle O^-}{\overset{\|}{P}}}\!-\!O\!\!\begin{array}{c} \\ \backslash \\ R_1 \end{array} \;+\; H_2O \qquad (4.44)
$$

4. Chemical recombination of vector—insertion into the host cell; recombining plasmid genes into the host genetic complement
5. Replication of host cell

There are many uses of recombinant DNA. As noted above, one technique that produces recombinant DNA is called cloning. In one cloning technique used for the production of the sheep Dolly in 1996, the DNA nucleus from a female's egg is replaced with the nucleus from another sheep. The egg is placed in the uterus of a third animal, known as the surrogate mother. Dolly is nearly genetically identical to the animal from which the nucleus was obtained but not genetically related to the surrogate mother.

Recombinant DNA has been used in a variety of ways. The growth hormone gene of rainbow trout has been transferred into carp eggs resulting in the transgenic carp producing larger fish. The milk production of dairy cows has been increased by cloning and introducing into the cows the cattle growth hormone bovine somatotropin.

Transgenic strawberry and potato plants have been produced that are frost resistant. Cotton, corn, and soybean plants have been produced with an increased resistance to herbicides allowing herbicide use without killing the transgenic crop-producing plants. Larger and smaller varieties of other food-producing plants have been produced using recombinant DNA as have plants that produce certain amino acids needed for our nutrition.

Transgenic bacteria have been produced that can metabolize petroleum products including certain synthetic polymers.

Along with the production of insulin, many other medical uses have been achieved for recombinant DNA. This includes the production of *erythropoetin*, a hormone used to stimulate the production of red blood cells in anemic people; tissue *plasminogen activator* an enzyme that dissolves blood clots in heart attack victims; and *antihemophilic human factor VIII*, used to prevent and control bleeding for people with hemophilia. These three important genetically engineered proteins were all cloned in hamster cell cultures.

Gene engineering is the basis of gene therapy where genes are removed, replaced, or altered producing new proteins for the treatment of such diseases as muscular dystrophy, some cancers, adenosine deaminase deficiency, cystic fibrosis, and emphysema.

4.14   DNA PROFILING

DNA profiling is also referred to as DNA fingerprinting and DNA typing. It is used in paternity identification, classification of plants, criminal cases, identification of victims, heredity (of living, recently deceased, and anciently deceased), etc. DNA profiling is a tool that allows a comparison of DNA samples. There are two basic types of DNA profiling; one that uses **polymerase chain reaction (PCR) enzymes** and the second employs the **restriction fragment length polymorphism**

(RFLP) enzymes. The PCR approach utilizes a sort of molecular copying process where a specific region is selected for investigation. The PCR approach requires only a few nanograms of DNA. The DNA polymerase makes the copies of DNA strands in a process that mimics the way DNA replicates naturally within the cell. Segments are selected for special study and the results used to identify the DNA pattern.

With the exception of identical twins, each individual has a DNA profile that is unique. In excess of 99% of the over 3 billion nucleotides in human DNA are the same. But, for every 1000 nucleotides there is one site of variation or polymorphism. These DNA polymorphisms change the length of the DNA fragments produced by certain restriction enzymes. The resulting fragments are called restriction fragments length polymorphisms, or RFLPs. Gel electrophoresis is typically employed to separate the sizes and thus create a pattern of RFLLPs. The number and size of the fragments is used to create the DNA profile.

Several steps are involved in creating the genetic fingerprint. First, a sample of cells is obtained from a person's blood, bone, semen, hair roots, or saliva. The individual cells from the sample are split open, and the DNA is isolated. The DNA is treated with restriction enzymes that cleave the DNA strands at specific locations creating the fragments of varying lengths and compositions. The resulting fragments undergo electrophoresis using a gel, which allows the separation of the fragmented DNA. Because the gel is fragile, a thin nylon membrane, covered by a towel, is laid over the gel. As moisture is drawn to the towel from the electrophoresis gel, the DNA fragments arc transferred to the nylon membrane. This process is called blotting. The DNA bands that binds to select fragments are visible to the eye but they are too numerous to be useful. Thus, a radioactive solution is washed over the nylon membrane that binds to like fragments, generally to only 6–20 of the DNA clusters. A sheet of photographic film is placed on the top of the nylon membrane that records these cluster sites. The film is then developed producing a pattern of thick-and-thin bands. This pattern is the genetic pattern for that particular sample. This process can take a month or more at commercial labs for routine analysis, but when needed, the analysis can be made in only a day or two.

There are different restriction enzymes that cut DNA at different sites. The previous sequence can be repeated several times for the same DNA sample. From a study of each restriction enzyme, a probability that another person will have the same profile is assigned. Thus, one restriction enzyme may have the possibility that another person has the same match of 1 in 100 or 1%. A second restriction enzyme may have the probability of 1 in 1000 or 0.1%. A third restriction enzyme may have a probability for a match being 1 in 500 or 0.2%. If there were a match with all three restriction enzymes, the probability would be $0.01 \times 0.001 \times 0.002$ or 0.00000002 or 0.000002% or 1 part in 50,000,000. There is a caution to using the "multiplication rule," in that DNA sequences are not totally random. In fact, DNA sequence agreements generally diverge as one's ancestors are less closely related.

The RFLP method requires a sample about 100 times larger than required for the PCR approach, but with repeated sequences using different restriction enzymes, RFLP is more precise.

It must be noted that factors leading to DNA degradation, such as moisture, chemicals, bacteria, heat, and sunlight, will impact negatively on DNA profiling since the precise sequences and length of the DNA and DNA fragments may be changed. While DNA, in general, is robust and can exist "alive" over thousands of years (such as the germination of seeds found in the pyramids of Egypt), DNA degradation decreases the probability of precise matches. Also, DNA contamination due to the addition of DNA from another source greatly confuses the final results.

DNA sequencing has found importance in a wide range of areas. It is being used to identify individuals at greater risk for having certain diseases, such as breast cancer. It has allowed taxonomists to determine evolutionary relationships among plants, animals, and other life forms. Currently, it is a basis for the so-called Eve theory that says that all of us are related to a common woman, called Eve after the Biblical Eve. It is also being used to trace the (ancient) movement of people about Earth. It is used for the screening of certain diseases such as the presence of the sickle-cell gene.

DNA profiling was used to determine whether unearthed bones that were said to be from Jesse James were in fact his. DNA samples were taken from grandchildren and compared to those obtained from the bone material and shown to be similar, so that while it cannot be absolutely said the bones were from Jesse James, DNA evidence was consistent with them being his bones.

4.15 ASPHALT

Nearly 100 billion pounds of asphalt are used annually in the United States. It is the viscous material that is not easily distilled as are more volatile fractions such as gasoline and kerosene are removed from crude oil. They are also found in natural deposits such as the La Bera Tar Pits in California. It is sometimes referred to as bitumen, covered in Section 10.14. Here, we will cover additional material related to asphalt.

About 80% of asphalt is employed along with rocks to form our roadways. The asphalt acts as the binder for the rock aggregates similar to concrete. Asphalt concrete is the most widely used recycled material, on a weight and percentage basis, in the United States. Over 80% of asphalt from road surfaces removed from resurfacing projects is reused as a component of the new construction. Roofing shingles are also a major use of asphalt.

Asphalt is a viscoelastic material and this viscoelasticity must be present over the temperature range where the material is being used. Unfortunately, the asphalt often softens during the summer days allowing rutting or permanent deformation to occur. During the winter, the asphalt chains tend to become more organized resulting in a more brittle material that can undergo thermal and fatigue cracking. With time, the more volatile, shorter chains leave the mixture causing the material to lose some of its flexibility. This process is called aging.

Asphalt emulsions are made by mixing the asphalt generally with the petroleum solvents but some formulations include the more environmentally acceptable water. These emulsions are often employed to resurface concrete and asphalt surfaces such as driveways.

Along with short-chained polymers with up to 150 carbons, asphalt contains a variety of other elements. The composition varies as to the source and the treatment of the crude oils. Crude oil from Venezuela contains a particularly high amount of metals, such as iron, nickel, and vanadium, in comparison to crude oil from the Middle East. Some of the asphalts contain polar functional groups such as alcohols, amines, thiols, and carboxylic-containing units. This allows the asphalt chains to aggregate similar to many detergents with the aliphatic units reaching outward from a polar core. For rocks with a mildly polar surface, this results in some binding between the rock surfaces and the asphalt resulting in a more secure material.

4.16 SUMMARY

1. Physically, there is little difference in the behavior, study, or testing of natural and synthetic polymers. The fundamental principles that underpin the behavior of macromolecules apply equally to both synthetic and natural polymers. Both groupings contain members that exhibit characteristics that are unique to that grouping. Even so, differences within even these groupings are by degree rather than kind, with the fundamental laws continuing to be applicable.
2. Contributions from studying both the natural and synthetic polymers are being used to forward the science of both sets of macromolecules.
3. Organic polymers are responsible for the very life—both plant and animal—that exists. Their complexity allows for the variety that is necessary for life to occur, to reproduce, and to adapt.

    The structures of largely linear natural and synthetic polymers can be divided into primary structures used to describe the particular sequence of (approximate) repeat units. Secondary structure is used to describe the molecular shape or conformation of the polymer. Tertiary structure describes the shaping or folding of macromolecules. Quaternary structure gives the overall shape to the groups of tertiary-structured macromolecules. The two basic secondary structures are helixes and sheets.

4. Polysaccharides are the most abundant, on a weight basis, naturally occurring organic polymer. They are truly complex molecules with most structures representable on only an "average" basis. They are diverse with respect to both occurrence and size. The most important polysaccharides, on a knowledge, weight, and use basis are cellulose and starch. Cellulose is composed largely of D-glucose units that are associated with other materials such as lignin. Cotton is one of the purest forms of cellulose, though we get most of our cellulose from wood pulp. Cellulose becomes soluble in aqueous solutions only when the hydrogen bonding within the cellulose is broken.

5. While cellulose is used extensively itself in the form of paper and paper-based products, it also forms the basis of a number of synthetic products including a variety of acetylated products such as rayon.

6. Starch is composed of two major components—amylopectin and amylose. It is the second most abundant polysaccharide and found largely within plants.

7. Proteins are composed of 20 different α-amino acids and contain peptide linkages similar to those present in polyamides. With the exception of glycine, all the amino acids contain a chiral carbon. The geometrical shape and behavior of the giant proteins is a product of the various preferred geometries that allow the molecules to balance factors such as preferred bond angle, secondary bonding forces—emphasizing hydrogen bonding, size, shape, hydrophobic/hydrophilic interactions, external/internal chemical environments, and cross-linking. The small chains of amino acids are referred to as peptides.

8. Secondary structures for proteins are generally fibrous and globular. Proteins such as keratins, collagen, and elastin are largely fibrous and have secondary structures of sheets and helixes. Many of the globular proteins are composed of protein chains present in secondary structures approximating helixes and sheets.

9. Enzymes are one important group of proteins. They serve as natural catalysts immobilizing various components that will be later joined or degraded.

10. The two major types of nucleic acids are deoxyribonucleic acid, DNA, and ribonucleic acid, RNA. Nucleic acids are polyphosphate esters containing the phosphate, the sugar, and the base moieties. Nucleic acids contain one of the five purine/pyrimidine bases that are coupled within double-stranded helixes. DNA, which is an essential part of the chromosome of the cell, contains the information for the synthesis of protein molecules. For double-stranded nucleic acids, as the two strands separate, they act as a template for the construction of a complementary chain. The reproduction or duplication of the DNA chains is called replication. The DNA undergoes semiconservative replication where each of the two new strands contains one of the original strands.

11. The flow of biological genome knowledge is from DNA to RNA via transcription and from RNA to direct protein synthesis via translation.

12. Polyisoprenes form the basis of many natural polymers including the two most widely known natural plastics, gutta percha and balata, and an elastomer known as *Hevea brasiliensis*, or natural rubber. The hard plastics, balata and gutta percha, are *trans* isomers while natural rubber is the *cis* isomer of 1,4-polyisoprene. The polymer chain of amorphous natural rubber and other elastomers uncoils during stretching and returns to its original low-entropy form when the material is above its T_g. Chain slippage is minimized through the presence of cross-links. Unlike other solids, stretched rubber contracts when heated. The long-range elasticity is dependent on the absence of strong intermolecular forces. Eventually natural rubber formed the basis of the rubber industry through gaining knowledge of factors such as cross-linking that allowed the material to remain coherent but elastomeric even when exposed to moderate temperatures.

13. Genetic engineering is based on chemical manipulations that are exactly analogous to those carried out by chemist in basic chemistry laboratories, but it involves the use of biological agents.

14. Melamines are polymeric agents that play many roles in nature including forming our skin pigmentation and with its growth through an exposure to sunlight, acting to protest us from the harmful effects of the sunlight.

GLOSSARY

Accelerator: Catalyst for the vulcanization of rubber.

Alkali cellulose: Cellulose that has been treated with a strong basic solution.

α-Cellulose: Cellulose that is not soluble in 17.5% basic solution.

α-Helix: Right-handed helical conformation.

Amylopectin: Highly branched starch polymer with branches or chain extensions on carbon 6 of the anhydroglucose repeating unit.

Amylose: Linear starch polymer.

Antioxidant: Compound that retards polymer degradation.

Balata: Naturally occurring *trans*-1,4-polyisoprene.

β-Arrangement: Pleated sheetlike conformation.

β-Cellulose: Cellulose soluble in 17.5% basic solution but not soluble in 8% caustic solution.

Boeseken–Haworth projections: Planar hexagonal rings used for simplicity instead of staggered chain forms.

Carbohydrate: Organic compound often with an empirical formula CH_2O; sugar, starch, and cellulose are carbohydrates.

Carbon black: Finely divided carbon used for the reinforcement of rubber.

Carboxymethylhydroxyethylcellulose (CMHEC): Made from the reaction of sodium chloroacetate with hydroxethylcellulose.

Carrageenin: Mixture of several polysaccharides containing D-galactose units; obtained from seaweed.

Casein: Milk protein.

Cellobiose: Repeat unit in cellulose.

Cellophane: Sheet of cellulose regenerated by the acidification of an alkaline solution of cellulose xanthate.

Celluloid: Product from a mixture of CN and camphor.

Cellulose: Linear polysaccharide consisting of many anhydroglucose units joined by β-acetal linkages.

Cellulose acetate: Product from the acetylation of cellulose.

Cellulose nitrate: Made from the reaction of cellulose and concentrated nitric acid; also known as gun cotton.

Cellulose xanthate: Product of soda cellulose and carbon disulfide.

Chitin: Polymer of acetylated glucosamine present in the exoskeletons of shellfish.

Collagen: Protein present in connective tissue.

Compounding: Processing of adding essential ingredients to a polymer such as rubber.

C-terminal amino acid: Amino acid with a carboxylic acid end group.

Cyclodextrins: Oligomeric cyclic products formed from the reaction of starch treated with a certain enzyme.

Degree of substitution (DS): Number that designates the average number of reacted hydroxyl groups in each anhydroglucose unit in cellulose or starch.

Denaturation: Change in conformation of a protein resulting from heat or chemicals.

Deoxyribonucleic acid (DNA): Nucleic acid in which deoxyribose unit are present; compose the human genome.

Dextran: Branched polysaccharide synthesized from sucrose by bacteria.

DNA profiling: Identification method based on variations between the DNA of individuals.

Drying: Jargon used to describe the cross-linking of unsaturated polymers in the presence of air and a heavy metal catalyst (drier).

Elastin: Protein that is the major material of arterial blood vessels and ligaments that is noted for its flexibility.

Enzyme: Protein with specific catalytic activity.

Ester gum: Ester of rosin and glycerol.

Ethylhydroxyethylcellulose (EHEC): Nonionic mixed ether formed from HEC and ethyl chloride.

Fibrillar protein: Hairlike, insoluble, intermolecularly hydrogen-bonded protein.

Fibrils: Threadlike strands or bundles of fibers.

Fossil resins: Resins obtained from the exudate of prehistoric trees.

Galalith: Commercial casein plastics.

γ-Cellulose: Cellulose soluble in 8% caustic solution.

Genome: Total genetic information contained in a cell, an organism, or a virus.

Globular proteins: Proteins with an overall globular structure formed from the contributions of secondary structures including sheets and helices.

Glycine: Simple and only nonchiral α-amino acid.

Glycogen: Highly branched polysaccharide, which serves as the reserve carbohydrate in animals.

Guayule: Shrub that produces *cis*-1,4-polyisoprene rubber.

Gutta percha: Naturally occurring *trans*-1,4-polyisoprene.

***Hevea Brasiliensis*:** Natural rubber, NR.

Humic acid: Polymeric aromatic carboxylic acid found in lignite.

Hydrocellulose: Cellulose degraded by hydrolysis.

Hydroxyethylcellulose (HEC): Produced from alkaline cellulose and ethylene oxide.

Hydroxypropylcellulose: Thermoplastic cellulose ether formed from alkaline cellulose and propylene oxide.

Isoelectric point: pH at which an amino acid does not migrate to either the positive pole or the negative pole in a cell.

Ketatin: Fibrillar protein.

Latex: Stable dispersion of polymer particles in water.

Lignin: Noncellulosic resinous component of wood.

Mercerized cotton: Cotton fiber that has been immersed in caustic solution, usually under tension, and washed with water removing the excess caustic.

Methylcellulose (MC): Formed from alkaline cellulose and chloromethane.

Myosin: Protein present in muscle.

Native cellulose: Naturally occurring cellulose; like cotton.

N-terminal amino acid: Amino acid with an amino end group

Nucleoside: Contains a pentose and a base.

Nucleotide: Contains a phosphate, a pentose, and a base.

Oligosaccharide: Low-molecular weight polysaccharide.

Polynucleotide: Nucleic acid.

Polypeptide: Protein; often used for low molecular weight proteins.

Polysaccharide: Polymer containing many sugar units; like cellulose and starch.

Primary structure: Term used to describe the primary configuration present in a protein chain.

Prosthetic group: Nonprotenious group conjugated (connected) to a protein.

Protein: Polyamide in which the building blocks are α-amino acids joined by peptide (amide) linkages.

Purine base: Compounds consisting of two fused heterocyclic rings, namely, a pyrimidine and an imidazole ring; essential part of nucleic acids.

Pyrimidine: A 1,3-diazine; essential part of nucleic acids.

Racked rubber: Stretched rubber cooled below its T_g.

Rayon: Cellulose regenerated by acidification of a cellulose xanthate (viscose) solution.

Recent resins: Resins obtained from the exudate of living trees.

Regenerated cellulose: Cellulose obtained by precipitation from solution.

Replication: Term used to describe duplication such as the duplication of DNA.

Retrogradation: Process whereby irreversible gel is produced by the aging of aqueous solutions of amylose starch.

Ribonucleic acid (RNA): Nucleic acid in which ribose units are present. Essential units of life and replication.

Schweitzer's solution: Ammoniacal solution of copper (II) hydroxide.

Secondary structure: Term used to describe the conformation of a protein molecule such as a helix.

Shellac: Natural polymer obtained from the excreta of insects in Southeast Asia.

Starch: Linear or branched polysaccharides of many anhydroglucose units joined by an α linkage; amylose is the linear fraction and amylopectin is the branched fraction.

Tenacity: Term used for the tensile strength of a polymer.

Terpene: Class of hydrocarbons having the empirical formula C_5H_8.

Tertiary structure: Term used to describe the shape or the folding of a protein.

Transcription: Term used to describe the transfer of information from DNA to RNA.

Translation: Term used to describe the transfer of information from RNA to protein synthesis.

Viscose: Alkaline solution of cellulose xanthate.

Zwitterion: Dipolar ion of an amino acid.

EXERCISES

1. Why is starch digestible by humans? Why is cellulose not digestible by humans?
2. How does cellobiose differ from maltose?
3. Why is cellulose stronger than amylose?
4. How does the monosaccharide hydrolytic product of cellulose differ from the hydrolytic product of starch?
5. What is paper made from?
6. How many hydroxyl groups are present on each anyhydroglucose unit in cellulose?
7. Which would be more polar—tertiary or secondary cellulose acetate?
8. Why would you expect chitin to be soluble in hydrochloric acid?
9. Which is more apt to form a helix: (a) amylose or (b) amylopectin?
10. Why is amylopectin soluble in water?
11. Define a protein in polymer science language.
12. Which α-amino acid does not belong to the L-series of amino acids?
13. To which pole will an amino acid migrate at a pH above its isoelectric point?
14. Why is collagen stronger than albumin?
15. What are the requirements for a strong fiber?
16. Which protein would be more apt to be present in a helical conformation: (a) a linear polyamide with small pendant groups, or (b) a linear polyamide with bulky pendant groups?
17. What is the difference between the molecular weight of (a) ribose and (b) deoxyribose?
18. What is the repeating unit in the polymer DNA?
19. Which is more acidic: (a) a nucleoside or (b) a nucleotide?
20. What base found in DNA is not present in RNA?
21. Why would you predict helical conformations for RNA and DNA?
22. If the sequence of one chain of a double helix of DNA is ATTACGTCAT, what is the sequence of the adjacent chain?

23. Why is it essential to have trinucleotides rather than dinucleotides as codons for directing protein synthesis?
24. How do the configurations differ for (a) gutta percha and (b) natural rubber?
25. The formation of what polymer is responsible for tanning?
26. Will the tensile force required to stretch rubber increase or decrease as the temperature is increased?
27. Does a stretched rubber band expand or contract when heated?
28. List three requirements for an elastomer.
29. Why is there an interest in the cultivation of guayule?
30. Are the polymerization processes for synthesis and natural *cis*-polyisoprene (a) similar or (b) different?
31. What does the presence of C_5H_8 units in NR indicate?
32. Why is *E. coli* most often used in gene splicing?
33. Why does a rubber band become opaque when stretched?
34. What is the most important contribution to retractile forces in highly elongated rubber?
35. What is present in so-called vulcanized rubber compounds?
36. When a rubber band is stretched, what happens to its temperature?
37. Why are natural plastics not used more?
38. What type of solvent would you choose for shellac?
39. Why is lignin sometimes referred to as being a two-dimensional polymer?
40. Which is a polymer: (a) rosin or (b) ester gum?
41. If the annual production of paper is over 100 million ton, how much lignin is discarded from paper production annually?
42. Might an article molded from Galalith be valuable?

FURTHER READINGS

Bloomfield, V., Crothers, D., and Tinoco, I. (2000): *Nucleic Acids*, University Science Books, Sausalito, CA.
Carraher, C. and Sperling, L. H. (1983): *Polymer Applications of Renewable-Resource Materials*, Plenum, New York.
Chiellini, E. (2001a): *Biorelated Polymers*, Kluwer, New York.
Chiellini, E. (2001b): *Biomedical Polymers and Polymer Therapeutics*, Kluwer, New York.
Coates, J. and Carraher, C. (1983): *Polym. News*, 9:77.
Dumitriu, S. (2001): *Polymeric Biomaterials*, Dekker, New York.
Ferre, F. (1997): *Gene Quantification*, Springer-Verlag, New York.
Gar, H., Cowman, M., and Hales, C. (2008): *Carbohydrate Chemistry: Biology and Medical Applications*, Elsevier, New York.
Gebelein, C. and Carraher, C. (1995): *Industrial Biotechnological Polymers*, Technomic, Lancaster, PA.
Kaplan, D. (1994): *Silk Polymers*, ACS, Washington, DC.
Kennedy, J., Mitchell, J., and Sandford, P. (1995): *Carbohydrate Polymers*, Elsevier, New York.
Malsten, M (2003): *Biopolymers at Interfaces*, Dekker, New York.
McGrath, K. and Kaplan, D. (1997): *Protein-Based Materials*, Springer-Verlag, New York.
Merrifield, R. B. (1975): Solid phase peptide synthesis, *Polym. Prep.*, 16:135.
Paulsen, B. (2000): *Bioactive Carbohydrate Polymers*, Kluwer, New York.
Scheirs, J. and Narayan, R. (2010): *Biodegradable Plastics: A Practical Approach*, Wiley, Hoboken, NJ.
Scholz, C. and Gross, R. (2000): *Polymers from Renewable Resources*, ACS, Washington, DC.
Steinbuckel, A. (2001a): *Lignin, Humic, and Coal*, Wiley, New York.
Steinbuchel, A. (2001b): *Polyisoprenoides*, Wiley, New York.
Steinbuchel, A. (2003): *Biopolymers, Miscellaneous Biopolymers and Biodegradation of Synthetic Polymers*, Wiley, Hoboken, NJ.
Tsuji, H. (2008): *Degradation of Poly(Lactide)*, Nova, Hauppauge, NY.
Wuisman, P. and Smit, T. (2009): *Degradable Polymers for Sketal Implants*, Nova, Hauppauge, NY.
Yu., L. (2008): *Biodegradable Polymer Blends and Composites from Renewable Resources*, Wiley, Hoboken, NJ.

5 Step-Reaction Polymerization (Polycondensation Reactions)

In Chapter 2, we noted that there is a direct relationship between polymer structures and their physical properties. In this chapter we examine condensation polymers that primarily have relatively strong attractions, secondary bonding, between and within polymer chains. Because of these strong secondary-bonding attractions, these polymers are typically strong requiring relatively a lot of force to elongate them. Many of them are synthesized so that there is good symmetry allowing the chains to more closely pack together, further enhancing the magnitude of the secondary bonding and increasing the glass transition of the materials. When these materials are allowed to form mainly crystals, they can act as fibers and when they are processed so that the noncrystalline fraction is large, they can act as plastics.

5.1 COMPARISON BETWEEN THE POLYMER TYPE AND THE KINETICS OF POLYMERIZATION

There is a large, but not total, overlap between the terms condensation polymers and stepwise kinetics and the terms addition (or vinyl) polymers and chain kinetics. In this section, we will look at each of these four terms and illustrate the similarities and differences between them.

The terms addition and condensation polymers were first proposed by Carothers and are based on whether the repeat unit of the polymer contains the same atoms as the monomer. An addition polymer has the same atoms as the monomer in its repeat unit:

$$H_2C{=}CH \quad \longrightarrow \quad -(-CH_2{-}CH-)_n- \qquad (5.1)$$

with X substituents shown above each carbon

where "X" is CH_3 for polypropylene, H for polyethylene, and C_6H_5 for polystyrene.

The atoms in the backbone of addition polymers are almost always only carbon.

Condensation polymers generally contain fewer atoms in the polymer backbone than in the reactants because of the formation of by-products during the polymerization process, and the backbone contains noncarbon atoms:

$$X{-}A{-}R{-}A{-}X + Y{-}B{-}R'{-}B{-}Y \quad \longrightarrow \quad -(-A{-}R{-}A{-}B{-}R'{-}B{-})- + XY \qquad (5.2)$$

where A–X can be most Lewis bases such as $-NH_2$, $-SH$, $-OH$ and B–Y can be Lewis acids such as

$$-\overset{O}{\underset{}{\overset{\|}{C}}}{-}OH, -\overset{O}{\underset{}{\overset{\|}{C}}}{-}Cl, -\overset{O}{\underset{R}{\overset{\|}{P}}}{-}Cl, -\overset{O}{\underset{O}{\overset{\|}{\underset{\|}{S}}}}{-}Cl$$

The corresponding polymerizations to form these polymers are typically addition and condensation polymerizations.

The term *stepwise kinetics*, or *step-growth kinetics*, refers to polymerizations in which the molecular weight of the polymer increases in a slow, stepwise manner as reaction time increases. Figure 5.1 contains a depiction of the stepwise polymerization process. Initially, only monomers R and R_1 are present (far left). After some time, all the monomer has reacted, yet no high molecular weight chains are found with the greatest $\overline{DP}$ being 2.5 (middle). As polymerization continues, longer chains grow with a high $\overline{DP}$ of 10 (far right) for this depiction. Eventually, all the chains will connect finally giving a polymer. Figure 5.2 contains a representative plot of $\overline{DP}$ as a function of time for this process.

The formation of polyesters from a dialcohol (diol) and a dicarboxylic acid (diacid) will be used to illustrate the stepwise kinetic process. Polymer formation begins with one diol molecule reacting with one diacid forming one repeat unit of the eventual polyester (Equation 5.3):

$$\text{HO}-\overset{\overset{\text{O}}{\|}}{\text{C}}-\text{R}-\overset{\overset{\text{O}}{\|}}{\text{C}}-\text{OH} + \text{HO}-\text{R}'-\text{OH} \rightleftarrows \text{HO}-\overset{\overset{\text{O}}{\|}}{\text{C}}-\text{R}-\overset{\overset{\text{O}}{\|}}{\text{C}}-\text{O}-\text{R}'-\text{OH} + \text{H}_2\text{O} \qquad (5.3)$$

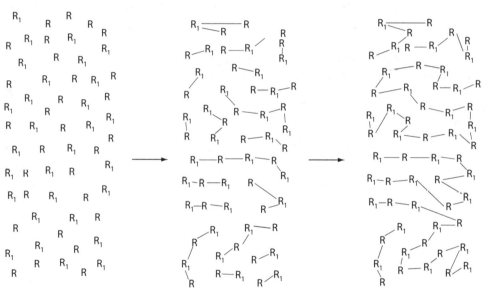

FIGURE 5.1 Depiction of stepwise chain growth for monomers R and R_1 as the polymerization begins (far left) and progressing toward the right.

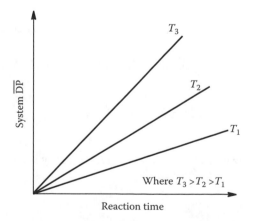

FIGURE 5.2 System molecular weight for stepwise kinetics as a function of reaction time and reaction temperature, T.

This ester unit can now react with either an alcohol or acid group producing chains ending with either two active alcohol functional groups or two active acid groups (Equation 5.4):

$$+ \text{HO—R'—OH} \rightleftharpoons \text{HO—R'—O—C(O)—R—C(O)—O—R'—OH} + H_2O$$

HO—C(O)—R—C(O)—O—R'—OH (5.4)

$$+ \text{HO—C(O)—R—C(O)—OH} \rightleftharpoons \text{HO—C(O)—R—C(O)—O—R'—O—C(O)—RC(O)—OH} + H_2O$$

The chain with two alcohol ends can now condense with a molecule containing an acid end and the chain with two acid ends can now condense with a molecule containing an alcohol group resulting in molecules that contain one acid and one alcohol group. This reaction continues through the reaction matrix whenever molecules with the correct functionality, necessary energy of activation, and correct geometry collide. The net effect is the formation of dimers, trimers, etc., until the polymer is formed:

$$\text{HO—R'—O—C(O)—R—C(O)—O—R'—OH} + \text{HO—C(O)—R—C(O)—OH}$$
$$\rightleftharpoons \text{HO—R'—O—C(O)—R—C(O)—O—R'—O—C(O)—R—C(O)—OH} + H_2O \quad (5.5)$$

$$\text{HO—C(O)—R—C(O)—O—R'—O—C(O)—RC(O)—OH} + \text{HO—R'—OH}$$
$$\rightleftharpoons \text{HO—C(O)—R—C(O)—O—R'—O—C(O)—RC(O)—O—R'—OH} + H_2O \quad (5.6)$$

Thus, the reactants are consumed with few long chains formed until the reaction progresses toward total reaction of the chains with themselves. Thus, polymer formation occurs one step at a time, hence the name "stepwise" kinetics.

Chain-growth reactions require initiation to begin chain growth. We will look at the free radical growth forming polystyrene to illustrate chain growth. Here, the initiation of a styrene molecule, R, will illustrate the chain-growth process. The initiator reacts with a styrene monomer creating a free radical active chain end. This free radical chain end then reacts with another styrene monomer that, in turn, reacts with another styrene monomer, etc., until termination stops chain growth with the formation of a polystyrene polymer chain (Figure 5.3). Polystyrene chains are formed from the beginning of the polymerization process with growth occurring through the formation of polymer

FIGURE 5.3 Molecular weight for chain-growth kinetics as a function of reaction time beginning with only monomer (R; far left) with reaction progressing toward the right.

chains. As the first chain is formed (Figure 5.3; middle) only polymer and monomer are present. As additional polymer chains are formed (Figure 5.3; right), the polymer mix still consists of only polymer chains and unreacted monomer. This is shown in the following sequences:

$$\text{(5.7)}$$

$$\text{(5.8)}$$

$$\text{(5.9)}$$

$$\text{(5.10)}$$

The average $\overline{DP}$ for the entire system, neglecting unreacted monomer, does not markedly change as additional polymer is formed (Figure 5.4). Moreover, unlike the stepwise process, average chain length decreases with increase in reaction temperature. More information about this is presented in Chapters 6 and 7.

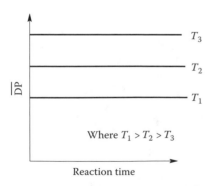

FIGURE 5.4 Idealized average molecular weight of formed polymer as a function of reaction time and temperature, T, for chain-type polymerization.

Both the stepwise and chain-wise polymerizations produce polymers that are polydisperse with respect to chain lengths.

Most addition polymers are formed from polymerizations exhibiting chain-growth kinetics. This includes the typical polymerizations, via free radical or some ionic mode, of the vast majority of vinyl monomers such as vinyl chloride, ethylene, styrene, propylene, methyl methacrylate, and vinyl acetate. By comparison, most condensation polymers are formed from systems exhibiting stepwise kinetics. Industrially this includes the formation of polyesters and polyamides (nylons). Thus, there exists a large overlap between the terms stepwise kinetics and condensation polymers and chain-wise kinetics and addition (or vinyl) polymers. A comparison of the two types of systems is given in Table 5.1.

TABLE 5.1

Comparison between Stepwise and Chain-Wise Polymerizations

Chain	Step
Growth occurs by addition of one unit at a time	Any two unlike molecular units can react
Monomer concentrations decrease steadily throughout the polymerization	Monomer disappears early in the reaction
Polymer chains are formed from the beginning of the polymerization and throughout the process	Polymer chain length increases steadily during the polymerization
Average chain length for reacted species remains approximately constant throughout the polymerization	Average molecular weight for the reaction (for the reacted species) increases throughout the polymerization
As reaction time is increased polymer yield increases, but molecular weight remains about the same	High "extents" of reaction are required to obtain high chain lengths
Reaction mixture contains almost only unreacted monomer, polymer, and very little growing polymer chains	Reaction system contains various stages, chain lengths, of product present in a calculable distribution

Even so, there is not a total overlap between the various characteristics of vinyl-chain kinetics and condensation-step kinetics. Following are examples illustrating the lack of adherence to this overlap:

1. The formation of polyurethanes and polyureas typically occur through stepwise kinetics with the polymer backbones clearly containing noncarbon atoms. Yet, no by-product is released through the condensation process because condensation occurs through an internal rearrangement and shift of the hydrogen—neither step resulting in the expulsion of a by-product.

$$\text{OCN—R—NCO} + \text{HO—R'—OH} \rightarrow \text{—(—}\overset{\overset{\displaystyle O}{\|}}{C}\text{—NH—R—NH—}\overset{\overset{\displaystyle O}{\|}}{C}\text{—O—R'—O—)—} \qquad (5.11)$$

Polyurethane

2. Internal esters (lactones) and internal amides (lactams) are readily polymerized through a chain-wise kinetic process forming polyesters and polyamides, clearly condensation polymers with respect to having noncarbons in the backbone, but without the expulsion of a by-product.

(5.12)

Nylon-6

A similar ring opening polymerization with ethylene oxide forming poly(ethylene oxide) forms a noncarbon backbone via a chain-wise kinetic process.

(5.13)

Poly(ethylene oxide)

3. Interfacially formed condensation polymers such as polyesters, polycarbonates, nylons, and polyurethanes are typically formed on a microscopic level in a chain-growth manner due to comonomer migration limits and largely because of the highly reactive nature of the reactants employed for such interfacial polycondensations.

5.2   INTRODUCTION

While condensation polymers account for only a modest fraction of all synthetic polymers, most natural polymers are of the condensation type. The first all-synthetic polymer, Bakelite, was produced by the stepwise polycondensation of phenol and formaldehyde.

As shown by Carothers in the 1930s, the chemistry of condensation polymerizations is essentially the same as classic condensation reactions leading to the formation of monomeric esters, amides, and so on. The principle difference is that the reactions used for polymer formation are bifunctional instead of monofunctional.

Table 5.2 contains a listing of a number of industrially important synthetic condensation polymers.

5.3   STEPWISE KINETICS

The kinetics for stepwise polycondensation reactions and the kinetics for monofunctional aminations and esterifications, for example, are similar. Experimentally, both kinetic approaches are essentially identical. Usual activation energies (120–240 kJ/mol) require only about one collision in 10^{12}–10^{15} to be effective in producing polymer at 100°C, whereas for the vinyl reactions, the activation energies are much smaller (8–20 kJ/mol), with most collisions of proper orientation being effective in lengthening the chain. This is in agreement with the slowness of the stepwise process in comparison to chain polymerizations.

While situations that are more complicated can occur, we will consider only the kinetics of simple polyesterification. The kinetics of most other common polycondensations follows an analogous pathway.

For uncatalyzed reactions where the diacid and diol are present in equimolar amounts, one diacid is experimentally found to act as a catalyst. The experimental expression dependencies are described in the usual manner as follows:

$$\text{Rate of polycondensation} = -\frac{d[A]}{dt} = k[A]^2[D] \tag{5.14}$$

where
    [A] is the concentration of diacid
    [D] is the diol concentration

Since [A] = [D], we can write

$$-\frac{d[A]}{dt} = k[A]^3 \tag{5.15}$$

Rearrangement gives

$$-\frac{d[A]}{[A]^3} = k \; dt \tag{5.16}$$

TABLE 5.2
Properties and Uses of Some Important Synthetic Condensation-Type Polymers

Type (Common Name)	Typical Properties	Typical Uses
Polyamides (nylons)	Good balance of properties; high strength, good elasticity and abrasion resistance, good toughness; favorable solvent resistance, outdoor weathering, moisture resistance	Fibers—about half of all nylon fiber goes into tire cord; rope, cord, belting, fiber cloths, thread, hose, undergarments, dresses; plastics—use as an engineering material, substitute for metal bearings, bearings, cams, gears, rollers, jackets on electrical wire
Polyurethanes	Elastomers—good abrasion resistance, hardness, resistance to grease, elasticity; fibers—high elasticity, excellent rebound; coatings—good resistance to solvents and abrasion; good flexibility, impact resistance; foams—good strength per weight, good rebound, high impact strength	Four major forms used—fibers, elastomers, coatings, foams; elastomers—industrial wheels, heel lifts; fibers—swimsuits, foundation garments; coatings–floors where impact and abrasion resistance are required, bowling pins; foams—pillows, cushions
Polyureas	High T_g, fair resistance to greases, oils, solvents	Not widely used
Polyesters	High T_g and T_m; good mechanical properties, resistance to solvents and chemicals; good rebound low moisture absorption, high modulus; film—high tensile strength (about that of steel), stiff, high resistance to failure on repeated flexing, high impact strength, fair tear strength	Fibers—garments, permanent press and "wash and wear" garments, felts, tire cord; film—magnetic recording tape, high grade film
Polyethers	Good thermoplastic behavior, water solubility; moderate strength and stiffness	Sizing for cotton and synthetic fibers; stabilizers for adhesives, binders, and film-formers in pharmaceuticals; thickeners; production of films
Polycarbonates	Crystalline with good mechanical properties, high impact strength; good thermal and oxidative stability, transparent, self-extinguishing, low moisture absorption	Machinery and business
Phenol–formaldehyde resins	Good heat resistance, dimensional stability, resistance to cold flow, solvent, dielectric properties	Used in molding applications; appliances, TVs, automotive parts, filler, impregnating paper, varnishes, decorative laminates, electrical parts, countertops, toilet seats; adhesive for plywood, sandpaper, brake linings, abrasive wheels
Polyanhydrides	Intermediate physical properties; medium to poor T_g and T_m	No large industrial applications
Polysulfides	Outstanding oil and solvent resistance; good gas impermeability, resistance to aging, ozone, bad odors, low tensile strength, poor heat resistance	Gasoline hoses, tanks, gaskets, diaphragms
Polysiloxanes	Available in a wide range of physical states—from liquids to greases, to waxes, to resins, to elastomers; excellent high and moderate low temperature physical properties; resistant to weathering and lubricating oils	Fluids—cooling and dielectric fluids, in waxes and polishes; as antifoam and mold release, for paper and textile treatment: elastomers—gaskets, seals, cable, wire insulation, hot Liquids and gas movement, surgical and prosthetic devices, sealing compounds; resins—varnishes, paints, encapsulating and impregnating agents
Polyphosphate and poly-phosphonate esters	Good fire resistance, fair adhesion, moderate moisture stability, fair temperature stability	Additives promoting flame retardance; adhesive for glass (since it has a similar refractive index), pharmaceuticals, surfactant

Integration of Equation 5.16 over the limits of $A = A_0$ to $A = A_t$ and $t = 0$ to $t = t$ gives

$$2kt = \frac{1}{[A_t]^2} - \frac{1}{[A_0]^2} = \frac{1}{[A_t]^2} + \text{constant} \qquad (5.17)$$

It is usual to express Equation 5.17 in terms of the extent of reaction, p, where p is the defined as the fraction of functional groups that have reacted at time t. Thus, $1 - p$ is the fraction of unreacted groups. A_t is in turn $A_0 \times (1 - p)$, that is,

$$A_t = A_0(1 - p) \qquad (5.18)$$

Substitution of the expression for A_t from Equation 5.18 into Equation 5.17 and rearrangement gives

$$2A_0^2 kt = \frac{1}{(1 - p)^2} + \text{constant} \qquad (5.19)$$

which is the equation of a straight line, that is,

$$mx = y + b$$

where
  $m = 2A_0^2 k$
  $x = t$
  $y = 1/(1 - p)^2$
  b is the constant

A plot of $1/(1 - p)^2$ as a function of time should be linear with a slope $2A_0^2 k$ from which k is determined. The determination of k at different temperatures allows the calculation of activation energy. Thus, one definition of the specific rate constant is $k = Ae^{-E_a/RT}$. The log of both sides gives $\log k = \log A - E_a/RT$, which again is the equation of a straight line where $y = \log k$, $b = \log A$, $m = -E_a/R$ and $x = 1/T$ where E_a is the activation energy, A is a constant, k is the specific rate constant, and T is the temperature.

The number-average degree of polymerization can be expressed as

$$\overline{DP}_n = \frac{\text{Number of original molecules}}{\text{Number of molecules at time } t} = N_0/N = \frac{A_0}{A_t} \qquad (5.20)$$

Thus,

$$\overline{DP}_n = \frac{A_0}{A_t} = \frac{A_0}{A_0(1 - p)} = \frac{1}{1 - p} \qquad (5.21)$$

The relationship given in Equation 5.21 is called the ***Carothers' equation*** because it was first found by Carothers while working with the synthesis of polyamides (nylons). For an essentially quantitative synthesis of polyamides where p is 0.9999, the DP is approximately 10,000, the value calculated using Equation 5.21:

$$\overline{DP}_n = \frac{1}{1 - p} = \frac{1}{1 - 0.0000} = \frac{1}{0.0001} = 10,000 \qquad (5.22)$$

Thus, Carothers' equation allows a determination of maximum $\overline{DP}$ as a function of extent of polymerization and the purity of reactants. This value of 10,000 Da is sufficient to produce polyesters that will give strong fibers. The value of p is decreased, as is the $\overline{DP}$, if impurities are present or if some competing reaction, such as cyclization, occurs. Since the values of k at any temperature can be determined from the slope $(2kA_0^2)$ of when $1/(1-p)^2$ is plotted against t, $\overline{DP}_n$ at any time t can be determined from the expression

$$(\overline{DP}_n)^2 = 2kt[A_0]^2 + \text{constant} \qquad (5.23)$$

Much longer times are required to effect the formation of high polymer polyesters in uncatalyzed esterifications than for acid- or base-catalyzed systems. For catalyzed systems, since the added acid or base is a catalyst, its apparent concentration does not change with time; thus, it is not included in the kinetic rate expression. In such cases, the reaction follows the rate expression

$$\text{Rate of polycondensation} = -\frac{d[A]}{dt} = k[A][B] \qquad (5.24)$$

For $[A] = [B]$ we have

$$-\frac{d[A]}{dt} = k[A]^2 \qquad (5.25)$$

and rearrangement gives

$$-\frac{d[A]}{[A]^2} = kt \qquad (5.26)$$

which on integration and subsequent substitution gives

$$kt = \frac{1}{A_t} - \frac{1}{A_0} = \frac{1}{A_0(1-p)} - \frac{1}{A_0} \qquad (5.27)$$

Rearrangement gives

$$A_0kt = \frac{1}{(1-p)} - 1 = \overline{DP}_n - 1 \qquad (5.28)$$

which predicts a linear relationship of $1/(1-p)$ with reaction time. This is shown in Figure 5.5

Useful polymers are not obtained unless the value for the fractional conversion p is at least 0.990, that is, a $\overline{DP}$ greater than 100.

We can also calculate the weight average chain length as follows. We have that

$$\overline{DP}_w = \frac{(1+p)}{(1-p)} \qquad (5.29)$$

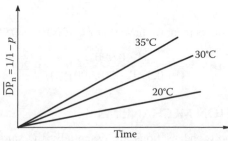

FIGURE 5.5 Plot of chain length, $\overline{DP}_n$, as a function of reaction time for the acid-catalyzed condensation of ethylene glycol with terephthalic acid producing poly(ethylene terephthalate), PET.

The molecular weight distribution is defined as M_w/M_n or $\overline{DP}_w/\overline{DP}_n$ and

$$\frac{\overline{DP}_w}{\overline{DP}_n} = \frac{(1+p)}{(1-p)} \text{ divided by } \frac{1}{(1-p)} = 1 + p \tag{5.30}$$

Where p is near or equal to 1, that is, for a high molecular weight situation,

$$\frac{\overline{DP}_w}{\overline{DP}_n} = 1 + 1 = 2 \tag{5.31}$$

for condensation polymerizations.

Of note, the molecular weight distribution for condensation polymers formed from stepwise kinetics is generally small in comparison to vinyl polymers formed from the chain-wise kinetics where $\overline{DP}_w/\overline{DP}_n$ values of 4–20 are not unusual.

It is important to note that the rate constant k for reactions for similar monofunctional compounds is essentially the same as for the difunctional compounds used in the formation of polycondensation polymers. Likewise, as in the case of reactions of small molecules, the rate constant k increases with temperature in accordance with the Arrhenius equation (5.32).

$$k = A\,e^{-E_a/RT} \tag{5.32}$$

As noted in Chapter 3, solution and bulk viscosity, resistance to flow, increases as molecular weight increases. Energy and time are required to process polymers into useful items. Increases in energy and time are reflected in an increase in the price of the item and decrease in the lifetime of the equipment used to process the polymers. There is need to produce polymeric materials that exhibit the needed physical properties and yet have a reasonable viscosity to minimize the energy and time required to produce the particular product. Thus, molecular weight control is important.

We have already looked the major indicator of molecular weight control. It is Carothers' relationship $\overline{DP}_n = 1/(1-p)$ so that chain length can be controlled through the extent of reaction and the purity of reactants. This was initially one of the least understood relationships and early in the commercial production of polymers cost industry a lot of time and money because of a lack of understanding of its importance. Its importance is now well understood. Thus, the extent of reaction can be controlled through simply stopping the reaction by cooling it. More typically, it is controlled using an excess of one or the other reactant or the introduction of some monofunctional reactant that will halt chain growth when it is added to a growing polymer chain.

For the situation when an excess of one reactant is employed chain growth continues until all of the reactant that is present in the smaller molar amount reacts leaving chains with end groups derived from the reactant that is present in excess. For a reaction where the number of moles of "B" are in excess, $N_B > N_A$, the extent of reaction for A is p and for B it is "rp" where r is the ratio of N_A/N_B. The following relationship can be derived:

$$\overline{DP}_n = \frac{(1+r)}{(1+r-2rp)} \tag{5.33}$$

For an excess in N_B of one mole percent we have that $r = 100/101 = 0.9901$ and

$$\overline{DP}_n = \frac{(1+0.9901)}{(1+0.9901-2(0.9901)(1))} = 201 \tag{5.34}$$

5.4 POLYCONDENSATION MECHANISMS

Proposed mechanisms for polycondensations are essentially the same as those proposed in the organic chemistry of smaller molecules. Here, we will briefly consider several examples to illustrate this similarity between reaction mechanisms for small molecules and those forming polymers.

For instance, the synthesis of polyamides (nylons) is envisioned as a simple S_N2 type Lewis acid–base reaction with the Lewis base nucleophilic amine attacking the electron-poor, electrophilic carbonyl site followed by loss of a proton.

$$\text{(5.35)}$$

$$\text{(5.36)}$$

A similar mechanism is proposed for most polyesterifications (5.37).

$$\text{(5.37)}$$

In Equation 5.38, we see a number of resonance forms for the isocyanate moiety. Because it is surrounded on both sides by atoms, N and O, that have greater electronegativities than the carbon atom, the carbon is electron poor and the site for attack by amines (polyureas) and alcohols (polyurethanes).

$$\text{(5.38)}$$

Polyurethane formation occurs with the attack of the nucleophilic alcohol at the electron-poor isocyanate carbon with a proton shift followed by rearrangement to the urethane structure.

$$\text{(5.39)}$$

Polyether formation from ring opening forming poly(ethylene oxide) occurs through acid or base catalysis as shown in 5.40 and 5.41.

$$(5.40)$$

$$(5.41)$$

Ethylene oxide is a polar molecule with an excess of electron density on the oxygen making it the site for acid attack whereas the ethylene moiety is electron poor and the site for nucleophilic attack by the base.

Again, what we have learned from smaller molecule chemistry applies to polymer reactions typically with only modifications that consider the larger size of the polymer.

5.5 POLYESTERS AND POLYCARBONATES

Carothers and his research group at Dupont began to investigate the formation of polymers from the reaction of aliphatic diacids with diols, generally adipic acid and ethylene glycol (derived from reaction of ethylene oxide with water; major ingredient in most antifreeze), in search of materials that would give them fibers. They were only able to form syrupy mixtures. This is because unlike reactions with diamines (Section 4.7), the equilibrium reaction greatly disfavors ester formation. Further, the ability to have almost equal amounts of functional groups is easily achieved with the amines through the formation of salts with the amines as shown in Equation 5.53 but diols do not form such salts. The critical need to have the reactants present in equal molar amounts for equilibrium-determined reactions is clearly seen in Equation 5.21. Carothers' group understood the principle of "driving" an equilibrium reaction so sought to remove water thus forcing the reaction toward ester formation. For this they developed a so-called molecular still, which was simply heating the mixture and applying a vacuum coupled with a "cold-finger" that allowed evacuated water to condense and be removed from the reaction system. Since the fractional conversion (p) was only 0.95 the average chain length of these polyesters was less than 20.

| Ethylene glycol | Adipic acid | Poly(ethylene adipate) | Water |

$$(5.42)$$

The Dupont research team turned from the synthesis of polyesters to tackle, more successfully, the synthesis of the first synthetic fiber material, nylon, that approached, and in some cases exceeded, the physical properties of natural analogs (Section 5.6). The initial experience with polyesters was put to use in the nylon venture. Today, we know that Carothers would have had greater success in producing high molecular weight polyesters had his group employed transesterification, ring opening (of the "diol" derived moiety), or the reaction of diols with acyl dichlorides (Schotten–Baumann reaction) or anhydrides. These techniques then compose the major techniques employed in the synthesis of polyesters. Each of these reactions involves the nucleophilic addition to the carbonyl group as shown in Section 5.5 and are illustrated in Equations 5.5 and 5.6. Focusing on the carbonyl-containing reactant, polyester formation employing direct esterification (reaction between an acid and an alcohol) and transesterification is relatively slow with each step reversible. Reaction rates are increased using acid catalysts that coordinate the carbonyl oxygen enhancing the electrophilic nature of the carbonyl carbon. Weakly basic catalysts are often employed in transesterification reactions probably to increase the nucleophilicity of the alcohol through the formation of an alkoxide ion. The reaction with anhydrides and acid chlorides are more rapid and can occur in an essentially non-reversible fashion. Moreover, anhydrides and acid chlorides are considered so-called high-energy reactants since they often involve additional energy requiring steps in their production and thus are less suitable for the large-scale production of materials. The activity energies for direct esterification and transesterification are about 30 kcal/mol (120 kJ/mol) while the activation energies for anhydride and acid chloride reaction with alcohols are about 15–20 kcal/mole (60–80 kJ/mol).

The initial polyester formation actually occurred much earlier and is attributed to Gay Lussac and Pelouze in 1833 and Berzelius in 1847. These polyesters are called *glyptals* and *alkyds*, and they are useful as coating materials and not for fiber production. While these reactions had low fractional conversions, they formed high molecular weight materials because they had funtionalities (i.e., the number of reactive groups on a single reactant) greater than two resulting in cross-linking.

When the average functionality is greater than 2, cross-linking occurs. Glyptal polyesters were produced in 1901 by heating glycerol and phthalic anhydride. Since the secondary hydroxyl is less active than the terminal primary hydroxyls in glycerol, the first product formed at conversions of less than 70% is a largely linear polymer. A cross-linked product is formed by further heating because the functionality of glycerol is three (Equation 5.43).

Kienle synthesized alkyds in the 1920s from trifunctional alcohols and dicarboxylic acids. Unsaturated oils, called drying oils, were transesterified with the phthalic anhydride in the reaction so that an unsaturated polymer was obtained which could later be reacted producing a cross-linked product.

The term alkyd is sometimes used to describe all polyesters produced from polybasic acids and a polyhydric alcohol. The terms saturated and unsaturated polyesters have been used to distinguish between those alkyd polyesters that are saturated from those that contain unsaturation.

(5.43)

Phthalic anhydride + glycerol ⟶ Cross-linked polyester

Unsaturated polyesters have been produced from the reaction of ethylene glycol with phthalic anahydride or maleic anhydride (Equation 5.44). These polyesters may be dissolved in organic solvents and used as cross-linking resins for the production of fibrous glass-reinforced composites.

$$\text{Maleic anhydride} \; + \; \text{Ethylene glycol} \; \longrightarrow \; \text{Unsaturated polyester}$$

(5.44)

Aromatic polyesters had been successfully synthesized from the reaction of ethylene glycol and various aromatic diacids but commercialization awaited a ready inexpensive source aromatic diacides. An inexpensive process was discovered for the separation of the various xylene isomers by crystallization. The availability of inexpensive xylene isomers allowed the formation of terephthalic acid through the air oxidation of the p-xylene isomer. Dupont, in 1953, produced polyester fibers from melt spinning, but it was not until the 1970s that Dupont produced polyester fibers became commercially available.

Expanding on the work of Carothers and Hill on polyesters, Whinfield and Dickson, in England, overcame the problems of Carothers and coworkers by employing an ester interchange reaction between ethylene glycol and the methyl ester of terephthalic acid forming the polyester poly(ethylene terephthalate) with the first plant coming online in 1953. This classic reaction producing Dacron, Kodel, and Terylene fibers is shown in 5.45:

Dimethyl terephthalate   Ethylene glycol   Poly(ethylene terephthalate)
                                                                    PET

(5.45)

While PET is normally made as described in Equation 5.45, it can also be made from the ring opening reaction with ethylene glycol as shown in 5.46:

Ethylene   Terephthalic acid   Poly(ethylene terephthalate)

(5.46)

Polyester fibers (poly(ethylene terephthalate), PET; Equations 5.45 and 5.46) are the world's leading synthetic fibers produced at an annual rate of over 1.5 million ton in the United States. Fibers are produced if the product is pushed through a small hole. As the polyester emerges from the hole, tension is applied assisting the polymer chains to align, giving the fiber additional strength in the direction of pulling. The crystallization of polyester resin can be achieved through heating to about 190°C followed by slow cooling. Rapid cooling, called quenching, produces a more amorphous material from which transparent film is made. Film strength is increased through the application of heat and pulling of the film. Biaxially oriented PET film is one of the strongest films available. Thicker film, because of its low carbon dioxide permeability, is used in the manufacture of containers for carbonated drinks through injection molding. It is also used as magnetic films in x-ray and other photographic applications. Thinner film is used in food applications such as food packaging including the boil-in-a-bag foods.

PET is difficult to mold because of its high melting, T_m 240°C, point. PET also crystallizes relatively slowly so that extra care must be exercised to ensure that the PET molded products are fully crystallized or the partial crystallized portions will be preferred sites for cracking, crazing, shrinkage, etc. Thus, nucleating agents and crystallization accelerators have been used to improve the crystallization rate. Post annealing has been used where appropriate.

PBT (5.47), because of the addition of two methylene units in the diol-derived portion, is lower melting than PET with a T_g of about 170°C. The moldability of aryl polyesters has been improved through the use of poly(butylene terephthalate) (PBT) instead of PET or by the use of blends of PET and PBT. These aryl polyesters are used for bicycle wheels, springs, and blow-molded containers.

Poly(butylene terephthalate), PBT

(5.47)

In comparison to PET, PBT melts at a lower temperature, crystallizes more rapidly, and is often employed as a molding compound. PBT offers a balance of properties between nylons and acetals with low moisture absorption, good fatigue resistance, good solvent resistance, extremely good self-lubrication, and good maintenance of physical properties even at relatively high use temperatures. Fiber-reinforced PBT molding compound is sold under the trade name Celanex. A PBT molding compound was first sold under the trade name Valox. Today, there are many PBT molding compounds available.

Poly(dihydroxymethylcyclohexyl terephthalate), 5.48, was introduced by Eastman Kodak as Kodel in 1958. In comparison to PET and PBT, the insertion of the cyclohexyl moiety gives a more hydrophobic material as well as a more moldable product that can be injected molded. The sole raw material is again dimethyl terephthalate. The reduction of the dimethyl terephthalate gives the dialcohol cyclohexanedimethanol. (Notice the reoccurring theme of reusing or using in a modified form the same readily available and inexpensive materials.) This material, along with blends and mixtures, is often extruded into film and sheeting for packaging. Kodel type materials are used to package hardware and other heavy items as well as blow molded to produce packaging for shampoos, liquid detergents, etc.

$$(5.48)$$

Poly(dihydroxymethylcyclohexyl terephthalate)

The hard–soft block copolymer approach employed to produce segmental polyurethanes (Sections 5.9 and 8.4) has also been used with polyesters with the hard block formed from 1,4-butadienediol and terephthalic acid while the soft block is provided from oligomeric (approximate molecular weight of 2000 Da) poly(tetramethylene glycol) and is sold under the trade name of Hytrel.

Along with nylons, polyester fibers approach and exceed common natural fibers such as cotton and wool in heat stability, wash-and-wear properties, and wrinkle resistance. Blended textiles from polyester, cotton, and wool can be also made to be permanent press and wrinkle resistant. The fibers are typically formed from melt or solvent spinning. Chemical and physical modification is often employed to produce differing fiber appearances from the same basic fiber material. Self-crimping textiles are made by combining materials with differing shrinkage properties. Different shaped dyes produce materials with varying contours and properties including hollow fibers.

Several "wholly" aromatic polyesters are available. As expected, they are more difficult to process and stiffer, less soluble, but are employed because of their good high thermal performance. Ekonol is the homopolymer formed from p-hydroxybenzoic acid (Equation 5.49). Ekonol has a T_g in excess of 500°C. It is highly crystalline and offers good strength.

$$(5.49)$$

p-Hydroxybenzoic acid ester   Poly-p-benzoate

It is not unexpected that such aromatic polyesters have properties similar to polycarbonates because of their structural similarities.

Polycarbonates were originally produced from the interfacial polymerization reaction of phosgene with bisphenol A (2,2′-bis(4-hydroxyphenyl) propane) (Equation 5.50). It is an unusually tough and transparent plastic available under the trade names of Lexan (General Electric) and Merlon (Mobay). Polycarbonates can also be made from the ester interchange between diphenyl carbonate and bisphenol A. The melting point of polycarbonates is decreased from 225°C to 195°C when propyl groups replace the methyl pendant groups.

Bisphenoyl A   Phosgene   Polycarbonate

$$(5.50)$$

Polycarbonates and polycarbonate–polyester copolymers are used for glazing sealed beam head-lights, door seals, popcorn cookers, solar heat collectors, and appliance housings.

Nonrecordable compact disks, CDs, are made of rigid, transparent polycarbonates (such as Equation 5.50) with a reflective metal coating on top of the polycarbonate. A laser is used to encode information through the creation of physical features sometimes referred to as "pits and lands" of different reflectivities at the polycarbonate–metal interface.

Recordable CDs contain an organic dye between the polycarbonate and the metal film. Here, a laser creates areas of differing reflectiveness in the dye layer through photochemical reactions.

A beam from a semiconductor diode laser "interrogates" the undersides of both types of CDs seeking out areas of reflected, corresponding to the binary "one," and unreflected, corresponding to the binary "zero" light. The ability to "read" information is dependent on the wavelength of the laser. Today, most of the CD players use a near-infrared laser because of the stability of such lasers. Efforts are underway to develop stable and inexpensive lasers of shorter wavelengths that will allow the holding of more information within the same space.

There is a concern that bisphenol A (BPA) may cause neural and behavioral changes in infants and children. A major line of exposure to infants is in baby bottles. In our modern age, the composition of baby bottles had evolved from glass to polyethylene to polycarbonates and now back to glass and other newer materials such as Tritan, which is a proprietary copolyester developed by Eastman with properties similar to polycarbonate. While the use of most polycarbonates will continue, the "better to be safe than sorry" motto is best used for infants that are more susceptible than adults are to various agents.

5.6   SYNTHETIC POLYAMIDES

Wallace Hume Carothers was brought to Dupont because his fellow researchers at Harvard and the University of Illinois called him the best synthetic chemist they knew. He started a program aimed at understanding the composition of natural polymers such as silk, cellulose, and rubber. Many of his efforts related to condensation polymers were based on his belief that if a monofunctional reactant reacted in a certain manner forming a small molecule, 5.51, those similar reactions except employing reactants with two reactive groups would form polymers, 5.52.

$$R-OH + HOOC-R' \rightleftharpoons R-O-\overset{\overset{\displaystyle O}{\|}}{C}-R' + HOH \tag{5.51}$$

Small ester

$$HO-R-OH + HOOC-R'-COOH \rightleftharpoons -(-O-R-O-\overset{\overset{\displaystyle O}{\|}}{C}-R'-\overset{\overset{\displaystyle O}{\|}}{C}-) + H_2O \tag{5.52}$$

Polyester

While the Carothers' group had made both polyesters and polyamides, they initially emphasized work on the polyesters since they were more soluble and easier to work with. One of Carothers' coworkers, Julian Hill, noticed that he could form fibers if he took a soft polyester material on a glass stirring rod and pulled some of it away from the clump. Because the polyesters had too low softening points for use as textiles, the group returned to work with the polyamides. They found that fibers could also be formed by the polyamides similar to those formed by the polyesters. These polyamides allowed the formation of fibers that approached, and in some cases surpassed, the strength of natural fibers. This new miracle fiber was introduced at the 1939 New York World's Fair in an exhibit that announced the synthesis of this wonder fiber from "coal, air, and water"—an exaggeration—but nevertheless eye catching. When the polyamides and nylons were first offered for sale in New York City, on May 15, 1940 nearly one million pairs were sold in the first few hours. Nylon sales took a large drop when it was noted that nylon was needed to produce the parachute material so critical to World War II.

The first polyesters produced by Carothers had relatively low molecular weights because of low fractional conversions. Carothers was successful in producing higher molecular weight polymers by shifting the equilibrium by removing the water produced. Equation 5.52 is an equilibrium process with the removal of water driving it toward polymer formation. However, these aliphatic polyesters, which he called "super polymers," lacked stiffening groups in the chain and thus had melting points that were too low for laundering and ironing.

Carothers' next step was to move from polyesters to nylons and to increase the fractional conversion (p) by making salts by the equivalent reaction of 1,6-hexanediamine (hexamethylenediamine) and adipic acid. These salts were recrystallizable from ethanol giving essentially a true 1:1 ratio of reactants. Thus, a high molecular weight polyamide generally known as nylon, in this case nylon-66, with a melting point of 265°C was produced from the thermal decomposition of this equimolar salt as shown in Equation 5.53.

Since the molecular weight of this nylon-66, produced by Carothers, was higher than he desired, he added 1% acetic acid (a monofunctional acid that acted to terminate chain growth) that produced a lower molecular weight nylon-66 with a melting point of about 200°C.

Molded nylon-66 is used for lawnmower blades, bicycle wheels, tractor hood extensions, skis for snowmobiles, skate wheels, motorcycle crankcases, bearings, and electrical connections. Fiber nylon-66 is used in clothing, fabrics, and rugs.

(5.53)

Nylon-66

In the early 1950s, George deMestral was walking in the Swiss countryside. When he got home, he noticed that his jacket had many cockleburs on them. For some reason he examined the cockleburs and noticed that they had a lot of tiny "hooks." His cotton jacket had loops that "held" the cockleburs. He began putting into practice his observations, making combinations of materials with rigid hooks and flexible loops or eyes. The initial hook and eye for commercial use was made in France. Today, Velcro, the name given to the hook-and-eye combination, is often based on nylon as both the hook and eye material. Remember that nylon can be made to behave as both a fiber and as a plastic. Polyester is blended with the nylon to make it stronger. Polyesters have also been employed to make hook-and-eye material. The hook-and-eye material is used to fasten shoes, close space suits, and in many other applications.

The general structure for aliphatic nylons for naming purposes is

$$-(-NH-(CH_2)_a-NH-C(=O)-CH_2)_b-C(=O)-)-$$

(5.54)

where a is the number of carbons derived from the amine-associated portion, and b is the number of carbons, including the carbonyl carbon atoms, associated with the diacid. Thus, nylon-66 has 6 carbons derived from hexamethylenediamine, a is 6, and 6 carbons derived from adipic acid.

Nylon-610 is derived from hexamethylenediamine and sebacic acid (dodeconic acid) and is more resistant to moisture and more ductile than nylon-66 because of the presence of the additional flexible nonpolar methylenes. The properties of these polyamides are improved by the formation of polyether blocks (NBC) and by blending with thermoplastics such as EPDM, PET, PBT, and TPE. NBC (Nyrim) is more expensive than reaction injection molded (RIM) polyurethane, but it can be heated to 200°C without melting. NBC moldings are produced from the RIM of poly(propylene glycol) and caprolactam. The tendency of these moldings to swell in water is reduced by reinforcing them with fibrous glass.

Nylon-6 (5.55) structurally quite similar to nylon-66, was initially produced in Germany by the ring-opening polymerization of caprolactam partly as a way to avoid the patents established by Dupont a decade before. The copolymer of nylon-6 and nylon-66 has a smoother surface than either of the homopolymers.

$$-(-NH-(CH_2)_5-\overset{\overset{\displaystyle O}{\|}}{C}-)- \tag{5.55}$$

Nylon-6

Nylon-6,6 is the dominant (sales-wise) nylon in the United States while nylon-6 is the dominant nylon in Europe.

The ability to form strong hydrogen bonding is reduced, and the flexibility increased, by placing bulky methoxymethyl pendent onto nylons (5.56).

$$-(-NH-(CH_2)_6-\underset{\underset{\displaystyle CH_2-O-CH_3}{|}}{N}-\overset{\overset{\displaystyle O}{\|}}{C}-CH_2)_4-\overset{\overset{\displaystyle O}{\|}}{C}-)- \tag{5.56}$$

While aliphatic-containing polyamides are given the name nylons, those where at least 85% of the amide groups are attached to an aromatic are called ***aramids***. Aramids are stronger and tougher than nylons but they are also more difficult to solubilize and fabricate. Because the presence of the aromatic groups causes the aramids to be stiff, they often form liquid crystals that are present in a nematic liquid crystal state in concentrated solution.

Aramids are generally prepared by the solution or interfacial polycondensation of meta and para substituted diacid chlorides and/or diamines. In some systems, synthesis is achieved under rapidly stirred conditions where the polymer is quasisoluble in the system. The polymer mixture is forced through a small opening into a nonsolvent forming a fiber without the need to dissolve the polymer before fiber formation.

Poly(*m*-phenylene isophthalamide) (5.57), sold under the trade name of Nomex, exhibits good thermal stability decomposing above 370°C.

$$\tag{5.57}$$

Isophthaloyl chloride   *m*-Phenylenediamine   Poly(*m*-phenylene isophthalamide)

Nomex is sold as both a fiber and a sheet. The sheet material is often a laminate with paper and used in electrical applications such as circuit boards and transformer cores. Firefighter gear generally includes a Nomex hood to protect the face not protected by the helmet. Race car drivers are required to wear flame-resistant underwear, balaclava, socks, shoes, and gloves that are typically made from Nomex fiber. Military pilots also wear flight suits that are made from Nomex.

Because of the meta orientations of the reactants, Nomex chains do not realign on fiber formation resulting is a material that is weaker than the corresponding para-oriented polymer, Kevlar.

The corresponding aramid produced using the para reactant in place of the meta gives poly(p-phenylene terephthalamide), PPT, produced under the trademark of Kevlar and developed by Dupont by Stephanie Kwolek and Roberto Berendt in 1965 (5.58). Like Nomex, Kevlar exhibits good thermal stability decomposing above about 500°C. Because of its outstanding strength, it was used as the skin covering of the Gossamer Albatross, which was flown across the English Channel using only human power. Aramids are also used as fiber reinforcement in composites and as tire cord.

$$(5.58)$$

Terephthaloyl Chloride p-Phenylenediamiine   Poly(p-phenylene terephthalamide) (PPT)

Kevlar fiber has a high tensile strength (about 3000 MPa) and a relatively low density (about 1.4 g/mL) leading to the often used statement that Kevlar is five times stronger than steel (on an equal weight basis). It gains part of its strength from the interchain hydrogen bonds formed between the carbonyl groups and hydrogen atoms on amines on neighboring chains and on the partial stacking of the phenylene rings allowing pi–pi interactions between the members of the stacking units. The molecularly rigid chains tend to form sheet-like structures similar to those of silk.

Kevlar is used in the manufacture of so-called bulletproof clothing used by the military, policemen, and SWAT teams. In truth, most bulletproof clothing is bullet-resistant and unless quite bulky not able to "stop" most rifle bullets or high caliber hand guns. It is also used in the construction of bullet resistant face masks by military personnel and motorcycle riders and to protect against abrasion by motorcycle riders.

There are many other commercial uses for Kevlar. It is used as the inner liner for some bicycle tires to protect against puncture. It is also being used for bowstrings in archery. Drumheads have been made from Kevlar. Kevlar is widely employed as the protective outer sheath for fiber-optic cable and as the reinforcing layer in rubber bellows, expansion joints, and hoses for high temperature applications.

Aramid materials are also employed in the United States' space program along with other "space-age" materials. The thermal Micrometeroid Garmet on the Extravehicular Mobility Unit, Advanced Crew Escape Suit, thermal blankets, and the fuselage, bay doors, upper wind surfaces of the Space Shuttle Orbiter employs Nomex, Kevlar, and Gore-Tex materials. The airbags for the Mars Pathfinder and MER rovers, the Galileo atmospheric probe, and the new Crew Exploration Vehicle all have aramid materials. Aramid materials are also used in the form of thin pads to protect sintered silica-fiber mats from stress and vibrations during the flight of the space shuttle.

Aramid fibers are widely employed in the construction of composites where high strength is required. The continuous phase is often an epoxy resin. Applications include cricket bats, helicopter rotor blades, bodies for formula one race cars, kayaks, tennis rackets, lacrosse sticks, and ice hockey sticks.

Several so-called semi-aromatic nylons have been produced. Nylon-6,T is produced from the condensation of terephthalic acid and 1,6-hexanediamine (5.59). Both reactants are readily available and inexpensive and the resulting materials offer greater strength than simply wholly aliphatic nylons such as nylon-6,6. Nylon-6,T has a very high T_m of 370°C and a T_g of 180°C. The high T_m results in the need for a high temperature to be employed in processing so that a third reactant is often introduced to lower the T_m and the processing temperature. "Third reactants" often used are adipic acid, caprolactam, isophthalic acid, and 1,5-hexyl diamine. These materials are sold under the trade names of Zytel HTN, Ultramid T, and Amodel R.

(5.59)

Nylon-6,T

Nylons offered new challenges to the chemical industry. Because of the presence of polar groups the attractive forces between chains was high in comparison to vinyl polymers. Nylons are generally semicrystalline, meaning they have a good amount of order. Thus, while they have a T_m, the main physical transition is the T_g so that they undergo a sharper transition from solid to melt in comparison to many of the vinyl polymers discussed in the next three chapters. Thus, the processing temperature window is narrower. If melt flow is required for processing, then the temperature must be sufficient to allow ready flow but low enough so as not to break primary bonds within the processed material. Even so, processing techniques have been developed that allow nylons to be readily processed using most of the standard techniques.

Since chains of nylons having an even number of carbon atoms between the amide and acid groups pack better, some of their melting points are higher than comparable nylons with odd numbers of carbon atoms. Further, the melting points decrease and the water resistance increases as the number of methylene groups between the amide and acid groups increases.

The presence of the polar groups results in materials with relatively high T_g and T_m values so that unlike most vinyl polymers that must be above their T_g to allow needed flexibility, nylons, and many condensation polymers function best where strength, and not flexibility, is the desired behavior. Because of the presence of these polar groups, that also allow for hydrogen bonding, nylons and most condensation polymers are stronger, more rigid and brittle, and tougher in comparison to most vinyl polymers. Nylons are also "lubrication-free" meaning they do not need a lubricant for easy mobility so that they can be used as mechanical bearings and gears without need for periodic lubrication.

In general, more crystalline nylons are used as fibers while less crystalline nylon materials are more used as plastics. The amount of crystallinity is controlled through a variety of means including the introduction of bulky groups and asymmetric units, rapid cooling of nonaligned melts, and introduction of plasticizing materials. The theme of using asymmetric units was used by Grace and Company in developing Trogamid T, an amorphous transparent nylon, from the condensation of terephthalic acid with a mixture of 2,2,4- and 2,4,4-trimethylhexamethylene diamines.

5.7　POLYIMIDES

General Electric made the initial announcement for the commercial preparation of polyetherimides, PEIs, in 1982 under the trade name of Ultem. The final reaction involves the imidization of a diacid anhydride through reaction with a diamine, here *m*-phenylenediamine (5.60). The "ether" portion of the polymer backbone results from the presence of ether linkages within the diacid anhydride.

Ultem is used in medical and chemical instrumentation. Products made from Ultem are high melting, offer good stiffness, transparency, impact and mechanical strength, high flame resistance, low smoke generation, and broad chemical resistance. Some of these properties are expected. The high flame resistance is at least in part derived from the presence of already partially or largely oxidized atoms in the product. The low smoke generation is partially derived from the largely cyclic structure with other cyclic structures predictable from the product structure if exposed to sufficient heat. These cyclic structures often give products that are not evolved with good char formation when the material is exposed to ordinary flame conditions.

(5.60)

Polyetherimide

The general good mechanical properties are a result of the presence of strong double bonds present within polycyclic structures composing the polymer backbone and the presence of strongly polar bonding units that allow the formation of good interactions between chains. Further, the structure is largely rigid with good dimensional stability along the polymer backbone. Any flexibility is gained largely from the presence of the ether linkages for the polyetherimides and the presence of methylene units for the polyimides. These products offer good stable melt viscosities even after recycling several times. They can be processed using a variety of techniques including the formation of sections as thin as 5 mil.

While Ultem is an example of thermoplastic polyimides there are a number of thermoset polyimides. These are known for good chemical resistance and mechanical properties as well as good thermal stability. They are employed in electronic applications as insulating films on magnetic wires, for flexible cables, and for medical tubings. On a laptop computer, the cable connecting the main logic board to the display is often a polyimide with copper conductors. Polyimides are also used as high temperature adhesives in the semiconductor sector.

5.8　POLYBENZIMIDAZOLES AND RELATED POLYMERS

Many heterocyclic polymers have been produced in an attempt to develop high-temperature-resistant polymers for aerospace applications. Among these polymers are the polybenzimidazoles (PBIs), which, as shown in Equation 5.61, are prepared from aromatic tetramines and esters of dicarboxylic acids. In standardized procedures, the reactants are heated below 300°C forming soluble prepolymer, which is converted to the final insoluble polymer by further heating.

Tetraaminobiphenyl   Dicarboxylic acid ester   Polybenzimidazole   (5.61)

PBI fiber has an extremely high melting point and does not ignite. They were initially made under the supervision of Carl Shipp "Speed" Marvel before 1983 as part of an overall project for the U.S. Air Force to create aerospace material. Celanese first produced PBI fibers (5.62) in 1983 from the reaction between tetra-aminobiphenyl and diphenyl isophthalate.

3.3′,4,4′-Biphenyltetramine   Diphenyl isophthalate   PBI   (5.62)

PBI fibers are employed in the production of firefighter coats and suits, high-temperature gloves, and welders', astronaut space, and race driver suits.

Polymers, such as PBI, have a weak link in them since a single covalent bond connects the phenyl rings in biphenyl. This weakness is overcome by the synthesis of ladder polymers, such as polyquinoxaline (Equation 5.63), which have two covalent bonds throughout the chain. Thus, the integrity of the polymer is maintained even if one bond is broken.

(5.63)

Polyquinoxaline

5.9 POLYURETHANES AND POLYUREAS

Polyurethanes (PUs), or polycarbamates, were first made by Bayer reacting diols and diisocyanates. The PU shown in Equation 5.64 is sold under the trade name Perlon U. As in the case with nylons and polyesters, higher melting products are formed when the number of carbon atoms is even since this allows a closer packing of the chains.

(5.64)

Isocyanates react with water producing unstable carbamic acids that decompose forming diamines and carbon dioxide. This forms one basis for the production of PU foams. Thus, isocycnate-terminated PUs are formed which, when exposed to moisture, form foams.

Many PUs are based on aromatic diisocyanates such as toluene diisocyanate (TDI; 5.65) that impart chain stiffness into the PU. PU foams are used in bedding, furniture, automotive seating, and carpet backing.

Cross-linked PU coatings, elastomers, and foams can be produced using an excess of the diisocyanate, which reacts with the urethane hydrogen producing an allophanate, or by incorporating polyols such as glycerol in the reaction mixture. Larger polyols called macroglycols, are used to form segmented products including Spandex used in clothing and undergarments. These products are one basis for segmented elastomers where aromatic polyurethanes that form hydrogen bonding act as the "hard" segment, and the polyols such as poly(ethylene glycol) form the "soft" segment. These segmented PUs can be thermoplastics or thermosets depending on whether or not cross-linking is introduced.

(5.65)

TDI

Reaction injection molding (RIM) is increasing in importance and emphasizes the production of thermoset polyurethanes. Here, liquid monomers are mixed together under high pressure prior to injection into the mold. Polymerization occurs within the mold. Most automotive dash panels are RIM-produced. Notice the tough surface and semiflexible underbelly of the dash. The amount of "foam" formation is controlled to give the finished product.

PUs are also widely used as coating materials sold as finished polymers, two-part systems, and prepolymer systems. Water-based PU systems are now available allowing easy home use. Aromatic diisocyanate-derived coatings generally offer poor external light stability while aliphatic-derived systems offer good light stability.

Polyureas (Equation 5.66) are made similar to PUs, except diamines are employed in place of alcohols.

$$(5.66)$$

The diisocyanate can be aliphatic or aromatic and it can be a monomer, oligomer, or polymer where the amine links together the isocyanate functional groups. A number of polyureas are actually a resin blend that can contain amine-terminated polymers or amine-terminated chain extenders.

Polyureas are used where toughness is needed such as in truck bed liners. Truck bed liners can be also be sprayed onto the bed surface. Bridges and pipes are increasingly being coated with polyureas since these last longer than simply painting these surfaces. Tanks, sewers, manholes, and roofs are often coated with polyureas. Railcars are increasingly being coated with polyureas for abrasion protection. In all of these applications, corrosion protection is essential and polyurethanes offer good stability to weather and many chemicals. It is also finding application where water retention is needed such as in wastewater treatment, and landscape and water containments. It is also being sprayed over foam creating architectural designs. Polyureas are used for flooring and parking decks because they can be rapidly applied and offer good protection when the floors are cleaned. Some automotive objects are formed from polyurea RIM systems replacing other molded pieces.

5.10 POLYSULFIDES

The first synthetic elastomer, Thiokol (Equation 5.67), was synthesized by Patrick in the 1920s, by the condensation of alkylene dichlorides and sodium polysulfides. These solvent-resistant elastomers have limited use because of their foul odor. They can be reduced to liquid polymers that can be reoxidized to solid elastomers used in caulking materials and some rocket propellant formulations.

$$(5.67)$$

Because natural sulfur contains eight sulfur atoms within each molecule, the number of sulfur atoms is generally variable, being 1–8.

Poly(phenylene sulfide) (PPS; Ryton) is a solvent-resistant plastic that is useful in high-temperature service (Equation 5.68). PPS is used for pumps, sleeve bearings, cookware, quartz halogen lamp parts, and electrical appliance housings.

$$(5.68)$$

5.11 POLYETHERS

Hay, in 1956, discovered an oxidative coupling catalyst that allowed the production of polymeric aromatic ethers. The hope was to make polymers from readily available starting materials, mainly phenol. The main aromatic polyether today is derived not from phenol but rather from the catalytic coupling of 2,6-dimethylphenol. The resulting polymer is referred to as poly(phenylene ether), PPE, or poly(phenylene oxide), PPO; 5.69. While called poly(phenylene oxide), it is actually composed of phenylene oxide units containing methyl groups in the 2 and 6 positions of the phenylene moiety. PPO is made by a room temperature oxidation brought about by bubbling oxygen through a solution of the phenol in the presence of copper (I) chloride and pyridine. Initially, there was not a ready, inexpensive source of 2,6-dimethylphenol but because of early found positive properties of PPO, an inexpensive source of 2,6-dimethylphenol was found and both the monomer and the polymer became commercial in 1964.

(5.69)

PPO has a very high T_g of 215°C, a T_m of about 270°C, exhibits good hydrolytic stability, but it has a very high melt viscosity and a tendency to oxidize and gel at processing temperatures. In spite of these negative processing features, PPO showed good compatibility with polystyrene and so has found a place in the marketplace.

The Noryl trade name covers a variety of PPO-intense materials and the most important being with polystyrene. While the methyl groups discouraged good interactions between PPO chains, the aromatic character positively interacts with the phenyl group on the polystyrene and the methyl groups interact positively with the aliphatic polystyrene backbone. Unlike most polymer mixtures, the combination of PPO and polystyrene forms a miscible blend. These blends and alloys with polystyrene raise the heat distortion temperature to over 100°C allowing production of materials that can be boiled. Combinations with PS, in particular high-impact polystyrene, HIPS, are more easily processed and the PPO imparts needed flame resistance. The PS adds flexibility and lowers the melting point. The combinations also offer good hydrolytic stabilities and electrical properties, and they are relatively lightweight. Noryl is widely used in switch boxes because of its good electrical resistance. It can also be modified by addition to glass and other mineral fillers and are especially adaptable to metallizing. PPO-extruded sheet is being used for solar energy collectors, lifeguards on broadcasting towers, airline beverage cases, and window frames.

Aliphatic polyethers are also referred to as polyacetals. Polyoxymethylene, POM, precipitates spontaneously from uninhibited aqueous solutions of formaldehyde and was isolated by Butlerov in 1859. Staudinger, in the 1920s and 1930s, experimented with the polymerization of formaldehyde but failed to produce chains of sufficient length to be useful. While pure formaldehyde readily polymerized, it also spontaneously depolymerizes, un-zippers. In 1947, Dupont began a program to make useful polymers from formaldehyde since formaldehyde is inexpensive and readily available. After 12 years, they announced the commercialization of the polymer from formaldehyde, polyoxymethylene, under the trade name of Delrin. The "secret" was capping the end groups by the acetylation of the hydroxyl end groups, thus preventing the ready unzipping of the polymer chain (Equation 5.64). POM has a T_g of −75°C and a T_m of 180°C.

(5.70)

Celanese came out a year later with a similar product under the trademark of Celcon. Celanese circumvented Dupont's patent on the basis of employing a copolymer variation that allowed enhanced stabilization against thermal depolymerization (Equation 5.71). The copolymer has a T_m of 170°C.

$$(5.71)$$

POMs are also employed in plumbing and irrigation because they resist scale accumulation, have good thread strength, torque retention, and creep resistance. POMs have been employed to assist in the flow of water in fire hoses, water displays, and for some large ships, it is "squirted" from the front of the ship to cut down on friction because they help to align the water allowing an increased fuel efficiency.

Commercially the most important polyether is known as poly(ethylene glycol), PEG, which is also known as poly(ethylene oxide), PEO, or polyoxyethylene, POE. While these names are synonymous, historically PEG was used for lower chained products, PEO for longer chained products, and POE for both. Here we will simply use the term PEG. PEG comes in a variety of chain lengths from low that are liquids to high that are solids.

PEG is generally synthesized from the ring opening polymerization of poly(ethylene oxide) in water (5.72). Along with water, ethylene glycol and ethylene glycol oligomers are often used.

$$(5.72)$$

PEGs are generally designated as to some average molecular weight. Thus, PEG4000 corresponds to $n =$ about 90 (that is 4000/44 amu/unit).

PEGs come with a variety of end groups. Thus, the monofunctional methyl ether of PEG, methoxypoly(ethylene glycol) has one of the end groups a methoxy rather than the usual hydroxyl. They also come in a variety of geometries. Star PEGs have 10–100 PEG units radiating from a central core. Branched PEGs have 3–10 PEG chains radiating from a central core. Comb PEGs have multiple PEG chains grafted to a polymer backbone.

PEGs have a low toxicity and are employed in a number of products. It forms the basis for a number of laxatives such as MiraLax, GlycoLax, and Movicol. It is used in a number of sexual lubricants and skin creams. Bowel irrigations often employ PEG with some electrolytes to be used before surgery, colonoscopy, or in drug overdoses. It is sold under trade names such as Colyte, TriLyte, Fortrans, GoLately, and GlycoLax. It is attached to selected protein drugs to allow a longer acting medicinal effect. PEG-interferon alpha is used to treat hepatitis C and PEG-filgrastim is used to treat neutropenia.

It is employed as the soft segment in polyurethanes. PEG is used in toothpaste as a dispersant and as a lubricant in eye drops. It is also being used in certain body armor. Dr. Pepper adds PEG as an antifoaming agent. It has been used to help to preserve wood replacing water giving the wood increasing dimensional stability. Certain gene therapy vectors such as viruses can be coated with PEG to protect them from inactivation by immune systems and to de-target them from organs where they could buildup causing a toxic effect.

Because of their ready water solubility, PEGs are often used to impart water solubility to a polymer. This approach has been used to make electrically conductive polymers water soluble, as well as a number of drugs water soluble.

5.12 POLYSULFONES

Polysulfones exhibit excellent thermal oxidative resistance, and resistance to hydrolysis and other industrial solvents, and creep. The initial commercial polysulfones were synthesized by the nucleophilic replacement of the chloride on bis(*p*-chlorophenyl) sulfone by the anhydrous sodium salt of bisphenol A. It became commercially available in 1966 under the trade name Udel (5.73). It exhibits a reasonably high T_g of 190°C.

(5.73)

Udel

Union Carbide, in 1976, made available a second-generation polysulfone under the trade name of Radel. Radel (5.74) was formed from the reaction of a bisphenol and bis(*p*-chlorophenyl) sulfone. This polysulfone exhibits greater chemical/solvent resistance, a greater T_g of 220°C, greater oxidative stability, and good toughness in comparison to Udel.

(5.74)

Radel

Polysulfones can also be made from the Friedal–Crafts condensation of sulfonyl chlorides.

Polysulfones are used for ignition components, hair dryers, cookware, dielectric in capacitors, and structural foams. Because of their good hydrolytic stability, good mechanical properties, and high thermal endurance they are good candidate materials for hot water and food handling equipment, alkaline battery cases, surgical and laboratory equipment, life support parts, autoclavable trays, tissue culture bottles, and surgical hollow shapes, and film for hot transparencies. Membranes for use in hemodialysis, gas separation, food and beverage processing, and wastewater recovery have also been produced from polysulfones. Their low flammability and smoke production, again because of their tendency for polycyclic formation on thermolysis and the presence of moieties that are already partially oxidized, makes them useful as materials for the aircraft and automotive industries.

5.13 POLY(ETHER ETHER KETONE) AND POLYKETONES

Aromatic polyketones are semicrystalline materials that contain both ketone groups generally flanked by aromatic units. Many also have included within them ether moieties that allow for some flexibility and better processing. They have good thermal stabilities, as well as offering good mechanical properties, flame resistance, impact resistance, and resistance to the environment.

Polyetherketone, PEK, was introduced by Raychem in the 1970s (Equation 5.75). It is made by the Friedel–Crafts reaction requiring good solvents or an excess of aluminum chloride to keep the polymer in solution allowing polymer growth to occur. Most polymerizations require that the reactants remain mobile, through solution or being melted, so that the individual units involved in the reaction can get together. Rapid precipitation of growing polymer chains often results in the formation of only oligomeric to small chains.

(5.75)

Polyetherketone

ICI has introduced a new crystalline poly(ether ether ketone) (PEEK) (5.76). Applications include compressor plates, valve seats, thrust washers, bearing cages, and pump impellers. In the aerospace industry they are employed as aircraft fairings, fuel valves, and ductings. They are also used in the electrical industry as wire coatings and semiconductor wafer carriers.

(5.76)

Aromatic polyketone

Aliphatic polyketones are made from the reaction of olefin monomers and carbon monoxide using a variety of catalysts. Shell commercialized a terpolymer of carbon monoxide, ethylene, and a small amount of propylene in 1996 under the trade name of Carilon (Equation 5.77). They have a useful range between the T_g (15°C) and T_m (200°C) that corresponds to the general useful range of use temperatures for most industrial applications. The presence of polar groups causes the materials to be tough with the starting materials readily available.

(5.77)

5.14 PHENOLIC AND AMINO PLASTICS

Baekeland found that a relatively stable resole prepolymer could be obtained by the controlled condensation of phenol and formaldehyde under alkaline conditions. These linear polymers (PF) may be converted to infusible cross-linked polymers called resites by heating or by the addition of mineral acids. As shown in Equation 5.78, the initial products produced when formaldehyde is condensed with phenol are hydroxybenzyl alcohols. The linear resole polymer is called an A-stage resin, and the cross-linked resite is called a C-state resin.

C-stage (resite resin)   (5.78)

Baekeland recognized that the trifunctional phenol would produce network polymers and thus used difunctional ortho- or para-substituted phenols to produce linear paint resins. Linear thermoplastic products are formed by alkaline or acid condensation of formaldehyde with phenol derivatives such as p-cresol (Equation 5.79).

(5.79)

Since the acid condensation of 1 mol of phenol with 1.5 mol of formaldehyde produced thermoset C-stage products, Baekeland reduced the relative amount of formaldehyde used and made useful novolac resins in a two-step process. Thus, stable A-stage novolac resin is produced by heating 1 mol of phenol with 0.8 mol of formaldehyde in the presence of acid. After removal of water by vacuum distillation, the A-stage resin produced is cooled and then pulverized. The additional formaldehyde required to convert this linear polymer into a thermoset resin is supplied by hexamethylenetetramine. The latter, which is admixed with the pulverized A-stage resin, is produced by the condensation of formaldehyde and ammonia. Other ingredients such as fillers, pigments, and lubricants are mixed in with the resin and hexamethylenetetramine. The A-stage resin is further polymerized. The term phenol-molding compound is applied to the granulated B-stage novolac resin.

While the condensation of urea and formaldehyde was described in 1884, urea–formaldehyde resins (UF) were not patented until 1918. Comparable products, based on the condensation of formaldehyde and melamine (2,4,6-triamino-1,3,5-triazine), were not patented until 1939. The term melamine–formaldehyde (MF; Equations 5.80 and 5.81) is used to describe these products.

(5.80)

Melamine   Cross-linked MF   (5.81)

Urea and melamine are tetra- and hexa-functional molecules. However, the formation of a network polymer is prevented by adding alcohols such as *n*-butanol and by condensing with formaldehyde at low temperatures under basic conditions. While phenol resins have better moisture and weather resistance than urea resins, the latter are preferred for light-colored objects. For example, the interior layers of laminated countertops are bonded together by phenolic resins, but either urea or melamine resins are used for the decorative surface. Melamine plastics are more resistant to heat and moisture than UF and thus, are used for decorative surfaces and for dinnerware.

5.15 SYNTHETIC ROUTES

The previous sections describe the synthesis of a number of important condensation polymers. Here, we will briefly consider the three main synthetic techniques utilized in the synthesis of these polymers.

The *melt technique* is also known by other names to describe the same or similar processes. These names include high melt, bulk melt, and simply bulk or neat. The melt process is an equilibrium-controlled process in which the polymer is formed by driving the reaction toward completion, usually by the removal of the by-product. For polyesterifications involving the formation of hydrogen chloride or water, the driving force is the removal of the hydrogen chloride or water. Reactants are introduced along with any added catalyst to the reaction vessel. Heat is applied to melt the reactants allowing them to condense together. The heat is maintained or increased above this melt temperature. Pressure is reduced to remove the condensate. Typical melt polymerizations take several hours to several days before the desired polymer is produced. Yields are necessarily high.

Solution condensations are also equilibrium processes, with the reaction often driven by the removal of the by-product by distillation or by salt formation or precipitation. Many solution condensations are run near room temperature. Solvent entrapment is a problem, but since a reaction may occur under considerably reduced temperatures, compared to the melt process, thermally induced side reactions are minimized. Side reactions with the solvent may be a problem. Because the reactants must be energetic, many condensations are not suitable for the solution technique.

The *interfacial technique* (IF), while old, gained popularity with the work of Morgan and Carraher in the 1960s and 1970s. Many of the reactions can be carried out under essentially non-equilibrium conditions. The technique is heterophasic, with two fast-reacting reactants dissolved in a pair of immiscible liquids, one of which is usually water. The aqueous phase typically contains the Lewis base such as diol, diamine, or dithiol. The organic phase contains the Lewis acid, generally an acid halide, dissolved in a suitable organic solvent such as hexane. Reaction occurs near the interface, hence the name. With all the potential that the interfacial system offers, it has not attracted wide industrial use because of the high cost of the necessarily reactive monomers and the cost of solvent removal. One commercial use for the IF system is the production of polycarbonates. Another use involves the synthesis of aramids. Morgan and others noted that some polymers formed with rapid stirring would remain in solution for a while before they precipitated. The problem with aramids was the need to form fibers from their solutions. Thus, the aromatic nylons had to be redissolved after formation. Today, aramids are synthesized using rapidly stirred systems where the polymer solution is sent through a small hole into a nonsolvent. This allows fibers to be produced without needing to redissolve the polymer.

Table 5.3 contains a comparison of these three major polycondensation processes.

TABLE 5.3
Comparison of Requirements for Different Polycondensation Techniques

Requirement	Melt	Solution	Interfacial
Temperature	High	Limited only by the MP and BP of the solvent used—generally about room temperature	Generally room temperature
Stability to heat	Necessary	Unnecessary	Unnecessary
Kinetics	Equilibrium, stepwise	Equilibrium, stepwise	Generally nonequilibrium, chain-wise
Reaction time	1 h to several days	Several minutes to 1 h	Several seconds to 1 h
Yield	Necessarily high	Less necessary high	Low to high
Stoichiometric equivalence	Necessarily high	Less necessary high	Less necessary
Purity of reactants	Necessary	Less necessary	Less necessary
Equipment	Specialized, often sealed	Simple, open	Simple to complex, can be open
Pressure	High, low	Atmospheric	Atmospheric

5.16   LIQUID CRYSTALS

Everyday of our lives we "run across" liquid crystals. They are commonly found in computer monitors, digital clocks, other "read-out" devices, etc.

Reintzer, in 1888, first reported "liquid crystal" behavior. In working with chloesteryl esters, he observed that the esters formed opaque liquids, which on heating turned clear. We now know, that as a general rule, that many materials are clear if they are anisotropic, random or if the materials are composed of ordered molecules or segments of molecules, whereas they are opaque if there exists a mixture of ordered and disordered regions. Lehmann interpreted this behavior as an evidence of a "third" phase that exists "between" the solid and isotropic liquid states. Lehman named this new phase as the liquid crystal phase. Friedel called this phase the mesophase after the Greek word "mesos" meaning intermediate. The initial molecules investigated as liquid crystals were large monomeric molecules.

Flory, in 1956, predicted that solutions of rod-like polymers could also exhibit liquid crystal behavior. The initial synthetic polymers found to exhibit liquid crystal behavior were concentrated solutions of poly(gamma-benzyl glutamate) and poly(gamma-methyl glutamate). These polymers exist in a helical form that can be oriented in one direction into "ordered groupings" giving materials with anisotropic properties.

Liquid crystals, LCs, are materials that undergo physical reorganization where at least one of the rearranged structures involve molecular alignment along a preferred direction causing the material to exhibit nonisotropic behavior and associated birefringent properties; that is, molecular asymmetry.

Liquid crystalline materials can be divided into two large groupings—thermotropic and lyotropic. Thermotropic liquid crystals are formed when "pure" molecules such as cholesteryl form ordered structures upon heating. When liquid crystals occur through mixing with solvents they are called lyotropic liquid crystals.

Thermotropic liquid crystals can be further divided into (a). enantiotropic materials where the liquid crystal phases are formed on both heating and cooling cycles and (b). mesotropic materials where the liquid crystals are stable only on supercooling from the isotropic melt. The mesotropic liquid crystals have been further divided into three groupings as follows:

- Smectic meaning "soap"
- Nematic meaning "thread"
- Cholesteric derived from molecules with a chiral center

Liquid crystal polymers are typically composed of materials that are rigid and rod-like with a high length to breath ratio or materials that have a disc shape. The smaller groupings that give the material liquid crystal behavior are called "mesogens." These mesogens are simply portions of the overall polymer that are responsible for forming the anisotropic liquid crystal and that, in fact, form the liquid crystal segments. Such mesogens can be composed of only segments from the backbone of the polymer, segments from the side chain or segments from both the backbone and the side chain.

Mesogens form the ordered structures necessary to give the overall material anisotropic behavior. There have been identified a number of different mesogen groupings. Chains are arranged so that the mesogen portions are aligned in one preferred direction with the ordering occurring in a three-dimensional layered fashion compose one group of arrangements called smectic arrangements. Here, the lateral forces between the mesogen portions are relatively higher than the lateral forces between the nonmesogen portions allowing a combination of segments that permit "flowing" (the passage of nonmesogen portions) and segments that retain contact (mesogen portions) as the material flows imparting a "memory"-type behavior of the material. A number of different "packings" of the mesogens have been identified. The most ordered of the mesogenic groupings is called "smectic B," which is a hexagonally, close-packed structure present in a three-dimensional arrangement. A much less ordered grouping is called the "Smectic A" phase. Here there is a somewhat random distribution of the mesogens between the layers.

Nematic liquid crystals offer much less order in comparison to smectic arrangements. Here, the directional ordering of the mesogen portions along one axis is retained, but the centers of gravity of the mesogen portions are no longer "coupled." Thus, the forces between the chains are less resulting in a generally greater fluidity for nematic liquid crystals in comparison with smectic structures. Nematic liquid crystals also offer nonlinear behavior.

The chiral nematic assembly is formed by materials that have chiral centers and that form a nematic phase. Here, a "chiral-imposed twist" is imparted to the linear chains composing each layer resulting in a three-dimensional helical arrangement.

The molecular asymmetry typically occurs not because of intermolecular interaction, but because two molecules cannot occupy the same space at the same time. Molecular chains can exist in a random arrangement until a given concentration is exceeded causing the molecules to rearrange in a more ordered fashion to accommodate the larger number of molecules within the same volume. Often, this occurs such that there is an ordered phase and a more random phase. As the concentration of polymer increases, the ordered phase becomes larger at the expense of the disordered phase. This increase in polymer concentration can occur via several routes such as the addition of more polymers, the addition of a solution containing a higher concentration of polymer, and the evaporation of the solvent.

For crystalline polymer systems, transition from the crystalline structure to a mesophere structure occurs whereas for amorphous polymer systems, the mesophase occurs after the T_g has occurred. Some polymer liquid crystal systems form several mesophases. Mesophases can be detected using DSC, x-ray diffraction, and polarizing microscopy.

The introduction of flexible "spacer" units such as methylene, methylene oxide, and dimethylsiloxane groups lowers the melting point and increases the temperature range within which the mesophase is stable. Often these spacer units are introduced by copolymerization. Thus, preformed p-acetoxybenzoic acid is reacted with poly(ethylene terephthalate), PET (5.82), introducing a mesogenic unit (5.83) in a polymer that has flexible spacer units (from the ethylene glycol) in it.

Poly (ethylene terephthate), PET   (5.82)   Mesogenic unit   (5.83)

Along with the mesogen units contained within the polymer backbone, the mesogen units can occur as side chains. These mesogen units can be introduced either through reaction with monomers that contain the mesogen unit or through introduction with already formed polymers.

LC materials have also been employed as films, plastics, and resins. Poly(1,4-benzoate) has been marketed under the name Ekonol. It decomposes before it melts, hence it does not form LC melts. Copolymerization with 4,4′-biphenol and terephthalic acid gives Ekkcel. Ekkcel does melt before it decomposes. Certain forms can be compression molded and others injected molded. The reaction of poly(1,4-benzoate) with PET gives a material that can be injected molded. These LCs are chemical resistant, with high tensile strength. LC films with mesogenic side chains can be used in information storage devices.

5.17 MICROFIBERS

Microfibers are not a new chemically distinct fiber, but rather the term refers to fibers that have smaller diameters. Dupont first introduced microfibers in 1989. Microfibers have diameters that are less than typical fibers. Microfibers are half the diameter of fine silk fiber, one-quarter the diameter of fine wool, and one hundred times finer than human hair. Denier, the weight in grams of 9000 m length of a fiber, is the term used to define the diameter or fineness of a fiber. While the definition for the thickness of microfibers is varied, a typical definition is that microfibers have a denier that is 0.9 denier or less. For comparison, the nylon stocking is knit from 10 to 15 denier fiber.

Microfibers allow a fabric to be woven that is lightweight and strong. Microfibers can be tightly woven so that wind, rain, and cold do not easily penetrate. Rainwear manufacturers use microfibers for this reason. They also have the ability to allow perspiration to pass through them. Thus, so-called microfiber athletic-wear is becoming more commonplace. Microfibers are also very flexible because the small fibers can easily slide back and forth on one another. The first fabric made from microfiber was Ultrasuede™ where short polyester microfibers were imbedded into a polyurethane base. Today microfibers are made mainly from polyesters, nylon, acrylic, and rayon fibers.

The use of the term microfibers is now extended to glass and Teflon-related materials. Here we will restrict ourselves to only fabric applications.

In 1970, Toray Industries scientist Miyoshi Okamoto created the first microfiber. A few months later, his colleague Toyohiko Hikota developed a process that allowed the production of fabric that was later trademarked as Ultrasuede™. Ultrasuede was produced from PET polyester fibers and they were so thin that a pound of them laid end-to-end would reach from the earth to the moon and back. Ultrasuede is soft and supple, resistant to stains and discoloration, and machine-washable and dry-cleanable.

Because of progress made in spinning and fiber processing, smaller fibers can be routinely made with varying configurations, coatings, etc. Microfiber production is mainly achieved using three techniques. The first technique will be illustrated using the processing technique employed to manufacture Ultrasuede.

The processing to form Ultrasuede is complex. First, ultra-microfibers are spun that are so light and fine that a single 50 mile long strand would weight less than one gram. These ultra-microfibers are then extruded through spinnerets creating a configuration that is similar to that present in tendons and hair bundles. These strands are then transformed by ironing, curling, cutting, and needle-punching into a felt-like material that is then impregnated with a special adhesive binder. The material is then formed into the desired contours, the protective coating is dissolved with a solvent, and the material undergoes further processing creating the nonwoven fabric.

Microfibers are also made by simple extrusion through a spinneret with a smaller hole than normally employed for fiber production. The third method involves spinning a bicomponent fiber and using a solution to split the fiber into smaller pieces. Initially, bicomponent fibers in the range of 2–4 denier are spun after which the fibers are split into microfibers. If a 32 segment pie of nylon/polyester fiber is used, the final denier is in the range of 0.1 denier. Brushing and other techniques can be used to enhance the effects. Splittable, hollow fibers are also being used to achieve fiber splitting. For instance, for polyester/polypropylene fiber combinations, "natural" splitting occurs after passage through the spinneret. After mechanical drawing, the yarn has a denier of about 0.2 well within that described for microfibers. More recently, tipped fibers are being studied. Here, a bicomponent fiber is made except the second polymer is placed on the tip of the fiber. After spinning, the fibers are twisted and then wet heat is applied resulting in the tips of the fibers breaking apart into microfibers with a denier of about 0.2. Because the two different polymers have different heats of elongation, the physical changes cause the two polymer components to break apart.

The shape of the microfiber influences the end properties. For instance, Tomen has developed Technofine™, which is a polyester microfiber with a "W" shape cross section. The increased surface allows a more rapid transport of water away from the skin and also increases the evaporation rate making garments made from it to dry quicker and more adsorbent.

Most textiles have limited elongation but many of the microfiber textiles have elongations of 100%.

Currently, no industry regulations exist that describe the amount of microfiber that must be present to call it a microfiber material. Even so, typically industrial use calls for a material to be at least 35%–40% microfiber to be called a microfiber product. Microfibers are most commonly used in dress and blouse garments. They are also used to make hosiery, ties, scarves, intimate apparel, curtains, draperies, sheets, towels, rainwear, swimwear, blankets, tents, sleeping bags, track and jogging wear, as well as many other items. The greater surface area results in a fabric with deeper, richer, and brighter colors possible.

The care of microfiber products is similar to that of the normal-fiber materials made from the same polymer. One caution is heat sensitivity. Because the fibers are so fine, heat penetrates more quickly causing them to scorch or glaze more quickly than "normal" fibers if too much heat is applied or heat is applied over too long a period. Typically, microfibers are wrinkle resistant, but if ironing is done than it should be accomplished using lower temperatures and only as directed.

5.18 SUMMARY

1. Many naturally occurring and some synthetic polymers are produced by condensation reactions many of which are described kinetically by the term stepwise polymerization. A high fractional conversion is required to form linear polymers such as polyesters, nylons, polysulfides, polyurethanes, polycarbonates, polysulfones, polyimides, polybenzimidazoles, and polyethers. However, a high fractional conversion is not required for the production of network, cross-linked, products such as epoxy, phenol, urea, formaldehyde, and melamine resins. One major exception to the production of condensation polymers through the stepwise kinetic process is the use of the interfacial reaction system employing reactive reactants that follows a chain-wise kinetic process. The interfacial system is employed to produce polycarbonates and some aramids. The remainder of the condensation polymers are generally produced using the melt and solution techniques.
2. The rate expressions and values, mechanisms, and the activation energies for the condensation reactions forming polymers are similar to those of small molecule reactions. Reaction rate increases with temperature in accordance with the Arrhenius equation. Average $\overline{DP}$ also increases as the reaction temperature increases to the ceiling temperature where polymer degradation occurs. Long chains are only formed at the conclusion of classical polycondensation processes.
3. The average $\overline{DP}$ for the formation of linear condensation polymers can be calculated using the Carothers' equation, average $\overline{DP} = 1/(1 - p)$.
4. Cross-linked products are formed when the functionality of either reactant is greater than two. Linear products are formed when the functionality of both reactants is two.
5. Condensation polymers tend to exist below their T_g at room temperature. They typically form fairly ordered structures with lots of strong interactions between the various chains giving strong materials with some, but not much, elongation when stretched. They are normally used as fibers and plastics. They have high stress/strain ratios.

GLOSSARY

Alkyds: Term originally used to describe oil-modified polyester, but now used for many polyester plastics and coatings.

Allophanates: Reaction product of a urethane and an isocyanate.

Amino resins: Urea and melamine–formaldehyde resins.

Aramids: Aromatic polyamides.

A-stage: Linear prepolymer of phenol and formaldehyde.

Bakelite: Polymer produced by the condensation of phenol and formaldehyde first by Leo Baekeland.

Bifunctional: Molecule with two active functional groups.

Bisphenol A: 2,2'-Bis(4-hydroxphenol)propane.

B-stage: Advanced A-stage resin.

Carbamate: A urethane.

Carbamic acids: Unstable compounds that decompose spontaneously giving amines and carbon dioxide.

Carothers, W. H.: Inventor of nylon who also standardized much of the polymer nomenclature we use today.

Carothers' equation: $\overline{DP}_n = 1/(1 - p)$

Condensation reaction: Reaction in which two molecules react producing a third molecule and a by-product such as water.

Cyclization: Ring formation.

Dacron: Trade name for PET fiber.

Drying: cross-linking of an unsaturated polymer generally in the presence of oxygen.

Drying oil: An unsaturated oil like tung oil.

Engineering plastic: Plastics whose physical properties are good enough to permit their use as structural materials; generally they can be cut, sawn, and drilled.

Epoxy resin: Polymer produced by the condensation of epichlorohydrine and a dihydric alcohol or by the epoxidation of an unsaturated molecule.

Ester interchange: Reaction between an ester of a volatile alcohol and a less volatile alcohol in which the lower boiling alcohol is removed by distillation.

Filament: Extrudate when a polymer melt or solution is forced through a hole in a spinneret.

Functionality: Number of active functional groups present in a molecule.

Functionality factor: Average number of functional groups present per reactive molecule in a mixture of reactants.

Furan resin: Resin produced form furfuryl alcohol or furfural.

Gel point: Point at which cross-linking begins to produce polymer insolubility.

Glyptals: Polyesters, usually cross-linked by heating.

Incipient gelation: Point where the $\overline{DP}$ reaches infinity.

Interfacial polymerization: One in which the polymerization reaction occurs at the interface of two immiscible liquids.

Kodel: Trade name for a PET fiber.

Ladder polymer: Double-chained temperature-resistant polymer.

Laminate: Layers of sheets or paper or wood or other material adhered by resins and pressed together like plywood.

Long oil alkyd: One obtained in the presence of 65%–80% of an unsaturated oil.

Medium oil alkyd: Alkyd obtained in the presence of 50%–65% of an unsaturated oil.

Melamine–formaldehyde resin: Resin produced by the condensation of melamine and formaldehyde.

Molding compound: Name given to describe a mixture of a resin and essential additives.

Nonoil alkyd: An oil-free alkyd containing no unsaturated oils.

Novolac: Polymers prepared by the condensation of phenol and formaldehyde under acidic conditions.

Nylon: Synthetic polyamide.

Oil length: Term used to indicate the relative percentage of unsaturated oils used in the production of alkyds.

Phenoxy resin: Polymer with hydroxyl pendent groups resembling an epoxy resin without epoxy groups.

Poly(ethylene terephthalate) (PET): Linear polyester used to produce fibers and for the blow-molding preparation of soft drink bottles; produced from terephthalic acid and ethylene glycol.

Prepolymer: Low molecular weight material (oligomer) capable of further polymerization.

Resite: Cross-linked resole.

Resole: Linear polymer prepared by condensation of phenol and formaldehyde under alkaline conditions.

Schotten–Baumann reaction: Traditionally the reaction between an acid chloride and a Lewis base.

Short oil alkyd: An alkyd obtained in the presence of 30%–50% of an unsaturated oil.

Step-reaction polymerization: Polymerization in which polyfunctional reactants react to produce larger units in a continuous stepwise manner.

Thiokol: Trade name for a polysulfide elastomer.

Unsaturated polyester: Term used to describe alkyds with unsaturated chains, particularly those produced by the condensation of maleic anhydride and ethylene glycol.

Urea–formaldehyde resin: Resin produced by condensation of urea and formaldehyde.

Wasted loops: Formation of cyclic compounds instead of polymer chains.

EXERCISES

1. Which of the following will give a polymer when condensed with adipic acid: (a) ethanol, (b) ethylene glycol, (c) glycerol, (d) aniline, or (e) ethylenediamine?
2. Could Carothers have produced strong polyester fibers by ester interchange or Schotten–Baumann reactions using aliphatic reactants?
3. Which would be useful as a fiber: (a) poly(ethylene terephthalate) or (b) poly(hexylene terephthalate)?
4. If the fractional conversion in an ester interchange reaction is 0.99999, what would be the average degree of polymerization of the polyester produced?
5. Use the logarithmic form of the Arrhenius equation to show that the value of the rate constant k increases as the temperature increases.
6. What is the first product produced when a molecule of sebacyl chloride reacts with a molecule of ethylene glycol?
7. What is the next product formed in question 6?
8. How would you improve the strength of the filament produced in the nylon rope trick without changing the reactants?
9. Name the product produced by the condensation of adipic acid and 1,4-tetramethylenediamine?
10. In which reaction would you expect the more "wasted loop": the reaction of oxalyl chloride with (a) ethylenediamine or (b) 1,6-hexanediamine?
11. Which system would be more apt to produce "wasted loops: (a) a dilute solution or (b) a concentrated solution?
12. If the values of A_0 and k are 10 mol/L and 10^{-3} L mol/s, respectively, how long would it take to obtain an average $\overline{DP}$ of 37?
13. Which will give the lower index of polydispersity: (a) $p = 0.000$ or (b) $p = 0.90$?
14. If you used a 2% molar excess of bisphenol A with TDI, what would be the maximum average degree of polymerization obtainable assuming $p = 1$?
15. Why would you predict the product obtained in question 14 be a useful fiber assuming an average $\overline{DP}$ of 100?
16. Name the product formed from reacting a phenol with formaldehyde.
17. Give Carothers' equation. What is its significance?
18. What is the product of polymerized formaldehyde? Is it stable?
19. What is the product of a diol and diisocyanate? What are general uses of this product?

20. Which would be the better or stronger fiber: one made from an ester of (a) terephthalic acid or (b) phthalic acid?
21. What would be the deficiency of a nylon film that was stretched in one direction only?
22. Which would be more flexible: (a) poly(butylene terephthalate) or (b) poly(hexylene terephthalate)?
23. Which would be more apt to deteriorate in the presence of moisture: (a) Lexan molding powder or (b) Lexan sheet?
24. What reactants are typically employed to make a polycarbonate?
25. How would you prepare a nylon with greater moisture resistance than nylon-66?
26. How would you prepare a nylon that would be less "clammy" when used as clothing?
27. Which would be higher melting: (a) a polyamine or (b) a polyester with similar members of methylene groups in the repeat unit?
28. Why is a methoxymethylated nylon more flexible than nylon?
29. Why is Bakelite not used in forming molded objects?
30. Isn't it wasteful to decompose a diisocyanate by hydrolysis to produce foams?
31. How would you prepare a hydroxyl-terminated polyester?
32. For the acid-catalyzed condensation of ethylene glycol with terephthalic acid, what is the relationship between $\overline{DP}$ and extent of reaction?
33. For stepwise kinetics what happens to chain length as time increases?
34. Why do polyurethanes and epoxy resins have good adhesive properties?
35. Why are furan resins relatively inexpensive?
36. To produce cross-linked polyesters what is normally present?
37. Could you produce a soluble novolac resin from resorcinol?
38. Can you explain why there are so many terms used, such as novolac, resole, etc., in phenolic resin technology?
39. Why is Bakelite not used for dinnerware?
40. Which of the following could be a nonpetrochemical plastic: (a) Bakelite, (b) urea plastics, or (c) melamine plastics?
41. Which would produce the better fiber: the reaction product of phthalic acid and (a) 1,4-butanediol, or (b) 2-hydroxylbutanol?
42. Why is PEG widely used in drug delivery?
43. What are aramids?
44. What are general properties of aramids?

FURTHER READINGS

Amato, I. (1997): *The Materials the World Is Made of*, Harper Collins, New York.
Bayer, A. (1878): Phenol-formaldehyde condensates, *Ber. Bunsenges. Phys. Chem.*, 5:280, 1094.
Binder, W. (2007): *Hydrogen Bonded Polymers*, Springer, New York.
Carothers, W. H. (1929): An introduction to the general theory of condensation polymers, *J. Am. Chem. Soc.*, 51:2548.
Carothers, W. H. (1938): US Patent 2,130,947.
Carraher, C. and Preston, J.: (1982): *Interfacial Synthesis*, vol. 3, Marcel Dekker, New York.
Carraher, C. and Swift, G. (2002): *Functional Condensation Polymers*, Kluwer, New York.
Carraher, C., Swift, G., and Bowman, C. (1997): *Polymer Modification*, Plenum, New York.
Collins, P. J. and Hird, M. (1997): *Introduction to Liquid Crystals*, Taylor & Francis, London, U.K.
Craver, C. and Carraher, C. (2000): *Applied Polymer Science*, Elsevier, New York.
Donald, A. (2005): *Liquid Crystalline Polymers*, Cambridge University Press, Cambridge, U.K.
Hill, J. W. and Carothers, W. H. (1932, 1933): Polyanhydrides, *J. Am. Chem. Soc.*, 54:1569; 55:5023.
Imrie, I. (2004): *Introduction to Liquid Crystalline Polymer*, Taylor & Francis, New York.
Kadolph, S. and Langford, A. (2001): *Textiles*, Prentice Hall, Englewood Cliffs, NJ.
Martin, S. M. and Patrick, J. C. (1936): Thiokol, *Ind. Eng. Chem.*, 28:1144.
Marvel, C. S. (1959): *An Introduction to the Organic Chemistry of High Polymers*, Wiley, New York.

Millich, F. and Carraher, C. (1977): *Interfacial Synthesis*, vols. 1 and 2, Dekker, New York.

Morgan, P. W. (1965): *Condensation Polymers by Interface and Solution Methods*, Wiley, New York.

Morgan, P. W. and Kwolek, S. L. (1959): The nylon rope trick, *J. Chem. Ed.*, 36:182.

Pascault, J., Sautereau, H., Verdu, J. and Williams, R. (2002): *Thermosetting Polymers*, Dekker, New York.

Rogers, M. (2003): *Synthetic Methods in Step-Growth Polymers*, Wiley, Hoboken, NJ.

Sadler, S. R. and Karo, W. (1998): *Polymer Synthesis (Three Volumes)*, Academic Press, Orlando, FL.

Seavey, K. and Liu, Y. (2008): *Step-Growth Polymerization Process Modeling and Product Design*, Wiley, Hoboken, NJ.

Yamashito, H. and Nakano, Y. (2008), *Polyester: Properties, Preparation and Applications,* Nova Science Publishers, Hauppauge, NY.

6 Ionic Chain Reaction and Complex Coordination Polymerization (Addition Polymerization)

In the following three chapters, we will examine, in comparison to condensation polymers, polymers that have low seconding bonding between polymer chains. The presence of these low-energy bonding forces allows for less energy being required for elongation. Many of these polymers have glass transition temperatures below room temperatures allowing relatively easy "sliding" of chains past one another resulting in a flexible material. Many of these materials are employed in the production of elastomers and plastics. Polymers with polar pendent groups are used as coatings and adhesives.

In contrast to the slow step-reaction polymerizations discussed in Chapter 5, chain polymerizations are usually rapid, and the initiation species continues to propagate until termination. Thus, in the extreme case, a single initiation species, which would produce one high molecular chain, leaving all of the other monomer molecules unchanged, could be produced. In any case, the concentration of monomer, which is usually a derivative of ethylene, decreases continuously throughout the reaction. In contrast to stepwise polymerization, the first species produced is a high molecular polymer.

A kinetic chain reaction usually consists of at least three steps: (1) initiation, (2) propagation, and (3) termination. The initiator may be an anion, a cation, a free radical, or a coordination catalyst. While coordination catalysts are the most important commercially, the ionic initiators will be discussed first in an attempt to simplify the discussion of chain-reaction polymerization.

In almost all of the polymerizations described in this chapter, there is a sensitive and critical balance between the catalyst activity and polymerization. For instance, if the catalyst is too active it may bind at unwanted sites, including the solvent. If the catalyst complex is not sufficiently active, then ready initiation does not occur. The choice of the solvent is also important. Some solvents will react with the catalysts binding it rather than allowing the catalyst to initiate the desired polymerization. Others may "hold" the catalyst complex together rather than allowing the catalyst to initiate polymerization. Still others may not allow the catalyst complex to form. As in much of science, the precise ingredients and conditions were developed through a combination of intuition, science, art, and research. This process still continues.

While most vinyl monomers undergo free radical polymerization (Chapter 7), a smaller number undergo ionic polymerization. Cationic polymerizations require monomers that have electron-releasing groups such as an alkoxy, phenyl, or vinyl group. Anionic polymerization occurs with monomers containing electron-withdrawing groups such as carboxyl, nitrile, or halide. This selectivity is due to the strict requirements for the stabilization of anionic and cationic species.

Compared with free radical polymerizations, ionic polymerizations are not as well defined. Reactions can use heterogeneous initiators and they are usually quite sensitive to the presence of impurities. Thus, kinetic studies are difficult and the results sensitive to the particular reaction conditions. Further, the rates of polymer formation are more rapid.

TABLE 6.1

Major Techniques Used in the Production of Important Vinyl Polymers

Free radical	Low-density polyethylene, LDPE
	Poly(vinyl chloride)
	Poly(vinyl acetate)
	Polyacrylonitrile and acrylic fibers
	Poly(methyl methacrylate)
	Polyacrylamide
	Polychloroprene
	Poly(vinyl pyridine)
	Styrene–acrylonitrile copolymers, SAN
	Polytetrafluoroethylene
	Poly(vinylene fluoride)
	Acrylonitrile–butadiene–styrene copolymers, ABS
	Ethylene–methacrylic acid copolymers
	Styrene–butadiene copolymers, SBR
	Nitrile rubber, NBR
	Polystyrene
Cationic	Polyisobutylene
	Butyl rubber
	Polyacetal
Anionic	Thermoplastic olefin elastomers (copolymers of butadiene, isoprene, and styrene)
	Polyacetal
Complex	High-density polyethylene, HDPE
	Polypropylene
	Polybutadiene
	Polyisoprene
	Ethylene–propylene elastomers

Cationic and anionic polymerizations are similar. Both involve the formation and propagation of ionic species. While high-energy, low-stability ions would be expected to react with most double bonds, ionic species that are stable enough to propagate are difficult to form and are easily destroyed. The "energetic-window" that allows the formation of such charged species is narrow. While polar solvents might be desirable to solvate the ions, and hence help to stabilize them, they often cannot be used. Some polar solvents, such as water and alcohols, react with and destroy most ionic initiators. Other polar solvents, such as ketones, prevent initiation because of the formation of stable complexes with the initiators. Ionic polymerizations are therefore conducted in low or moderately polar solvents, such as hexane and ethylene dichloride.

By bulk, almost all vinyl polymers are made by four processes (Table 6.1)—free radical (>50%), complex coordinate (12%–15%), anionic (10%–15%), and cationic (8%–12%). Three of these techniques are covered in this chapter.

6.1 CHAIN GROWTH POLYMERIZATION—GENERAL

The next three chapters will deal with polymers formed from chain growth polymerization. Chain growth polymerization is also called addition polymerization and is based on free radical, cationic, anionic, and coordination reactions where a single initiating species causes the growth of a polymer chain.

The kinetic chain reaction typically consists of three steps—initiation, propagation, and termination. The initiators for free radical, anionic, and cationic polymerizations are organic radicals, carbanions, and carboniums. Chain growth is exothermic with the polymerization mainly controlled by the steric and resonance factors associated with the monomer. Generally, the less resonance stabilization in the growing chain the more exothermic the reaction. In addition, the greater the steric factors for the reactants, the less exothermic the polymerization.

6.2 CATIONIC POLYMERIZATION

The art of cationic polymerization, like that of many other types of polymerization, is at least a century old. However, the mechanisms for these reactions have only recently become better understood.

The first species produced in cationic polymerizations are carbocations, and these were unknown as such prior to World War II. It is now known that pure Lewis acids, such as boron trifluoride and aluminum chloride, are not effective as initiators. A trace of a proton-containing Lewis base, such as water, is also required. The Lewis base coordinates with the electrophilic Lewis acid, and the proton is the actual initiator. Since cations cannot exist alone, they are accompanied by a ***counterion***, also called a ***gegenion***: (Equation 6.1).

$$BF_3 \quad + \quad H_2O \quad \rightleftharpoons \quad H^+, BF_3OH^-$$

<table>
<tr><td>Lewis acid
catalyst</td><td>Lewis acid
cocatalyst</td><td>Catalyst–cocatalyst
complex</td></tr>
</table>

(6.1)

Since the required activation energy for ionic polymerization is small, these reactions may occur at very low temperatures. The carbocations, including the macrocarbocations, repel one another; hence, chain termination does not occur by combination but is usually the result of reaction with impurities.

Both the initiation step and the propagation step are dependent on the stability of the carbocations. Isobutylene (the first monomer to be polymerized commercially by ionic initiators), vinyl ethers, and styrene have been polymerized by this technique. The order of activity for olefins is $Me_2C = CH_2 > MeCH = CH_2 > CH_2 = CH_2$, and for para-substituted styrenes the order for the substituents is Me-O > Me > H > Cl. The mechanism is also dependent on the solvent as well as the electrophilicity of the monomer and the nucleophilicity of the gegenion. Rearrangements may occur in ionic polymerizations.

The rate of initiation (R_i) for typical cationic reactions is proportional to the concentration of the monomer [M] and the concentration of the catalyst—cocatalyst complex [C] as follows:

Isobutylene	Catalyst–cocatalyst complex	Carbocation	Gegenion
(M)	(C)	(M^+)	

(6.2)

$$R_i = k_i[C][M]$$

(6.3)

Propagation, or chain growth, takes place in a head-to-tail configuration as a result of resonance stabilization and steric factors by carbocation (M^+) addition to another monomer molecule. The head stabilizes the cation best, so it is the growing site while the least sterically hindered site is the site for attack by the cation resulting in the head-to-tail arrangement. The rate constant for growth is essentially independent of chain length so is the same for all propagation steps and is influenced by the dielectric constant of the solvent. The rate is fastest in solvents with high dielectric constants, promoting the separation of the carbocation—gegenion pairs. The chemical and kinetic equations for this are as follows:

Carbocation (Gegenion)   Isobutylene   Macrocarbocation   Gegenion
(M+)   (M)   (M+)

$$(6.4)$$

$$R_p = k_p[M][M^+] \tag{6.5}$$

The termination rate R_t, assumed to be a first-order process, is simply the dissociation of the macrocarbocation–gegenion complex forming BF_3 and H_2O and the now neutral "dead" polymer chain. This is expressed as follows:

$$R_t = k_t[M^+] \tag{6.6}$$

Termination may also occur by chain transfer, where a proton is transferred to a monomer molecule, leaving a cation that can serve as an initiator. The $\overline{DP}$ is equal to the kinetic chain length (v) when chain transfer occurs. The chemical and kinetic equations for termination via chain transfer are as follows:

$$(6.7)$$

$$R_{tr} = k_{tr}[M][M^+] \tag{6.8}$$

It is experimentally found that the rate of initiation equals the rate of termination and since the propagation step is so rapid, then the number of growing chains is constant. Since it is difficult to determine values for some members of the kinetic expressions, including $[M^+]$ the following approach is normally taken to eliminate the need for determining $[M^+]$. Since there is a steady state of growing chains, the rate of initiation is equal to the rate of termination, giving $R_i = R_t$, and solving for $[M^+]$ gives the following:

$$k_i[C][M] = k_t[M^+] \tag{6.9}$$

therefore

$$[M^+] = \frac{k_i[C][M]}{k_t} \tag{6.10}$$

This expression of $[M^+]$ is substituted into the propagation rate expression (Equation 6.5) giving Equation 6.11.

$$R_p = k_p[M][M^+] = \frac{k_p k_i[C][M]^2}{k_t} = k'[C][M]^2 \tag{6.11}$$

For termination by chain transfer we have

$$k_i[C][M] = k_t[M][M^+] \tag{6.12}$$

and

$$[M^+] = \frac{k_i[C][M]}{k_t[M]} = \frac{k_i[C]}{k_t} \tag{6.13}$$

giving

$$R_p = k_p[M][M^+] = \frac{k_p k_i[C][M]}{k_t} = k'[C][M] \tag{6.14}$$

The $\overline{DP}$ can also be described when internal dissociation is the dominant termination step as follows:

$$\overline{DP} = \frac{R_p}{R_t} = \frac{k_p[M][M^+]}{k_t[M^+]} = k''[M] \tag{6.15}$$

But if chain transfer is the dominant termination step then

$$\overline{DP} = \frac{R_p}{R_{tr}} = \frac{k_p[M][M^+]}{k_{tr}[M][M^+]} = \frac{k_p}{k_{tr}} = k'' \tag{6.16}$$

It is important to note that regardless of how termination occurs, the molecular weight is independent of the concentration of the initiator. However, the rate of ionic chain polymerization is dependent on the dielectric constant of the solvent, the resonance stability of the carbonium ion, the stability of the gegenion, and the electropositivity of the initiator.

The rates of all single-step reactions increase as the temperature increases. This may not be true for the 'net' effect for multistep reactions such as those involved with multistep polymerizations, here the cationic polymerization. For cationic polymerizations the activation energies are generally of the order $E_{tr} > E_i > E_p$. Remembering that the description of the specific rate constant is

$$k = A e^{-Ea/RT} \tag{6.17}$$

the overall or "net" activation energy for chain growth from Equation 6.11 is

$$E_{(overall)} = E_p + E_i - E_{tr} \tag{6.18}$$

and for chain length from Equation 6.16 it is

$$E_{(overall)} = E_p - E_{tr} \qquad (6.19)$$

For many cationic polymerizations, the net activation is negative, using the relationships given in Equation 6.11, so that the overall rate of polymerization decreases, for these cases, as the temperature is increased. Further, using Equation 6.16 and since $E_{tr} > E_p$, the overall degree of polymerization does decrease as the temperature is increased. This is pictured in Figure 5.4.

The cationic polymerization of vinyl isobutyl ether at −40°C produces stereoregular polymers. Poly(vinyl isobutyl ether) has a low T_g and is used as an adhesive and as an impregnating resin. This production of stereoregular structures has been known for some time and is especially strong for vinyl ethers. Several general observations have been noted. First, the amount of stereoregularity is dependent on the nature of the initiator. Second, stereoregularity increases with a decrease in temperature. Third, the amount and type (isotactic or syndiotactic) is dependent on the polarity of the solvent. For instance, the isotactic form is preferred in nonpolar solvents, but the syndiotactic form is preferred in polar solvents.

Commercial polymers of formaldehyde are also produced using cationic polymerization. The polymer is produced by ring opening of trioxane. Since the polyacetal, polyoxymethylene (POM), is not thermally stable, the hydroxyl groups are esterified (capped) by acetic anhydride. The commercial polymer is a strong engineering plastic. Engineering plastics typically have higher modulus and higher heat resistance than general-purpose plastics.

Another stable polyacetal (polyoxymethylene [POM]; Celcon) is produced by the cationic copolymerization of a mixture of trioxane and dioxolane.

$$(6.20)$$

| Trioxane | Polyacetal | Capped polyoxymethylene, POM |

Many cyclic ethers have been polymerized using cationic polymerization. The tendency for ring cleavage decreases as the size of the ring increases. In fact, the six-membered ring oxacyclohexane is so stable that it does not polymerize even in the presence of a promoter. Ethylene oxide (also called oxirane) polymerizes forming poly(ethylene oxide) (PEO; also known as poly(ethylene glycol)) (Equation 6.21) in the presence of acids such as sulfuric acid, producing a wide range of chain sized polymers sold under various trade names including Carbowax and Polyox. PEO is also used in cosmetics and pharmaceuticals (as water-soluble pill coatings and capsules).

$$(6.21)$$

Polyterpenes, coumarone–indene resins, and so-called petroleum resins are produced commercially using cationic polymerization. These are used as additives for rubber, coatings, floor coverings, and adhesives.

6.3 ANIONIC POLYMERIZATION

Anionic polymerization was used to produce synthetic elastomers from butadiene at the beginning of the twentieth century. Initially, alkali metals in liquid ammonia were used as initiators, but by the 1940s they were replaced by metal alkyls such as n-butyllithium. In contrast to vinyl monomers with electron-donating groups polymerized by cationic initiators, vinyl monomers with electron-withdrawing groups are more readily polymerized using anionic polymerization. Accordingly, the order of activity using an amide ion initiator is acetonitrile > methyl methacrylate > styrene > butadiene.

The chemical and kinetic relationships for the anionic polymerization of acrylonitrile follow the same three major steps found for cationic polymerizations—namely initiation, propagation, and termination:

$$
\begin{array}{ccc}
& \overset{\displaystyle CN}{\underset{\displaystyle |}{}} & \overset{\displaystyle CN}{\underset{\displaystyle |}{}} \\
:NH_2^- \ + \ H_2C{=}CH & \longrightarrow & H_2N{-}CH_2{-}CH{:}^-
\end{array}
\qquad (6.22)
$$

$$
\text{Amide ion} \qquad \text{Acrylonitrile} \qquad\qquad \text{Carboanion}
$$

$$
R_i = k_i[C][M] \qquad (6.23)
$$

where $C = :NH_2^-$

$$
\begin{array}{cccc}
\overset{\displaystyle CN}{\underset{\displaystyle |}{}} & \overset{\displaystyle CN}{\underset{\displaystyle |}{}} & \overset{\displaystyle CN}{\underset{\displaystyle |}{}} & \overset{\displaystyle CN}{\underset{\displaystyle |}{}} \\
H_2N{-}CH_2{-}CH{:}^- + H_2C{=}CH & \longrightarrow \longrightarrow \longrightarrow & H_2N{-}({-}CH_2{-}CH{-})_n{-}CH_2{-}CH{:}^-
\end{array}
\qquad (6.24)
$$

$$
\text{Carboanion} \qquad \text{Acrylonitrile} \qquad\qquad \text{Macrocarbanion}
$$

$$
R_p = k_p[M][M^-] \qquad (6.25)
$$

with termination occurring through solvent transfer gives

$$
\begin{array}{cccc}
\overset{\displaystyle CN}{\underset{\displaystyle |}{}} & \overset{\displaystyle CN}{\underset{\displaystyle |}{}} & \overset{\displaystyle CN}{\underset{\displaystyle |}{}} & \overset{\displaystyle CN}{\underset{\displaystyle |}{}} \\
H_2N{-}({-}CH_2{-}CH{-})_n{-}CH_2{-}CH{:}^- + NH_3 & \rightarrow & H_2N{-}({-}CH_2{-}CH{-})_n{-}CH_2CH_2 \ + \ :NH_2^-
\end{array}
\qquad (6.26)
$$

$$
R_{tr} = k_{tr}[NH_3][M^-] \qquad (6.27)
$$

As in the case with cationic polymerizations, the number of growing chains is constant so that a steady state exists such as the $R_i = R_{tr}$. This is useful because it is difficult to determine the concentration of $[M^-]$ so that it can be eliminated as follows:

$$
k_i[C][M] = k_{tr}[NH_3][M^-] \qquad (6.28)
$$

and

$$
[M^-] = \frac{k_i[C][M]}{k_{tr}[NH_3]} \qquad (6.29)
$$

Substituting into Equation 6.25 gives

$$
R_p = k_p[M][M^-] = \frac{k_p[M]k_i[C][M]}{k_{tr}[NH_3]} = \frac{k'[M]^2[C]}{[NH_3]} \qquad (6.30)
$$

$\overline{DP}$ can be described as follows:

$$\overline{DP} = \frac{R_p}{R_{tr}} = \frac{k_p[M][M^-]}{k_{tr}[NH_3][M^-]} = \frac{k'[M]}{[NH_3]} \qquad (6.31)$$

Using the same approach we did with the cationic polymerization, we have for the rate of propagation

$$E_{(overall)} = E_p + E_i - E_{tr} \qquad (6.32)$$

and for the dependence of chain length we have

$$E_{(overall)} = E_p - E_{tr} \qquad (6.33)$$

Thus, the rate of propagation and the molecular weight are both inversely related to the concentration of ammonia. The activation energy for chain transfer is larger than the activation energy for propagation. The overall activation energy is about 160 kJ/mol. The reaction rate increases and the molecular weight decreases as the temperature is increased as shown in Figure 5.4. The reaction rate is dependent on the dielectric constant of the solvent and the degree of solvation of the gegenion. Weakly polar initiators, such as Grignard's reagent, may be used when strong electron-withdrawing groups are present on the monomer, but monomers with less electron-withdrawing groups require more highly polar initiators such as *n*-butyllithium.

Synthetic *cis*-1,4-polyisoprene (Equation 6.34) is produced at an annual rate of about 100,000 ton by the anionic polymerization of isoprene when a low dielectric solvent, such as hexane, and *n*-butyllithium is used. But, when a stronger dielectric solvent, such as diethylether, is used along with *n*-butyllithium, equal molar amounts of *trans*-1,4-polyisoprene and *cis*-3,4-polyisoprene units are produced.

$$(6.34)$$

The formation of stable macroanions produces what Szwarz has called "living polymer." These, macroanions or macrocarbanions have been used to produce block copolymers such as Kraton. Kraton is an ABA block copolymer of styrene (A) and butadiene (B). Termination is brought about by addition of water, ethanol, carbon dioxide, or oxygen.

Living polymers are generally characterized by (1) an initiation rate that is much larger than the polymerization rate; (2) polymer molecular weight related to [monomer]/[initiator]; (3) linear molecular weight-conversion relationship; (4) narrow molecular weight range; and (5) stabilization of the living end groups allowing the formation of telechelics, macromers, block copolymers, and star polymers.

Group-transfer polymerizations make use of a silicon-mediated Michael addition reaction. They allow the synthesis of isolatable, well-characterized living polymers whose reactive end groups can be converted into other functional groups. It allows the polymerization of alpha, beta-unsaturated esters, ketones, amides, or nitriles through the use of silyl ketenes in the presence of suitable nucleophilic catalysts such as soluble Lewis acids, fluorides, cyanides, azides, bifluorides, and HF_2^-.

As the polymerization occurs, the reactive ketene silyl acetal group is transferred to the head of each new monomer as it is added to the growing chain (Equation 6.35). Similar to anionic polymerization, the molecular weight is controlled by the ratio of the concentration of monomer to initiator. Reactions are generally carried out at low temperatures (about 0°C–50°C) in organic liquids such as THF. Compounds with "active" hydrogen atoms such as water and alcohols will stop the polymerization and their presence will curtain polymer chain length. Under the right conditions, polymerization will continue until all the monomer has been used up.

$$MeOOC—C(Me)_2—(—CH_2—\underset{\underset{O}{\overset{\|}{C}}—O—Me}{\overset{Me}{\overset{|}{C}}—)—CH_2—\underset{O—SiMe_3}{\overset{Me\ \ O—Me}{\overset{|}{C}=C}} \qquad \overset{Acid}{\longrightarrow}$$

(6.35)

$$MeOOC—C(Me)_2—(—CH_2—\underset{\underset{O}{\overset{\|}{C}}—O—Me}{\overset{Me}{\overset{|}{CH}}—)—CH_2—\underset{C—O—Me}{\overset{Me}{\overset{|}{C}}—H}$$

The trimethylsiloxy end group is a "living" end that continues to add units as long as the monomer is available or until it is neutralized.

6.4 STEREOREGULARITY AND STEREOGEOMETRY

As noted in Chapter 2, there exists stereogeometry and stereoregularity in polymers. These differences have profound effects on the physical properties and, to a lesser degree, chemical properties of the polymers produced from the same monomer. Three possible units can be formed from the polymerization of butadiene as shown in Equation 6.36.

| 1,3-Butadiene | 1,2- | *Cis*-1,4- | *Trans*-1,4- |

(6.36)

For isoprene, there are four possible units formed (Equation 6.37).

As shown in Figure 2.5, there are three possible stereoregular forms for mono-substituted vinyl polymers. These are isotactic, all of the pendant groups are on one side of the chiral carbon; syndiotactic, the pendant groups appear on alternate sides of the chiral carbon; and atactic, some mixture of geometries about the chiral carbon.

| Isoprene | 1,2- | 3,4- | *Cis*-1,4 | *Trans*-1,4- | (6.37) |

It is important to realize that polymer configuration and conformation are related. Thus, there is a great tendency for isotactic polymers (configuration) to form helical structures (conformation) in an effort to minimize steric constrains brought about because of the isotactic geometry.

6.5 POLYMERIZATION WITH COMPLEX COORDINATION CATALYSTS

Prior to 1950, the only commercial polymer of ethylene was a highly branched polymer called high-pressure polyethylene where extremely high pressures were used in the polymerization process. The technique for making linear polyethylene was discovered by Marvel and Hogan and Banks in the 1940s and 1950s and by Nobel Laureate Karl Ziegler in the early 1950s. Ziegler prepared high-density polyethylene (HDPE) by polymerizing ethylene at low pressure and ambient temperatures using mixtures of triethylaluminum and titanium tetrachloride. Another Nobel Laureate, Giulio Natta, used Ziegler's complex coordination catalyst to produce crystalline, stereoregular polypropylene. These catalysts are now known as Ziegler–Natta (or Natta–Ziegler) catalysts.

In general, a Ziegler–Natta catalyst is a combination of a transition metal compound from Groups IVB (4) to VIIIB (10) and an organometallic compound of a metal from Groups IA (1) to IIIA (13) in the periodic table. It is customary to refer to the transition metal compounds as the catalyst (because reaction occurs at the transition metal atom site) and the organometallic compound as the cocatalyst.

Here we will use titanium to illustrate the coordination polymerization process. Several exchange reactions between catalyst and cocatalyst occur with Ti(IV) reduced to Ti(III). The extent and kind of stereoregulation can be controlled through a choice of reaction conditions and catalyst/cocatalyst. The titanium salt is present as a solid. The precise mechanism probably varies a little depending on the catalyst/cocatalyst and reaction conditions. Here we will look at the polymerization of propylene using titanium chloride and triethylaluminum. In general, a monomeric molecule is inserted between the titanium atom and the terminal carbon atom in a growing chain. Propagation occurs at the solid titanium salt surface—probably at defect, corners, and edges. The monomer is always the terminal group on the chain. Triethylaluminum reacts with the titanium-containing unit producing ethyltitanium chloride as the active site for polymerization:

Titanium	Triethyl	Ethyltitanium	Diethylaluminum
chloride	aluminum	chloride (active	chloride
		site)	

(6.38)

The propylene forms a pi-complex with the vacant d-orbital of titanium as shown in Equation 6.39.

(6.39)

The ethyl groups transfer to the propylene, opening up a new active site. The growing chain transfers to the site vacated by the ethyl group creating a new active site that attracts, through pi-interactions, another propylene monomer. This sequence is shown in Equation 6.40. The edges of the solid titanium salt are believed to help provide the contour necessary to form the stereoregular chains.

$$(6.40)$$

Most vinyl monomers give a predominance of the isotactic product. Typically, the more exposed the catalytic site, the less the stereoregularity of the resulting polymer. Isotactic polypropylene is produced using this technique, as is HDPE.

The versatility of such stereoregulating systems is demonstrated in the polymerization of 1,3-butadiene where all four of the potential structures—isotactic-1,2-, syndiotactic-1,2-, *trans*-1,4- and *cis*-1,4, can be synthesized in relatively pure form using different catalysts systems.

6.6 SOLUBLE STEREOREGULATING CATALYSIS

The 1940s were a time of studying the kinetics and mechanism of the production of vinyl polymers that took "center stage" in the 1950s. The 1950s incubated the solid-state stereoregulating catalysis that spawned a chemical revolution with the synthesis of stereoregular vinyl polymers in the 1960s. The 1980s and early 1990s served as a foundational time for soluble stereoregulating catalysis spawning another revolution related to the production of vinyl polymers with enhanced properties.

The solid-state stereoregulating catalysts "suffered" from at least three problems. First, while stereoregular polymers were formed with good control of the stereo-geometry, polymer properties still fell short of predicted (upper limit) values. This was probably due to the presence of the associated solid catalyst structure that accompanies the active catalytic site. This "excess baggage" restricts the motion of the growing chains so that stereoregular control was good, but the tendency to form good secondary structures was interrupted.

Second, in many cases the solid-state catalysis were incorporated, as contaminants, within the growing polymer making an additional purification step necessary in the polymer processing to rid the polymer of this undesired material.

Third, many solid-state catalysts offered several "active polymerization sites" due to differences in the precise structure at and about the active sites. This resulted in an average stereoregular product formed.

The new soluble catalysts offer a solution to these three problems. First, the "smaller" size of the active site, and associated molecules, allows the growing chains to "take advantage" of a natural tendency for the growing polymer chain to form a regular helical structure (in comparison to polymers formed from solid-state catalysts).

Second, the solution catalysts allow the synthesis of polymers that contain little or no catalytic agents, allowing the elimination of the typical additional "clean-up" steps necessary for polymers produced from solid-state catalysts.

Third, the newer soluble catalytic sites are homogeneous offering the same electronic and stereo-structure allowing the synthesis of more stereoregular-homogeneous polymers.

The new soluble stereoregulating polymerization catalysts require three features:

- A metal atom (active) site
- A cocatalyst or counterion
- A ligand system

While the major metal site is zirconium, other metals have been successfully used including Ti, Hf, Sc, Th, and rare earths (such as Nd, Yb, Y, Lu, and Sm). Cyclopentadienyls (Cps) have been the most commonly used ligands though a number of others have been successfully employed including substituted Cps and bridged Cps. The most widely used metal–ligand grouping is zirconocene dichloride (zironocene dichloride has a distorted tetrahedral geometry about Zr).

Methylalumoxane (Equation 6.41), MAO, is the most widely used counterion. MAO is an oligomeric material with the following approximate structure:

$$
\begin{array}{c}
H_3C \\
\backslash \\
Al{-}O{-}[{-}\overset{\displaystyle CH_3}{\underset{\displaystyle |}{Al}}{-}O{-}]_n{-}\overset{\displaystyle CH_3}{\underset{\displaystyle }{Al}} \\
/ \qquad\qquad\qquad \backslash \\
H_3C \qquad\qquad\qquad\quad CH_3
\end{array}
\qquad \text{where } n = 4\text{--}20 \qquad\qquad (6.41)
$$

Methylalumoxane

It is believed that MAO plays several roles. MAO maintains the catalyst complex as a cation, but doing so without strongly coordinating to the active site. It also alkylates the metallocene chloride, replacing one of the chloride atoms with an alkyl group and removing the second chlorine, thus creating a coordinately unsaturated cation complex, Cp_2MR^+. As an olefin approaches the ion pair containing the active metal, a metallocene–alkyl–olefin complex forms. This complex is the intermediate stage for the insertion of the monomeric olefin into a growing polymer chain.

The structure of the catalyst complex controls activity, stereoselectivity, and selectivity toward monomers. The catalyst structure is sensitive to Lewis bases such as water and alcohols encouraging the use of strongly oxyphilic molecules, such as MAO, to discourage the inactivation (poisoning) of the catalyst.

These soluble catalysts are able to give vinyl polymers that have increased stereogeometry with respect to tacticity as well as allowing the growing chains to form more precise helical structures. Further, the homogeneity of the catalytic sites also allows for the production of polymers with narrow molecular weight "spreads."

The summation of these affects is the production of polymers with increased strength and tensile properties. For polyethylene, PE, the use of these soluble catalysts allows the synthesis of PE-chains with less branching compared to those produced using solid-state catalysts such as the Ziegler–Natta catalysts, ZNCs. PE produced using soluble catalysts also show increased properties in comparison to PE produced by solid catalysts. Table 6.2 gives some comparisons of the PEs produced using the ZNCs with those produced with soluble catalysts.

TABLE 6.2
Comparison of Properties of Polyethylene Using Solid (ZNC) and Soluble Catalysts

Property	Unit	Soluble	ZNC
Density	g/cc	0.967	0.964
Melt index		1.3	1.1
Haze		4.2	10.5
Tensile yield	psi	800	750
Tensile brake	psi	9400	7300
Elongation break	%	630	670

Source: Pain, C.F., *Proceedings Worldwide Metallocene Conference (Met Con '93)*, Catalyst Consultant, Inc., Houston, TX, May 26–28, 1993.

Values of M_w/M_n of 2 or less are common for the soluble catalyst systems whereas values of 4–8 are usual for ZNC systems. The soluble catalyst systems also are able to polymerize a larger number and greater variety of vinyl monomers to form homogeneous polymers and copolymers in comparison to solid catalyst systems.

The active site is a cationic metallocene-alkyl generated by the reaction of a neutral metallocene formed from the reaction with excess MAO or other suitable cocatalysts such as a borane Lewis acid. This sequence is shown in Figure 6.1 using MAO with ethylene to form polyethylene. Initiation and propagation occur through the precoordination and the insertion of the ethylene into the alkyl group-polymer chain. Here termination occurs through beta-hydride elimination producing a zirconium hydride and a long-chain alpha-olefin. These long-chain alpha-olefins can form linear high-density polyethylene or be used as comonomers with monomers such as 1-propylene, 1-hexene, or 1,5-hexadiene to give a variety of branched and linear products. These Group IV B metallocene catalysts are very active producing yields in excess of 1 ton of polyethylene per gram of catalyst per hour with a total efficiency on the order of 25 ton of PE per gram of catalyst.

These catalysis systems are also used to form other hydrocarbon polymers such as a variety of polypropylenes.

A major limitation of such Group IV B metallocene catalysts is that they are very air and moisture sensitive and not tolerant of heteroatom-containing monomers. In the case of heteroatom-containing monomers, the unbonded electron pairs on the heteroatom, such as oxygen, preferentially coordinate to the Lewis acid metal center in place of the carbon–carbon double bond. Some so-called middle and late transition metal organometallics are more tolerant to the presence of such heteroatoms and can be used as effective cocatalysts. These include some palladium, iron, cobalt, and nickel initiators.

FIGURE 6.1 Proposed mechanism for soluble stereoregulating catalyst polymerizations.

The use of transition and selected main group metal catalysis is increasing with the ability to design special catalytic systems for special polymer architecture and property production. These catalysis systems involve the transition metal as a site for active polymer growth. The new soluble stereoregulating catalysts are one example of these systems. These growing sites may be more or less ionic/covalent depending upon the catalyst used and such sites are not generally appreciably dissociated, as is the case in classical cationic and anionic systems. The metal's ligands can provide both electronic and steric structural control and are generally more robust in comparison to the anionic/cationic systems. Along with many advantages, there are some challenges. Because of their very nature, transition metal initiators can be very complex requiring several synthetic steps; they may be expensive and/or require costly cocatalysts; and control of the particular reaction conditions is very important since small, seemingly subtle changes can be magnified into larger polymer structural changes.

There is an increasingly large number of metal-catalyzed polymerizations including olefin metathesis reactions, which includes ring-opening metathesis polymerizations (ROMPs), formation of polyketones from the copolymerization of carbon monoxide, group transfer polymerizations, and step-growth addition/elimination (coupling) polymerizations. The study of metal catalytic sites is a vigorous area of ongoing research.

Polymers produced from single-site catalysts are increasingly being used in the marketplace. As noted above, the strength of the materials is increased because of the greater order in the individual polymer chains. For polyethylene this means the number of branches is less and for substituted polymers such as polypropylene this means that the order about the substituted-carbon is increased allowing for a denser, tighter fit of the individual polymer chain segments resulting in increased overall polymer strengths and less permeability for materials.

The use of materials produced from single-site catalysts in areas using thin films is increasing. For instance, bananas are generally produced at one location, and shipped and stored to other locations for sale. Even when picked green, they ripen rapidly when exposed to oxygen. Regular LLDPE is generally employed as a thin film to protect bananas for shipment and storage. Regular LLDPE permits some transfer of oxygen and because of the somewhat pointed nature of bananas, the film may be punctured. Single-site metallocene-based LLDPE is less permeable and less apt to tear and is now replacing regular LLDPE in this use. Its use is also increasing in the containment of heavier materials such as topsoil and water-purification salt using films. In both cases, thinner films, and consequently less film material is necessary to give an equal or better job performance. Single-site produced materials also offer better clarity, toughness, and easy sealability.

6.7 POLYETHYLENES

Tupperware was the idea of Earl Silas Tupper, a New Hampshire tree surgeon and plastics innovator. He began experimenting with polyethylene during the early part of World War II. In 1947, he designed and patented the famous "Tupper seal" that "sealed in" freshness. In order to close the container it had to be "burped" to remove air. Tupperware was also bug proof, spill proof, did not rot or rust, and did not break when dropped. Even with all of these advantages, few were sold. Brownie Wise was a divorced single mother from Detroit who desperately needed to supplement her income as a secretary. Her innovative idea was "Tupperware Parties." By 1951, Tupper had withdrawn all of the Tupperware from the stores and turned over their sales to Brownie Wise with the only source of the ware being through the Tupperware Parties.

The initial synthesis of polyethylene is attributed to many. In 1898, Han von Pechmann prepared it by accident while heating diazomethane. His colleagues Eugen Bamberger and Friedrich Tschirner characterized the white solid as containing methylene units and called it polymethylene. Polyethylene, from the ethylene monomer, was probably initially synthesized by M. E. P. Friedrich while a graduate student working for Carl S. Marvel in 1930 when it was an unwanted by-product from the reaction of ethylene and a lithium alkyl compound. In 1932, British scientists at the Imperial

Chemical Industries, ICI, accidentally made polyethylene while they were looking at what products could be produced from the high-pressure reaction of ethylene with various compounds. On March 1933, they found the formation of a white solid when they combined ethylene and benzaldehyde under high pressure (about 1400 atm pressure). They correctly identified the solid as polyethylene. They attempted the reaction again, but with ethylene alone. Instead of again getting the waxy white solid, they got a violent reaction and the decomposition of the ethylene. They delayed their work until December 1935 when they had better high-pressure equipment. At 180°C, the pressure inside of the reaction vessel containing the ethylene decreased consistently with the formation of a solid. Because they wanted to retain the high pressure, they pumped in more ethylene. The observed pressure drop could not be totally due to the formation of polyethylene, but something else was contributing to the pressure loss. Eventually they found that the pressure loss was also due to the presence of a small leak that allowed small amounts of oxygen to enter into the reaction vessel. The small amounts of oxygen turned out to be the right amount needed to catalyze the reaction of the additional ethylene that was pumped in subsequent to the initial pressure loss (another "accidental" discovery). The ICI scientists saw no real use for the new material. By chance, J. N. Dean of the British Telegraph Construction and Maintenance Company heard about the new polymer. He had needed a material to encompass underwater cables. He reasoned that polyethylene would be water resistant and suitable to coat the wire protecting it from the corrosion caused by the salt water in the ocean. In July of 1939, enough polyethylene was made to coat 1 nautical mile of cable. Before it could be widely used, Germany invaded Poland and polyethylene production was diverted to making flexible high frequency insulated cable for ground and airborne radar equipment. Polyethylene was produced, at this time, by ICI, Dupont, and Union Carbide for the United States.

Polyethylene did not receive much commercial use until after the war when it was used in the manufacture of film and molded objects. Polyethylene film displaced cellophane in many applications being used for packaging produce, textiles, and frozen and perishable foods, etc. This polyethylene was branched and had a relatively low softening temperature, below 100°C, preventing its use for materials where boiling water was needed for sterilization. The branched polyethylene is called low-density, high-pressure polyethylene because of the high pressures usually employed for its production and because of the presence of the branches, the chains are not able to closely pack, leaving voids and subsequently producing a material that had a lower density in comparison to low-branched polyethylene.

Karl Ziegler, director of the Max Planck Institute for Coal Research in Muelheim, Germany was extending early work on polyethylene attempting to get ethylene to form polyethylene at lower pressures and temperatures. His group found that certain organometallics prevented the polymerization of ethylene. He then experimented with a number of other organometallic materials that inhibited polyethylene formation. Along with finding compounds that inhibited polyethylene formation, they found compounds that allowed the formation of polyethylene under much lower pressures and temperatures. Further, these compounds produced a polyethylene that had fewer branches and higher softening temperatures.

Giulio Natta, a consultant for the Montecatini company of Milan, Italy applied the Ziegler catalysts to other vinyl monomers such as propylene and found that the polymers were higher density, higher melting, and more linear than those produced by the then classical techniques such as free radical initiated polymerization. Ziegler and Natta shared the Nobel Prize in 1963 for their efforts in the production of vinyl polymers using what we know to day as solid-state stereoregulating catalysts.

While many credit Natta and Ziegler as first having produced so-called high-density PE and stereoregular polyolefins, Phillips' scientists first developed the conditions for producing stereospecific olefin polymers and high-density PE. In 1952, J. Paul Hogan and Robert Banks discovered that ethylene and propylene polymerized into what we today know as high-density PE and stereoregular PP. As with many other advancements, their initial studies involved other efforts, here to improve fuel yields by investigating catalysts that converted ethylene and propylene to higher molecular weight products. They found that chromium trioxide supported on a silica–alumina catalyst produced a hard solid rather than the usual waxy-like PE. They quickly looked at other olefins and soon discovered a crystalline PP, namely a stereoregular PP, specifically i-PP.

While the common name for the monomer is ethylene, the official name is ethane so that what we know as polyethylene is often referred to as polyethene or polythene. Even so, the name polyethylene is so entrenched in our common vocabulary it will be employed here. Polyethylene can be produced using radical, anionic, cationic, and ion coordination polymerization. This is a result of the lack of substituents on the ethylene monomer. Each of these different polymerizations results in a different type of polyethylene. Today there exists a wide variety of "polyethylenes" that vary in the extent and length of branching as well as molecular weight, molecular weight distribution, and amount of crystallinity. Some of these are pictured in Figure 6.2. Commercial low-density polyethylene, LDPE, typically has between 40 and 150 short alkyl branches for every 1000 ethylene units. It is produced using high pressure (15,000 to 50,000 psi) and temperatures (to 350°C). It has a density of about 0.912–0.935. Because of the branching, the LDPE is amorphous (about 50%) and sheets can allow the flow-through of liquids and gasses. Because of the branching and low amount of crystallinity, LDPE has a low melting point of about 100°C making it unsuitable for uses requiring sterilization using boiling water. LDPE has a combination of short to long branches, at a rate of about 10 short branches to every long branch.

High-density polyethylene, HDPE, produced using organometallic catalysts such as the Ziegler–Natta or Phillips catalysts, have less than 15 (normally within the range of 1 6) short alkyl branches (essentially no long branches) for 1000 ethylene units. Because of the regular structure of the ethylene units themselves and the low extent of branching, HDPE chains can pack more efficiently

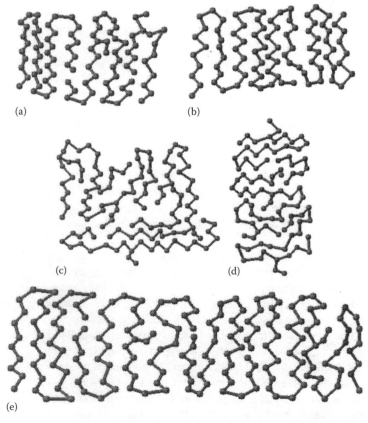

(a)   (b)

(c)   (d)

(e)

FIGURE 6.2   (a) Ball-and-stick models of HDPE, (b) UHMWPE, (c) LDPE, (d) LLDPE, and (e) ULPE.

resulting in a material with greater crystallinity (generally up to 90%), higher density (0.96), with increased chemical resistance, hardness, stiffness, barrier properties, melting point (about 130°C), and tensile strength. Low molecular weight (chain lengths in the hundreds) HDPE is a "wax" while "typical" HDPE is a tough plastic.

Linear low-density polyethylene, LLDPE, can be produced with less than 300 psi and at about 100°C. It has a density between 0.915 and 0.925 g/mL. It is actually a copolymer of ethylene with about 8%–10% of an alpha-olefin such as 1-butene, 1-pentene, 1-hexene, or 1-octene. Through control of the nature and amount of alpha-olefin, we are able to produce materials with densities and properties between those of LDPE and HDPE. LLDPE does not contain the long branches found in LDPE. Because of its toughness, transparency, and flexibility it is used in film applications as packaging for cables, toys, pipes, and containers.

Very low-density polyethylene, VLDPE, has a density range between 0.88 and 0.915 g/mL. It is largely linear chains with high amount of short-chain branching generally made by copolymerization of the ethylene with short-chain alpha olefins such as 1-butene, 1-hexene, and 1-octene so that it is structurally similar to LLDPE except the alpha-olefins are generally longer resulting is a lower density. VLDPE is generally made using metallocene catalysts.

Ultrahigh molecular weight polyethylene, UHMWPE, is a high-density polyethylene with chain lengths over 100,000 ethylene units. Because of the great length of the chains, they "inter-tangle" causing physical cross-links, increasing the tensile strength, and related properties of these materials. (By comparison, HDPE rarely is longer than 2000 ethylene units.) UHMWPE is about 45% crystalline and offers outstanding resistance to corrosion and environmental stress-cracking, outstanding abrasion resistance and impact toughness, and good resistance to cyclical fatigue and radiation failure, and with a low surface friction. It is produced using catalysts systems similar to those used for the production of HDPE (i.e., Ziegler–Natta and Phillips catalysts). It has a density of about 0.93.

Ultralinear polyethylene, ULPE, has recently become available using soluble stereoregulating catalysts. Along with a decreased amount of short-chained alkyl branching, ULPE has a narrower molecular weight spread.

At extremely high chain lengths, fibers can be made of largely linear polyethylene producing a material that is extremely strong and resistant to punctures and cutting even by scissors. One such product is "Spectra" composed of high molecular weight, about 70,000 units long, polyethylene fibers stretched about 100% to align the chains. Spectra is used in the construction of bullet-resistant vests and surgical gloves where the nicking and cutting of the surgeon's hands by the sharp instruments is unwelcome. Spectra is said to be 10 times stronger than steel on a weight basis and 35% stronger than armid fibers such as Kevlar. Like other woven materials, these gloves are vulnerable to punctures by sharp objects such as needles.

Cross-linked polyethylene, PEX or XLPE, is a medium to high-density PE that contains cross-links resulting in a thermoset material. The cross-linking causes a reduced flow for the material and permeability and increased chemical resistance. Because of the good retention and stability toward water, some water-pumping systems use PEX tubing since its cross-linking causes the tubing, once expanded over a metal nipple, to return to its original shape resulting in a good connection.

Medium-density polyethylene, MDPE, has a density between about 9.3 and 9.4 g/mL. It is mainly produced using ion coordination polymerization. It has good shock and drop resistance.

Polymethylene can be produced through several routes including the use of diazomethane or a mixture of carbon monoxide and hydrogen. This polymer has only a little branching.

It is well accepted that the history of a polymer, including polymer processing, influences polymer behavior. Some of these influences are just becoming known. Interestingly, as in much of science, once the critical parameters are known on a macro-level, we are able to better understand them on a molecular and conceptual level. An example that illustrates this involves the processing of annealed polyethylene. In general, for most linear polyethylene, micelles and associated spherulites are formed when it is melted and then slowly cooled. If no force is applied during the annealing process (simple melt crystallization), a high amount of force and large deformation is required to breakdown the initial spherulite structures with reformation occurring along the axis of the pull when high-strength polyethylene rod, film, and sheet is produced. However, if the polyethylene is crystallized under pressure applied in one direction, less energy and lower deformation is required to align the polyethylene spherulites since the spherulites are already partly aligned. In both cases, stretching of the molecular network is required. For the simple melt crystallized polyethylene, the original spherulite structure is destroyed during the deformation followed by the formation of new fibrillar structures. For the pressure-associated annealing process, elongated micelles are formed that largely remain after the deformation process.

LDPE films are nearly clear even though they contain a mixture of crystalline and amorphous regions. This is because the crystalline portions are space filling and not isolated spherulites allowing a largely homogeneous structure with respect to refractive index resulting in a transparent material. In fact, the major reason that LDPE films appear hazy or not completely transparent is because of the roughness of the surface and is not due to the light scattering of the interior material.

Space-filling models of amorphous and crystalline linear polyethylene are given in Figures 6.3 and 6.4.

Typical uses of the various polyethylenes include

1. UHMWPE-battery separators, lightweight fibers, permanent solid lubricant materials in railcar manufacture, automobile parts, truck liners; liners to hoppers, bins, and chutes; farm machinery as sprockets, idlers, wear plates, and wear shoes; moving parts of weaving machines and can and bottle handling machines, artificial joints including hip and knee replacements, gears, butcher's chopping boards, sewage-treatment-bearings, sprockets, wear shoes; lumbering-chute, sluice and chain-drag liners; neutron shields; and also as components in bullet-resistant wear.
2. "Typical" HDPE-blow-molded products—bottles, cans, trays, drums, tanks, and pails; injection-molded products—housewares, toys, food containers, cases, pails, and crates; films, pipes, bags, conduit, wire and cable coating, foam, insulation for coaxial and communication cables.
3. Low molecular weight HDPE-spray coatings, emulsions, prinking inks, wax polishes, and crayons.

FIGURE 6.3   Space-filling structure of a portion of a linear amorphous polyethylene region.

FIGURE 6.4   Space-filling structure of a portion of a linear crystalline polyethylene region.

4. LDPE-packaging products, bags, industrial sheeting, and piping and tubing, films, garbage cans, industrial containers, and household items.
5. LLDPE-telephone jacketing, wire and cable insulation, piping and tubing, drum liners, bottles, and films.
6. MDPE-shrink film, packaging film, gas pipes, and fittings.
7. VLDPE-frozen food and ice bags, food packaging, and stretch wrap.

Recently, efforts are underway to form ethylene, and hence polyethylene, from green chemistry feedstocks—sugarcane-derived ethanol and other natural feedstocks.

Plastomers is the name given to copolymers of ethylene that have a little crystallinity, but are largely amorphous. They are also called very low-density polyethylene, VLDPE. They are more elastic than LLDPE but less stiff. They are used as a sealing layer in film applications and controlled permeation packaging for vegetables and fruits.

Most shopping bags are made from PE, PP, or paper. Today, there is a move to remove the so-called nongreen plastic bags. A brief comparison of the two is in order. In comparison to paper bags, plastic bags are lighter, stronger, lower cost, and water and chemical resistant. They cost less energy to produce, transport, and recycle. When disposed of improperly plastic bags are unsightly and represent a hazard to wildlife and infants. Neither paper (because of lack of oxygen) nor plastic bags decompose in landfills. Both can be recycled. Paper is made from a renewable resource, cellulose from trees. With the move toward PLA bags, long-term disposal and degradation will become less of a problem. Thus, the answer is not apparent but what is apparent is that we must increase our efforts to recycle and properly dispose of both plastic and paper bags.

6.8   POLYPROPYLENE

Polypropylene, PP, is one of the three most heavily produced polymers (Section 1.3). The abundance of PP is the result of the variety of PP produced, its versatility allowing a wide variety of products to be produced, the availability of a large feedstock, and its inexpensiveness. Today, PP is used in such diverse applications as a *film* in disposable diapers and hospital gowns to geotextile liners; *plastic* applications as disposable food containers and automotive components; and *fiber* applications such as in carpets, furniture fabrics, and twine.

While PP was produced in 1951 by Karl Rehn, it only became commercially available in the late 1950s with the production by Natta and coworkers at Phillips of somewhat stereoregular PP. The first PP was not highly crystalline because the tacticity, a measure of the stereoregularity, was only approximate. But, with the use of the Natta–Ziegler and Phillips catalysts systems PP with greater stereoregularity was produced giving PP with enhanced physical properties such as increased stiffness, better clarity, and a higher distortion temperature. Today, with better catalysts, including the soluble metallocene catalysts, the tacticity has been increased so that PP with 99% isotacticity can be produced. The more traditional Natta–Ziegler catalysts have high catalystic efficiencies with one gram of catalyst producing 1 kg of PP. This high catalytic efficiency eliminates the need for catalyst removal. Most iPP is made using bulk propylene, either as a gas or as a liquid.

A brief side trip allows some insight into industrial workings. Phillips Petroleum Company was busily working on ways to improve refinery processes because they were in the 1950s almost solely a fuel company. Two young chemists, J. Paul Hogan and Robert L. Banks, were trying to develop catalysts that would act as high-performance gasoline additives when one of the catalysts, mixed with the petroleum propylene present in a pipe, plugged up the pipe with a whitish, taffy-like material. While many companies might have told Hogan and Banks to get back to their original efforts that were the "bread-and-butter" of the company, instead they were told to investigate the formation of this off-white material. At this time, most of the known plastics were either too brittle or softened at too low a temperature for most practical uses. This off-white material, produced from propylene, could be hardened giving a flexible material that melted above 100°C. After some effort, it was discovered that this catalyst could also be used to give a polyethylene that was superior to the "old" polyethylene. Eventually, the "old" polyethylene would be given the name of low-density polyethylene, LDPE, and the new polyethylene the name of high-density polyethylene, HDPE.

In 1953, a patent was applied for covering both the synthesis of polypropylene and polyethylene under the trade name of Marlex. Even after a material is discovered, many steps are needed before a product becomes available to the public. Management was getting glowing reports about the potential for these new materials. Based on these assessments, $50 million was committed for the construction of a large-scale polyethylene plant at Adams Terminal.

Several things were in operation that were not immediately apparent. First, the great investment of money and time needed to bring a product into the marketplace. At that time $50 million was a lot of money, about half a billion dollars today. Second, up to that time, technicians with little training were used in much of the chemical industry. The "educated" chemists were kept behind the research benches discovering "new" things. Most of the PhD chemists were, in fact, in universities rather than in industry. (Today, only about 11% of the professional chemists are teaching, the others are in industry and government.) Because of the sensitivity of the catalyst system used to produce the new polymers, substantially greater training had to be given to workers who dealt with the production of these new polymers. Third, this was a venture into a new arena by Phillips, up to now solely a gasoline-producing company. Finally, while not initially apparent, the catalysts used to produce these new polymers were part of a new group of catalysts being investigated internationally by many companies that allowed the production of so-called stereoregular polymers.

As noted previously, polypropylene, and other alpha-olefins, can reside in three main stereoregular forms differentiated from one another because of the precise geometry of the methyl group as the polymer is formed. When the methyl groups reside on the same side as one looks down a barrow of the chain, as below, it is called isotactic-polypropylene or simply i-PP.

(6.42)

The methyl group can also exist in alternate positions as shown below. This form is called syndiotactic-polypropylene or simply s-PP.

$$
\tag{6.43}
$$

The third structure consists of mixtures of the syndiotactic and isotactic structures favoring neither structure. This mixture of structures is called the atactic form, a-PP, the "a" meaning having nothing to do with. Space-filling models of all three different tactic forms are given in Figure 6.5.

Each particular tactic form has its own physical properties. The syndiotactic and isotactic forms are referred to as stereoregular forms and allow the chains to better fit together giving a material that is crystalline while the a-PP is amorphous. Being crystalline causes the polymer to be stronger, denser, less porous to small molecules such as water and oxygen, and to have a higher melting temperature. The old polypropylene is of the atactic form and the form synthesized by Hogan and Banks is i-PP. Only recently, using soluble stereoregulating systems has s-PP become commercially available.

Other groups were working on developing these stereoregulating catalysts. In fact, more research dollars and effort was spent during the 1950s developing these stereoregulating catalysts than that spent on cancer research during that time. The competition was fierce and monetary stakes high. The author was present at one of the initial presentations of the results from a number of different groups. Each group believed their catalyst system gave the best results and each believed that the product they obtained was the same product that other companies obtained. There were several times during the presentations where the groups would yell at one another and physically wrench the microphone from another speaker calling one another various unkind words. It was later learned that each of the groups had different catalytic systems that gave different products. Thus, the presenters were most probably giving accurate information but based on slightly differing catalyst systems.

Jumping from a laboratory-scale production to mass-scale production is difficult and made especially difficult because of the need to control the structure of the catalyst because only one form of

FIGURE 6.5 Space-filling models of syndiotactic (top), isotactic (middle), and atactic (bottom) forms of polypropylene. (My children and now grandchildren believe that the isotactic polypropylene is really a chorus-line of dancing teddy bears.)

the catalyst gave the desired i-PP and PE while other forms would give mixtures of polypropylene and PE structures. The first Marlex pellets came off the production line at the Adams plant and they were varied in color and size, and off-specification. It was hard for the sales staff to convince buyers that this was the miracle material that they had promised. The realization that the catalysts form was so important became painfully evident. The material was better than the old PE and PP but it was not as good as that obtained in the laboratory. Warehouse after warehouse of somewhat inferior material was produced with few buyers.

Relief was spelled, not "R-E-L-I-E-F," but rather "H-U-L-A H-O-O-P" that was reinvented by Richard Knerr and Arthur "Spud" Melin who founded the Wham-O company that also reinvented another old toy, the Frisbee. American children fell in love with the hula hoop and the somewhat inferior Phillips material was good enough to give good hula hoops. Demand for these plastic rings was sufficient to take the Adams plant output for a half year and turn it into hula hoops. Phillips president, Paul Endacott, was so pleased that he kept a hula hoop in his office to remind him of his "savior." By the time that particular wave of hula hoop mania wound down in 1959, the problems in the product line at the Adam terminal were overcome, and Marlex had found new markets. One of the first markets was the use of PE and PP baby bottles in hospitals to replace the old glass bottles. The Marlex bottles were less expensive, could be sterilized (remember, both the new PE and PP could be heated to boiling, 100°C, without melting), and would not shatter when dropped. The manufacture of Chiffon household liquid detergent took a chance on using plastic bottles rather than glass to hold their detergent. Their success caused other manufacturers to change to the lighter, nonbreakable, and less expensive plastic containers. Today, PP and PE holds the major share of the container market.

Copying the tubes used to make the hula hoop, plastic tubing for many varied application such as connecting air conditioners and ice makers, took off. Today, the three most used synthetic polymers are LLDPE, HDPE, and PP. New catalysts and production procedures has allowed the physical properties and varied uses of these "big three" synthetic polymers to be continually increased.

While it was possible to produce syndiotactic PP, s-PP, using the Ziegler–Natta solid systems, commercial s-PP has only recently become commercial using the soluble metallocene catalysts. These materials have a similar T_g as i-PP but they have a different balance between stiffness and toughness.

Atactic or amorphous forms of PP are also used. Initially atactic PP, a-PP, was obtained as a by-product of the production of i-PP. As an inexpensive by-product, it is used as a modifier for asphalt for roofing and in adhesives. As the effectiveness of catalyst systems becomes better, less a-PP is available so that today some a-PP is intentionally made for these applications.

While PP is often synthesized using Ziegler–Natta catalyst, today there exists many other catalyst systems, including the soluble catalysts systems described in Section 6.7. In 1980, Walter Kaminsky and coworkers developed many of these systems. These catalysts contain a Group IVB metallocene coupled with methylaluminoxane, MAO. Through varying the structure of the metallocene cocatalyst it is possible to tailor the tacticity of the product. Thus, for PP, the atactic product is formed when the simple zironocene and hafnocene dichlorides are employed; syndiotactic PP is produced when a bridged catalyst is employed; and isotactic PP is produced when another bridged catalyst is employed.

Stretching a film tends to orientate the polymer chains. Biaxially stretched or oriented polypropylene, BOPP, is strong and clear resulting in its wide use producing clear bags and packaging materials. PP is often used in the construction of cold-weather clothing under-layers such as long-sleeved shirts and underwear. It is also used in warm-weather wear because of its ability to remove perspiration from the skin. It is used in ropes that can float because of the low density of PP. In low-ventilation areas, it often replaces PVC because it emits less smoke and no dangerous chlorine-containing fumes. It serves as the suture material in Prolene. Because of its ability to retain color and to form thin sheets, it is being used to make stationary folders, storage boxes, and the cube stickers for the Rubik's cubes. It is also used to construct trading card holders (such as baseball cards) with each sheet containing three or nine pockets for the trading cards. In medicine, it is also used in hernia repair operations with a small sheet or a film of PP placed over the spot of the hernia to prevent the reformation of the surgical hernia.

Expanded polypropylene, EPP, foam is used for radio-controlled model aircraft, automobiles, and trucks by hobbyists because of the ability of the foam to absorb impacts and its low density.

6.9 POLYMERS FROM 1,4-DIENES

There are three important 1,4-dienes employed to produce commercially important polymers. These monomers possess a conjugated pi-bond sequence of $-C=C-C=C-$ that readily forms polymers with an average energy release of about 20 kcal/mol (80 kJ/mol) with the conversion of one of the double bonds into a lower (potential energy wise; generally more stable) energy single bond. For all of these products, cross-linking and grafting sites are available through the remaining double bond.

1,4-Butadiene can form three repeat units as described in Equation 6.36, the 1,2; *cis*-1,4; and *trans*-1,4. Commercial polybutadiene is mainly composed of the 1,4-*cis* isomer and is known as butadiene rubber, BR. In general, butadiene is polymerized using stereoregulating catalysts. The composition of the resulting polybutadiene is quite dependent on the nature of the catalyst such that almost total *trans*-1,4 units, or *cis*-1,4 units, or 1,2 units can be formed as well as almost any combination of these units. The most important single application of polybutadiene polymers is its use in automotive tires where over 10^7 ton are used yearly in the manufacture of automobile tires in the United States. BR is usually blended with natural rubber, NR, or styrene–butadiene rubber, SBR, to improve tire tread performance, particularly wear resistance.

The second use of butadiene is in the manufacture of ABS copolymers where the stereogeometry is also important. A polybutadiene composition of about 60% *trans*-1,4; 20% *cis*-1,4; and 20% 1,2 configuration is generally employed in the production of ABS. The good low temperature impact strength is achieved in part because of the low T_g values for the compositions. For instance, the T_g for *trans*-1,4-polybutadiene is about −14°C while the T_g for *cis*-1,4-polybutadiene is about −108°C. Most of the ABS rubber is made using an emulsion process where the butadiene is initially polymerized forming submicron particles. The styrene–acrylonitrile copolymer is then grafted onto the outside of the BR rubber particles. ABS rubbers are generally tougher than HIPS rubbers but are more difficult to process. ABS rubbers are used in a number of appliances including luggage, power tool housings, vacuum cleaner housings, toys, household piping, and automotive components such as interior trim.

Another major use of butadiene polymer is in the manufacture of high impact polystyrene, HIPS. Most HIPS has about 4%–12% polybutadiene in it so that HIPS is mainly a polystyrene intense material. Here the polybutadiene polymer is dissolved in a liquid along with styrene monomer. The polymerization process is unusual in that both a matrix composition of polystyrene and polybutadienes is formed as well as a graft between the growing polystyrene onto the polybutadiene is formed. The grafting provides the needed compatibility between the matrix phase and the rubber phase. The grafting is also important in determining the structure and size of rubber particles that are formed. The grafting reaction occurs primarily by hydrogen abstraction from the polybutadiene backbone either by growing polystyrene chains or alkoxy radicals if peroxide initiators are employed.

Interestingly isoprene, 2-methyl-1,3-butadiene, exists as an equilibrium mixture of cis and trans isomers.

$$(6.44)$$

Polyisoprene is composed of four structures as shown in Equation 6.37. As in the case of polybutadiene, it is the *cis*-1,4 structure that is emphasized commercially. The *cis*-1,4-polyisoprene is similar to the *cis*-1,4-polybutadiene material except it is lighter in color, more uniform, and less expensive to process. Polyisoprene is analogous to natural rubber in composition. The complete *cis*-1,4 product has a T_g of about −71°C. Interestingly, isomer mixtures generally have higher T_g values. Thus an equal molar product containing *cis*-1,4; *trans*-1,4; and 3,4 units has a T_g of about −40°C.

As with many polymers, polyisoprene exhibits non-Newtonian flow behavior at shear rates normally used for processing. The double bond can undergo most of the typical reactions such as carbene additions, hydrogenation, epoxidation, ozonolysis, hydrohalogenation, and halogenation. As with the case of the other 1,4-diene monomers, many copolymers are derived from polyisoprene or isoprene itself.

Polyisoprene rubbers are used in the construction of passenger, truck, and bus tires and inner liners as well as sealants and caulking compounds, sporting goods, gaskets, hoses, rubber sheeting, gloves, belts, and footwear.

Polychloroprene was the first commercially successful synthetic elastomer introduced in 1932 under the trade names of DuPrene and Neoprene by Dupont. Carothers and coworkers discovered it. Because of its early discovery, good synthetic routes were worked out prior to the advent of good stereoregulating catalytic systems. Thus, polychloroprene is largely manufactured by emulsion polymerization using both batch and continuous systems. Free radical products contain mainly 1,4-*trans* units. Along with the four "main" structural units analogous to those of polyisoprene, sequence distributions are available for both polyisoprene and polychloroprene. Polymerization can occur with the growing end being the 4 end or the 1 end (Equation 6.45). Generally the 1,4-polymerization sequence is favored with the growing end being carbon 4.

(6.45)

Structural regularity for the inclusion of the 1,4-*trans* unit is inversely proportional to temperature. Thus, at 90°C the product contains about 85% of the *trans*-1,4 units while this increases to almost 100% at −150°C. Both uncured and cured polychloroprene exists as largely crystalline materials because of the high degree of stereoregularity. Cured polychloroprene has good high tensile strength because of this and the application of stress to the material, either before or after curing, increases the tensile strength. The *trans*-1,4-polychloroprene has a T_g of about −49°C while 1,4-*cis*-polychloroprene has a T_g of about −20°C.

The compounding of polychloroprene is similar to that of natural rubber. Vulcanizing is achieved using a variety of agents including accelerators. Because of its durability, polychloroprene rubber is often used where deteriorating effects are present. It offers good resistance to oils, ozone, heat, oxygen, and flame (the latter because of the presence of the chlorine atom). Automotive uses include as hoses, V-belts, and weather stripping. Rubber goods include gaskets, diaphragms, hoses, seals, conveyer belts, and gaskets. It is also used in construction for highway joint seals, bridge mounts, expansion joints, and soil-pipe gaskets. Further, it is also used for wet-laminating and contact-bond adhesives, in coatings and dipped goods, as modifiers in elasticized bitumens and cements, and in fiber binders.

As noted above, the presence of the chlorine atom in polychloroprene makes it less apt to burn in comparison to materials such as hydrocarbon-only elastomers. Thus, it is used in fire doors, some combat related attire such as gloves and face masks and other similar applications. It is also used in the construction of objects that come into contact with water such as diving suits and fishing wader boots.

6.10 POLYISOBUTYLENE

Polyisobutylene, PIB; Equation 6.46, was initially synthesized in the 1920s but was developed by William Sparks and Robert Thomas at Standard Oil's (to become Exxon) Linden, NJ laboratory. It is one of the few examples of the use of cationic catalysis to produce commercial scale polymers. PIB and various copolymers are also called butyl rubber (IIR). Low molecular weight (about 5,000 Da) PIB can be produced at room temperature but large chains (over 1,000,000 Da) are made at low temperatures where transfer reactions are suppressed.

$$(6.46)$$

Because of the symmetry of the monomer, it might be expected that the materials would be quite crystalline like linear polyethylene. While PIB does crystallize under stress, it is does not crystallize under nonstressed conditions. This is because the geminal dimethyl groups on alternating carbons in the backbone cause the bond angles to be distorted from about the usual tetrahedral bond angle of 109.5° to 123° forcing the chain to straighten out. As a consequence of this the chains pack efficiently giving a relatively high-density material (density of 0.917 g/cc compared to densities of about 0.85 g/cc for many amorphous polymers) even when amorphous. This close packing reduces the incentive for crystallization, accounts for its low permeability, and produces an unusually low T_g of −60°C for such a dense material.

Polyisobutylene is often produced as a copolymer containing a small amount (1%–10%) of isoprene. Thus, the random copolymer chain contains a low concentration of widely spaced isolated double bonds that are later cross-linked when the butyl rubber is cured.

Butyl rubbers have lower permeability (including being essentially impermeable to air) and higher damping than other elastomers making them ideal materials for tire innerliners, basketball inside coatings, and engine mounts. It is used as an adhesive, caulking compound, chewing gum base, and an oil additive. Its use as an oil additive is related to its change in shape with increasing temperature. Since lubricating oil is not a good solvent for polyisobutylene, polyisobutylene is present as a coil at room temperature. However, as the temperature increases it begins to uncoil acting to counteract the decrease in the viscosity of the oil as the temperature is increased. PIB is also used in sealing applications such as in roof repairs.

It is being used in a number of green chemistry applications. Polyisobutylene (in the form of polyisobutylene succinimide) is added in small amounts to lubricating oils reducing the creation of an oil mist, thus lowering the workers' exposure to the oil mist. It is part of Elastol employed to clean up waterborne oil spills. It increases the crude oil's viscoelasticity resulting in an increased ability of the oil to remain together as it is vacuumed from the water's surface.

Polyisobutylene acts as a detergent and is used as a fuel additive in diesel fuel reducing fuel injector fouling. This in turn results in increased mileage and lowered unwanted emissions. It is also employed as part of a number of detergent packages that are blended into gasoline and diesel fuels.

6.11 METATHESIS REACTIONS

Chauvin, Grubbs, and Schrock won the 2005 Nobel Prize for developing metathesis reactions. Olefin metathesis is a catalytically induced reaction wherein olefins, such as cyclobutene and cyclopentene, undergo bond reorganization resulting in the formation of so-called polyalkenamers. Because the resulting polymers contain double bonds that can be subsequently used to introduce cross-linking, these materials have been used to produce elastomeric materials as well as plastics. Transition metal catalysts are required for these reactions. Catalysts include typical Natta–Ziegler types and other similar catalysts–cocatalysts combinations. The reactions can be run at room temperature and the stereoregularity can be controlled through the choice of reaction conditions and catalysts. For instance, the use of a molybdenum-based catalyst with cyclopentene gives the cis product whereas the use of a tungsten-based catalyst gives the trans product.

Trans-polypentenamer	*Cis*-polypentenamer
(6.47)	(6.48)

As expected, the methathesis polymerization of more strained cycloalkenes, such as cyclobutene, occurs more rapidly than less strained structures such as cyclopentene.

It is believed that polymerization occurs via a chain polymerization where ring opening occurs via the complete scission of the carbon–carbon double bond through the reaction with metal carbene precursors giving an active carbene species (Equation 6.49).

$$RCH = ML + CH = CH \longrightarrow \qquad \rightleftarrows \quad \rightleftarrows \quad \rightleftarrows \quad RH = CH - CH = ML \qquad (6.49)$$

where "L" is the ligand attached to the metal.

6.12 ZWITTERIONIC POLYMERIZATION

While most polymerizations require an initiator, catalyst, or some other forms of activation, zwitterionic copolymerizations do not. These copolymerizations require a specific combination of one monomer that is nucleophilic and a second that is electrophilic in nature. The interaction of these two comonomers gives a zwitterion that is responsible for both the initiation and the propagation.

Initiation $\qquad MN + ME \longrightarrow {}^+MN - ME^- \qquad (6.50)$

Propagation $\qquad {}^+MN - ME + {}^+MN - ME \longrightarrow {}^+MN - ME - MN - ME \longrightarrow \longrightarrow \longrightarrow \longrightarrow$

$$\longrightarrow {}^+MN - (-ME - MN) - ME \qquad (6.51)$$

If growth involves only addition and condensation reactions, then an alternating copolymer is formed. Sometimes a lateral reaction occurs where the zwitterion interacts with one of the monomers giving a product that is a statistical copolymer.

6.13 ISOMERIZATION POLYMERIZATION

Isomerization polymerizations are polyaddition reactions where the propagating species rearranges to energetically preferred structures prior to subsequent chain growth:

$$nA \longrightarrow -(-B-)_n{}^- \tag{6.52}$$

In 1962, Kennedy reported the first isomerization polymerization using 3-methyl-1-butene to give a 1,1-dimethyl polypropylene as below:

3-Methyl-1-butene   Poly-1,1-dimethylpropylene

$$\tag{6.53}$$

Isomerization polymerizations can be associated with coordination catalysts systems, ionic catalysts systems, and free-radical systems. The cationic isomerization polymerization of 4-methyl-1-pentene is of interest because the product can be viewed as an alternating copolymer of ethylene and isobutylene. This structure cannot be obtained by conventional approaches.

In the presence of certain Ziegler–Natta catalysts, an equilibrium exists between *cis*- and *trans*-1,3-pentadiene. Here, the *cis*-1,4-polypentadiene is formed from *trans*-1,3-pentadiene or from a mixture of the cis and trans isomers.

Such isomerizations are sometimes desired and sometimes are the cause of or explanation for unwanted structures. In the cationic polymerization forming poly(1-butene) nine different structural units have been found. Classical 1,2-hydride and 1,2-methide shifts, hydride transfer, and proton elimination account for these structures.

Unwanted branching of many polymers probably occurs through such isomerizations. Polypropylene, formed using cationic polymerization, has methyl, ethyl, *n*-propyl, *n*-butyl, isopropyl, gem-dimethyl, isobutyl, and *t*-butyl groups connected to the main chain.

6.14 PRECIPITATION POLYMERIZATION

Precipitation polymerization, also called slurry polymerization, is a variety of solution polymerization where the monomer is soluble but the polymer precipitates as a fine flock. The formation of olefin polymers via coordination polymerization occurs by a slurry process. Here the catalyst is prepared and polymerization is carried out under pressure and at low temperatures, generally less than 100°C. The polymer forms a viscous slurry. Care is taken so that the polymer does not cake-up on the sides and stirrer.

6.15 SUMMARY

1. Chain reactions, including ionic chain polymerization reactions, consist of at least three steps: initiation, propagation, and termination. Termination generally occurs through chain transfer producing a new ion and the "dead" polymer.

2. Cationic polymerizations occur with vinyl compounds that contain electron-donating groups using Lewis acids along with a cocatalyst as the initiator. Polymerizations generally occur at low temperatures in solvents with high dielectric constants. The $\overline{DP}$ is proportional to the concentration of monomer and the overall rate of polymerization is proportional to the square of the monomer concentration. In general, the rate of polymerization is dependent on the dielectric constant of the solvent, the resonance stability of the carbocation, the degree of solvation of the gegenion, and the electropositivity of the initiator.

3. Monomers with electron-withdrawing groups can undergo anionic polymerization in the presence of anionic initiators. The rate of polymerization is dependent on the dielectric constant of the solvent, the stability of the carbanion, the electronegativity of the initiator, the degree of solvation of the gegenion, and the strength of the electron-withdrawing substituents.

4. Stereoregular polymers can be formed. These polymers can be divided into three general stereoregular unit combinations. When the pendent groups, such as methyl for polypropylene, are all on one side of the polymer chain the polymer combination is isotactic; when the methyl groups alternate from one side to the other, the polymer combination is syndiotactic; and when the position of the methyl group is somewhat random it is atactic. The tacticity of polymers influence the physical properties of the products. In general, polymers with greater tacticity (order) have higher glass transition and melting temperatures, have a greater tendency to form crystalline products, are stronger, and are denser. Stereoregular polymers are produced at low temperatures in solvents that favor the formation of ion pairs between the carbocation and the gegenion. One of the most widely used stereoregulating systems is called the Ziegler–Natta catalyst system that generally uses a transition metal salt such as titanium chloride and a cocatalyst such as alkylaluminum. A proposed mechanism involves a reaction on the surface of $TiCl_3$, activated by the addition of an alkyl group from the cocatalyst. The monomer adds to this active site producing a pi complex, which forms a new active center by the insertion of the monomer between the titanium and carbon atoms. This step is repeated in the propagation reactions in which the alkyl group from the cocatalyst is the terminal group. Stereospecific polymers are also produced using the alfin and chromia on silica initiators. The alfin system consists of allyl sodium, sodium isopropoxide, and sodium chloride.

5. Soluble stereoregulating systems have been developed using an organometallic transition complex such as Cp_2TiCl_2 and a cocatalyst often methylalumoxane. This system has advantages over the Ziegler–Natta and similar systems in that the polymers produced are more stereoregular, a wider range of monomers can be used, and little or no catalyst incorporated into the polymers allowing the polymer to be directly used without having to undergo a procedure to remove the catalyst.

6. Two of the highest volume polymers are made using ionic polymerization—high-density polyethylene and i-PP. There exist a number of commercially available polyethylenes that vary in extent and kind of branching, chain length, and amount of crystallinity.

7. A number of polymers have been made using ring-opening polymerizations. Nylon-6, similar to structure and properties to nylon-6,6, is made from the ring opening of the lactam caprolactam. Poly(ethylene oxide) is made from the ring opening of ethylene oxide and is made more stable by capping the ends preventing ready depolymerization.

GLOSSARY

Alfin catalyst: Complex catalyst system consisting of allyl sodium, sodium isopropoxide, and sodium chloride.

Anionic polymerization: A polymerization initiated by an anion.

Butyl rubber (IIR): Copolymer of isobutylene and isoprene.

Capping: Reacting the end groups to produce a stable polymer.

Carbanion: Negatively charged organic ion.

Carbocation: Positively charged organic ion, that is, one lacking an electron pair on a carbon atom.

Cationic polymerization: Polymerization initiated by a cation and propagated by a carbonium ion.

Ceiling temperature: Threshold temperature above which a specific polymer is unstable and decomposes.

Celcon: Trade name of a copolymer of formaldehyde and dioxolane.

Chain-reaction polymerization: A rapid polymerization based on initiation, propagation, and termination steps.

Chain transfer: Process in which a growing chain becomes a dead polymer by abstracting a group from some other compounds, thereby generating another active site.

Copolymer: Polymer chain composed of units from more than one monomer.

Copolymerization: Polymerization of a mixture of more than one monomer.

Coupling: Joining of two active species.

Gegenion: A counterion.

Initiation: Start of a polymerization.

Isomerization polymerization: Polyaddition reaction in which the propagation species rearranges to energetically preferred structures prior to chain growth.

Kraton: Trade name for an ABA block copolymer of styrene–butadiene–styrene.

Lactom: Heterocyclic amide with one nitrogen atom in the ring.

Leuchs' anhydride: Cyclic anhydride that decomposes to carbon dioxide and an amino acid.

Living polymers: Macroanions or macrocarbanions.

Macroions: Charged polymer molecules.

Metathesis reaction: Catalytically induced reaction wherein olefins undergo bond reorganization resulting in the formation of polyalkenamers.

Monadic: Polyamide produced from one reactant.

Natta, Giulio: Discoverer of stereospecific polymers.

Oxirane: Ethylene oxide.

Polyacetal: Polyoxymethylene.

Polychloral: Polymer of trichloroacetaldehyde.

Promoter: Strained cyclic ethers that are readily cleaved.

Propagation: Continuous successive chain extension in a chain reaction.

Soluble stereoregulating catalyst: Soluble catalysts requiring a metal active site, cocatalyst or counterion, and a ligand system; capable of producing polymers with high stereoregularity and a minimum of branching.

Termination: The destruction of active-growing chain in a chain reaction.

Trioxane: Treimer of formaldehyde.

Ziegler, Karl: Discoverer of complex coordination catalysts.

Ziegler–Natta (or Natta–Ziegler) catalyst: Able to produce stereoregular polymers.

Zwitterionic polymerization: Copolymerization between nucleophilic and electrophilic comonomers.

EXERCISES

1. Describe the contents of the reaction flask 10 min after the polymerization of (a) reactants in a stepwise polymerization such as dimethyl terephthalate and ethylene glycol, and (b) monomer in chain reactions, such as isobutylene.
2. What is the initiator in the polymerization of isobutylene?
3. What is the general name of the product produced by cationic initiation?
4. What reactant besides the monomer is present in cationic chain propagation reactions?
5. What name is used to describe the negatively charged counterion in cationic chain-reaction polymerizations?
6. Is a Lewis acid (a) an electrophile or (b) a nucleophile?
7. Is a Lewis base (a) an electrophile or (b) a nucleophile?
8. Why is coupling not a preferred termination step in the cationic chain polymerization of a pure monomer?
9. Is the usual configuration of polymers produced by ionic chain polymerization (a) head-to-tail or (b) head-to-head.
10. Which condition would be more apt to produce stereoregular polymers in ionic chain polymerizations: (a) high temperatures or (b) low temperatures?
11. Name (a) a thermoplastic, (b) an elastomer, and (c) a fiber that is produced commercially by ionic chain polymerization.
12. Which technique would you choose for producing a polymer of isobutyl vinyl ether: (a) cationic or (b) anionic.
13. Which technique would you choose for producing a polymer of acrylonitrile: (a) cationic or (b) anionic?
14. Which of the following could be used to initiate the polymerization of isobutylene: (a) sulfuric acid, (b) boron trifluoride etherate, (c) water, or (d) butyllithium?
15. Which of the following could be polymerized by cationic chain polymerization?

(a)   (b)   (c)   (d)   (e)

16. Which polymer is more susceptible to oxidation: (a) HDPE or (b) PP?
17. When termination is by chain transfer, what is the relationship of average $\overline{DP}$ and the kinetic chain length?
18. What would be the composition of the product obtained by the cationic low-temperature polymerization of a solution of isobutylene in ethylene?
19. What is the relationship between the rates of initiation to the monomer concentration in ionic chain polymerization?
20. What effect will the use of a solvent with a higher dielectric constant have on the rate of propagation in ionic chain polymerization?
21. How does the rate constant k_p change as the yield of polymer increases?
22. Which will have the higher T_g value: (a) polystyrene or (b) polyisobutylene?
23. Which of the following could serve as an initiator for an anionic chain polymerization? (a) $AlCl_3$-H_2O, (b) BF_3-H_2O, (c) butyllithium, or (d) sodium metal?

24. What species, in addition to a dead polymer, is produced in a chain transfer reaction with a macrocarbocation in cationic chain polymerization?
25. What is the relationship between R_i and R_t under steady-state conditions?
26. What is the relationship between average $\overline{DP}$ and R_p and R_t?
27. Draw a structure of what i-PP looks like.
28. What percentage of polymer is usually found when a polymer produced by chain-reaction polymerization is heated above its ceiling temperature?
29. What is the relationship between the average degree of polymerization and initiator concentration in cationic chain polymerization.
30. Can the polymers found in the bottom of a bottle of insolubilized formaldehyde solution be useful?
31. How would you prepare stable polymers from formaldehyde?
32. Why is the thermal decomposition of polymers of formaldehyde called unzipping?
33. Discuss the advantages of the soluble stereoregulating catalysts in comparison to the Natta–Ziegler catalysts.
34. Why are there so many widely used polyethylenes?
35. How would you increase the flow rate of water in a fire hose?
36. Why is poly-3,3-bischloromethyloxybutylene crystalline?
37. Why are PP and PE the most widely used polymers?
38. What kind of polymers are made from lactones?
39. How could you remove unsaturated hydrocarbons from petroleum or coal tar distillates?
40. What species is produced by the reaction of an anionic chain polymerization initiator and the monomer?
41. What are the propagation species in anionic chain polymerizations?
42. Why are polymers produced by the anionic polymerization of pure monomers called "living polymers?"
43. Using the symbols A and B for repeating units in the polymer chain, which of the following is a block copolymer: (a) ABAABABBABAAB, (b) ABABABABAB, and (c) AAAAAABBBBBB?
44. What is the most widely used monadic nylon?
45. What is the repeating unit for nylon-4?
46. What is the catalyst and cocatalyst in the most widely used Ziegler–Natta catalyst?
47. Name two structures that are possible from the polymerization of 1,3-butadiene.
48. What is the principal difference between propagation reactions with butyllithium and a Ziegler–Natta catalyst?
49. List some physical properties that i-PP would have in comparison to a-PP.
50. Show the skeletal structures of *cis-* and *trans-*polyisoprene.
51. Write formulas for repeating units in the chains of (a) poly-1,4-isoprene and (b) poly-1,2-isoprene.
52. What is the most widely used catalyst for the production of HDPE?
53. What elastomer is produced by anionic chain polymerization?
54. What elastomer is produced by use of a Ziegler–Natta catalyst?
55. Why are there so many different kinds of PE?

FURTHER READINGS

Dragutan, V. and Dragutan, I. (2009): *Polymers from Cycloolefins*, Taylor & Francis, Boca Raton, FL.
Dubois, P., Coulembier, O., Raquez, J., and Degee, P. (2009): *Modern Arylation Methods,* Wiley, Hoboklen, NJ.
Gaylord, N. G. and Mark, H. F. (1959): *Linear and Stereoregular Addition Polymers*, Interscience, New York.
Giarrusso, A. (2004): *Stereospecific Polymerization and Stereoregular Polymers*, Wiley, Hoboken, NJ.
Hogan, J. P. and Banks, R. L. (1955): Philips process, U.S. Patent 2,717,888.

Moore, E. P. (1998): *The Rebirth of Polypropylene*, Hanser Gardner, Cincinnati, OH.

Morton, A. and Bolton, F. H. (1952): Alfin catalysis, *Ind. Eng. Chem.*, 40:2876.

Natta, G. and Danusso, F. (1967): *Stereoregular Polymers and Stereospecific Polymerization*, Pergamon, New York.

Oyama, S. T. (2008): *Mechanisms in Homogeneous and Heterogeneous Epoxidation Catalysis*, Elsevier, Amsterdam, the Netherlands.

Pain, C. F. (1993): *Proceedings Worldwide Metallocene Conference (Met Con '93)*, Catalyst Consultant, Inc., Houston, TX, May 26–28, 1993.

Scheirs, J. (2000): *Metallocene-Based Polyolefins*, Wiley, New York.

Severn, J. and Chadwick, J. (2008): *Tailor-Made Polymers: Via Immobilization of Alpha-Olefin Polymerization Catalysts*, Wiley, Hoboken, NJ.

Szwarc, M. (1968): *Carbanions, Living Polymers and Electron-Transfer Processes*, Wiley, New York.

Ziegler, K. (1952, 1955): Ziegler catalysts, *Angew. Chem.*, 64:323; 67:541.

7 Free Radical Chain Polymerization (Addition Polymerization)

Since many synthetic plastics and elastomers and some fibers are prepared by free radical polymerization, this method is important. Table 7.1 contains a listing of commercially important addition polymers including those that will be emphasized in this chapter because they are prepared using the free radical process.

As with other chain reactions, free radical polymerization is a rapid reaction consisting of the characteristic steps of initiation, propagation, and termination. Free radical initiators are produced by the homolytic cleavage of covalent bonds as well as numerous radiation-associated methods.

7.1 INITIATORS FOR FREE RADICAL CHAIN POLYMERIZATION

Free radical initiation can occur through application of heat (thermal), ultraviolet and visible light (photochemical), ionizing light, redox reagents, electricity (electrochemical), etc., that is, any process that creates the essential free radicals.

Light in the ultraviolet and visible range can disrupt selected bonds, thereby forming free radicals. Such disruption occurs as we are exposed to sunlight. Suntan treatments often contain certain compounds that can accept this damaging radiation. Related compounds are also used in foods to give them longer shelf life. They are generally known as antioxidants. Synthetic antioxidants include benzophenones, benzils, and certain organic ketones. Thus, diphenylketone decomposes on exposure to ultraviolet radiation of the appropriate wavelength forming two free radicals (Equation 7.1).

$$\text{Ph-}\overset{\overset{\displaystyle O}{\|}}{\text{C}}\text{-Ph} \xrightarrow{\text{UV}} \text{Ph-}\overset{\overset{\displaystyle O}{\|}}{\text{C}}{}^{\bullet} + \text{Ph}^{\bullet} \tag{7.1}$$

The advantage of using such photochemical initiations is that polymerization can be carried out at room temperature.

When molecules are exposed to light of higher energy, shorter wavelength, or higher frequency, electrons can be removed or added depending on the specific conditions. Usual forms of ionizing radiation employed industrially and experimentally include neutrons, x-rays, protons, and α and β particles. Oxidation–reduction, redox, reactions are also often employed to initiate free radical polymerizations in solution or heterogeneous organic-aqueous systems. Free radicals can be created by passing a current sufficient to initiate free radical polymerizations through a reaction system.

While application of heat or some other methods could rupture the pi bond in the vinyl monomer causing the formation of a two-headed free radical that can act as a free radical initiator, peroxides and dinitriles are generally employed as initiators. This is a consequence of the general bond dissociation energy trend of C–H > C–C > C–N > O–O. Dinitrile or azo compounds such as 2,2′-azobisisobutyronitrile, AIBN, require temperatures of about 70°C–80°C to produce decomposition with free radical formation. Peroxides such as benzoyl peroxide, BPO, require temperatures

TABLE 7.1
Industrially Important Addition Polymers

Polymer Name	Typical Properties	Typical Uses
Polyacrylonitrile	High strength; good stiffness; tough, abrasion resistant; good flex life; good resistance to moisture, stains, fungi, chemicals, insects; good weatherability	Carpeting, sweaters, skirts, socks, slacks, baby garments
Poly(vinyl acetate)	Water sensitive with respect to physical properties such as adhesion and strength; good weatherability, fair adhesion	Lower mol. wt. used in chewing gum, intermediate in production of poly(vinyl alcohol); water-based emulsion paints
Poly(vinyl alcohol)	Water soluble; unstable in acid and base systems; fair adhesion	Thickening agent for various suspension and emulsion systems; packaging film, wet-strength adhesive
Poly(vinyl butyral)	Good adhesion to glass; tough, good stability to sunlight; good clarity; insensitive to moisture	Automotive safety glass as the interlayer
Poly(vinyl chloride) and poly(vinylidene chloride); (called "the vinyls or vinyl resins"	Relatively unstable to heat and light; resistant to fire, insects, fungi, moisture	Calendered products such as film, Sheets, floor coverings; shower curtains, food covers, rainwear, handbags, coated fabrics, insulation for electrical cable and wire; old records
Polytetrafluoroethylene	Insoluble in most solvents, chemically inert, low dielectric loss, high dielectric strength, uniquely nonadhesive, low friction, constant electrical and mechanical properties from 20°C–250°C; high impact strength	Coatings for frying pans, wire, cable; insulation for motors, oils, transformers, generators; gaskets; pump and valve packings; nonlubricated bearings, biomedical
Polyethylene (LDPE)	Good toughness and pliability over wide temp. range; outstanding electrical properties; good transparency in thin films; resistant to chemicals, acids, bases; ages poorly in sunlight and oxygen; low density, flexible, resilient, high tear strength, moisture resistant	Films, sheeting used as bags, textile materials, pouches, frozen foods, produce wrapping, etc.; drapes, table cloths, covers for ponds, greenhouses, trash can liners, etc.; electrical wire and cable insulator; coating for foils, papers, other films, squeeze bottles
Polypropylene	Lightest major plastic; i-PP is major form sold; high tensile strength, stiffness, hardness, resistance to marring; good gloss, high T_g allows it to be sterilized; good electrical properties, chemical inertness, moisture resistance	Filament—rope, webbing, cordage, carpeting, injection molding applications in appliances, small house wares, and automotive fields
Polyisoprene (*cis*-1,4-polyisoprene)	Structurally close to natural rubber; properties similar to those of natural rubber; good elasticity, rebound	Replacement of natural rubber; often preferred because of greater uniformity and cleanliness

TABLE 7.1 (continued)
Industrially Important Addition Polymers

Polymer Name	Typical Properties	Typical Uses
SBR (styrene–butadiene rubber)	Random copolymer; generally slightly poorer physical properties than those of NR	Tire treads for cars; inferior to NR with respect to heat buildup and resilience, thus not used for truck tires; belting, molded goods, gum, flooring, rubber shoe soles, hoses, electrical insulation
Butyl rubber (copolymer of isobutylene)	Amorphous isoprene-largely 1,4 isomer; good chemical inertness, low gas permeability, high viscoelastic response to stresses, less sensitive to oxidative aging than most isoprene rubbers; better ozone stability than NR; good solvent resistance	About 60%–70% used for inner tubes for tires
Polychloroprene (mostly 1,4 isomer)	Outstanding oil and chemical resistance; high tensile strength; outstanding resistance to oxidative degradation, aging; good ozone and weathering response; dynamic properties same or better than most synthetic rubbers	Can replace NR in most applications; gloves, coated fabrics, cable and wire coatings, hoses, belts, shoe heels, solid tires
Polystyrene	Clear, easily colored; easily fabricated; transparent; fair mechanical and thermal properties; good resistance to acids, bases, oxidizing and reducing agents; readily attacked by many organic solvents; good electrical insulator	Production of ion-exchange resins, heat- and impact-resistant copolymer, ABS, resins, etc., foams, toys, plastic optical components, lighting fixtures, house wares, packaging, home furnishings
Poly(methyl methacrylate)	Clear, transparent, colorless, good weatherability, good impact strength, resistant to dilute basic and acidic solutions; easily colored; good mechanical and thermal properties; good fabricability; poor abrasion resistance compared to glass	Used in cast sheets, rods, tubes, molding, extrusion compositions, tail- and signal-light lenses, dials, medallions, brush backs, jewelry, signs, lenses, skylight "glass"; generally used where good light transmission is needed

in the range of 60°C–140°C for decomposition and free radical formation. While the dissociation bond energy for C–N is generally greater than for O–O, the formation of a stable N_2 molecule is the thermodynamic driving force due to an entropy effect allowing dissociation to occur at typically lower temperatures.

Here, we will focus on the simple heat-induced decomposition of classical free radical initiators.

The rate of decomposition of initiators usually follows first-order kinetics and is dependent on the solvent present and the temperature of polymerization. The rate is usually expressed as a half-life time ($t_{1/2}$) where $t_{1/2} = \ln 2/k_d = 0.693/k_d$. The rate constant ($k_d$) changes with temperature in accordance with the Arrhenius equation as shown in Equation 7.2:

$$k_d = Ae^{-E/RT} \tag{7.2}$$

Typical equations for the dissociation of AIBN and BPO are shown below. It should be pointed out that because of recombination, which is solvent dependent, and other side reactions of the created free radical (R·), the initiator efficiency is seldom 100%. Hence, an efficiency factor (f) is employed to show the fraction of effective free radicals produced.

The decomposition of AIBN and BPO to form radicals is given below:

$$(7.3)$$

BPO   Free radicals

$$(7.4)$$

AIBN   Free radicals

The precise structure of the initiating agent and initial addition to the monomer varies according to the reaction conditions, monomer, and initiator. For illustration, the peroxide bond in benzoyl peroxide, BPO, breaks forming two benzoyl free radicals. A small portion of the benzoyl free radical decomposes into the phenyl radical as shown in 7.3. For styrene, the major reaction (7.5) involves direct addition of the benzoyl radical to the tail end of the styrene monomer creating a molecule where the radical resides at the head or more sterically demanding and more radical stabilizing site:

$$(7.5)$$

The benzoyl peroxide decomposes with a specific rate constant of about 10^{-8} s^{-1}, an Arrhenius constant, A, of about 10^{16}, and an activation energy of about 28 kcal/mol (about 115 kJ/mol). As noted before, not all radicals initiate new chains. Some terminate prior to initiation forming inactive products, mainly phenyl benzoate (7.6). Thus, the efficiency constant, f, that reflects the ratio of benzoyl peroxide that actually forms chains is used:

$$(7.6)$$

Phenyl benzoate

While initiation can occur via a number of routes, as noted before, we will emphasize the use of chemical initiators for the formation of the free radicals necessary to begin the free radical polymerization process.

7.2 MECHANISM FOR FREE RADICAL CHAIN POLYMERIZATION

In general, the decomposition of the initiator (I) may be expressed by the following equation in which k_d is the specific rate or decay constant:

$$I \rightarrow 2R^\bullet \tag{7.7}$$

$$R_d = -\frac{d[I]}{dt} = k_d[I] \tag{7.8}$$

where R_d is the rate of decomposition.

The initiation of a free radical chain takes place by the addition of a free radical ($R^\bullet$) to a vinyl monomer (Equation 7.9). Polystyrene (PS) will be used to illustrate the typical reaction sequences. (Styrene, like many aromatic compounds, is toxic and concentrations that come into contact with us should be severely limited.) It is important to note that the free radical ($R^\bullet$) is a companion of all polymerizing species and is part of the polymer chain acting as an end group and hence should not be called a catalyst even though it is often referred to as such.

$$(7.9)$$

$$R^\bullet + M \rightarrow RM^\bullet \tag{7.10}$$

$$R_i = \frac{d[RM^\bullet]}{dt} = k_i[R^\bullet][M] \tag{7.11}$$

where
R_i is the rate of initiation
$R^\bullet$ is the free radical from benzoyl peroxide

The rate of decomposition of the initiator (I) (Equation 7.7) is the rate-controlling step in the free-radical polymerization as well as the formation of growing chains. Thus, the overall expression describing the rate of initiation can be given as

$$R_i = 2k_d f[I] \tag{7.12}$$

where f is the efficiency factor and is a measure of the fraction of initiator radicals that produce growing radical chains, that is, are able to react with monomers.

A "2" is inserted in Equation 7.12 because, in this presentation, for each initiator molecule that decomposes, two radicals are formed. The "2" is omitted from Equation 7.7 because this rate expression describes the rate of the decomposition of the initiator, but not the rate of formation of free radicals $R^\bullet$. (Similarly, in Equations 7.17 and 7.19, each termination results in the loss of two growing chains, thus a "2" appears in the descriptions.)

Propagation is a bimolecular reaction (Equation 7.13), which takes place by the addition of a new free radical ($RM^\bullet$) to another molecule of monomer (M), and by many repetitions of this step. While there may be slight changes in the propagation rate constant (k_p) in the first few steps, the rate constant is generally considered to be independent of chain length. This is what is also experimentally found. Hence, the symbols $M^\bullet$, $RM^\bullet$, and $RM_nM^\bullet$ may be considered equivalent in rate equations for free radical polymerization.

$$(7.13)$$

Styrene macroradical

Since the specific rate constants are approximately independent of the length of the growing chain, one specific rate constant is used to represent all of the propagation steps, k_p.

The rate of demise of monomer with time is described as:

$$\frac{-d[M]}{dt} = k_p[M^\bullet][M] + k_i[R^\bullet][M]$$

$$(7.14)$$

that is, monomer consumption only occurs in reactions described by Equations 7.9 and 7.13.

For long chains, the consumption of monomer by the initiation step (Equation 7.13) is small and can be neglected allowing Equation 7.14 to be rewritten as

$$R_p = \frac{-d[M]}{dt} = k_p[M^\bullet][M]$$

$$(7.15)$$

The polarity of the functional group in the monomers polymerized by free radical chain polymerization is between the positively inclined monomers characterized by undergoing cationic polymerization, and the negatively inclined monomers characterized by undergoing anionic polymerization. Figure 7.1 contains a listing of the addition polymerization routes taken by various monomers. As is true for the configuration of ionic growing chains, free radical polymers are also formed so that the addition will give a head-to-tail configuration because of the functional groups on the vinyl monomers which are better stabilized than the hydrogen atom and because this will balance the steric requirements present as addition occurs.

Unlike ionic polymerizations, the termination of the growing free radical chains usually occurs by the coupling of two macroradicals. Thus, the kinetic chain length (v) is equal to $\overline{DP}/2$. The chemical and kinetic equations for bimolecular termination are shown below (7.16).

$$(7.16)$$

$$H_2C=C(CH_3)_2 \qquad H_2C=CH(O-CH_3)$$
$$\longleftarrow\text{--------------Cationic---------------}\longrightarrow$$
$$H_2C=CHCl \quad H_2C=CCl_2 \quad H_2C=CHF \quad F_2C=CF_2$$
$$\longleftarrow\text{----------------------Free radical ------------------}\longrightarrow$$

$$\overset{O}{\overset{||}{H_2C=COCCH_3}} \quad H_2C=CH_2 \quad H_2C=CH-CH=CH_2 \quad \text{Styrene} \quad \overset{O}{\overset{||}{H_2C=CHCCH_3}}$$

$$\longleftarrow\text{Cationic}\longrightarrow$$
$$\longleftarrow\text{----------------------------------Free radical ----------------------------}\longrightarrow$$
$$\longleftarrow\text{--------------------------------------Anionic ---------------------------------}\longrightarrow$$

$$\overset{O}{\overset{||}{H_2C=C(CH_3)C-O-CH_3}} \quad \overset{O}{\overset{||}{H_2C=CHC-O-CH_3}} \quad H_2C=H(CN) \quad H_2C=C(CN)_2 \quad H_2C=CH(NO_2)$$

$$\longleftarrow\text{----------------------------Free radical --------------------------}\longrightarrow$$
$$\longleftarrow\text{-----------------------------------Anionic--------------------------------------}\longrightarrow$$

FIGURE 7.1 Type of chain initiation for some common monomers in the order of general decrease in electron density associated with the double bond and their tendency to undergo chain polymerization where A, anionic; C, cationic; and F, free radical.

Termination is a head-to-head configuration at the juncture of the two macroradicals. The extent of coupling termination can then be obtained by determining the extent of head-to-head configuration in the product. The kinetic equation for coupling termination is shown in Equation 7.17.

$$R_t = \frac{-d[M^\bullet]}{dt} = 2k_t[M^\bullet][M^\bullet] = 2k_t[M^\bullet]^2 \tag{7.17}$$

In some situations, termination occurs by disproportionation. This termination process involves chain transfer to a hydrogen atom from one chain end to the free radical chain end of another growing chain, resulting in one of the "dead" polymer chains having an unsaturated chain end.

(7.18)

$$R_{td} = 2k_{td}[M^\bullet]^2 \tag{7.19}$$

The kinetic chain length for termination by disproportionation is $\overline{DP} = v$ (compared with the relationship for coupling of $\overline{DP} = 2v$. The extent of the two types of termination is experimentally found by determining the number of head-to-head sites and unsaturated end groups).

While styrene macroradicals typically terminate by coupling, methyl methacrylate macroradicals terminate by coupling at temperatures below 60°C, but by disproportionation at higher temperatures.

The kinetic chain length is described in Equation 7.20.

$$v = \frac{R_p}{R_i} = \frac{R_p}{R_{td}} = \frac{k_p[M][M^\bullet]}{2k_{td}[M^\bullet]^2} = \frac{k_p[M]}{k_{td}[M^\bullet]} = \frac{k'[M]}{[M^\bullet]} \tag{7.20}$$

Because it is experimentally found that the number of growing chains is constant, there exists a steady state in $M^\bullet$ so that $R_i = R_{td}$. (A similar scenario can be used to develop rate expressions for coupling.)

While equations such as 7.15, 7.17, 7.19, and 7.20 are theoretically important, they contain $[M^\bullet]$, which is difficult to determine experimentally and are thus, of little practical use. The following is an approach to render such equations more useful by generating a description of $[M^\bullet]$ that involves more easily experimentally accessible terms.

The rate of monomer-radical *change* is described by:

$$\frac{d[M^\bullet]}{dt} = [\text{monomer-radical formed}] - [\text{monomer-radical utilized}] = k_i[R^\bullet][M] - 2k_t[M^\bullet]^2 \quad (7.21)$$

As noted before, it is experimentally found that the number of growing chains is approximately constant over a large extent of reaction. This situation is referred to as a "steady state." For Equation 7.21 this results in $d[M^\bullet]/dt = 0$ and

$$k_i[R^\bullet][M] = 2k_t[M^\bullet]^2 \quad (7.22)$$

Additionally, a steady-state value for the concentration of R· exists giving

$$\frac{d[R^\bullet]}{dt} = 2k_d f[I] - k_i[R^\bullet][M] = 0 \quad (7.23)$$

Solving for $[M^\bullet]$ from Equation 7.22 gives

$$[M^\bullet] = \left(\frac{k_i[R^\bullet][M]}{2k_t} \right)^{1/2} \quad (7.24)$$

and $[R^\bullet]$ from Equation 7.23 gives

$$[R^\bullet] = \frac{2k_d f[I]}{k_i[M]} \quad (7.25)$$

Substitution into Equation 7.24 the expression for $[R^\bullet]$, we obtain an expression for $[M^\bullet]$, Equation 7.26, that contains somewhat easily experimentally determinable terms.

$$[M^\bullet] = \left(\frac{k_d f[I]}{k_t} \right)^{1/2} \quad (7.26)$$

Using this relationship for $[M^\bullet]$ we get more useful rate (Equation 7.27) and kinetic chain length (7.29) equations.

$$R_p = k_p[M][M^\bullet] = k_p[M] \left(\frac{k_p f[I]}{k_t} \right)^{1/2} = \left(\frac{k_p^2 k_d f}{k_t} \right)^{1/2} [M][I]^{1/2} = k''[M][I]^{1/2} \quad (7.27)$$

Thus, the rate of propagation, or polymerization, is directly proportional to the concentration of the monomer and square root concentration of the initiator.

In preparation of describing the kinetic chain length, we can also describe the rate of termination using the new description for $[M^\bullet]$.

$$R_t = 2k_t[M^\bullet]^2 = \frac{2k_t k_d f[I]}{k_t} = 2k_d f[I] \quad (7.28)$$

TABLE 7.2
Energies of Activation for Propagation (E_p) and Termination (E_t) in Free Radical Chain Polymerization

Monomer	E_p (kJ/mol)	E_t (kJ/mol)
Methyl acrylate	30	22
Acrylonitrile	17	23
Butadiene	39	—
Ethylene	34	—
Methyl methacrylate	26	12
Styrene	33	12
Vinyl acetate	31	22
Vinyl chloride	15	18

TABLE 7.3
Typical Free Radical Kinetic Values

Specific Rate Constant	Activation Energies (kJ/mol)	
k_d	10^{-3} s^{-1}	E_d 80–160
k_i	10^3 L/mol-s	E_i 20–30
k_p	10^3 L/mol-s	E_p 20–40
k_t	10^7 L/mol-s	E_t 0–20

$$\overline{DP} = \frac{R_p}{R_i} = k_p[M]\left(\frac{\left(k_d f[I]/k_t\right)^{1/2}}{2k_d[I]}\right) = \frac{k_p[M]}{2(k_d k_t f[I])^{1/2}} = \frac{k'[M]}{[I]^{1/2}} \tag{7.29}$$

Thus, chain length is directly proportional to monomer concentration and inversely proportional to the square root of initiator concentration.

Typical energies of activation for propagation and termination are given in Table 7.2 and typical free radical kinetic values in Table 7.3.

As done in Chapter 6, the effect of temperature can be determined using average activation of the various steps. Again, the rates of all single step reactions increase as the temperature increases but the overall result may be different for complex reactions. For free radical polymerizations, the activation energies are generally of the order $E_d > E_i = E_p > E_t$. Remembering that the description of the specific rate constant is

$$k = Ae^{-Ea/RT} \tag{7.30}$$

the overall or "net" activation energy is

$$E_{(overall)} = E_t + E_i + E_p + E_d \tag{7.31}$$

Using only the specific rate constants involved with propagation from Equation 7.27 we have

$$R_p \propto \frac{k_p^2 k_d}{k_t} \tag{7.32}$$

so that the overall activation energy using average values is a positive value (Equation 7.33) and the overall rate of polymerization increases as temperature increases.

$$E_{p(\text{overall})} \, \alpha \, 2E_p + E_d - E_t = 2 \times 30 + 120 - 10 = 170 \tag{7.33}$$

For chain length, from Equation 7.29 we have

$$\overline{\text{DP}} \, \alpha \, \frac{k_p}{2k_d k_t} \tag{7.34}$$

so that the overall activation energy using average values is

$$\overline{\text{DP}}(\text{overall}) \, \alpha \, E_p - E_d - E_t = 30 - 120 - 10 = -100 \tag{7.35}$$

so that $\overline{\text{DP}}$ decreases as temperature increases as pictured in Figure 5.4.

The Gibbs free energy relationship for a reversible process at constant temperature for polymerization is described by

$$\Delta G_p = \Delta H_p - T \Delta S_p \tag{7.36}$$

where ΔH_p is the heat of polymerization defined by

$$\Delta H_p = E_p - E_{dp} \tag{7.37}$$

where
   E_p is the activation energy for propagation
   E_{dp} is the activation energy for depolymerization

The entropy term is negative so that it is the enthalpy or energy term that "drives" the polymerization. At low temperatures, the enthalpy term is larger than the $T \Delta S_p$ term so that polymer growth occurs. At some temperature, called the **ceiling temperature**, the enthalpy term and entropy term are the same and $\Delta G_p = 0$. Above this temperature, depolymerization occurs more rapidly than polymer formation so that polymer formation does not occur. At the ceiling, temperature depolymerization and polymerization rates are equal. The ceiling temperature is then defined as

$$T_c = \frac{\Delta H_p}{\Delta S_p} \tag{7.38}$$

since $\Delta G_p = 0$.

The ceiling temperature for styrene is about 310°C, for ethylene it is 400°C, for propylene it is 300°C, for methyl methacrylate it is 220°C, for tetrafluoroethylene it is 580°C, and for α-methylstyrene it is only 61°C.

It is interesting to note that due to their industrial importance, free radical polymerizations are the most studied reactions in chemistry. Furthermore, the kinetic approaches taken in this chapter are experimentally verified for essentially all typical free radical vinyl polymerizations.

There is some tendency for the formation of stereoregular sequences, particularly at low temperatures, but ionic and coordination catalysts are far superior in this aspect and are used to create stereoregular macromolecules.

Several additional comments are appropriate concerning chain type polymerization and termination. First, since the slow step is the initiation step, and the other steps, especially termination, have very low energy of activations and therefore are very fast, how does polymer form? Consider the relative concentrations of the various species. The concentration of polymer monomer is very high relative to the concentration of the free radical species favoring the formation of chain growth. The rate of termination is proportional to the square of the concentration of growing chain for both coupling (7.17) and disproportionation (7.19) and because of the concentration of growing chain being relatively quite small it allows polymer formation. Thus, chain growth resulting in polymer formation is the consequence of the high concentration of monomer and low concentration of growing chains. Second, in general as the steric hindrance increases the tendency for termination occurring through disproportionation increases since coupling requires the approach of the ends of two growing chains. This approach of growing chain ends becomes, in general, less favorable as the steric hindrance increases.

7.3 CHAIN TRANSFER

Transfer of the free radical to another molecule serves as one of the termination steps for general polymer growth. Thus, the transfer of a hydrogen atom at one end of the chain to a free radical end of another chain is a chain-transfer process we dealt with in Section 7.2 under termination via disproportionation. When abstraction occurs intramolecularly or intermolecularly by a hydrogen atom some distance from the chain end, branching results. Each chain-transfer process causes the termination of one macroradical and produces another macroradical. The new radical sites serve as branch points for chain extension or branching. As noted earlier, such chain transfer can occur within the same chain as shown in 7.39.

(7.39)

Chain transfer can also occur between chains.

(7.40)

Chain transfer can also occur with initiator, impurity, solvent, or other additive present in the polymerization system. While, the average chain length is equal to R_p divided by the sum of all termination reactions, it is customary to control all termination steps except the one that is being studied. Chain transfer to all other molecules, except solvents or some special additives, is typically negligible.

The chain-transfer reaction decreases the average chain length. This decrease in chain length increases as the concentration of the chain-transfer agent (S) increases and as the tendency of the chain transfer agent to 'chain-transfer' is increased. The resulting $\overline{DP}$ is equal to that which would have been obtained without the solvent or additive in addition to a factor related to the product of the ratio of the rate of propagation (R_p) and the rate of chain transfer (R_{tr}) and the ratio of the concentration of the monomer [M] to the concentration of chain transfer agent [S].

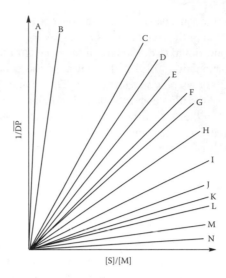

FIGURE 7.2 Molecular weight of polystyrene as a function of solvent and solvent concentration: A, *n*-butylmercaptan; B, carbon tetrabromide; C, carbon tetrachloride; D, *o*-cresol; E, *p*-cresol; F, *m*-cresol; G, phenol; H, *sec*-butylbenzene; I, cumene; J, ethylbenzene; K, chloroform; L, *n*-heptant; M, toluene; N, benzene. where [*S*], is the concentration of chain transfer agent and [M], is the concentration of styrene monomer.

The Mayo equation, Equation 7.41, that gives positive slops when the data are plotted, is the reciprocal relationship derived from the previously cited expression. The ratio of the rate of cessation or termination by transfer to the rate of propagation is called the chain transfer constant (C_s).

$$\frac{1}{\overline{DP}} = \frac{1}{\overline{DP_o}} + C_s \frac{[S]}{[M]}$$

(7.41)

The chain transfer constant is given as

$$C_s = \frac{k_{tr}}{k_p}$$

(7.42)

As shown in Figure 7.2, the molecular weight of polystyrene is reduced when it is polymerized in solvents, and the reduction or increase in slope is related to the chain transfer efficiency of the solvent. The slopes in this figure are equal to C_s.

Chain transfer constants of various solvents, including those given in Figure 7.2, are given in Table 7.4.

Chain transfer agents have been called regulators (of molecular weight). When used in large proportions, they are called telogens, since they produce low molecular weight polymers (telomers).

7.4 POLYMERIZATION TECHNIQUES

The principle free radical polymerization techniques are bulk, solution, and emulsion. Tables 7.5 and 7.6 briefly describe these techniques.

7.4.1 BULK POLYMERIZATION

Bulk polymerization of a liquid monomer such as methyl methacrylate is relatively simple in the absence of oxygen where small bottles or test tubes are used as the reaction vessel. The monomer can be heated in the presence of an initiator giving a clear plastic shaped like the container, but a

TABLE 7.4
Chain Transfer Constants of Solvent to Styrene
in Free Radical Chain Polymerization at 60°C

Transfer Agent	$C_s \times 10^4$	Transfer Agent	$C_s \times 10^4$
Acetic acid	2.0	1-Dodecanethiol	148,000
Benzene	0.01	Hexane	0.9
Butyl alcohol	0.06	N,N-Dimethylaniline	12
t-Butyl alcohol	6.7	1-Naphthalenethiol	1,500
Butyl disulfide	0.24	1-Octanethiol	190,000
Carbon tetrabromide	18,000	p-Methoxyphenol	260
Carbon tetrachloride	84	Phenol	8.1
Chloroform	0.5	Triethylamine	1.4
o-Chlorophenol	6.0	Toluene	0.105
2,6-Ditert-butylphenol	49	Water	0

TABLE 7.5
Types of Polymerization Systems

Monomer–Polymer Phase Relationship	Monomer Location	
	Continuous	Dispersed
Homogeneous (same phase)	Bulk, solid state, solution	Suspension
Heterogeneous (different phase)	Bulk with polymer precipitating	Emulsion; suspension with polymer precipitating

little smaller because of shrinkage. The volume of the monomers is generally larger than the final polymers; thus, the density of the polymer is greater than that of the original monomer. Poly(methyl methacrylate) (PMMA; 7.43) is sold under the trade names Lucite and Plexiglas as an atactic amorphous polymer.

$$(7.43)$$

Polymers such as the methacrylates play an essential role as photoresists. To make a photoresist, the methacrylate polymer is deposited onto silicon dioxide. A mask, that shields specific regions from subsequent exposure to light, is placed over the methacrylate polymer resist. The combination is exposed to light of such a strength as to induce methacrylate polymer bond breakage producing a somewhat degraded methacrylate product that is more soluble in organic liquids allowing the preferential removal of the exposed photoresist. These methacrylate polymers are especially designed to allow both easy degradation and subsequent easy removal. One of these polymers is a copolymer with adamantane and lactone-containing units (7.44).

Modified methacrylate copolymer

$$(7.44)$$

TABLE 7.6
Summary of Popular Polymerization Techniques

Bulk

Simplest of the techniques requiring only monomer and monomer-soluble initiators, and perhaps a chain-transfer agent for molecular weight control. Characterized, on the positive side, by the high polymer yield per volume of reaction, easy polymer recovery. Difficulty of removing unreacted monomer and heat control are negative features. Examples of polymers produced by bulk polymerization include poly(methyl methacrylate), polystyrene, and low-density (high pressure) polyethylene.

Solution

Monomer and initiator must be soluble in the liquid and the solvent must have the desired chain-transfer characteristics, boiling point (above the temperature necessary to carry out the polymerization and low enough to allow for ready removal if the polymer is recovered by solvent evaporation). The presence of the solvent assists in heat removal and control (as it also does for suspension and emulsion polymerization systems). Polymer yield per reaction volume is lower than for bulk reactions. Moreover, solvent recovery and removal (from the polymer) is necessary. Many free radical and ionic polymerizations are carried out utilizing solution polymerization including water-soluble polymers prepared in aqueous solution (namely poly(acrylic acid), polyacrylamide, and poly(N-vinylpyrrolidone). Polystyrene, poly(methyl methacrylate), poly(vinyl chloride), and polybutadiene are prepared from organic solution polymerizations.

Suspension

A water-insoluble monomer and an initiator are used. Again, a chain-transfer agent may be used to control chain size. Stirring is usual. Droplets of monomer containing the initiator and the chain-transfer agent are formed. A protective colloidal agent, often poly(vinyl alcohol), is added to prevent the coalescence of the droplets. Near the end, the particles become hard and are recovered by filtration. Because the liquid is water-based, solvent recovery and treatment problems are minimal. The products may contain a number of impurities including any of the agents added to assist in the polymerization process. Polymers produced by suspension polymerization include poly(vinyl chloride), polystyrene resins, and copolymers such as poly(styrene-co-acrylonitrile), SAN, and poly(vinyl chloride-co-vinylidene chloride).

Emulsion

The system usually contains a water-soluble initiator (in contrast to the requirement that the initiator must not be water soluble in suspension polymerizations), a chain-transfer agent, and a surfactant. The hydrophobic monomer forms large droplets that are stabilized by the surfactant. At a certain surfactant concentration, the surfactant molecules form micelles that contain 50–100 surfactant molecules. During the polymerization, the monomer, that has a small but real water solubility, migrate from the monomer droplets through the water and into these micelles. Polymerization begins when the water-soluble initiator enters into the monomer-containing micelle. Because the concentration of micelles (about 10^{21}/L) is high compared with the concentration of monomer droplets (about 10^{13}/L), the initiator is more likely to enter a micelle than a monomer droplet. As polymerization continues, monomer is transferred to the growing micelles. At about 50%–80% conversion the monomer droplets disappear and the micelles become large polymer-containing droplets. This suspension is called latex. The latex is stable and can be used as it is or the polymer recovered by coagulation. In inverse emulsion polymerization, the monomer, which is hydrophilic, is dispersed in an organic liquid. Here, the monomer is usually contained in an aqueous solution.

This type of photoresist is called a "positive photoresist" since it is the exposed area that is removed.

Negative photoresists are formed from polymers that undergo reactions that decrease their solubility when exposed to radiation. Thus, polymers such as *cis*-1,4-polyisoprene (7.45) cross-link when exposed to the appropriate radiation giving insoluble products.

$$\text{(structure)} \tag{7.45}$$

The rate of bulk polymerization can be followed by monitoring the change in volume or increase in viscosity. When the viscosity is high, the termination reaction is hindered since the macroradicals are unable to diffuse readily in the viscous medium. Thus, the number of growing chains increases. This autoacceleration, called the Norris–Trommsdorff effect, Tranmsdorff effect, or gel effect, causes the formation of unusually high molecular weight chains. Since vinyl polymerizations are exothermic, there is a buildup of heat, which further causes an additional autoacceleration of the reaction. If the temperature buildup is not controlled, it is possible that an explosion will occur. While the temperature can be easily controlled within a small test tube, it is more difficult in a large batch process. Stirring and external cooling are employed to control the polymerization process.

7.4.2 Suspension Polymerization

Water insoluble monomers such as vinyl chloride may be polymerized as suspended droplets (10–1000 nm in diameter) in a process called suspension (pearl) polymerization. Coalescence of droplets is prevented by the use of small amounts of water-soluble polymers, such as poly(vinyl alcohol) (PVA). The suspension process is characterized by good heat control and ease of removal of the discrete polymer particles.

Since PVC is insoluble in its monomer, it precipitates as formed in the droplet. This is actually advantageous, since it permits ready removal of any residual carcinogenic monomer from the solid beads by stripping under reduced pressure.

7.4.3 Solution Polymerization

Monomers may also be polymerized in solution using good or poor solvents for homogeneous and heterogeneous systems, respectively. In solution polymerizations, solvents with low chain transfer constants are used to minimize reduction in chain length.

Poly(vinyl acetate) (PVAc) may be produced by the polymerization of vinyl acetate (7.46). The viscosity of the solution continues to increase until the reaction is complete. Dilute polymer solutions are used to prevent the onset of autoacceleration because of the gel effect.

PVAc is used in adhesives and coatings and is hydrolyzed producing water-soluble poly(vinyl alcohol) (PVA; 7.46). The PVA may be reacted with butyraldehyde to produce poly(vinyl butyral) used as the inner lining of safety glass.

$$\text{(reaction scheme)} \tag{7.46}$$

When a monomer such as acrylonitrile is polymerized in a poor solvent, macroradicals precipitate, as they are formed (7.47). Since these are "living polymers," the polymerization continues as more acrylonitrile diffuses into the precipitated particles. This heterogeneous solution polymerization has been called **precipitation polymerization**. Polyacrylonitrile (PAN; produced under the trade name of Acrilan) cannot be colored, so small amounts of a comonomer such as styrene, vinyl chloride, or vinylidene chloride are added to allow the fabrics and textiles to be colored. These copolymers are called **modacrylics**. The use of vinylidene chloride and vinyl chloride gives a product, such as a carpet or wig, increased flame resistance.

$$(7.47)$$

7.4.4 EMULSION POLYMERIZATION

Many water-soluble vinyl monomers may be polymerized by the emulsion polymerization technique. This technique, which differs from suspension polymerization in the size of the suspended particles and in the mechanism, is widely used for the production of a number of commercial plastics and elastomers. While the particles in the suspension range from 10 to 1000 nm, those in the emulsion process range from 0.05 to 5 nm in diameter. The small beads produced in the suspension process may be separated by filtering, but the latex produced in emulsion polymerization is a stable system in which the charged particles cannot be recovered by ordinary separation procedures.

Since relatively stable macroradicals are produced in the emulsion process, the termination rate decreases and a high molecular weight product is rapidly produced. It is customary to use a water-soluble initiator such as potassium persulfate, an anionic surfactant such as sodium sterate, and to stir the aqueous mixture of monomer, initiator, and surfactant in the absence of oxygen at 40°C–70°C. When the concentration of soap exceeds the critical micelle concentration (CMC), the molecules are present as micelles in which the hydrophilic carboxylic acid ends are oriented toward the water–micelle interface, and the lyophilic hydrocarbon ends are oriented toward the center of the micelle. The micelles are present as spheres with a diameter of 5–10 nm when the soap concentration is less than 2%. However, with the higher concentrations typically employed, the micelles resemble aggregates of rods which are 100–300 nm in length.

As shown in Figure 7.3, the water-insoluble monomer, M, is attracted to the lyophilic ends in the micelles, causing the micelles to swell. The number of swollen micelles per milliliter of water is about 10^{18}. However, at the initial stages of polymerization (phase I) most of the monomers are present as globules that resemble those observed in suspension polymerization.

Since the initiation of polymerization takes place in the aqueous phase, essentially no polymerization occurs in the globules. Thus, they serve primarily as a reservoir of monomer supplied to the micelles to replace monomer converted to polymer. The number of droplets per milliliter of water is about 10^{11}. Hence, since there are 10 million times as many micelles as droplets, the chance of initiation of monomer in a droplet is very small and the chance that more than one growing chain occurs within the same droplet is very, very small.

The persulfate ion undergoes homolytic cleavage producing two sulfate ion radicals.

$$S_2O_8^{-2} \longrightarrow 2SO_4^{-}\cdot$$

$$(7.48)$$

The sulfate ion radical then initiates polymerization, here with a styrene monomer eventually forming a polystyrene oligomer radical and eventually a polystyrene radical (7.49).

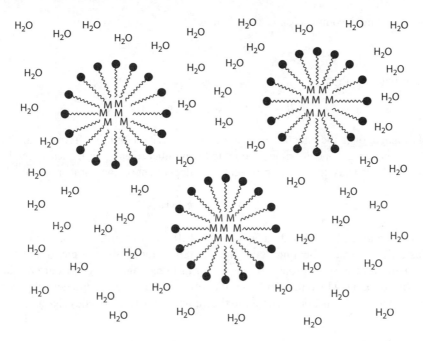

FIGURE 7.3 Micelles swollen with solubilized monomer. As the concentration increases, the micelles change from spherical, as shown, to more rodlike in appearance.

$$\text{(7.49)}$$

According to a theory proposed by Harkins and refined by Smith and Ewart, the first stages of propagation in an emulsion system also take place in the aqueous phase to produce a more lyophilic surface oligoradical. When the $\overline{DP}$ of the polystyrene oligoradical is 3–5, its solubility is much like that of styrene, and it migrates to the swollen micelle where propagation continues with the styrene molecules already present. According to the accepted theories, each micelle can accommodate only one free radical, as noted before, and until a second one enters and terminates the propagation reaction through coupling, propagation continues to take place in the micelles. From a statistical point of view, only one-half of the micelles (N/2) will contain growing chains at any one time. It should also be noted that since propagation occurs in the micelles, the rate of polymerization will be proportional to the number of micelles present, that is, the rate is proportional to the soap concentration.

As the micelles grow by absorption of more monomer and formation of polymer, they become relatively large particles that absorb soap from micelles that have not been inoculated or stung by oligoradicals. Thus, in stage II, when about 20% of the monomer has been converted to polymer, the micelles disappear and are replaced by large, but fewer, monomer–polymer particles.

Polymerization continues in stage II, and monomer continues to be supplied to the particles by the droplets in the aqueous phase. These droplets disappear when about 30% of the monomer has been converted to polymer. Polymerization continues in stage III after about 60% conversion, but all monomer must now be supplied to the macroradicals by a diffusion process in the micelles.

The rate of sulfate decomposition is described as

$$R_d = k_d[S_2O_8^{-2}]$$ (7.50)

The rate of initiation is then

$$R_i = k_i[SO_4^-][M] = 2k_d f[S_2O_8^{-2}]$$ (7.51)

The rate of propagation in the micelles is similar to that described for other free radical chain growth, but since the free radical concentration is equal to the number of active micelles, the value of N/2 is used instead of $[M^\bullet]$. Thus, the rate of propagation is dependent on the number of micelles present.

$$R_p = k_p[M][M^\bullet] = k_p[M](N/2)$$ (7.52)

The rate of production of free radicals at 50°C is about 10^{13} radicals/mL per 1 s. Thus, since there are 10^5 micelles for every free radical produced in one second, inoculation of any of the 10^{18} micelles/mL is infrequent. Hence, since propagation is a very fast reaction, long chains are produced before termination by coupling, which takes place as the result of the entrance of a new oligoradical in the active micelle. The $\overline{DP}$ is also proportional to the number of active micelles (N/2).

$$\overline{DP} = \frac{R_p}{R_i} = \frac{k_p\left(\dfrac{N}{2}\right)}{k_i[SO_4^-]}$$ (7.53)

7.5 FLUORINE-CONTAINING POLYMERS

Polytetrafluoroethylene, better known by its trade name of Teflon, was accidentally discovered by Roy J. Plunkett, a Dupont chemist who had just received his PhD from Ohio State 2 years before. He was part of a group searching for nontoxic refrigerant gases. On April 6, 1938, he and his assistant, Jack Rebok, had filled a tank with tetrafluoroethylene. After some time, they opened the valve but no gas came out. The tank weight indicated that there was no weight loss—so what happened to the tetrafluoroethylene? Using a hacksaw, they cut the cylinder in half and found a waxy white powder. He correctly surmised that the tetrafluoroethylene had polymerized. The waxy white powder had some interesting properties. It was quite inert toward strong acids, bases, heat, and was insoluble in any attempted liquid. It appeared to be quite "slippery."

Little was done with this new material until the military, working on the atomic bomb, needed a special material for gaskets that would resist the corrosive gas uranium hexafluoride that was one of the materials used to make the atomic bomb. General Leslie Groves, responsible for the U.S. Army's part in the atomic bomb project, had learned of Dupont's new inert polymer and had Dupont manufacture it for them.

Teflon was introduced to the public in 1960 when the first Teflon-coated muffin pans and frying pans were sold. Like many new materials, problems were encountered. Bonding to the surfaces was uncertain at best. Eventually the bonding problem was solved. Teflon is now used for many other applications including acting as a biomedical material in artificial corneas, substitute bones for nose, skull, hip, nose, and knees; ear parts, heart valves, tendons, sutures, dentures, and artificial tracheas. It has also been used in the nose cones and heat shield for space vehicles and for their fuel tanks.

Over one-half million vascular graft replacements are performed yearly. Most of these grafts are made of PET and PTFE. These relatively large diameter grafts work when blood flow is rapid, but they generally fail for smaller vessels.

Polytetrafluoroethylene is produced by the free-radical polymerization process. While it has outstanding thermal and corrosive resistance, it is a marginal engineering material because it is not easily machinable. It has low tensile strength, resistance to wear, and it has low creep resistance. Molding powders are processed by press and sinter methods used in powder metallurgy. It can also be extruded using ram extruder techniques.

PTFE is a crystalline polymer with melting typically occurring above 327°C. Because it is highly crystalline, it does not generally exhibit a noticeable T_g.

The C–F bond is one of the strongest single bonds known with a bond energy of 485 kJ/mol. While it is structurally similar to linear polyethylene, PE, it has marked differences. Because of the small size of hydrogen, PE exists as a crankshaft backbone structure. Fluorine is a little larger (atomic radius of F = 71 pm and for H = 37 pm) than hydrogen causing the teflon backbone to be helical forming a complete twist every 13 carbon atoms. The size of the fluorine is sufficient to form a smooth "protective" sheath around the carbon backbone. The concentration of F end groups is low in ultra high molecular weight PTFE contributing to its tendency to form crystals.

The electron densities of PE and PTFE are also different. The electronegativity value for C is 2.5, F = 4.0, and for H = 2.1. Thus, the electron density on the fluorine surface of PTFE is greater than that for PE.

For high molecular weight linear PE, the repeat unit length is about 0.254 nm forming crystalline portions with a characteristic thickness of about 10 nm. The chain length for tough solids from PE is about 4.5 times the crystalline thickness. Thus, tough solids occur at molecular weights greater than 5000 g/mol or chain lengths greater than about 45 nm. In comparison, the repeat unit length for PTFE is about 0.259 nm. The crystalline thicknesses for PTFE are about 100–200 nm or much thicker than for PE. Chain lengths for tough solids are about 4.5 times the crystalline thickness. Thus, much greater chain sizes, about 200,000–400,000 Da, are required to produce tough solids. The greater size of the crystalline portions also probably contributes to its higher T_m and greater difficulty in processing. The crystal thickness of PTFE is about 10–20 times the crystal thickness found for most other semicrystalline polymers such as PE.

At low molecular weights, PTFE is waxy and brittle. To achieve good mechanical properties ultrahigh molecular weights about 10 million dalton is usually needed. These long chains disrupt crystal formation because they are longer than a single crystal. However, the long chain lengths connect the crystals together adding to their strength. Moreover, these long chains result in extremely high viscosities so that ultrahigh molecular weight PTFE does not flow when melted and is thus, not melt processable. Form restrictive and costly methods are used to produce products from PTFE.

While vinyl fluoride was prepared about 1900, it was believed resistant to typical "vinyl" polymerization. German scientists prepared vinyl fluoride through the reaction of acetylene with hydrogen fluoride in the presence of catalysts in 1933 (Equation 7.54).

$$H—F \ + \ HC \equiv CH \longrightarrow H_2C = \overset{}{\underset{F}{\diagdown}} \tag{7.54}$$

It was not until 1958 that Dupont scientists announced the polymerization of vinyl fluoride forming poly(vinyl fluoride), PVF; Equation 7.55. Polymerization is accomplished using peroxide catalysts in water solutions under high pressure.

$$H_2C = \overset{}{\underset{F}{\diagdown}} \longrightarrow \tag{7.55}$$

In comparison to PTFE, PVF is easily processable using a variety of techniques used for most thermoplastic materials. It offers good flame retardancy, presumably due to the formation of HF that assists in the control of the fire. Thermally induced formation of HF is also a negative factor because of its toxicity. As in the case of poly(vinyl chloride), elimination of the hydrogen halide promotes formation of aromatic polycyclic products that themselves are toxic.

The difference in electronegativity between the adjacent carbons because of the differing electronegativities of H and F results in the C–F bond being particularly polar resulting in it being susceptible to attack by strong acids. The alternating bond polarities on the PVF chain give it a tight structure resulting in PVF films having a low permeability. This tight structure also results in good solvent resistance, resistance to cracking, and resistance to fading.

Friction and wear are important related characteristics. If a material has a high friction then it will generally have a shorter wear time because water or other friction event chemicals pass over the material with the higher friction causing greater wear. The friction eventually "wears" away polymer chains layer-by-layer. The engineering laws of sliding friction are simple. According to Amontons' laws, the friction F between a body (rain drop, wind, or board rubbing against the material) and a plane surface (the polymeric material) is proportional to the load L and independent to the area of contact A. The friction of a moving body is generally less than that of a static body. The kinetic friction is considered independent of the velocity. The coefficient of friction is defined as F/L. Polymers show a wide range of coefficients of friction so that rubbers exhibit relatively high values (BR = 0.4–1.5 and SBR = 0.5–3.0) whereas some polymers such as PTFE (0.04–0.15) and PVF (0.10–0.30) have low values.

The low coefficient of friction for PVF results in materials coated with it remaining somewhat free of dirt and other typical contaminants allowing PVF-coated materials to be less frequently cleaned. It is essentially self-cleaning as rain carries away dust and other particulates including bird droppings, acid rain, and graffiti. The low friction also results in longer lifetimes for materials coated with the PVF and for the PVF coating itself.

PVF has a T_g of about –20°C remaining flexible over a wide temperature range (from about –20°C to 150°C), even under cold temperatures. Because of its low coefficient of friction and tightly bound structure, it retains good strength as it weathers. Films, in Florida, retain much of their thickness even after about a decade losing less than 20% of their thickness. To increase their useful lifetimes, relatively thick films, such as 1 mil, are generally employed. The "slickness" also acts to give the material a "natural" mildew resistance.

Unlike PVC that requires plasticizers to be flexible, PVF contains no plasticizers and does not "dry out" like PVC. PVF, because of its higher cost in comparison to PE and PP, is used as a coating and selected "high end" bulk applications such as films. Films are sold by Dupont under the trade name of Tedlar. Tedlar is used in awnings, outdoor signs, roofing, highway sound barriers, commercial building panels, and solar collectors. It is used as a fabric coating, protecting the fabric from the elements. PVF is resistant to UV-related degradation and unlike PVC, it is inherently flexible. While transparent, pigments can be added to give films and coatings with varying colors. Protective coatings are used on plywood, automotive parts, metal sidings, lawn mower housings, house shutters, gutters, electrical insulations, and in packaging of corrosive chemicals. PVF has piezoelectric properties generating a current when compressed.

The processability of fluorine-containing polymers is improved by the replacement of one or more of the F atoms. Replacing one of the eight fluorine atoms with a trifluoromethyl group gives a product called FEP or Viton, actually a copolymer of tetrafluoroethylene and hexafluoropropylene (7.56). Polytrifluoromonochloroethylene (CTFE, Kel F; 7.57), in which one fluorine atom has been replaced by a chlorine atom has a less regular structure and is thus more easily processed. Poly(vinylidene fluoride) (Kynar) is also more easily processable but less resistant to solvents and corrosives.

(7.56)   (7.57)

7.6 POLYSTYRENE

The styrene monomer was discovered by Newman in 1786. The initial formation of polystyrene was by Simon in 1839. While polystyrene was formed almost 175 years ago, the mechanism of formation, described in Sections 7.1 through 7.3, was not discovered until the early twentieth century. Staudinger, using styrene as the principal model, identified the general free radical polymerization process in 1920. Initially, the commercialization of polystyrene, as in many cases, awaited the ready availability of the monomer. While there was available ethyl benzene, it underwent thermal cracking rather than dehydrogenation until the appropriate conditions and catalysts were discovered. Dow first successfully commercialized polystyrene formation in 1938. While most commercial polystyrene, PS, has only a low degree of stereoregularity, it is rigid and brittle because of the resistance of the more bulky phenyl-containing units to move in comparison, for example, to the methyl-containing units of polypropylene. This is reflected in a relatively high T_g of about 100°C for polystyrene. It is transparent because of the low degree of crystalline formation.

While PS is largely commercially produced using free radical polymerization, it can be produced by all four of the major techniques-anionic, cationic, free radical, and coordination-type systems. All of the tactic forms can be formed employing these systems. The most important of the tactic forms is syndiotactic PS, s-PS. Metallocene-produced s-PS is a semicrystalline material with a T_m of 270°C. It was initially produced by Dow in 1997 under the trade name of Questra. It has good chemical and solvent resistance in contrast to "regular" PS that has generally poor chemical and solvent resistance because of the presence of voids that are exploited by the solvents and chemicals.

The physical properties of PS are dependent on the molecular weight and presence of additives. While higher molecular weight PS offers better strength and toughness, it also offers poorer processability. Low molecular weight PS allows good processability but poorer strength and toughness. Generally, a balance is sought where intermediate chain lengths are used. Typically employed chain lengths are about 1500–3500 with standard molecular weight distributions of about 2.2–3.5. Small amounts of plasticizers are often used to improve processability.

Styrene is employed in the formation of a number of co and terpolymers. The best known is the terpolymer ABS.

PS is produced in three forms—extruded PS, expanded PS foam, and extruded PS foam. Expanded PS, EPS, was developed by the Koppers Company in Pittsburgh, PA in 1959. Extruded PS foam, XPS, and insulation were developed by Dow Chemical and sold under the trade name Styrofoam. This term is often used for many other expanded PS materials. EPS and XPS are similar and both generally contain a mixture of 90%–95% PS and 5%–10% gas (also referred to as a blowing agent), generally pentane, nitrogen, or carbon dioxide. The solid plastic is heated generally using steam followed by the introduction of the gas. EPS is produced from PS beads containing the entrapped blowing agents. When heated, the blowing agent turns to a gas expanding resulting in the individual beads expanding and fusing together. Extruded PS foam, XPS, is similarly formed except the blowing agents become gaseous as the mixture emerges from an extruder giving a more continuous product in comparison to EPS.

Extruded PS foam, XPS, is often known by its trade name Styrofoam. XPS has air inclusions giving the material some flexibility, low thermal conductivity, and a low density. It is widely used as commercial and residential insulations. In construction, it is also used to make ornamental pillars that are subsequently coated with a harder material. Under roads and buildings, it is employed to prevent soil disturbance due to weathering. Life rafts are generally made from XPS but can be also made from EPS.

Expanded PS foam, EPS, is incorrectly best known to the general public as Styrofoam cups, coolers, and egg containers. EPS can be generally distinguished from Styrofoam by the presence of small beads or spheres that are present in the EPS and missing in the XPS. It is also used to make a number of other products such as inexpensive surfboards and other water and pool floating devices. It is used to make packing peanuts and molded packing material for cushioning items such as TVs, computers, and stereo equipment.

Extruded PS is used for objects where a somewhat rigid inexpensive plastic material is needed such as plastic Petri dishes, plastic test tubes, plastic model kits, CD jewel cases, toys, house wares, and appliance parts. Medical products are sterilized after the product is made using irradiation or treatment with ethylene oxide.

Legislation was put in place in some states to insure the recycling of PS. Interestingly, some of this legislation was written such that all PS had to be recycled within some period of time such as a year. This legislation was changed to reflect the real concern of fast-food containers when it was pointed out that less than 10% PS is used in this manner and that well over twice as much was used as house insulation that should not be recycled every year or so.

7.7 POLY(VINYL CHLORIDE)

Poly(vinyl chloride), PVC, is one of the three most abundantly produced synthetic polymers. PVC is one of the earliest produced polymers. In 1835, Justus von Liebig and his research student Victor Regnault reacted ethylene dichloride with alcoholic potash forming the monomer vinyl chloride. Later, Regnault believed he polymerized vinyl chloride but later studies showed it to be poly(vinylidene chloride). In 1872, E. Baumann exposed vinyl chloride sealed in a tube to sunlight and produced a solid, PVC. Klasse, in Germany, found that vinyl chloride could be made by the addition of hydrogen chloride to acetylene in a system that could be scaled up for commercial production. (Today most vinyl chloride is made from the oxychlorination reaction with ethylene.) By World War I, Germany was producing a number of flexible and rigid PVC products and during World War I Germany used PVC as a replacement for corrosion-prone metals.

Waldo Semon was responsible for bringing many of the poly(vinyl chloride) products to market. As a young scientist at BF Goodrich, he worked on ways to synthesize rubber and to bind the rubber to metal. In his spare time, he discovered that PVC, when mixed with certain liquids, gave an elastic-like, pliable material that was rain proof, fire resistant, and that did not conduct electricity. Under the trade name Koroseal, the rubbery material came into the marketplace, beginning about 1926, as shower curtains, raincoats, and umbrellas. During World War II, it became the material of choice to protest electrical wires for the Air Force and Navy. Another of his inventions was the synthetic rubber patented under the name Ameripol that was dubbed "liberty rubber" since it replaced the natural rubber in the production of tires, gas masks, and other military equipment. Ameripol was a butadiene-type material.

As a side note, today there is a debate concerning the use of chlorine-containing materials and their effect on the atmosphere. This is a real concern and one that is being addressed by industry. PVC, and other chloride-containing materials have been in the past simply disposed of through combustion that often created unwanted hydrogen chloride. This practice has largely been stopped but care should be continued to see that such materials are disposed of properly. Further, simply outlawing all chloride-containing materials is neither possible nor practical. For instance, we need common table salt for life and common table salt is sodium chloride. Chlorine is widely used as a water disinfectant both commercially (for our drinking water) as well as for swimming pools. Further, PVC is an important material that is not easily replaced. Finally, the amount of chloride-containing residue that is introduced into the atmosphere naturally is large in comparison to that introduced by PVC. Even so, we must exercise care, as we want to leave a better world for our children and grandchildren; hence, a knowledge-based approach must be undertaken.

Another health concern is the presence of certain plasticizers. As noted later, PVC employs a large amount of additives including a variety of plasticizers. We do not live in a risk-free society but some risks should be eliminated or minimized. One group of plasticizers of interest is the phthalate plasticizers. As with other health concerns infant care is the greatest concern. Vinyl IV bags are often used in neonatal intensive care units. FDA has requested that manufacturers eliminate the use of questionable plasticizers for the production of these bags. Another possible solution is simply eliminating vinyl IV bags using another material for construction of these bags. Further, baby and children toys should be made from materials that do not contain these plasticizers. Another group that is at particular risk is the critically ill or injured patients. Again, special care should be exercised when dealing with special groups of people. As we are dealing with toxic materials we need to remember that all materials can be harmful if present in the wrong concentrations and/or locations. Even so, where we can we might minimize exposure to known potentially toxic materials.

Because of its versatility, some unique performance characteristics, ready availability, and low cost, PVC is now the third largest produced synthetic polymer behind polyethylene and polypropylene. PVC materials are often defined to contain 50% or more vinyl chloride units by weight. PVC is generally a mixture of a number of additives and often other units such as ethylene, propylene, vinylidene chloride, and vinyl acetate. Structurally similar products, but with differing properties, are made from the chlorination of polyethylene but almost all PVC is made from the polymerization of vinyl chloride. Typical homopolymers are about 400–1000 units long.

PVC is commercially produced by a number of techniques but mainly by suspension, emulsion, bulk, and solution polymerization. Typically, product properties and form can be tailored using a particular synthetic process and conditions. Particulate architecture is then controlled to achieve materials with specific sizes and distributions for specific uses and applications. Because of the tendency of PVC to split off hydrogen chloride, forming materials with high char at relatively low general processing temperatures, special care is taken with respect to temperature control and particulate architecture that allows ready processing of PVC by most of the common processing techniques.

Tacticity of the PVC varies according to the particular reaction conditions but generally, manufactures favor a syndiotactic form with many PVC materials being about 50% PVC. The reported amount of crystallinity is in the range of 5%–10%. This allows for a material with some strength, but one with sufficient amorphous regions to retain good flexibility.

PVC, in comparison to many other polymers, employs an especially wide variety of additives. For instance, a sample recipe or formulation for common stiff PVC pipe, such as used in housing and irrigation applications, may contain in addition to the PVC resin, tin stabilizer, acrylic processing aid, acrylic lubricant-processing aid, acrylic impact modifier, calcium carbonate, titanium dioxide, calcium sterate, and paraffin wax. Such formulations vary according to the intended processing and end use. In such nonflexible PVC materials, the weight amount of additive is about 5%–10%.

As noted before, there is a tendency for PVC to undergo elimination of hydrogen chloride when heated. The most labile chlorine atoms are those at tertiary or terminal sites. Once the initial chloride is eliminated, continued unzipping occurs with the formation of unsaturated backbone sites and the evolution of hydrogen chloride. The purpose of the stabilizer is to cap unzipping sites by substitution of more stable groups for the evolved chloride as depicted in 7.58.

$$\text{-----C--Cl} \ + \ \text{Sn--R} \ \longrightarrow \ \text{-----C--R} \ + \ \text{Sn--Cl} \tag{7.58}$$

Some of the tin stabilizers are based on oligomeric materials first made by Carraher and coworkers. These oligomeric materials are essentially "nonmigratable."

A sample formula for a flexible upholstery fabric covering might contain PVC resin, medium molecular weight polymeric plasticizer, stearic acid lubricant, calcium carbonate, pigment, antimony oxide, linear phthalate ester, epoxidized soybean oil, and linear phthalate ester. Here the weight amount of additive is in the range of 40%–70% by weight with the plasticizer often being about 60%.

PVC has a built in advantage over many other polymers in that it is itself flame resistant. About 50% of PVC is used as rigid pipe. About 70% of the water pipes in the United States are PVC. About 75% of the sewer pipes are PVC. Other uses of rigid PVC are as pipe fittings, electrical outlet boxes, and automotive parts. Uses of flexible PVC include in gasoline-resistant hose, hospital sheeting, shoe soles, electrical tape, stretch film, pool liners, vinyl-coated fabrics, roof coatings, refrigerator gaskets, floor sheeting, and electrical insulation and jacketing. A wide number of vinyl chloride copolymers are commercially used. Many vinyl floor tiles are copolymers of PVC.

Many flat sheet signs are made of PVC. Films are also formed from PVC. Many of these signs and films are simply referred to as vinyl. These films and sheets form many of our commercial signs and markings on vehicles. Unplasticized or rigid PVC is used in the construction industry as a siding simply known as vinyl siding. It is also used to repair window frames and sills and fascia. It is also widely used in the construction of plastic gutters, downpipes, and drainpipes.

The flame resistance of PVC is a mixed blessing. In a fire, the PVC emits hydrogen chloride with the chlorine scavenging free radicals helping eliminate the fire. Hydrogen chloride fumes present a health concern when we breathe them. In a fire, moisture helps dilute the hydrogen chloride causing it to settle on the cooler surfaces rather than remaining air borne. Even so, in closed structures such as tunnels alternative materials are advised.

7.8 POLY(METHYL METHACRYLATE)

Poly(methyl methacrylate), PMMA, (7.42) is also referred to as acrylic glass or fiber. It was initially developed in 1928 and became commercially available in 1933 by Rohm and Hass. It is a thermoplastic transparent plastic or fiber produced by the polymerization of methyl methacrylate using emulsion, solution, and bulk free radical polymerization. *Acrylic fiber* refers to both fibers made from PMMA and also those made from polyacrylonitrile. PMMA is often used as an alternative to glass and sold under a number of trade names including Lucite, Plexiglas, Acrylex, Acrylite, Oroglass, and Vitroflex. When completely burned, the products are carbon dioxide and water making it an environmentally positive material in this regard.

In comparison to glass, PMMA is lighter and has greater impact strength than glass or PS, but less than that of polycarbonate. It is more easily scratched than glass, but has optical properties similar to glass in that it transmits up to 98% of light with a low reflection. It is inferior to glass with respect to environmental stability and resistance to solvent, thought its environmental stability is better than most other plastics.

As already noted, PMMA is an often used as an alternative to glass. PMMA is used in the exterior lenses of automobiles and trucks. The spectator protection shield in ice hockey stadiums is made of PMMA. The windows of many aircraft and windows of police vehicles for riot control are PMMA as are many motorcycle helmet visors.

PMMA is also used in medicine. It has a good compatibility with human tissue and has been used for the replacement of intraocular lenses in eyes. Hard contact lenses are often composed of PMMA. Bone cement containing PMMA is used to connect bond implants and remodel lost bones. PMMA has a Young's modulus near that of bone, and so does a good job of load sharing with the native bone. Dentures are often PMMA. In cosmetic surgery, colored PMMA microspears are implanted to reduce wrinkles and scars.

Artistic uses of PMMA are many. Acrylic paints are largely PMMA suspended in water. Because PMMA is hydrophobic, materials are added to allow the suspension to occur. Salvador Dali has used PMMA plastic sheets to paint on.

PMMA has many other uses including as artificial fingernails, bodies of electric guitars and basses, light guide in electrical devices, short distance optical fibers, drums called Vistalites that were used by the Led Zeppelin drummer John Bonham, and extensively in the sign business as either the backside, backplate, letters or design, or faceplate. The high-heels of shoes are often made of PMMA.

PMMA is also used in the semiconductor industry as a resist in the electron beam lithography process. For instance, a semiconductive wafer is coated with a PMMA-containing solution. Patterns are then cut onto this surface using an electron beam, deep UV light, or x-rays. The areas exposed to this radiation contain degraded PMMA that is removed by a chemical developer making it a positive photoresist. PMMA allows precision etching with nanoscale patterns created.

7.9　POLYACRYLONITRILE

Polyacrylonitrile, PAN (7.47), forms the basis for a number of fibers and copolymers. As fibers, they are referred to as acrylics or acrylic fibers. The development of acrylic fibers began in the early 1930s in Germany but they were first commercially produced in the United States by Dupont (Orlon) and Monsanto (Acrilan) about 1950.

Because of the repulsion of the cyanide groups, the polymer backbone assumes a rodlike conformation. The fibers derive their basic properties from this stiff structure of PAN where the nitrile groups are randomly distributed about the backbone rod. Because of strong bonding between the chains, they tend to form bundles. Most acrylic fibers actually contain small amounts of other monomers such as methyl acrylate and methyl methacrylate. Because they are difficult to dye, small amounts of ionic monomers, such as sodium styrene sulfonate, are often added to improve their dyeability. Other monomers are also employed to improve dyeability. These include small amounts (about 4%) of more hydrophilic monomers such as N-vinyl-2-pyrrolidone (7.59), methacrylic acid, or 2-vinylpyridine (7.60).

(7.59)　　　　　　　　(7.60)

Acrylic fibers are used as an alternative to wool for sweaters. PAN is also used in the production of blouses, blankets, rugs, curtains, shirts, craft yarns, and pile fabrics used to simulate fur.

Because of the presence of the cyano group there can be concern that when burned HCN can be given off. Contrary to this, at temperatures above 160°C, PAN begins forming cyclic imines that dehydrogenate forming dark-colored heat-resistant fused ring polymers with conjugated C=C and C=N bonding. This is described in Section 10.12. Thus, rather than forming toxic HCN, polyacrylonitrile forms thermally stable intermediates that act to curtail burning.

Fibers with more than 85% acrylonitrile units are called acrylic fibers but those containing 35%–85% acrylonitrile units are referred to as modacrylic fibers. The remainder of the modacrylic fibers is derived from comonomers such as vinyl chloride or vinylidene chloride that are specifically added to improve flame resistance.

7.10　SOLID-STATE IRRADIATION POLYMERIZATION

There are numerous examples of solid-state polymerizations. Here, we will briefly describe examples based on addition polymers. Generally, the crystalline monomer is irradiated with electrons or some forms of high-energy radiation such as gamma or x-rays. Since many monomers are solids only below room temperature, it is customary to begin irradiation at lower temperatures with the temperature only raised after the occurrence of initial polymerization. (Some reactions are carried to completion at the lower temperature.) After polymerization, the monomer is removed. Table 7.7 contains a list of some of the common monomers that undergo solid-state irradiation polymerization.

TABLE 7.7
Monomers That Undergo Solid-State Irradiation Polymerization

Styrene	Formaldehyde
Acetaldehyde	Acrylic acid (and salts)
Methacrylic acid (and salts)	Trioxane
1,3-butadiene	3,3-Bischloromethylcyclooxabutane
Isoprene	Acrylonitrile
Acrylamide	β-propiolactone
Diacetylenes	

This approach can offer several advantages. First, polymers can be formed from monomers that do not give polymer under more typical reaction conditions. Second, under some cases, the crystalline structure acts as a template giving order that might be difficult to otherwise achieve. Third, removal and interference by solvents or additives is eliminated since they are not present. Fourth, the polymers produced by this technique are often different from those that are produced from the same monomer except when produced using typical reaction techniques.

7.11 PLASMA POLYMERIZATION

Organic and inorganic molecules can be placed in the vapor state either through heating, low pressure, simply spraying, or some combination of these processes. These molecules are then subjected to some ionizing energy that forms active species that react with one another eventually depositing themselves on a surface. Often the products are polymeric with complex structures. The term plasma polymerization is generally used to describe the process resulting in surface film formation while the term deposition is generally used to describe the deposition of powdery particles formed in the gas phase. Others describe plasma polymerization as the polymerization that occurs at high rates in the gas phase resulting in powder formation and deposition as any sorption occurring on the surface. In truth, it is difficult to separate the two reaction sequences because active molecules can react both in the gaseous phase upon collision and on the surface.

Plasma environments are often created using plasma jets, ion beams, glow discharges, corona discharges, laser-induced plasmas, and electron beams. Low temperature plasmas can also be created using radio frequency, audio frequency, microwave, or direct current energy sources. In general terms, the molecules enter the reactor as neutral species. They become reactive species as electronic energy is transferred to them. The reactive species can be ions, free radicals, or excited molecules. Reaction can occur in the gaseous phase and/or at the solid surface. Commercially, reactors often consist of a low-pressure glow discharge of reactive species. Because a small amount of electromagnetic radiation is emitted in the visible region, the term glow discharge was derived.

This approach allows the deposition of thin films at low temperatures. By comparison, polymer deposition generally requires very high temperatures. For instance, the chemical vapor deposition of silicon nitride requires a temperature of about 900°C whereas the plasma chemical deposition requires a temperature of only 350°C.

A number of typical polymer-forming monomers have been polymerized using plasma polymerization including tetrafluoroethylene, styrene, acrylic acid, methyl methacrylate, isoprene, and ethylene. Polymerization of many nontypical monomers has also occurred including toluene, benzene, and simple hydrocarbons.

Plasma films are usually highly cross-linked; resistant to higher temperatures abrasions, and chemical attacks; and its adhesion to the surface is high. Adhesion to the surface is generally high because the growing polymer complex can fit the surface contour and thus "lock-itself in" (physical adhesion) and because, in many instances, the species are active enough to chemically react with the surface molecules to chemically bond to the surface. The surface can be prepared so that the chemical reaction is enhanced.

Plasma surface treatment of many polymers, including fabrics, plastics, and composites, often occurs. The production of ultrathin films via plasma deposition is important in microelectronics, biomaterials, corrosion protection, permeation control, and for adhesion control. Plasma coatings are often about 1–100 nm thick.

7.12 SUMMARY

1. For classical free-radical polymerizations, the rate of propagation is proportional to the concentration of monomer and the square root of the initiator concentration. Termination usually occurs through a coupling of two live radical chains but can occur through disproportionation. The rate of termination for coupling is directly proportional to initiator concentration. The $\overline{DP}$ is directly proportional to monomer concentration and inversely proportional to the square root of the initiator concentration.
2. The first chains produce high molecular weight products. Within the polymerizing system, the most abundant species are the monomer and polymers chains.
3. Increasing the temperature increases the concentration of free radicals thus decreasing the chain length. Increasing the temperature increases the rate of polymer formation.
4. The rate-controlling step is the rate of initiation.
5. There is a steady state concentration of growing chains.
6. The kinetic chain length v is equal to $\overline{DP}$ for disproportionation termination, but $\overline{DP} = 2v$ for termination by coupling.
7. Chain transfer reactions usually decrease $\overline{DP}$ and often will introduce branching as it occurs within or between polymer chains. Solvent also can act as effective chain transfer agents lowering chain length.
8. Vinyl monomers can be polymerized using solution, bulk, suspension, and emulsion techniques. Each having its own characteristic strengths and weaknesses.
9. When polymerizations become viscous, termination slows allowing an increase in the number of growing chains and rate of polymerization. This is known as the gel of Trommsdorff effect. If such reactions are allowed to continue without cooling, explosions are possible.
10. Monomers may be polymerized using a water-soluble initiator while dispersed, by agitation, in a concentrated soap solution. In this emulsion system, initiation occurs in the aqueous phase and propagation occurs in the soap micelles. Since the growing macroradicals are not terminated until a new free radical enters the micelle, high molecular weight products are rapidly obtained. The rate of polymerization and $\overline{DP}$ are proportional to the number of activated micelles.
11. Polyfluorocarbons are resistant to heat, solvent, and corrosives. The resistance is greatest in the regularly structured PTFE and decreases as the geometry is upset by substitution of other atoms for fluorine.
12. PVC is an important polymer used for many commercial purposes including as pipes for water delivery systems within homes.

GLOSSARY

Backbiting: Hydrogen atom abstraction that occurs when a chain end of a macroradical doubles back on itself forming a more stable hexagonal conformation.

Branch point: Point on a polymer chain where additional chain extension occurs producing a branch.

Bulk polymerization: Polymerization of monomer without added solvents or water; also called neat.

Ceiling temperature (T_c): Characteristic temperature above which polymerization occurs but the polymer decomposes before it is recovered.

Chain stopper: Chain transfer agent that produces inactive free radicals.

Chain transfer: Process in which a free radical abstracts an atom or group of atoms from a solvent, telogen, or polymer.

Chain-transfer constant (C_s): Ratio of cessation or termination of transfer to the rate of propagation.

Critical micelle concentration: Minimum concentration of soap in water that will produce micelles.

Dead polymer: Polymer in which chain growth has been terminated.

Disproportionation: Process by which termination occurs as a result of chain transfer between two macroradicals yielding dead polymers.

Half-life time: Time required for half the reactants to be consumed (generally for a first order reaction).

Heterolytic cleavage: Cleavage of a covalent bond that leaves one electron with each of the two atoms. The products are free radicals.

Homopolymer: Polymer composed of only one repeating unit.

Kel F: Trade name for polytrifluorochloroethylene.

Kinetic chain length: Length of the polymer chain initiated by one free radical.

Macroradicals: Electron-deficient polymers having a free radical present on the chain.

Micelles: Ordered groups of soap molecules in water.

Oligoradical: Low molecular weight macroradical.

Piezoelectric: Conversion of mechanical force, such as pressure, into electrical energy.

Plasticizer: High-boiling compatible liquid that lowers T_g and flexibilizes the polymer.

Retarder: Additive that acts as a chain transfer agent producing less active free radicals.

Saran: Trade name for copolymers of vinyl chloride and vinylidene chloride.

SBR: Rubbery copolymer of styrene and butadiene.

Suspension polymerization: Process in which liquid monomers are polymerized in liquid droplets suspended in water.

Telogen: Additive that readily undergoes chain transfer with a macroradical.

Telomer: Low molecular weight polymer resulting from chain transfer of a macroradical with a telogen.

Telomerization: Process in which telomers are produced by chain transfer reactions.

Trommsdorff effect: Decrease in termination rate in viscous media that results in the formation of higher molecular weight polymers.

EXERCISES

1. Use a slanted line to show the cleavage of (a) boron trifluoride-water, (b) sodamide, and AIBN in cationic, anionic, and free radical initiations, respectively.
2. Which type of chain-reaction polymerization is most likely to terminate by coupling?
3. If an initiator has a half-life of 4 h, what percentage of this initiator will remain after 12 h?
4. If some heat-to-head configuration is detected in a polymer chain known to propagate by head-to-tail addition, what type of termination has occurred?
5. Which is the better way to increase polymer production rates: (a) increasing the temperature, or (b) increasing the initiator concentration?
6. Name three widely used thermoplastics produced by free radical chain polymerization.
7. What effect does the increase of polarity of the solvent have on free-radical polymerization?

8. Show the repeat units for (a) PS, (b) PVC, and (c) PMMA.
9. Can you think of any advantage of the Trommsdorff effect?
10. What is the limiting step in free radical chain polymerization?
11. In general, which is more rapid: (a) free radical chain reactions or (b) step-reaction polymerizations?
12. If one obtained a yield of 10% polymer after 10 min of polymerizing styrene by a free radical mechanism, what would be the composition of the other 90%?
13. Why is $t_{1/2}$ for all first-order reactions equal to $0.693/k_d$?
14. How could you follow the rate of decomposition of AIBN without directly measuring the rate of polymerization?
15. What is the usual value for the energy of activation of free radical initiation?
16. What is the advantage of producing free radicals by UV radiation?
17. Why is PVC considered negative by some environmentally?
18. If $[M^\bullet]$ is equal to 1×10^{-11} mol/L under steady-state conditions, what will $[M^\bullet]$ equal after (a) 30, (b) 60, and (c) 90 min?
19. In general, what is the activation energy in the free radical chain propagation of polymer chains?

relationship between the rate of propagation and the concentration of

transfer with solvent occurs, what effect does this have on the average degree ation?

adical polymerization name two steady-state assumptions.

ner is used to produce PVA?

advantage and one disadvantage for the bulk-batch polymerization of

ease or decrease when the average $\overline{DP}$ for a single growing chain goes from

26. Why is ethylene more readily polymerized by free radical chain polymerization than isobutylene?
27. What is the termination mechanism in free-radical polymerization is the average $\overline{DP} = v$?
28. The value of v increases as the polymerization temperature of a specific monomer is increased. What does this tell you about the termination process?
29. In general, what is the activation energy of termination?
30. Why would you not recommend the use of poly-α-methylstyrene for the handle of a cooking utensil?
31. Why is polyfluoroethylene generally known by the public as Teflon.
32. Which would you expect to have the higher chain transfer constant: (a) carbon tetrafluoride, or (b) carbon tetrachloride?
33. While the addition of dodecyl mercaptan to styrene causes a reduction in $\overline{DP}$, the rate of polymerization is essentially unchanged. Explain.
34. Would it be safe to polymerize styrene by bulk polymerization in a 55-gal drum?
35. How do the kinetics of polymerization differ in the bulk and suspension polymerization methods?
36. Since the monomers are carcinogenic, should the polymerization of styrene, acrylonitrile, and vinyl chloride be banned?
37. What are some unusual properties of polyfluoroethylene?
38. Why does polymerization not take place in the droplets instead of in the micelles in emulsion polymerization?
39. Why does initiation not occur in the micelles in emulsion polymerization?
40. What would happen if one added a small amount of an inhibitor to styrene before bulk polymerization?

41. Name several places you might "run across" polystyrene daily.
42. Why is the T_g of PTFE higher than that of FEP?
43. Why does an increase in soap concentration increase the average $\overline{DP}$ and R_p in emulsion polymerizations?
44. Which will have the higher specific gravity (density): (a) PVC or (b) PVDC?
45. Compare the properties of PMMA and glass.
46. Why is there a concern over the use of plasticizers?

FURTHER READINGS

Bamford, C. H., Barb, W. G., Jenkins, A. D., and Onyon, P. E. (1958): *Kinetics of Vinyl Polymerization by Radical Mechanism*, Butterworths, London, U.K.

Bhowmick, A. and Stephens, H. (2000): *Handbook of Elastomers*, Dekker, New York.

Buback, M. and van Herk, A. (2007): *Radical Polymerization: Kinetics and Mechanism*, Wiley, Hoboken, NJ.

Cazacu, M. (2008): *Advances in Functional Heterochain Polymers*, Nova, Hauppauge, NY.

Chern, C. (2008): *Principles and Applications of Emulsion Polymerization*, Wiley, Hoboken, NJ.

Dubois, P. (2000): *Advances in Ring Opening Polymerizations*, Wiley, New York.

Gaylord, N. G. and Mark, H. F. (1958): *Linear and Stereoaddition Polymers*, Interscience, New York.

Harkins, W. D. (1947, 1950): Emulsion polymerization, *J. Am. Chem. Soc.*, 69:1429; *J. Polymer Sci.*, 5:217.

Kissin, Y. and Centi, G. (2008): *Alkene Polymerization Reactions with Transition Metal Catalysts*, Elsevier, New York.

Korolyov, G. and Mogilevich, M. (2009): *Three-Dimensional Free-Radical Polymerization, Process, and Technology*, Springer, New York.

Mayo, F. R. (1943): Chain transfer, *J. Am. Chem. Soc.*, 65:2324.

Mishra, M. and Yagic, Y. (2008): *Handbook of Vinyl Polymers: Radical Polymerization, Process, and Technology*, CRC Press, Boca Raton, FL.

Niyazi, F. (2007): *Photochemical Reactions in Heterochain Polymers*, Nova, Hauppauge, NY.

Papaspyrides, C. and Vouviouka, S. (2009): *Solid-State Polymerization*, Wiley, Hoboken, NJ.

Peacock, A. (2000): *Handbook of Polyethylenes*, Dekker, New York.

Plunkett, R. J. (1941): Polytetrafluoroethylene, U.S. Patent 2,230,654.

Scheirs, J. (2003): *Modern Styrenic Polymers*, Wiley, Hoboken, NJ.

Smith, D. (2008): *Handbook of Fluoropolymer Science and Technology*, Wiley, Hoboken, NJ.

Smith, W. V. and Ewart, R. H. (1948) Emulsion polymerization, *J. Chem. Phys.*, 16:592.

Trommsdorff, E., Kohle, H., and Lagally, P. (1948): Viscous polymerization, *Makromol. Chem.*, 1:169.

Vasile, C. (2000): *Handbook of Polyolefins*, Dekker, New York.

Walling, C. (1957): *Free Radicals in Solution*, Wiley, New York.

8 Copolymerization

The previous two chapters dealt with homopolymers where the forces between the polymer chains are relatively small allowing their use as plastics. Here we will look at copolymers where the various units allow the copolymers to also act as elastomers because of the introduction of chemical cross-linking. To some extent, the introduction of physical cross-linking allows some homo- and copolymers to act as elastomers.

While the mechanism of copolymerization is similar to that discussed for the polymerization of one reactant (homopolymerization), the relativities of monomers may differ when more than one is present in the feed, that is, reaction mixture. Copolymers may be produced by step reaction polymerization or chain reaction polymerization. It is important to note that if the reactant species are M_1 and M_2 then the composition of the copolymer is not a physical mixture or blend, though the topic of blends will be dealt with in this chapter.

Many naturally occurring polymers are largely homopolymers, but proteins and nucleic acids are copolymers composed of a number of different mers. While many synthetic polymers are homopolymers, the most widely used synthetic rubber (SBR) is a copolymer of styrene (S) and butadiene (B), with the "R" representing "rubber." There are many other important copolymers. Here we will restrict ourselves to vinyl-derived copolymers.

Copolymers may be **random** in the placement of units,

$$-M_1M_2M_1M_2M_2M_1M_1M_2M_1M_1M_2M_1M_2M_2M_1M_1M_2M_2M_2M_1M_1M_2M_1M_1- \qquad (8.1)$$

alternating in the placement of units,

$$-M_1M_2M_1M_2M_1M_2M_1M_2M_1M_2M_1M_2M_1M_2M_1M_2M_1M_2M_1M_2M_1M_2- \qquad (8.2)$$

block in either or both of the different units in which there are long sequences of the same repeating unit in the chain,

$$-M_1M_1M_1M_1M_1M_1M_1M_2M_2M_2M_2M_2M_2M_2M_2M_1M_1M_1M_1M_1M_1M_1M_1M_1M_2M_2M_2M_2- \qquad (8.3)$$

or **graft** copolymers in which the chain extension of the second monomer is present as branches.

$$
\begin{array}{lll}
-M_1M_1M_1M_1M_1M_1M_1M_1M_1M_1M_1M_1M_1M_1M_1M_1M_1M_1M_1M_1- & & \\
\quad M_2 \qquad\qquad M_2 \qquad\qquad\qquad M_2 & & \\
\quad M_2 \qquad\qquad M_2 \qquad\qquad\qquad M_2 & & \\
\quad M_2 \qquad\qquad M_2 \qquad\qquad\qquad M_2 & & (8.4)\\
\quad M_2 \qquad\qquad M_2 \qquad\qquad\qquad M_2 & & \\
\quad M_2 \qquad\qquad M_2 & & \\
\qquad\qquad\qquad M_2 & &
\end{array}
$$

Each of these types of copolymers offers different physical properties for a particular copolymer combination. It is interesting to note that block copolymers may be produced from one monomer only if the arrangement around the chiral carbon atom changes sequentially. These copolymers are called **stereoblock** copolymers.

8.1 KINETICS OF COPOLYMERIZATION

Because of a difference in the reactivity of monomers, expressed as reactivity ratios (r), the composition of the copolymer (n) may be different from that of the reactant mixture or feed (x). When x equals n, the product is said to be an *azeotropic copolymer*.

In the early 1930s, Nobel Laureate Staundinger analyzed the product obtained from the copolymerization of equimolar quantities of vinyl chloride (VC) and vinyl acetate (VAc). He found that the first product produced was high in VC, but as the composition of the reactant mixture changed because of a preferential depletion of VC, the product was becoming higher in VAc. This phenomenon is called the *composition drift*.

Wall studied the composition drift and derived what is now called the Wall equation where n was equal to rx when the reactivity ratio r was equal to the ratio of the propagation rate constants. Thus, r was the slope of the line obtained when the ratio of monomers in the copolymer (M_1/M_2) was plotted against the ratio of monomers in the feed (m_1/m_2). The Wall equation is not general:

$$n = \frac{M_1}{M_2} = r\left(\frac{m_1}{m_2}\right) = rx \tag{8.5}$$

The copolymer equation that is now accepted was developed in the late 1930s by a group of investigators including Wall, Dostal, Lewis, Alfrey, Simha, and Mayo. These workers considered the four possible extension reactions when monomers M_1 and M_2 were present in the feed. As shown in the following text, two of these reactions are homopolymerizations or self-propagating steps (8.6 and 8.8), and the other two are heteropolymerizations or cross-propagating steps (8.7 and 8.9). The ratio of the propagating rate constants are expressed as *monomer reactivity ratios* (or simply reactivity ratios), where $r_1 = k_{11}/k_{12}$ and $r_2 = k_{22}/k_{21}$. $M_1^\bullet$ and $M_2^\bullet$ are used as symbols for the macroradicals with $M_1^\bullet$ and $M_2^\bullet$ terminal groups, respectively.

Reaction	Rate constant	Rate expression	
$M_1^\bullet + M_1 \rightarrow M_1 M_1^\bullet$	k_{11}	$R_{11} = k_{11}[M_1^\bullet][M_1]$	(8.6)
$M_1^\bullet + M_2 \rightarrow M_1 M_2^\bullet$	k_{12}	$R_{12} = k_{12}[M_1^\bullet][M_2]$	(8.7)
$M_2^\bullet + M_2 \rightarrow M_2 M_2^\bullet$	k_{22}	$R_{22} = k_{22}[M_2^\bullet][M_2]$	(8.8)
$M_2^\bullet + M_1 \rightarrow M_2 M_1^\bullet$	k_{21}	$R_{21} = k_{21}[M_2^\bullet][M_1]$	(8.9)

Experimentally, it is found that the specific rate constants for the various reaction steps described earlier are essentially independent of chain length, with the rate of monomer addition primarily dependent only on the adding monomer unit and the growing end. Thus, the four reactions between two comonomers can be described using only these four equations.

As is the case with the other chain processes, determining the concentration of the active species is difficult so that expressions that do not contain the concentration of the active species are derived. The change in monomer concentration, that is, the rate of addition of monomer to growing copolymer chains, is described in the following:

$$\text{Disappearance of } M_1: -\frac{d[M_1]}{dt} = k_{11}[M_1^\bullet][M_1] + k_{21}[M_2^\bullet][M_1] \tag{8.10}$$

$$\text{Disappearance of } M_2: -\frac{d[M_2]}{dt} = k_{22}[M_2^\bullet][M_2] + k_{12}[M_1^\bullet][M_2] \tag{8.11}$$

Since it is experimentally observed that the number of growing chains remains approximately constant throughout the duration of most copolymerizations (i.e., a steady state in the number of growing chains), the concentrations of $M_1^\bullet$ and $M_2^\bullet$ are constant, and the rate of conversion of $M_1^\bullet$ to $M_2^\bullet$ is equal to the conversion of $M_2^\bullet$ to $M_1^\bullet$. Solving for $M_1^\bullet$ gives

$$[M_1^\bullet] = \frac{k_{21}[M_2^\bullet][M_1]}{k_{12}[M_2]} \tag{8.12}$$

The ratio of disappearance of monomers M_1/M_2 is described by Equation 8.13 from Equations 8.10 and 8.11. Remember, that this is also the average composition of the growing chains and resulting polymer:

$$\frac{d[M_1]}{d[M_2]} = \frac{k_{11}[M_1^\bullet][M_1] + k_{21}[M_2^\bullet][M_1]}{k_{22}[M_2^\bullet][M_2] + k_{12}[M_1^\bullet][M_2]} = \frac{[M_1](k_{11}[M_1^\bullet] + k_{21}[M_2^\bullet])}{[M_2](k_{22}[M_2^\bullet] + k_{12}[M_1^\bullet])} \tag{8.13}$$

Substitution of $[M_1^\bullet]$ into Equation 8.13 gives

$$\frac{d[M_1]}{d[M_2]} = \frac{[M_1]}{[M_2]} \left\{ \frac{(k_{11}k_{21}[M_2^\bullet][M_1]/k_{21}[M_2]) + k_{21}[M_2^\bullet]}{(k_{12}k_{21}[M_2^\bullet][M_1]/k_{21}[M_2]) + k_{22}[M_2^\bullet]} \right\} \tag{8.14}$$

Division by k_{12} and cancellation of the appropriate k's gives

$$\frac{d[M_1]}{d[M_2]} = \frac{[M_1]}{[M_2]} \left\{ \frac{(k_{11}[M_2^\bullet][M_1]/k_{12}[M_2]) + [M_2^\bullet]}{([M_2^\bullet][M_1]/[M_2]) + k_{22}[M_2^\bullet]/k_{21}} \right\} \tag{8.15}$$

Substitution of $r_1 = k_{11}/k_{12}$ and $r_2 = k_{22}/k_{21}$ and cancellation of $[M_2^\bullet]$ gives

$$\frac{d[M_1]}{d[M_2]} = \frac{[M_1]}{[M_2]} \left\{ \frac{(r_1[M_1]/[M_2]) + 1}{([M_1]/[M_2]) + r_2} \right\} \tag{8.16}$$

Multiplication by $[M_2]$ gives what are generally referred to as the "copolymerization equations," Equations 8.17 and 8.19, which gives the copolymer composition without the need to know any free radical concentration, and which gives the composition of the growing polymer as a function of monomer feed (8.19):

$$n = \frac{d[M_1]}{d[M_2]} = \frac{[M_1](r_1[M_1] + [M_2])}{[M_2]([M_1] + r_2[M_2])} \tag{8.17}$$

Equation 8.17 is also presented in another form that allows greater ease of seeing the relationship between the monomer feed, x, and copolymer composition. This is achieved by the following steps.
  Multiplying through by $[M_1]$ and $[M_2]$ gives

$$n = \frac{d[M_1]}{d[M_2]} = \frac{[M_1]r_1[M_1] + [M_1][M_2])}{[M_2][M_1] + [M_2]r_2[M_2])} \tag{8.18}$$

Then divide both the top and bottom by $[M_1][M_2]$ to get the second form, 8.19, of the copolymerization equation but in terms of the composition of the feed (x) on the composition of the copolymer (n) as shown in the following.

$$n = \frac{d[M_1]}{d[M_2]} = \frac{r_1([M_1]/[M_2])+1}{r_2([M_2]/[M_1])+1} = \frac{r_1 x + 1}{r_2/x + 1}$$

(8.19)

The reactivity ratios are determined by an analysis of the change in the composition of the feed during the early stages of polymerization. Typical free radical chain copolymerization reactivity ratios are given in Table 8.1. The copolymer composition and type can be predicted by looking at

TABLE 8.1
Typical Free Radical Chain Copolymerization Reactivity Ratios at 60°C

M_1	M_2^a	r_1	r_2	$r_1 r_2$
Acrylamide	Acrylic acid	1.38	0.36	0.5
	Methyl acrylate	1.30	0.05	0.07
	Vinylidene chloride	4.9	0.15	0.74
Acrylic acid	Acrylonitrile (50°C)	1.15	0.35	0.40
	Styrene	0.25	0.50	0.04
	Vinyl acetate (70°C)	2	0.1	0.2
Acrylonitrile	Butadiene	0.25	0.33	0.08
	Ethyl acetate (50°C)	1.17	0.67	0.78
	Maleic anhydride	6	0	0
	Methyl methacrylate	0.13	1.16	0.15
	Styrene	0.04	0.41	0.16
	Vinyl acetate	4	0.06	0.24
	Vinyl chloride	3.3	0.02	0.07
Butadiene	Methyl methacrylate	0.70	0.32	0.22
	Styrene	1.4	0.78	1.1
Isoprene	Styrene	2	0.44	0.88
Maleic anhydride	Methyl acrylate	0	2.5	0
	Methyl methacrylate	0.03	3.5	0.11
	Styrene	0	0.02	0
	Vinyl acetate (70°C)	0.003	0.055	0.0002
Methyl acrylate	Acrylonitrile	0.67	1.3	0.84
	Styrene	0.18	0.75	0.14
	Vinyl acetate	9.0	0.10	0.90
	Vinyl chloride	5	0	0
Methyl methacrylate	Styrene	0.50	0.50	0.25
	Vinyl acetate	20	0.015	0.3
	Vinyl chloride	12.5	0	0
Styrene	p-Chlorostyrene	0.74	1.03	0.76
	p-Methoxystyrene	1.2	0.82	0.95
	Vinyl acetate	55	0.01	0.55
	Vinyl chloride	17	0.02	0.34
	2-Vinylpyridine	0.56	0.9	0.50
Vinyl acetate	Vinyl chloride	0.23	1.7	0.39
	Vinyl laurate	1.4	0.7	0.98
Vinyl chloride	Dimethyl maleate	0.77	0.009	0.007

Source: Brandrup and Immergut (1975).

[a] Temperatures other than 60°C are shown in parentheses.

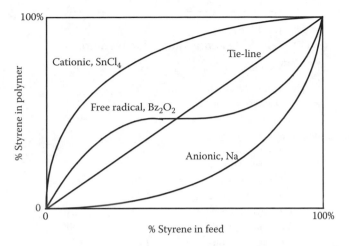

FIGURE 8.1 Instantaneous copolymer composition as a function of monomer composition and initiator employed for the comonomer system of styrene and methyl methacrylate using different modes of initiation. (From Landler, Y., *Compt. Rend.*, 230, 539, 1950.)

the values of r_1 and r_2. If the value of r_1 is greater than 1, then M_1 tends to react with itself producing homopolymers, or block copolymers in M_1. Preference for reaction with the unlike monomer occurs when r_1 is less than 1. When r_1 and r_2 are approximately equal to 1, the conditions are said to be ideal, with a random (not alternating) copolymer produced, in accordance with the Wall equation. Thus, a random copolymer (ideal copolymer) would be produced when chlorotrifluoroethylene is copolymerized with tetrafluoroethylene (Table 8.1).

When r_1 and r_2 are both approximately zero, as in the case with the copolymerization of maleic anhydride and styrene, an alternating copolymer is produced. In general, there will be a tendency toward alternation when the product $r_1 r_2$ approaches zero. In contrast, if the values of r_1 and r_2 are similar and the product $r_1 r_2$ approaches 1, the tendency will be to produce random copolymers. The value of $r_1 r_2$ for most copolymerizations is between 1 and 0, and thus this value may be used with discretion for estimating the extent of randomness in a copolymer.

The resonance stability of the macroradical is an important factor in polymer propagation. Thus, for free radical polymerization, a conjugated monomer such as styrene is at least 30 times as apt to form a resonance-stabilized macroradical as vinyl acetate, resulting in a copolymer being rich in styrene-derived units when these two are copolymerized.

Strongly electrophilic or nucleophilic monomers will polymerize exclusively by anionic or cationic mechanisms. However, monomers that are neither strongly electrophilic nor nucleophilic generally polymerize by ionic and free radical processes. The contrast between anionic, cationic, and free radical methods of addition copolymerization is clearly illustrated by the results of copolymerization utilizing the three modes of initiation (Figure 8.1). Such results illustrate the variations of reactivities and copolymer composition that are possible from employing the different initiation modes. The free radical "tieline" resides near the middle since free radical polymerizations are less dependent on the electronic nature of the comonomers relative to the ionic modes of chain propagation.

As noted before, the copolymerization can be controlled by the control of the monomer feed in accordance of Equations 8.17 and 8.19.

8.2 THE Q–e SCHEME

A useful scheme for predicting r_1 and r_2 values for free radical copolymerizations was developed by Alfrey and Price in 1947. The Alfrey–Price Q–e scheme is similar to the Hammett equation approach except that it is not primarily limited to substituted aromatic compounds. In the semiempirical

TABLE 8.2
Typical Q and e Values for Monomers

Monomer	Q	e
Benzyl methacrylate	3.64	0.36
Methacrylic acid	2.34	0.65
2-Fluoro-1,3-butadiene	2.08	−0.43
p-Cyanostyrene	1.86	−0.21
p-Nitrostyrene	1.63	0.39
2,5-Dichlorostyrene	1.60	0.09
Methacrylamide	1.46	2.24
p-Methoxystyrene	1.36	−1.11
2-Vinylpyridine	1.30	−0.50
p-Methylstyrene	1.27	−0.98
Methacrylonitrile	1.12	0.81
p-Bromostyrene	1.04	−0.32
Styrene	1.00	−0.80
m-Methylstyrene	0.91	−0.72
Methyl methacrylate	0.74	0.40
Acrylonitrile	0.60	1.20
Methyl acrylate	0.42	0.60
Vinylidene chloride	0.23	0.36
Vinyl chloride	0.044	0.20
Vinyl acetate	0.026	−0.22

Q–e scheme, the reactivities or resonance effects of the monomers and macroradicals are evaluated empirically by Q and e values. The polar properties of both monomers and macroradicals are designated by arbitrary "e" values. Thus, as shown in Table 8.2, Q_1 and Q_2 are related to the reactivity, and e_1 and e_2 are related to the polarity of monomers M_1 and M_2, respectively. Styrene is assigned the Q value of 1 and an e value of −0.80. Higher Q values indicate greater resonance stability or reactivity, and higher e values (less negative) indicate greater electron-withdrawing power of the α-substituents on the vinyl monomer (in comparison to the phenyl substituent in styrene).

8.3 COMMERCIAL COPOLYMERS

One of the first commercial copolymers, introduced in 1928, was made of vinyl chloride (87%) and vinyl acetate (13%) (Vinylite). Because the presence of the vinyl acetate mers disrupted the regular structure of PVC, the copolymer was more flexible and more soluble than PVC itself.

Copolymers of vinyl chloride and vinylidene chloride were introduced in the 1930s. The copolymer with a very high vinyl chloride content is used as a plastic film (Pliovic), and the copolymer with a high vinylidene chloride content is used as a film and filament (Saran).

Polybutadiene, produced in emulsion polymerization, is not useful as an elastomer. However, the copolymer with SBR and acrylonitrile (Buna-N) are widely used as elastomers.

Ethylene–propylene copolymers (EPMs) show good resistance to ozone, heat, and oxygen and are used in blends to make today's external automotive panels. Two general types of EPMs are commercially available. EPMs are saturated and require vulcanization if used as a rubber. They are used in a variety of automotive applications including as body and chassis parts, bumpers, radiator and heater hoses, seals, mats, and weather strips. EPMs are produced using Ziegler–Natta catalysts.

The second type of EPM is the ethylene–propylene-diene terpolymers (EPDMs). These are made by polymerizing ethylene, propylene, and a small amount (3 mol%–10 mol%) of nonconjugated diolefine employing Ziegler–Natta catalysts. The side chains allow vulcanization with sulfur. They are employed in the production of appliance parts, wire and cable insulation, coated fabrics, gaskets, hoses, seals, and high-impact polypropylene.

8.4 BLOCK COPOLYMERS

While block copolymers do not occur naturally, synthetic block copolymers have been prepared by all known classical polymerization techniques. The first commercial block copolymer was a surfactant (Pluronics) prepared by the addition of propylene oxide to polycarbanions of ethylene oxide. While neither water-soluble poly(ethylene oxide) nor water-insoluble poly(propylene oxide) exhibits surface activity, the ABA block copolymer consisting of hydrophilic and lyophilic segments, is an excellent surfactant. Each block has 20 plus repeat units of that variety (8.20).

$$(8.20)$$

The ethylene oxide block is hydrophilic, while the propylene oxide block is (relatively) hydrophobic. The copolymer forms micelles in aqueous solutions with the hydrophilic portions pointing outward, interacting with the water, while the hydrophobic portions form the inner core, shielded from the water by the ethylene oxide derived block. A micelle is also formed in organic liquids, but here the hydrophobic propylene oxide block "faces" outward while the ethylene oxide block acts as the inner core.

Polymers that contain two reactive end groups are referred to as *telechelic polymers*. Joseph Shivers, a Dupont chemist, invented Spandex in 1959 after about a ten-year search. It was first named Fiber K but Dupont chose the more appealing, smooth-sounding trade name of Lycra.

Elastomeric polyurethane fibers consisting of at least 85% segmented polyurethane are commercially available under the name Spandex. They are block copolymers and are among the first products specifically designed using the concept of soft and hard segments. The soft or flexible segment is composed of poly(ethylene oxide) chains that contain two hydroxyl end groups (one at each end of the poly(ethylene oxide) chain). The hard or stiff segment is derived from the reaction of the diisocyanate with the hydroxyl end groups forming polar urethane linkages that connect the poly(ethylene oxide) polymer segments through the urethane linkages. Such products are often referred to as segmented polyurethane fibers (8.21). These products have found wide use in cloths including bras. Similar products have also been formed using hydroxyl-terminated polyethylene in place of the poly(ethylene oxide).

$$(8.21)$$

Flexible polyester   Diisocyanate   Flexible block copolymer
  or polymer

A related product is formed from the analogous reaction using hydroxyl-terminated poly(ethylene oxide) with aromatic diacids to form a segmented aromatic polyester block copolymer that is sold under the trade name of Hytrel.

Poly(ethylene oxide) itself is made from ethylene oxide through a base catalyzed ring-opening polymerization. While the major use of poly(ethylene oxide) is in the manufacture of polyurethane products, including foams, it can be synthesized giving viscous liquids to solids depending on the molecular weight. It is used as surfactants, functional fluids, thickeners, lubricants, and plasticizers. Further, the circuit boards that hold the electronic components of computers are made from poly(ethylene oxide).

The most widely used chain reaction block copolymers are those prepared by the addition of a new monomer to a macroanion. AB and ABA block copolymers called Soprene and Kraton, respectively, are produced by the addition of butadiene to styryl macroanions or macrocarbanions (8.22). This copolymer is normally hydrogenated (8.23).

(8.22)

(8.23)

Today, new copolymers are making use of the hard–soft block strategy where the hard segment is a block portion as polyethylene that readily crystallizes forming a physical cross-link. The soft segment consists of blocks formed from α-olefins such as 1-butene, 1-hexene, and 1-octane where the substituted alkane-arm discourages crystallization.

8.5 GRAFT COPOLYMERS

The major difference between block and graft copolymers is the position of the second kind of unit. Thus, the information that applies to block copolymers can often be applied to graft copolymers. So, domains where physical cross-linking occurs via crystallization can occur either in block components or within graft copolymers where the necessary symmetry occurs.

Graft copolymers of nylon, protein, cellulose, or starch, or copolymers or vinyl alcohol have been prepared by reaction of ethylene oxide with these polymers. Graft copolymers are also produced when styrene is polymerized by Lewis acids in the presence of poly-p-methoxystyrene. The Merrifield synthesis of polypeptides is also based on graft copolymers formed from chloromethaylated polystyrene. Thus, the variety of graft copolymers is enormous.

The most widely used graft copolymer is the styrene-unsaturated polyester copolymer. This copolymer, which is usually reenforced by fibrous glass, is prepared by the free radical chain polymerization of a styrene solution of an unsaturated polyester.

The graft copolymers of acrylamide, acrylic acid, and cellulose or starch are used as water absorbents and in enhanced oil recovery systems.

8.6 ELASTOMERS

Elastomers typically contain chemical and/or physical cross-links. While there are thermoplastic elastomers such as SBS and block copolymers, most elastomers are thermosets. Elastomers are characterized by a disorganized (high-entropy) structure in the resting or nonstressed state. Application of stress is accompanied by a ready distortion requiring (relative to plastics and fibers) little stress to effect the distortion. This distortion brings about an aligning of the chains forming a structure with greater order. The driving force for such a material to return to its original shape is largely a return to the original less organized state. While entropy is the primary driving force for elastomers to return to the original resting state, the cross-links allow the material to return to its original shape thereby giving the materials a type of memory. Materials that allow easy distortion generally have minimal interactions between the same or different chains. This qualification is fulfilled by materials that do not bond by dipolar (or polar) or hydrogen bonding. Thus, the intermolecular and intramolecular forces of attraction are small relative to those present in fibers and plastics. Hydrocarbon-intense polymers are examples of materials that meet this qualification. Production generally requires the initial production of linear polymers, followed by the insertion of cross-links through a process called vulcanization or curing. Addition of fillers and other additives such as carbon black also is typical. Table 8.3 contains a listing of important elastomers.

The introduction of cross-links to inhibit chain slippage was discovered by Goodyear in 1839. He accomplished this through the addition of sulfur to natural rubber (NR). Shortly after this, an accelerator, zinc (II) oxide, was used to speed up the process. Other additives were discovered often through observation and trial and error so that today's elastomers often have a number of important additives that allow them to perform demanding tasks.

About 1915, Mote found that a superior abrasion-resistant elastomer was produced using carbon black (Chapter 10). Today, it is recognized that factors such as surface area, structure, and aggregate size are important features in the production of superior elastomers. For instance, high surface areas (small particle size) increase the reinforcement and consequently the tensile strength and improve the resistance to tearing and abrasion. Large aggregates give elastomers that have improved strength before curing, high modulus, and an improved extrusion behavior.

Rubbers typically have low hysteresis. *Hysteresis* is a measure of the energy absorbed when the elastomer is deformed. A rubber that absorbs a great amount of energy as it is deformed (such as a tire hitting bumps on the roadway) is said to have a high hysteresis. The absorbed energy is equivalent to the reciprocal of resilience such that a material with a low hysteresis has a high resilience. Rubbers with a particularly high hysteresis are used where heat buildup is desirable such as in tires to give the tread a better grip on the road and the tire a smoother ride.

Typical thermosetting elastomers are difficult to recycle because their cross-linking prevents them from being easily solubilized and reformed through the application of pressure and heat. Recycling can be accomplished through the particalizing (grinding into small particles) of the elastomeric material followed by a softening-up by the application of a suitable liquid and/or heat and, finally the addition of a binder that physically or chemically allows the particles to bind together in the desired shape.

8.6.1 THERMOPLASTIC ELASTOMERS

A number of thermoplastic elastomers have been developed since the mid-1960s. The initial thermoplastic elastomers were derived from plasticized PVC and are called plastisols. *Plastisols* are formed from the fusing together of PVC with a compatible plasticizer through heating. The plasticizer acts to lower the T_g to below room temperature. Conceptually, this can be through of as the plasticizer acting to put additional distance between the PVC chains, thus lowering the inter- and intrachain forces as well as helping to solubilize chain segments. The resulting materials are used in a number of areas including the construction of boot soles.

TABLE 8.3
Common Elastomers and Their Uses

Common Name (Chemical Composition	Abbreviation	Uses and Properties
Acrylonitrile–butadiene–styrene (terpolymer)	ABS	Oil hoses, fuel tanks, gaskets, pipe and fittings, appliance and automotive housings. Good resistance to oils and gas.
Butadiene rubber	BR	Tire tread, hose, and belts. Very low hysteresis; high rebound.
Butyl rubber (from isobutene and 0.5%–3% isoprene)	IIR	Innertubes, cable sheathing, tank liners, roofing, seals, and coated fabrics. Very low rebound; high hysteresis.
Chloroprene rubber (polychloroprene)	CR	Wire and cable insulation, hose footwear, mechanical automotive products. Good resistance to oil and fire, good weatherability.
Epichlorohydrin (epoxy copolymers)		Seals, gaskets, wire and cable insulation. Good resistance to chemicals.
Ethylene–propylene rubbers (random copolymers with 60%–80% ethylene	EP or EPM	Cable insulation, window strips. Outstanding insulative properties.
Ethylene–propylene–diene (random terpolymers)	EPDM	Good resistance to weathering, resistant to ozone attack.
Fluoroelastomers (fluorine-containing copolymers)		Wire and cable insulation, aerospace applications. Outstanding resistance to continuous exposure to high temperatures, chemicals, and fluids.
Ionomers (largely copolymers of ethylene and acid-containing monomers with metal ions)		Golf ball covers, shoe soles, weather stripping. Tough, flame-resistant, good clarity, good electrical properties, abrasion-resistant.
Natural rubber (polyisoprene)	NR	General-purpose tires, bushings, and couplings, seals, footwear, belting. Good resilience.
Nitrile rubber (random copolymer of butadiene and acrylonitrile)	NBR	Seals, automotive parts that are in contact with oils and gas, footwear, hose. Outstanding resistance to oils and gas, little swelling in organic liquids.
Polysulfide		Adhesive, sealants, hose binders. Outstanding resistance to oil and organic solvents.
Polyurethanes		Sealing and joining, printing rollers, fibers, industrial tires, footwear, wire and cable coverings.
Silicones (generally polydimethylsiloxane)		Medical applications, flexible molds, gaskets, seals. Extreme use temperature range.
Styrene–butadiene rubber (random copolymer)	SBR	Tire tread, footwear, wire and cable covering, adhesives. High hysteresis.

The hard/soft segment scenario is utilized in the formation of a number of industrially important thermoplastic elastomers. Thermoplastic elastomers contain two or more distinct phases and their properties depend on these phases being intimately mixed and small. These phases may be connected either chemically or physically. For a material to be a thermoplastic elastomer at least one phase must be soft or flexible under the operating conditions and at least one phase must be hard with the hard phase(s) becoming soft (or fluid) at higher temperatures. Often the hard segments or phases are crystalline thermoplastics while the soft segments or phases are amorphous. In continuous chains containing blocks of hard and soft segments, the molecular arrangement normally contains crystalline regions where there is sufficient length in the hard segment to form the crystalline regions or phases where the soft segments form amorphous regions.

Such hard/soft scenarios can also be achieved through employing grafts where the pendant group typically acts as the hard segment with the backbone acting as the soft segment. In 8.24 there is a representation of a typical graft copolymer chain. In order for an effective network to be formed each "A" chain needs to have at least two "B" grafts to allow for the formation of a continuous inter-linked network. While there has been a lot of research done with such graft materials, they have not yet become very important commercially.

$$-\text{A-A-A-A-A-A-A-A-A-A-A-A-A-A-A-A-A-A-A-A-A-A-A-A-}$$
$$| \qquad\qquad | \qquad\qquad |$$
$$(B)_n \qquad\qquad (B)_n \qquad\qquad (B)_n \qquad\qquad\qquad (8.24)$$

Thermoplastic elastomers can also be achieved through physical mixing of hard and soft segments. These are fine dispersions of a hard thermoplastic polymer and an elastomer. The two materials generally form interdispersed co-continuous phases. Often the physical combining is achieved through intense mechanical mixing but in some cases, such as with polypropylene and EPMs the effect of blending is achieved through polymerizing the finely dispersed elastomer phase (EPM) phase simultaneously with the hard polypropylene.

At times, the phases are cross-linked during the mechanical mixing. This process is referred to as "dynamic vulcanization" and results in a finely dispersed discontinuous cross-linked elastomer phase. The products are referred to as thermoplastic vulcanizates or dynamic vulcanizates. The products of this process have an insoluble elastomer phase giving the material greater oil and solvent resistance. The cross-linking also reduces or eliminates the flow of this phase at high temperatures and/or under high stress. This allows the material better resistance to compression set.

SBS block copolymers differ structurally from the random copolymer SBR. Because styrene and butadiene blocks are incompatible, they form separate phases joined at the junctions where the various blocks are connected. This activity results in the formation of an elastomeric material where the butadiene blocks form the soft segments and the styrene blocks form hard blocks.

The block copolymer made from connecting blocks of polystyrene with blocks of polybutadiene illustrates another use of soft and rigid or hard domains in thermoplastic elastomers. The polystyrene blocks give rigidity to the polymer while the polybutadiene blocks act as the soft or flexible portion. The polystyrene portions also form semicrystalline domains that add to the strength of the copolymer and these domains also act as "physical cross-links" allowing the soft portions to respond in an "elastomeric" manner, while the semicrystalline domains give the material the "elastomeric" memory. The polybutadiene blocks act as the continuous phase while the polystyrene blocks act as the discontinuous phase. Heating the material above the T_m of the polystyrene domains allows whole chain mobility allowing the processability of the virgin material and the subsequent reprocessability of used material. The rigid domains reform, upon cooling. Block polystyrene–polybutadiene copolymers are used in the soles of many of the athletic shoes.

Worldwide sales of thermoplastic elastomers are approximately 1.5 million ton with a value of about $5 billion.

8.7 BLENDS

There is an ongoing search for new materials and materials that exhibit needed properties. Blends are one of the major avenues of achieving these new materials without actually synthesizing new polymers. *Polymer blends* are physical mixtures of two or more polymers though sometimes the various phases are chemically bonded together. These blended mixtures may offer distinct properties, one set of properties related to the first member of the blend, and another set of properties related to the second member of the blend. The blended mixtures may also offer some averaging of properties. The property mix of polymeric blends is dependent on a number of factors, one of the major being the miscibility of the polymers in one another. This miscibility is in turn dependent on the nature of the polymers composing the blend and the amount of each component in the blend. Here polymer blends will be divided into miscible and immiscible polymer blends.

The extent of mixing is related to time since mixing requires sufficient time to allow the polymer chains to mix. Thus, for miscible blends particular structures can be "frozen-in" by rapid cooling when the desired mixing is achieved. Here, micelles of particular structures can cause the mixture to perform in a manner that is governed by the particular grouping that may not occur if more total mixing occurs.

Miscibility/immiscibility can be described in simple thermodynamic terms as follows, at constant temperature. Mixing occurs if the free energy of mixing is negative.

$$\Delta G_{\text{mixing}} = \Delta H_{\text{mixing}} - T \Delta S_{\text{mixing}} \tag{8.25}$$

Mixing is exactly analogous with polymer solubility. The driving force for mixing and solubility is the entropy or random-related term. The entropy-related term must overcome the opposing enthalpy energy term. In a more complete treatment, temperature and volume fraction must be considered.

8.7.1 IMMISCIBLE BLENDS

Immiscible combinations are all about us. Oil and water is an immiscible combination; as is the lava in the so-called lava-lamps; and chicken broth in chicken soup. Immiscible blends are actually a misnaming at the molecular level since they are not truly mixed together. However, at the macro-level they appear mixed, so the name immiscible blends.

Immiscible blends are said to be phase separated, that is, there are different phases mixed together. Both phases are solid in behavior.

Because PS is brittle with little impact resistance under normal operating conditions, early work was done to impart impact resistance. The best-known material from this work is called high-impact polystyrene or HIPS. HIPS is produced by dispersing small particles of butadiene rubber in with the styrene monomer. The bulk or mass polymerization of the styrene begins producing what is referred to as prepolymerization material. During the prepolymerization stage, styrene begins to polymerize with itself forming droplets of polystyrene with phase separation. When nearly equal phase volumes of polybutadiene rubber particles and polystyrene are obtained, phase inversion occurs and the droplets of polystyrene act as the continuous phase within which the butadiene rubber particles are dispersed. The completion of the polymerization generally occurs employing either bulk or aqueous suspension conditions.

Most HIPS has about 4%–12% polybutadiene in it so that HIPS is mainly a polystyrene-intense material. The polymerization process is unusual in that both a matrix composition of polystyrene and polybutadienes is formed as well as a graft between the growing polystyrene onto the polybutadiene being formed. Grafting provides the needed compatibility between the matrix phase and the rubber phase. Grafting is also important in determining the structure and size of rubber particles that are formed. The grafting reaction occurs primarily by hydrogen abstraction from the polybutadiene backbone either by growing polystyrene chains or alkoxy radicals if peroxide initiators are employed.

HIPS is an immiscible blend that is used in many applications and used to be employed as the material for many of the automotive bumpers. The polystyrene portion is strong and inflexible while the polybutadiene particles are flexible allowing an impact to be distributed over a larger area. The polybutadiene rubbery portion allows the bumper to bend and indent and protects the PS from fracturing while the PS phase resists further deformation. This combination gives a strong flexible material.

In general, the morphology on a molecular level varies with the fraction of each component in the mixture. In general terms, we can talk about a continuous and discontinuous phase. For a combination of polymers A and B, such as PS and polybutadiene, at low amounts of A, polymer A will typically act as a discontinuous phase surrounded by B. Thus, at low amounts of PS, the PS congregates as small particles in a "sea" or continuous phase of B, the polybutadiene. As the fraction of polymer A increases, the spheres eventually become so large as to join together forming a continuous phase, and so two continuous phases are present. As the fraction of A continuous to increase, polymer B becomes the discontinuous phase, being surrounded by the continuous phase polymer A.

The discontinuous phase generally takes the rough shape of a sphere to minimize surface area exposure to the other phase. The size of the spheres influences the overall properties and varies with concentration. In general, because of the affinity of like polymer chains, spheres tend to grow. Larger sphere sizes are promoted because they give less relative contact area with the other phase.

As noted earlier immiscible blends can exhibit different properties. If the domains are of sufficient size, they may exhibit their own T_g and T_m values. Many commercially used immiscible blends have two separate T_g and/or T_m values.

Blends can also offer variable physical properties as already noted for HIPS. Consider a blend of polymer PS and butadiene where butadiene is the major component. PS is the stronger material with the blend weaker than PS itself. In some cases, the blend can be stronger than the individual polymers. Heat and pressure can result in the change of the discontinuous phase becoming flattened out when pressed against a mold. The spheres can also be caused to elongate forming rodlike structures with the resulting structure being similar to composites where the rodlike structures strengthen the overall structure.

The strength can also be increased by using approximately the same amounts of the two polymers so that they form two continuous phases. Here, both phases can assist the blend to be strong. Another approach is to use compatibilizers. **Compatibilizers** are materials that help bind together the phases allowing stress–strain to be shared between the two phases. Many compatibilizers are block copolymers where one block is derived from polymers of one phase and the second block composed of units derived from polymers of the second phase. The two blocks get "locked" into the structures of the like phases and thus serve to connect the two phases.

Graft copolymers are also used as compatibilizers to tie together different phases. HIPS contains polystyrene grafted onto polybutadiene backbones. This allows stress–strain to be transferred from the PS to the polybutadiene phase transferring energy that might break the brittle PS to the more flexible polybutadiene phase. That is why HIPS is stronger than PS itself.

Compatibilizers also act to modify the tendency to form large spheres. The formation of large spheres is a result of the two polymer components trying to segregate. The compatibilizer causes the two phases to come together minimizing the tendency to form large spheres. For instance, for a mixture of 20:80 PS:polybutadiene the sphere size is about 5–10 μm whereas addition of about 9% PS-polyethylene (polyethylene is enough like polybutadiene to be incorporated into the polybutadiene phases) block copolymer results in PS spheres of about 1 μm. This increases the interface between the two phases resulting in better mechanical properties because stress–strain can be more effectively transferred from one phase to the other.

8.7.2 MISCIBLE BLENDS

Miscible blends are not as easy to achieve as immiscible blends. As noted earlier, entropy is the major driving force for materials to mix. Because polymer chains are already in a state of relatively high order, increases in randomness are not easily achieved so that immiscible blends are often more easily formed. To make matters worse, for amorphous polymers the amount of disorder in the unmixed polymer is often higher than for blends that tend to arrange the polymer chains in a more ordered fashion.

Sometimes special attractions allow polymers to mix. This is true for the mixture of PS and poly(phenylene oxide), PPO, where the interaction between the phenyl groups allow the miscibility of the two polymers. For many combinations, such preferential associations are not present.

Another approach is the use of copolymers. There are a number of variations to this. In some situations, polymer–copolymer combinations are used where the adage "the enemy of my enemy is my friend" comes into play. Thus, the random copolymer of styrene and acrylonitrile forms a miscible blend with poly(methyl methacrylate), PMMA. The copolymer is composed of nonpolar styrene units and polar acrylonitrile units that are incompatible with one another. These units will blend with PMMA in order to avoid one another.

Another approach is to use copolymers where the structures of the copolymer are similar to that of the other phase. This is what occurs for PE and copolymers of ethylene and propylene.

Generally, miscible blends will have properties somewhere between those of the unblended polymers. These properties will be dependent on the ratio of the two polymers and this ratio is often used to obtain a particular property. These properties include mechanical, chemical, thermal, weathering, etc. For instance, PPO is thermally stable with a high T_g, about 210°C. While this is good for some applications, it is considered too high for easy processing. Thus, PS, with a T_g of about 100°C, is added to allow a lower processing temperature. Noryl is blended with a second polymer that is generally PS or HIPS. Noryl is used in the construction of internal appliance components, brackets, and structural components in office products, large computer and printer housings, automotive wheel covers, and high tolerance electrical switch boxes and connectors.

Industrial companies have long-term strategies. For example, Exxon (now ExxonMobile) is the third largest chemical company in the United States. Some time ago, they made the decision to emphasize the ethylene and propylene monomers that are obtained from the petrochemical interests of ExxonMobile. Thus, ExxonMobile has a research emphasis on the commercialization of products from these monomers. The major materials made from ethylene and propylene are polymeric, either homopolymers or copolymers. Efforts include developing catalysts that allow the formation of polymeric materials from the ethylene and propylene monomers and the use of these catalysts to synthesize polymeric materials that have varying properties allowing their application in different marketplaces in society.

A driving force for the conversion of gasoline to polymeric materials is increased value in the products made from the polymers. The general trail is gasoline → ethylene, propylene monomers → raw polymers and copolymers → finished products.

As noted earlier, one major use of HIPS was in automotive bumpers. These have been largely replaced by another blend, but here it is a miscible blend of PE and PP. HIPS bumpers have a more rubbery feel to them while the PE–PP materials are more plastic in their behavior and feel. These PE and PP intense plastics are not only made into automotive bumpers but also as side and bottom panels. There are several processes employed to produce the raw materials used in the production of these automotive parts. One process developed by Exxon begins with the production of PP using a catalyst developed by Exxon. This catalyst system produces isotactic stereoregular polypropylene, i-PP, which is stronger and denser than atactic nonstereoregular polypropylene, a-PP. At some time during this polymerization, some of the liquid polyproplyene monomer is removed and ethylene monomer is added to the mix. Because of the continued presence of the catalyst that can also polymerize the ethylene monomer, copolymer containing ethylene and i-PP units is produced. This product can be roughly pictured as being formed from i-PP particles that have unreacted propylene monomer removed creating open-celled sponge-like particles. These i-PP particles then become impregnated with ethylene monomer eventually resulting in the formation of a copolymer that contains both i-PP and PE units with an i-PP outer shell. This product is eventually mixed with an ethylene(60%)-i-PP(40%) copolymer giving a blended material that has two continuous phases with a final i-PP content of 70%. This material is used in molding the automotive bumpers and panels. The i-PP units contribute stiffness and the PE units contribute flexibility to the overall product. The inner PE units in the impregnated particles allow stress to be rapidly distributed to the stiffer outer i-PP shell. The closeness of structure between the two components, the i-PP impregnated particles and the PE/i-PP copolymer, allows the miscible mixing of the two phases.

Research scientists such as Edward Kresge developed the actual conditions and concentrations and ratios of monomer over a long period of time.

8.8 POLYMER MIXTURES—IPNS AND ALLOYS

Today there exist a number of polymer mixtures including blends noted in the Section 8.7. Here we will briefly look at two more mixtures. A plastic or polymer alloy is a physical mixture of two or more polymers in a melt and is often structurally similar to blends, in fact, the terms blends and alloys are sometimes used to describe the same materials. While some cross-linking may occur, such materials are generally not viewed as copolymers, though they are covered in this chapter along with blends and dendrites. Many alloys contain a matrix that is a mixture of polymer types with many of them containing ABS. ABS thermoplastics have a two-phase morphology consisting of elastic particles dispersed in a styrene–acrylonitrile (SAN) copolymer matrix.

Interpenetrating polymer networks (IPNs) are described by Sperling to be an intimate combination of two polymers, both in network form, where at least one of the two polymers is synthesized and/or cross-linked in the immediate presence of the other. This is similar to taking a sponge cake; soaking in it warm ice cream, and refreezing the ice cream, resulting in a dessert (or mess) that has both spongy and stiff portions. Such IPNs, grafts, blocks, and blends can offer synergistic properties that are being widely exploited.

8.9 DENDRITES

Along with the varying structures given in the previous sections, there exist other structurally complex molecules called dendrites developed by a number of scientists including Tomali and Frechet. These molecules can act as "spacers," "ball-bearings," and as building blocks for other structures. Usually, they are either wholly organic or they may contain metal atoms. They may or may not be copolymers depending on the particular synthetic route employed in their synthesis.

While some make a distinction between dendrimers and hyperbranched polymers, we will not do so here. In essence, hyperbranched polymers are formed under conditions that give a variety of related but different structures while dendrimers are formed one step at a time giving a fairly homogeneous product.

Dendrites are highly branched, usually curved, structures. This name comes from the Greek name for tree "dendron." Another term often associated with these structures is "dendrimers" describing the oligomeric nature of many dendrites. Because of the structure, dendrites can contain many terminal functional groups for each molecule that can be further reacted. Moreover, most dendrites contain "lots" of unoccupied space that can be used to "carry" drugs, fragrances, adhesives, diagnostic molecules, cosmetics, catalysts, herbicides, and other molecules.

The dendrite structure is determined largely by the "functionality" of the reactants. The dendrite pictured in Figure 8.2, left, can be considered as being derived from a tetra-functional monomer formed from the reaction of 1,4-diaminobutane and acrylonitrile. The resulting polypropylenimine

(a) (b)

FIGURE 8.2 (a) Dendrite structure derived from the reaction of 1,4-diaminobutane and acrylonitrile, and (b) derived from "bent" difunctional reactants.

dendrimer has terminal nitrile groups that can be further reacted extending the dendrimer or terminating further dendrimer formation. The resulting molecule is circular with some three-dimensional structure. The dendrimer shown in Figure 8.2, right is derived from difunctional reactants that are "bent" to encourage "fanlike" expansion rather than the typical linear growth.

Numerous approaches have been taken in the synthesis of dendrites or dendrimers. These approaches can be divided into two groupings. In divergent dendrimer growth, growth occurs outward from an inner core molecule. In convergent dendrimer growth, developed by Frechet and others, various parts of the ultimate dendrimer are separately synthesized and then they are brought together to form the final dendrimer.

The somewhat spherical shape of dendimers gives them some different properties in comparison to more linear macromolecules. On a macroscopic level, dendimers act as ball-bearings rather than strings. In solution, viscosity increases as molecular weight increases for linear polymers. With dendimers, viscosity also increases with molecular weight up to a point after which viscosity decreases as molecular weight continues to increase.

Dendimers are used as host molecules, catalysts, self-assembling nanostructures; analogues of proteins, enzymes, and viruses; and in analytical applications including in ion-exchange displacement chromatography and electrokinetic chromatography.

We are continuing to recognize that polymer shape is important in determining material property. Another group of structurally complex shapes is referred to as stars. There are a number of synthetic routes to star polymers. Fetters and coworkers developed a number of star polymers based on chlorosilanes. For instance 3,12, and 18-arm star polymers can be formed. These arms are now reacted with other reactants such as living polystyrene or polybutadiene giving now the star polymers with the silicon-containing inner core and polymer outer core. Through the control of the length of the grafted polystyrene or other reactants, the size of the "star" can be controlled.

For dendrimers made using flexible arms, the core is mobile and depending upon the situation spends some time near the outer layer of the dendrimer sphere. Counter, stiff, rigid arms produce a dendrimer that "holds" its core within the interior of the dendrimer.

The term "generation" describes the number of times "arms" have been extended. The nature of each generation can be varied so that mixtures of steric requirements and hydrophobic/hydrophilic character can be introduced offering materials with varying structures and properties. By varying the hydrophobic and hydrophilic interactions and steric nature of the arms, secondary and tertiary structural preferences can be imposed on the dendrimer.

The dendrite structure can be used as a synthetic tool to construct a particular property into an overall structure.

8.10 IONOMERS

Ionomers are ion-containing copolymers typically containing over 90% (by number) ethylene units with the remaining being ion-containing units such as acrylic acid. These "ionic" sites are connected through metal atoms. Ionomers are often referred to as processable thermosets. They are thermosets because of the cross-linking introduced through the interaction of the ionic sites with metal ions. They are processable or exhibit thermoplastic behavior because they can be reformed through the application of heat and pressure.

As with all polymers, the ultimate properties are dependent upon the various processing and synthetic procedures that the material is exposed to. This is especially true for ionomers where the location, amount, nature, and distribution of the metal sites strongly determine the properties. Many of the industrial ionomers are made where a significant fraction of the ionomer is unionized and where the metal-containing reactants are simply added to the preionomer followed by heating and agitation of the mixture. These products often offer superior properties to ionomers produced from fully dissolved preionomers.

Bonding sites are believed to be of two different grouping densities. One of these groupings involves only a few or individual bonding between the acid groups and the metal atoms. The second bonding type consists of large concentrations of acids groups with multiple metal atoms (clusters). This metal–acid group bonding (salt formation) constitutes sites of cross-linking. It is believed that the "processability" is a result of the combination of the movement of the ethylene units and the metal atoms acting as "ball bearings." The "sliding" and "rolling" is believed to be a result of the metallic nature of the acid–metal atom bonding. (Remember that most metallic salts are believed to have a high degree of ionic, nondirectional bonding as compared with typical organic bonds where there exists a high amount of covalent, directional bonding.) Recently, Carraher and coworkers have shown that the ethylene portions alone are sufficient to allow ionomers to be processed through the application of heat and pressure.

Ionomers are generally tough and offer good stiffness and abrasion resistance. They offer good visual clarity, high melt viscosities, superior tensile properties, oil resistance, and are flame retarders. They are used in the automotive industry in the formation of exterior trim and bumper pads, in the sporting goods industry as bowling pin coatings, golf ball covers, in the manufacture of roller skate wheels, and ski boots. Surlyn (Dupont; poly(ethylene-*co*-methacrylic acid) is used in vacuum packaging for meats, in skin packaging for hardware and electronic items (such as seal layers and as foil coatings of multiwall bags), and in shoe soles.

Sulfonated EPDMs are formulated to form a number of rubbery products including adhesives for footwear, garden hoses, and in the formation of calendered sheets. Perfluorinated ionomers marketed as Nafion (Dupont) are used for membrane applications including chemical processing separations, spent acid regeneration, electrochemical fuel cells, ion-selective separations, electrodialysis, and in the production of chlorine. It is also employed as a "solid"-state catalyst in chemical synthesis and processing. Ionomers are also used in blends with other polymers.

8.11 FLUOROELASTOMERS

Fluoroelastomers are specialty copolymers containing a high amount of fluorine-containing units. These copolymers can be divided into three groups. Viton A (8.26) is a copolymer of vinylidene difluoride and hexafluoropropylene. It is used as a general purpose sealing material in the automotive industry and with aerospace fuels and lubricants.

$$\text{(8.26)}$$

Viton B (8.27) is a terpolymer produced from the polymerization of tetrafluoroethylene, vinylidene fluoride, and hexafluoropropylene. It is used in power utility seals and gaskets.

$$\text{(8.27)}$$

Viton F is also a terpolymer of tetrafluoroethylene, vinylidene fluoride, and hexafluoropropylene. It is used as seals for concentrated aqueous solutions of inorganic acids and for oxygenated automotive fuels. Finally, Viton Extreme is a copolymer of tetrafluoroethylene and propylene and the terpolymer of ethylene, tetrafluoroethylene, and perfluoromethylvinylether.

Viton O-rings are used in SCUBA diving when glass blends called Nitrox are employed. Viton has a low flammability even in the presence of high amounts of oxygen often found in Nitrox. It also stands up well in such increased oxygen conditions. As part of the move toward green fuels, Viton lined hoses are often employed when biodiesel fuels are used.

These copolymers are often employed as mixtures with other polymers. They are compatible with many hydrocarbons but not compatible with ketones and organic acids.

8.12   NITRILE RUBBER

A number of names including nitrile rubber, Buna-N, and nitrile butadiene rubber (NBR) refer to the copolymer of acrylonitrile and butadiene. The composition of the copolymer varies depending on what the intended end product is. Since butadiene is employed, the resulting copolymer has sites of unsaturation that are often subsequently reacted forming thermosets. The greater the amount of cross-linking the more rigid the product along with the corresponding decreased porosity and diffusion properties. The greater the proportion of acrylonitrile incorporated into the copolymer, the greater the resistance to oils, fuels, and other chemicals and the stiffer the material. Moreover, the form of the butadiene can vary from 1,2-butadienc and 1,3-butadiene.

Typically, NBR has a wide operational temperature range of about −40°C to 120°C making it useful for extreme automotive applications as well as cooling units. Its good resistance to oils and other chemicals allows its use around ketones, hydrocarbon liquids, esters, and aldehydes. In the lab, many of the gloves are nitrile. These gloves are also used in home and industrial cleaning and medically as examination and disposable gloves. Nitrile gloves have greater puncture resistance compared to rubber gloves. NBRs ability to withstand extreme temperatures and resistance to oils encourages their automotive uses as hoses, seals, belts, oil seals, and grommets. NBR is also used as adhesives, expanded foams, floor mats, and surface treatment of paper, synthetic leather, and footwear.

NBR is often synthesized using a radical initiating agent. Thus, emulsifier, acrylonitrile, and butadiene, a catalyst, and radical initiators are added to the reaction vessel. The vessel is heated to 30°C–40°C for the formation of so-called hot NBR aiding in the polymerization process and promoting branch formation. Reaction continues for about 5–12 h to about 70% conversions when a terminating agent such as dimethhldithiocarbamate or diethylhydroxylamine is added. The production of cold NBR is similar except the polymerization temperature is in the range of 5°C–15°C. Lower temperature NBR contains less branching generally producing a somewhat stiffer product. Unreacted monomer is removed and reused in a subsequent reaction. The latex is filtered, removing unwanted solids, and then sent to blending tanks where an antioxidant is added. The resulting latex is coagulated by the addition of aluminum sulfate, calcium chloride, or other coagulating compound. The latex is dried giving a flaky crumb-like product that is then used to produce the desired product. Because of the variation in reactants and conditions of polymerization there exists a variation in product structure.

8.13   ACRYLONITRILE BUTADIENE STYRENE TERPOLYMERS

The terpolymer formed from the reaction of acrylonitrile, butadiene, and styrene is referred to by the first letters of the three monomers, ABS. It is made by polymerizing acrylonitrile and styrene in the presence of polybutadiene. The proportions of reactants vary widely but are generally in the range of 15%–35% acrylonitrile, 5%–30% butadiene, and 40%–60% styrene. The resulting product contains long butadiene chains captured by shorter chains of poly(acrylonitrile-co-styrene). The polar nitrile groups attract one another binding the mixture together resulting in a material that is stronger than simply polystyrene. The "plastic" styrene contributes a shiny product with surface that is largely impervious. The "rubbery" polybutadiene contributes resilience and a wide range of operating temperatures between −25°C and 60°C. This combination results in what is referred to as "rubber toughening" where the rubbery polybutadiene is dispersed within a plastic or more rigid styrene matrix bound together by the acrylonitrile units. Typically, impacts are transferred from the more rigid styrene-rich portions to the rubbery butadiene-rich regions that are able to absorb the impact through segmental chain movement. Impact resistance can be increased by increasing the amount of polybutadiene to a limit. Aging is also dependent on the ABS composition and is generally dependent on the amount of butadiene since the unsaturation is typically responsible for limiting the use time because of the increased cross-linking with time. Thus, it is customary to add an antioxidant to curtail aging.

The properties of the end product are dependent on the processing conditions. For instance, processing the ABS at higher temperatures increases the gloss and heat resistance of the material while increased impact resistance and strength result in employing lower processing temperatures.

The electrical properties of ABS are relatively independent over a wide range of applied frequencies making it a good material where varied electrical frequencies are present.

ABS is used to make rigid molded products such as piping and fittings, fuel tanks, automotive body parts, toys, wheel covers, enclosures, and where good shock absorbance is needed such as golf club heads and protective head gear. In fact, our ever-present Lego building blocks are made from ABS plastic.

8.14   EPDM RUBBER

EPDM rubber's name is derived from E for ethylene, P from propylene, D from diene, and M from the ASTM rubber classification. The dienes currently employed are dicyclopendadiene (8.28), ethylidene norbornene (8.29), and vinyl norbornene (8.30). Along with supplying the needed double bond for subsequent cross-linking (curing), the dienes also supply steric hindrance encouraging amorphous formation.

Dicyclopentadiene (8.28)   Ethylidene Norbornene (8.29)   5-Vinyl-2-norbornene (8.30)

The ethylene content is about 45%–75%, diene content is about 2.5%–12%, with the remainder being propylene. The greater the ethylene content the higher the loading capability and better mixing and extrusion of the material. The polymers are cured employing a peroxide agent.

EPDM rubbers have excellent ozone, weather, and heat properties. While it is compatible with fireproofing hydraulic fluids, water, and bases, it is not compatible with most gasoline and other hydrocarbon liquids, concentrated acids, and halogen-intense solvents. The automotive industry uses EPDM rubbers as weather seals including door, trunk, hood, and window seals. The noise in automobiles may be the result of the friction between the EPDM seals and the mated surface. This is normally corrected by using a special coating that increases the chemical resistance of the rubber. EPDM is also used to waterproof roofs. As a green material application, it allows the harvesting of rainwater for other uses since it does not introduce pollutants into the rainwater.

EPDM rubber is also used to produce garden hoses, washers, belts, tubing, electrical insulation, mechanical goods, motor oil additive, pond liner, RV roofs, and as an impact modifier.

8.15   SUMMARY

1. Unlike homopolymers, which consist of chains with identical repeating units, copolymers are macromolecules that contain two or more unlike units. These repeat units may be randomly arranged, or may alternate in the chain. Block copolymers are linear polymers that consist of long sequences of repeating units in the chain, and graft copolymers are branch polymers in which the branches consist of sequences of repeating units that differ from those present in the backbone. These different copolymers give materials with differing properties, even when synthesized using the same monomers.

2. The copolymerization between two different monomers can be described using only four reactions, two homo-polymerizations and two cross-polymerization additions. Through appropriate arrangements, equations that allow copolymer composition to be determined from the monomer feed ratio are developed.

3. The product of the reactivity ratios can be used to estimate the copolymer structure. When the product of the reactivity ratios is near one, the copolymer arrangement is random; when the product is near zero, the arrangement is alternating; when one of the reactivity ratios is large, blocks corresponding to that monomer addition will occur.

4. Block and graft copolymers may differ from mixtures by having properties derived from each component. Block copolymers can be used as thermoplastic elastomers and graft copolymers with flexible backbones can be used for high-impact plastics. Block and graft copolymers can be produced by step reaction polymerization and chain reaction polymerization. The principle block copolymers are thermoplastic elastomers and elastic fibers such as ABS. The principal graft copolymers are grafted starch and cellulose.

5. Blends are physical mixtures of polymers. Depending on the extent and type of blend, the properties may be characteristic of each blend member or may be some "blend" of properties. Immiscible blends are phase separated with the phases sometimes chemically connected. They are generally composed of a continuous and discontinuous phase. HIPS is an example of an immiscible blend. Miscible blends occur when the two blended materials are compatible. Often the properties are a mixture of the two-blended materials. The plastic automotive panels and bumpers are generally made from a miscible blend of polyethylene and a copolymer of polyethylene and polypropylene.

6. Dendrites are complex molecules formed from the building up of the dendrite by individual steps (divergent growth) or from bringing together the units already formed (convergent growth).

7. Ionomers are often referred to as processable thermosets. They are cross-linked using metal ions such that application of heat and pressure allow them to be reformed.

GLOSSARY

AB: Block copolymer with two separate mers.
ABA: Block copolymer with three sequences of mers in the order shown.
ABS: Three-component copolymer of acrylonitrile, butadiene, and styrene.
Alloy: Rubber-toughened materials in which the matrix can be a mixture of polymer types.
Alternation copolymer: Ordered copolymer in whichever other building is a different mer.
Azeotropic copolymer: Copolymer in which the feet and composition of the copolymer are the same.
Blends: Mixtures of different polymers on a molecular level; may exist in one or two phases.
Block copolymer: Copolymer that contains long sequences or runs of one mer or both mers.
Buna-N: Elastomeric copolymer of butadiene and acrylonitrile.
Buna-S: Elastomeric copolymer of butadiene and styrene.
Butyl rubber: Elastomeric copolymer of isobutylene and isoprene.

Charge-transfer complex: Complex consisting of an electron donor (D) and an electron acceptor (A) in which an electron has been transferred from D to A, resulting in the charge-transfer agent $D^+ A^-$.

Composites: Mixtures of different polymers, one forming a continuous phase (the matrix), and one the discontinuous phase (often a fiber).

Compositional drift: Change in composition of a copolymer that occurs as copolymerization takes place with monomers of different reactivities.

Copolymer: Macromolecule containing more than one type of mer in the backbone.

Dendrites: Complex molecules that are highly branched.

Domains: Sequences in a block copolymer or polymer.

Graft copolymer: Branched copolymer in which the backbone and the branches consist of different mers.

Homopolymer: Macromolecule consisting of only one type of mer.

Hytrel: Trade name for a commercial TEP.

Ideal copolymer: Random copolymer.

Interpenetrating polymer network (IPN): Intimate combination of two polymers, both in network form, where at least one of the two polymers is synthesized and or cross-linked in the immediate presence of the other.

Ionomers: Copolymers typically containing mostly ethylene units, with the remaining units being ion containing that are neutralized through reaction with metals.

Kraton: Trade name for an ABA block copolymer of styrene and butadiene.

Living polymers: Macrocarbanions.

Polyallomers: Block copolymers of ethylene and propylene.

Q–e scheme: A semiempirical method for predicting reactivity ratios.

Random copolymer: Copolymer in which there is no definite order for the sequence of the different mers.

Reactivity ratio: Relative reactivity of one monomer compared to another monomer.

Saran: Trade name for copolymer of vinyl chloride and vinylidene chloride.

SBR: Elastomer copolymer of styrene and butadiene.

Spandex: Elastic fiber consisting of a block copolymer of a polyurethane, hard segment, and a polyester or polyether, soft segment.

Wall equation: Predecessor to the copolymer equation.

EXERCISES

1. Draw representative structures for (a) homopolymers, (b) alternation copolymers, (c) random copolymers, (d) AB block copolymers, and (e) graft copolymers of styrene and acrylonitrile.
2. If equimolar quantities of M_1 and M_2 are used in an azeotropic copolymerization, what is the composition of the feed after 50% of the copolymer has formed?
3. Define r_1 and r_2 in terms of rate constants.
4. Do the r_1 and r_2 values increase or decrease during copolymerization?
5. What is the effect of temperature on r_1 and r_2?
6. What will be the composition of copolymers produced in the first part of the polymerization of equimolar quantities of vinylidene chloride and vinyl chloride?
7. What monomer may be polymerized by anionic, cationic, and free radical chain techniques?
8. Which chain polymerization technique would you select to polymerize (a) isobutylene, (b) acrylonitrile, and (c) propylene?
9. If $r_1 r_2$ is about zero, what type of copolymer would be formed?
10. Show a structure for an AB block copolymer.
11. What is the value of $r_1 r_2$ for an ideal random copolymer?

12. Which would polymerize more readily?

(a)   (b)

13. Why are most elastomers difficult to recycle?
14. What is the composition of the first copolymer chains produced by the copolymerization of equimolar quantities of styrene and methyl methacrylate in (a) free radical, (b) cationic, and (c) anionic copolymerization?
15. What is the composition of the first copolymer butyl rubber chains produced form equimolar quantities of the two monomers?
16. What is the composition of the first copolymer butyl rubber chains produced from a feed containing 9 mol of isobutylene and 1 mol of isoprene?
17. How would you ensure that the production of butyl rubber of uniform composition occurs in question 16.
18. For a graft copolymer with a backbone derived from monomer A and the graft B what often occurs when B is long?
19. What is the composition of the first polymer chains produced by the copolymerization of equimolar quantities of vinyl chloride and vinyl acetate?
20. What are the advantages, if any, of the vinyl chloride–vinyl acetate copolymer over PVC itself?
21. Why are ionomers superior to LDPE?
22. What is the difference between buna-S, GRS, and SBR?
23. How can you connect two different phases in polymer blends?
24. Which sequence in the ABA block copolymer of ethylene oxide and propylene oxide is more lyophilic?
25. When will a blend exhibit two distinct properties?
26. Are dendrites really polymeric?
27. What product is obtained if 1.5 mol of styrene is copolymerized with 1 mol of maleic anhydride in benzene?
28. Why are ionomers referred to as processable thermosets?
29. Of what use is the copolymerization equation?
30. How could you use graft copolymerization techniques to reduce the water solubility of starch?
31. What is the end group when azobiscyanopentanoic acid is used as an initiator?

FURTHER READINGS

Alfrey, T., Bohrer, J., and Mark, H. (1952): *Copolymerization*, Interscience, New York.
Bhattacharya, A., Ray, P., and Rawlins, J. (2008): *Polymer Grafting and Crosslinking*, Wiley, Hoboken, NJ.
Brandrup, J. and Immergut, E. (1975). *Polymer Handbook*, Wiley, New York.
Calleja, F. and Roslaniec, Z. (2000): *Block Copolymers*, Dekker, New York.
Carraher, C. and Moore, J. A. (1984): *Modification of Polymers*, Plenum, New York.
Frechet, J. (2002): *Dendrimers and Other Dendritic Polymers*, Wiley, New York.
Frechet, J. and Tomalia, D. (2002): *Dendrimers and Other Dendritic Polymers*, Wiley, New York.
Hadjichristidis, N., Pispar, S., and Floudas, G. (2007): *Block Copolymers: Synthetic Strategies, Physical Properties and Applications*, Wiley, Hoboken, NJ.
Ham, G. E. (1964): *Copolymerization*, Interscience, New York.
Hamley, I. (2004): *Developments in Block Copolymer Science and Technology*, Wiley, Hoboken, NJ.

Hamely, I. (2008): *Block Copolymers in Solution: Fundamentals and Applications*, Wiley, Hoboklen, NJ.

Landler, Y. (1950): *Comptes Rendus*, 230:539.

Lipatov, Y. and Alekseeva, T. (2007): *Phase-Separated Interpenetrating Polymer Networks*, Springer, New York.

Mark, J., Etman, B., and Bokobza, L. (2007): *Rubberlike Elasticity*, Cambridge University, Ithaca, NY.

Mishra, M. and Kobayashi, S. (1999): *Star and Hyperbranched Polymer*, Dekker, New York.

Newkome, G., Moorefield, C., and Vogtle, F. (2001): *Dendrimers and Dendrons*, Wiley, New York.

Nwabunma, D. (2008): *Polyolefin Blends and Composites*, Wiley, Hoboken, NJ.

Pascaullt, J. (2003): *Polymer Blends*, Wiley, Hoboken, NJ.

Robeson, L. (2007): *Polymer Blends: An Introduction*, Carl Hanser GmbH, New York.

Shonaike, G. and Simon, G. (1999): *Polymer Blends and Alloys*, Dekker, New York.

Stann, M. (2000): *Blends*, Wiley, New York.

Stauart, G., Spruston, N., and Hausser, M. (2007): *Dendrites*, Oxford University, Ithaca, NY.

Tomali, D. (2002): *Dendrimers and Other Dendric Polymers*, Wiley, New York.

Utracki, L. (2007): *Polymer Blends Handbook*, Springer, New York.

Wall, F. T. (1944): The structure of copolymers, *J. Am. Chem. Soc.*, 66:2050.

Walling, C. (1957): *Free Radicals in Solution*, Wiley, New York.

Yan, D., Frey, H., and Gao, C. (2008): *Hyperbranched Polymers: Synthesis, Properties, and Applications*, Wiley, Hoboken, NJ.

9 Organometallic and Metalloid Polymers

9.1 INTRODUCTION

Classical polymer chemistry emphasizes materials derived from about a dozen elements (including C, H, O, N, S, P, Cl, and F). Chapters 9 and 10 deal with polymers containing additional elements. This chapter focuses on inorganic and metal-containing polymers containing organic units.

Elements such as silicon, sulfur, and phosphorus catenate similar to the way carbon does, but such catenation generally does not lead to (homo) chains with high degrees of polymerization. Further, such products might be expected to offer lower thermal stabilities and possibly lower strengths than carbon-based polymers since their bond energies are generally lower (Table 9.1). The alternative of using heteroatomed backbones is attractive, since the resultant products can exhibit greater bond energies (Table 9.1).

One common misconception concerns the type of bonding that can occur between inorganic and organic atoms. With the exception of the clearly ionic bonding, many of the inorganic–organic bonding is of the same general nature as that present in organic compounds. The percentage contribution of the organic–inorganic bonding due to covalent contributions is typically well within that found in organic acids, alcohols, and thio and nitro moieties (for instance, the usual limits are about 5% ionic character for the B–C bond to 55% ionic for the Sn–O, and both are clearly directional bonding in character). Thus, the same spatial, geometrical rules apply to these polymers as to the more classical polymers such as PE, PS, nylons, polyesters, and PP. The exception is the ionomers where the metals are bonded through ionic bonding to the oxygen atoms.

The number of potential inorganic–organic polymers is great. The inorganic portions can exist as oxides, salts, in different oxidation states, different geometries, etc. The importance of these inorganic–organic polymers can be appreciated by considering the following. First, photosynthesis, the conversion of carbon dioxide and water by sunlight to sugars is based on a metal-containing polymer—chlorophyll. In addition, a number of critical enzymes, such as hemoglobin, contain a metal site as the key site for their activities. Second, the inorganic—organic polymers produced so far exhibit a wide range of properties, such as electrical conductivity, specific catalytic operations, wide operating temperatures, greater strengths, and greater thermal stabilities, not common to most organic polymers (Table 9.2). Third, inorganic—organic polymers form the basis for many insulators and building materials. Fourth, inorganic elements are present in high abundances in Earth's crust (Table 9.3).

The topic of metal- and metalloid-containing polymers can be divided by many means. Here the topic will be divided according to the type of reaction employed to incorporate the inorganic atom into the polymer chain. While many other types of reactions have been employed to produce metal and metalloid-containing polymers including redox, coupling, ring-opening polymerizations, this chapter will focus on addition, condensation, and coordination reactions. Emphasis is given to unifying factors.

TABLE 9.1
General Magnitude of Bonds

Bond	General Bond Energy (kJ/mol)	Ionic Character[a] (%)	Bond	General Bond Energy (kJ/mol)	Ionic Character (%)
Al–O	560	60	P–O	400	40
B–C	360	5	P–S	320	5
B–N	440	20	S–S	240	0
B–O	460	45	Si–Si	220	0
Be–O	500	65	Si–C	300	10
C–C	340	0	Si–N	420	30
C–H	400	5	Si–O	440	50
C–N	300	5	Si–S	240	10
C–O	340	20	Sn–Sn	160	0
C–S	260	5	Sn–O	520	55
P–P	200	0	Ti–O	640	60
P–N	560	20			

[a] Based on Pauling electronegativity values. The percentage of ionic bonding should be less when pi-bonding occurs. Given to nearest 5%.

TABLE 9.2
Actual and Potential Applications for Organometallic and Metalloid Polymers

Biological	Anticancer, antiviral, treatment of arthritis, antibacterial, antifungal, antifouling, treatment of Cooley's anemia, algicides, molluscidides, contrast agents, radiology agents
Electrical/optical	Nonlinear optics, lithography, conductors, semiconductors, piezoelectronic, pyroelectronic, solar energy conversion, electrodes, computer chip circuitry
Analytical, catalytic, building	UV absorption, smart materials, nanocomposites, laser, sealants, paints, caulks, lubricants, gaskets

TABLE 9.3
Relative Abundances of Selected Elements in Earth's Upper (10 Miles) Crust

Element	Weight (%)	Element	Weight (%)
Oxygen	50	Titanium	0.4
Silicon	26	Fluorine	0.3
Aluminum	7.3	Chlorine	0.2
Iron	4.2	Carbon	0.2
Calcium	3.2	Sulfur	0.1
Sodium	2.4	Phosphorus	0.1
Potassium	2.3	Barium	0.1
Magnesium	2.1	Manganese	0.1
Hydrogen	0.4		

9.2 INORGANIC REACTION MECHANISMS

Many of the polymerizations and monomer syntheses are simply extensions of known inorganic, organometallic, and organic reactions. The types and language used to describe inorganic–organic reaction mechanisms are more diversified than those employed by classical organic chemists.

The majority of inorganic reactions can be placed into one of two broad classes—oxidation/reduction (redox) reactions including atom and electron transfer reactions, and substitution reactions. Terms such as inner sphere, outer sphere, and photo-related reactions are employed to describe redox reactions. Such reactions are important in the synthesis of polymers and monomers; in the use of metal-containing polymers as catalysts; and in applications that involve the transfer of heat, electricity, and light. They will not be dealt with to any appreciable extent in this chapter.

Terms such as liability, inertness, ligand, associative, interchange, and dissociative are important when discussing substitution reactions. The ligand is simply (typically) the Lewis base that is substituted for and is also the agent of substitution. Thus, in the reaction between tetrachloroplatinate and diamines forming the anticancer and antiviral platinum II polyamines, the chloride is the leaving group or departing ligand, while the amine-functional group is the ligand that is the agent of substitution.

$$(9.1)$$

There is a difference between the thermodynamic terms stable and unstable, and the kinetic terms labile and inert. Furthermore, the difference between the terms stable and unstable, and the terms labile and inert are relative. Thus $Ni(CN)_4^{-2}$ and $Cr(CN)_6^{-3}$ are both thermodynamically stable in aqueous solution, yet kinetically the rate of exchange of radiocarbon-labeled cyanide is quite different. The half-life for exchange is about 30 s for the nickel complex and 1 month for the chromium complex. Taube has suggested that those complexes that react completely within approximately 60 s at 25°C be considered labile, while those that take a longer time be considered inert. This rule of thumb is often given in texts but is not in general use in the literature. Actual rates and conditions are superior tools for the evaluation of the kinetic/thermodynamic stabilities of complexes.

The term "D mechanism" (dissociation) is loosely comparable to $S_N 1$-type reaction mechanisms, but it does not imply an observed rate law. Here, a transient intermediate is assumed to live long enough to be able to differentiate between various ligands, including the one just lost, and between solvent molecules. Thus, the overall rate expression may be dependent on the nature of LL′, solvents, or some combinations as pictured below where S = solvent, L is the leaving ligand, and L′ is the incoming ligand.

$$ML_4 \rightleftharpoons ML_3 + L \qquad\qquad (9.2)$$

$$ML_3 + L' \rightarrow ML_3L' \qquad\qquad (9.3)$$

$$ML_3 + S \rightleftharpoons ML_3S \qquad\qquad (9.4)$$

In the I_d mechanism, dissociative interchange, the transition state involves extensive elongation of the $M\cdots L$ bond, but not rupture.

$$ML_4 + L' \rightarrow [L\cdots ML_3 \cdots L'] \rightarrow ML_3L' + L \qquad\qquad (9.5)$$

The ML_3L' species is often called an outer sphere complex, or, if ML_4 is a cation and L an anion, the couple is called an ion pair.

For the I_a mechanism, an associative interchange, the interaction between M and L′, is more advanced in the transition state than in the case of the I_d. The M···L′ bonding is important in defining the activated complex. Both of these interchange mechanisms are loosely connected to the S_N 2-type mechanism.

For the A mechanism, associative, there is a fully formed intermediate complex ML_4L', which then dissociates, being roughly analogous to the E_1-type reaction mechanism.

It is important to remember that the same electronic, steric, mechanistic, kinetic, and thermodynamic factors that operate in regard to smaller molecules are in operation during a polymerization process.

9.3   CONDENSATION ORGANOMETALLIC POLYMERS

Condensation reactions exhibit several characteristics such as (typically) the expulsion of a smaller molecule on reaction leading to a repeat unit containing fewer atoms than the sum of the two reactants, and most reactions can be considered in terms of polar (Lewis acid–base; nucleophilic–electrophilic) mechanisms. The reaction site can be at the metal atom (that is adjacent to the atom)

$$R_2MX_2 + H_2N-R-NH_2 \rightarrow -(-MR_2-NH-R-NH-)- + HX \tag{9.6}$$

or it can be somewhat removed from the metal site.

$$\tag{9.7}$$

Research involving condensation organometallic polymers was catalyzed by the observation that many organometallic halides possess a high degree of covalent character within their composite structures and that they can behave as organic acid chlorides in many reactions, such as hydrolysis

$$\underset{\substack{\| \\ R-C-Cl + H_2O}}{O} \rightarrow \underset{\substack{\| \\ R-C-OH + HCl}}{O} \tag{9.8}$$

$$-\underset{|}{\overset{|}{M}}-Cl + H_2O \rightarrow -\underset{|}{\overset{|}{M}}-OH + HCl \tag{9.9}$$

and polyesterfication

$$\underset{\substack{\| \\ Cl-C-R-}}{O}\underset{\substack{\| \\ C-Cl}}{O} + HO-R'-OH \rightarrow -\underset{\substack{\| \\ (C-R-}}{O}\underset{\substack{\| \\ C-O-R'-O)}}{O}- + HCl \tag{9.10}$$

$$Cl-\underset{|}{\overset{|}{M}}-Cl + HO-R'-OH \rightarrow -(-\underset{|}{\overset{|}{M}}-O-R'-O-)- + HCl \tag{9.11}$$

Thus, many of the metal-containing polycondensations can be considered as extensions of organic polyesterfications, polyamination, etc. reactions.

The most important organometalloid polymers are the polysiloxanes based on the siloxane Si–O linkage found in glass and quartz (Chapter 10). The polysiloxanes were incorrectly named silicones by Kipping in the 1920s, but this name continues to be widely used. Originally, it was wrongly believed to have a structure similar to a ketone, hence, silicone.

The production of silicate glass is believed to be a transcondensation of the siloxane linkages in silica. A comparable poly(silicic acid) is produced when silicon tetrachloride is hydrolyzed:

$$SiCl_4 \rightarrow Si(OH)_4 \rightarrow -(-\underset{\underset{OH}{|}}{\overset{\overset{OH}{|}}{Si}}-O-)- \tag{9.12}$$

Silicic acid   Poly(silicic acid)

The poly(silicic acid) further condenses producing a cross-linked gel. This cross-linking can be prevented by replacing the hydroxyl groups in silicic acid with alkyl groups. Ladenburg prepared the first silicone polymer in the nineteenth century by the hydrolysis of diethyldiethoxysilane. Kipping, in the early 1940s, recognized that these siloxanes could also be produced by the hydrolysis of dialkyldichlorosilanes giving a poly(silicic acid)-like structure where the hydroxyl groups are replaced by the alkyl groups.

In 1945, Rochow discovered that a silicon–copper alloy reacts with organic chlorides forming a new class of compounds called organosilanes.

$$CH_3Cl + Si(Cu) \rightarrow (CH_3)_2SiCl_2 + Cu \tag{9.13}$$

Dimethyldichlorosilane

These compounds react with water forming dihydroxylsilanes,

$$(CH_3)_2SiCl_2 + H_2O \rightarrow (CH_3)_2Si(OH)_2 + HCl \tag{9.14}$$

and eventually forming dimeric, oligomeric, and finally polysiloxanes (9.15). Because of the toxicity of HCl, the chloride groups are replaced by acetate groups, leading to the familiar vinegar smell for many silicon caulks and sealants. Branching and cross-linking were introduced through the use of methyltrichlorosilane. Modern caulks and sealants are made using the tetrafunctional group tetraethoxysilane to introduce cross-linking into the resin.

Polysiloxanes are characterized by combinations of chemical, mechanical, and electrical properties, which, taken together, are not common to any other commercially available class of polymers. They exhibit relatively high thermal and oxidative stabilities, low power loss, high dielectric strength, and unique rheological properties, and are relatively inert to most ionic reagents. Almost all of the commercially utilized siloxanes are based on polydimethylsiloxane with trimethylsiloxy end groups. They have the widest use of temperature range for commercial polymers suitable for outdoor applications from the winter of Nome Alaska to the summer of south Florida (about −60°C to 40°C). The first footprints on the moon were made with polysiloxane elastomeric boots.

$$\tag{9.15}$$

The reason for the low temperature flexibility is because of a very low T_g, about −120°C, which is the result of the methyl groups attached to the silicon atoms that are free to rotate causing the oxygen and other surrounding atoms to "stay away" creating a flexible chain.

(9.16)

Polysiloxanes degrade by an unzipping mechanism forming six- and eight-membered rings. They also form the same rings when polymerized forming what is referred to as "wasted loops" because they must be removed before the polysiloxane is useful. This tendency to form six- and eight-membered rings is based on the good stability of such siloxane rings.

(9.17)

Six-membered loop

(9.18)

Eight-membered loop

Polysiloxanes are used in a wide variety of applications. The viscosity or resistance to flow increases as the number of repeat units increases but physical properties, such as surface tension and density, remain the same after a $\overline{DP}$ of about 25. The liquid surface tension is lower than the critical surface tension of wetting resulting in the polymer spreading over its own absorbed films. The forces of attraction between polysiloxane films is low resulting in the formation of porous films that allow oxygen and nitrogen to readily pass through but not water. Thus, semipermeable membranes, films, have been developed that allow drivers to "breath air under water" for short periods.

As noted above, viscosity increases with $\overline{DP}$ allowing many of the uses to be grouped according to chain length. Low viscosity fluids with $\overline{DP}$'s of 2–30 are used as antifoams and in the flow control of coatings applications. These applications are the direct consequence of the low attractions between polysiloxane chains which, in turn, is responsible for their low surface tension. Thus, they encourage a coatings material to flow across the surface filling voids, corners, and crevices. Their good thermal conductivity and fluidity at low temperatures allow their use as low temperature heat exchangers and in low temperature baths and thermostats.

Viscous fluids correspond to a $\overline{DP}$ range of about 50–400. These materials are employed as mold release agents for glass, plastic, and rubber parts. They are good lubricants for most metal to nonmetal contacts. They are used as dielectric fluids (liquids) in a variety of electrical applications including transformers and capacitors; as hydraulic fluids in the vacuum and hydraulic pumps; in delicate timing and photographic devices; as antifoam agents; components in protective hand creams; toners in photocopiers; in oil formulations when mixed with thickeners; and in inertial guidance systems. High-performance greases are formed by mixing polysiloxane fluids with polytetrafluoroethylene or molybdenum disulfide. Brake fluids are formulated from polydimethylsiloxane fluids with $\overline{DP}$s approximately 50. High viscosity fluids with $\overline{DP}$s approximately 700–6000 are used as damping fluids for weighting meters at truck stops. They act as liquid springs in shock absorbers. The longer chained fluids are used as impact modifiers for thermoplastic resins and as stationary phases in gas chromatography.

As with the alkanes, even longer chains form the basis for solid polysiloxanes that, according to design, can be classified as thermoplastics, engineering thermoplastics, elastomers, and when cross-linked as thermosets. Solid polysiloxanes are used in a variety of applications including as sealants, thermostripping, caulking, dampening, O-rings, and window gaskets. Weather-stripping on cooling units, trucks, and automobiles is often made of polysiloxanes.

Room temperature–vulcanizing (RTV) silicon rubbers make use of the room temperature reaction of certain groups, which can be placed on polydimethylsiloxanes that react with water. When exposed to water, such as that normally present in the atmosphere, cross-links are formed creating an elastomeric product.

The first contact lenses were based on poly(methyl methacrylate). While they could be polished and machined, they did not permit gas exchange and were rigid. By the early 1970s, these were replaced by soft contact lenses containing cross-linked poly(2-hydroxyethyl methacrylate) HEMA. These so-called disposable lenses do permit gas exchange. More recently, Salamone and coworkers developed contact lenses based on the presence of siloxane units. Polysiloxanes have good gas permeability. These polymers are referred to as Tris materials and are generally copolymers containing units as shown below:

(9.19)

Polysiloxanes are widely employed as biomaterials. Artificial skin can be fabricated from a bilayer fabricated from a cross-linked mixture of bovine hide, collagen, and chondroitin sulfate derived from shark cartilage with a thin top layer of polysiloxane. The polysiloxane acts as a moisture and oxygen-permeable support, and protects the lower layer from the "outer world." A number of drug delivery systems use polysiloxanes because of the flexibility and porous nature of the material.

The first Silly Putty™, also called Nutty Putty, was made over 50 years ago by mixing together silicone oil with boric acid. That original formula has changed only a little, though colorants have been added to give the material a brighter color and some the ability to "glow-in-the-dark." Today, the formula contains about 70% dimethylsiloxane and boric acid, 17% quartz, 9% Thixatrol ST (a commercial rheology modifier that is a derivative of caster oil), and several other minor constituents. Silly Putty is a dilatant material (Section 13.1) meaning it has an inverse thioxtropy (a thioxotropic liquid is one whose viscosity decreases with time). In essence, the resistance of flow increases faster than the increase in the rate of low. Thus, under short interaction times (Section 13.1) it behaves as a solid where the various molecular components resist ready movement, acting as a solid and under sharp impact like hitting it with a hammer or rapidly "snapping" it, it will act as a brittle material. Under a relatively long interaction time, the molecular chains are able to yield and the material acts as a liquid. Under moderate interaction times, there is segmental movement and the material acts as a rubber.

Silly Putty is one of the several materials whose discovery occurred at about the same time by different individuals. In this case, the two individuals are Earl Warrick, working for Dow Corning, and James Wright, a researcher working for General Electric. The discoveries occurred in 1943 during searches for synthetic rubber during World War II. Initially no practical use was found. By 1949, it was found in a local toy store as a novelty item. Despite its good sales, the store dropped it after one year. The next year Peter Hodson began packaging it in the now familiar plastic egg. Today, it sells for about the same price it did in 1950. Silly Putty sells at a rate of about 6 million eggs, or 90 ton, yearly.

Carraher and coworkers have produced a wide variety of organometallic condensation polymers based on the Lewis acid–base concept. Polymers have been produced from Lewis bases containing amine, alcohol, acid, thiol, and related units including a number of drugs such as ciprofloxacin and acyclovir. Lewis acids contain metals and metalloids such as Ti, Zr, Hf, V, Nb, Si, Ge, Sn, Pb, As, Sb, Bi, Mn, Ru, P, Co, Fe, and S. These compounds have potential uses in the biomedical arena as antifungal, antibacterial, anticancer, Parkinson's treatment, and antiviral drugs. These polymers show promise in a wide variety of other areas including electrical, catalytic, and solar energy conversions. Polymers referred to as polydyes, because of the presence of dye moieties in the polymer backbones, have impregnated paper products, plastics, rubber, fibers, coatings, and caulks giving the impregnated material color, (often) added biological resistance, and special photo properties.

Organotin compounds are the most widely used metal-containing compounds. The bulk of these materials are employed to improve the heat stability of PVC as it is formed into piping. Today, because polymers leach much slower than small molecules, organotin polymers initially synthesized by Carraher and coworkers are being used in PVC piping. Organotin polyamine containing the plant growth hormone kinetin (9.20) increases the germination of damaged seeds and thus may help in providing food in third world countries. It also increases the germination rate of saw grass seed from about 0% to over 50% and is an important agent in replacing the "sea of grass" (actually saw grass) in the Everglades. Some of the organotin polymers are found to be effective anticancer, antibacterial, and antiviral agents (one such polymer is shown in 9.21). They show a good inhibition of several ovarian cancer test lines and inhibition of a wide variety of viruses including those responsible for small pox, herpes simplex, chickenpox, and shingles.

(9.20)

(9.21)

Some of the polymers, including the metallocene-containing products, show the ability to control laser radiation. Depending on the range of radiation, laser energy can be focused allowing the material to be cut readily or it can be dispersed imparting to the material, containing the polymer, added stability toward the radiation.

A number of these polymers exhibit a phenomenon called "anomalous fiber formation," reminiscent of "metallic whiskers."

9.4   COORDINATION POLYMERS

Coordination polymers have served humankind since before recorded history. The tanning of leather and the generation of selected colored pigments depend on the coordination of metal ions. A number of biological agents, including plants and animals, owe their existence to coordinated polymers such as hemoglobin. Many of these coordination polymers have unknown and/or irregular structures.

The drive for the synthesis and characterization of synthetic coordination polymers was catalyzed by work supported and conducted by the United States Air Force in a search for materials, which exhibited high thermal stabilities. Attempts to prepare highly stable, tractable coordination polymers were disappointing. Typically, only oligomeric products were formed, and the monomeric versions were often more stable than the polymeric versions.

Bailar listed a number of principles that can be considered in designing coordination polymers. Briefly these are as follows. (1) Little flexibility is imparted by the metal ion or within its immediate environment; thus, flexibility must arise from the organic moiety. Flexibility increases as the covalent nature of metal–ligand bond increases. (2) Metal ions only stabilize ligands in their immediate vicinity; thus, the chelates should be strong and close to the metal ions. (3) Thermal, oxidative, and hydrolytic stabilities are not directly related; polymers must be designed specifically for the properties desired. (4) Metal–ligand bonds have sufficient ionic character to permit them to rearrange more readily than typical "organic bonds." (5) Polymer structure (such as square planar, octahedral, linear, network) is dictated by the coordination number and the stereochemistry of the metal ion or the chelating agent. (6) Lastly, employed solvents should not form strong complexes with the metal or chelating agent, or they will be incorporated into the polymer structure and/or prevent reaction from occurring.

Coordination polymers can be prepared by a number of routes, with the three most common being (1) preformed coordination metal complexes polymerized through functional groups where the actual polymer-forming step may be a condensation or addition reaction;

$$(9.22)$$

(2) reaction with polymer-containing ligands; or

$$(9.23)$$

(3) polymer formation through chelation.

$$(9.24)$$

Carraher and coworkers employed the last two processes to recover the uranyl ion. The uranyl ion is the natural water-soluble form of uranium oxide. It is also toxic, and acts as a heavy metal toxin. Through the use of salts of dicarboxylic acids and poly(acrylic acid), the uranyl ion was removed to 10^{-5} M with the resulting product being much less toxic and convertible to uranium oxide by heating.

Many of the organometallic polymers are semiconductors with bulk resistivities in the range of 10^3–10^{10} Ω cm suitable for specific semiconductor use. Further, some exhibit interesting photoproperties.

In 1964, Rosenberg and coworkers found that bacteria failed to divide but continued to grow. After much effort, they found that the cause of this anomalous growth was a broken electrode and eventually identified the chemical as *cis*-dichlorodiamineplatinum II. This compound is now licensed under the name Platinol and is also known as cisplatin. Cisplatin is the most widely used anticancer drug. Carraher, Rosenberg, Allcock, and others have reduced the toxic effects of cisplatin through the placement of the platinum moiety into various coordination platinum polymers. Some of these polymers inhibit various cancer growths with much less toxic effects. Carraher and coworkers also found that many of these are also very active antiviral agents, and some are able to prevent the onset of virally related juvenile diabetes in test animals.

Simple chelation polymers are all about us, but they are not always recognized as such. During the winter, the north experiences freezing temperatures and the associated ice. Sodium chloride and calcium chloride are the most used freezing point-lowering agents. They are inexpensive and readily available. On the negative side, they adversely affect surrounding plant life and must be reapplied generally after each rain–ice cycle.

Recently magnesium acetate has begun to be used for an application in especially dangerous sites such as bridges. The acetate can be either internally bridged, where it is not polymeric,

(9.25)

or it can be bridged forming a complex linear polymeric material.

(9.26)

Hellmuth believes that the polymeric material forms. If the magnesium acetate were simply the internally chelated material then it should be quickly washed away. The combination of calcium and magnesium acetates is known as CMA. CMA is more expensive than sodium chloride or calcium chloride, but it does not damage plant life and it has a much longer effective life. The polymer seeks the cracks and the crevices in and around the pavement and remains until it is needed again to lower the freezing point of water. Often, one application of CMA is sufficient for a winter season.

9.5 ADDITION POLYMERS

A large number of vinyl organometallic monomers have been prepared, homopolymerized, and copolymerized with classic vinyl monomers by Pittman and others. These include polymers containing Mo, W, Fe, Cr, Ir, Ru, and Co.

The effect that the presence of the organometallic function exerts in vinyl polymerizations is beginning to be fully understood. A transition metal may be expected, with its various readily available oxidation states and large steric bulk, to exert unusual electronic and steric effects during polymerization. The polymerization of vinylferrocene will be employed as an example. Its homopolymerization has been initiated by radical, cationic, coordination, and Ziegler–Natta initiators. Unlike the classic organic monomer styrene, vinylferrocene undergoes oxidation at the iron atom when peroxide initiators are employed. Thus, azo initiators (such as AIBN) are typically used.

Here, we see one difference between an organic monomer and an organometallic monomer in the presence of peroxide initiators. The stability of the ferricinium ion makes ferrocene readily oxidizable by peroxides, whereas styrene, for example, undergoes polymerization in their presence. Unlike most vinyl monomers, the molecular weight of polyvinylferrocene does not increase with a decrease in initiator concentration because of the unusually high chain-transfer constant for vinylferrocene. Finally, the rate law for vinylferrocene homopolymerization is first order in initiator in benzene. Thus, intramolecular termination occurs. The Mossbauer studies support a mechanism involving electron transfer from iron to the growing chain radical giving rise to a zwitterion that terminates polymerization.

The high electron richness of vinylferrocene as a monomer is illustrated in its copolymerization with maleic anhydride, where 1:1 alternation copolymers are formed over a wide range of monomer feed ratios and $r_1 r_2 = 0.003$. Subsequently, a large number of detailed copolymerization studies have been undertaken using metal-containing vinyl monomers.

Other inorganic and metal-containing polymers have been formed using the addition approach. These include polyphosphazenes, polyphosphonitriles, and poly(sulfur nitride). Phosphonitrilic polymers have been known for many years, but since they lacked resistance to water they were not of interest as commercial polymers. However, when the pendant chlorine groups are replaced by fluorine atoms, amino, alkoxy, or phenyoxy groups, these polymers are more resistant to hydrolysis. Allcock and coworkers have pioneered these efforts. Phosphonitrile fluoroelastomers are useful throughout a temperature range of −56°C to 180°C. Phosphazene is produced by the thermal cleavage of a cyclic trimer obtained from the reaction between phosphorus pentachloride and ammonium chloride.

$$(9.27)$$

Poly(phosphonitrile chloride)

Amorphous elastomers are obtained when phosphazene is refluxed with nucleophiles, such as sodium trifluoroethoxide or sodium cresylate, and secondary amines. Difunctional reactants, such as dihydroxybenzenes (hydroquinone), produce cross-linked phosphazenes.

$$(9.28)$$

Polyphosphazenes generally exhibit very low T_g values consistent with the barriers to internal rotation being low, and this property indicates the potential of these polymers for elastomer applications. In fact, theoretical calculations based on a rotational isomeric model assuming localized pi bonding predict the lowest (400 J/mol repeating unit) known polymer barrier to rotation for the skeletal bonds for polydifluorophosphazene. Temperature intervals between T_g and T_m are unusually small, and generally fall outside the frequently cited relationship. This behavior may be related to complications in the first order transition generally found for organo-substituted phosphazenes and not common to other semicrystalline polymers. Two first-order transitions are generally observed for organo-substituted phosphazenes with a temperature interval from about 150°C–200°C. The lower first-order transition can be detected using DCS, DTA, and TMA. An examination by optical microscopy reveals that the crystalline structure is not entirely lost but persists throughout the extended temperature range to a higher temperature transition, which appears to be T_m, the true melting temperature. The nature of this transitional behavior resembles the transformation to a mesomorphic state similar to that observed in nematic liquid crystals. It appears from the relationship between the equilibrium melting temperature (heat and entropy of fusion; $T_m = H_m/S_m$) and the low value of H_m at T_m compared with the lower transition temperature that the upper transition, T_m, is characterized by a very small entropy change. This may be due to an onset of chain motion between the two transitions leading to the small additional gain in conformational entropy at T_m. The lower transition is believed to correspond to T_g.

Allcock and coworkers have employed polyphosphazenes in a variety of uses including the broad areas of biomedical and electrical applications. From a practical point of view, polyphosphazenes are usually soft just above the lower transition so that the compression molding of films can be carried out. This suggests that the lower transition temperature represents the upper temperature for most useful engineering applications of polyphosphazenes in unmodified forms.

Sulfur nitride polymers (polythiazyls) $[-(-S=N-)-]$, which have optical and electrical properties similar to those of metals, were first synthesized in 1910. These crystalline polymers, which are superconducive at 0.25 K, may be produced at room temperature using the solid-state polymerization of the dimer (S_2N_2). A dark blue–black amorphous paramagnetic form of poly(sulfur nitride) is produced by quenching the gaseous tetramer in liquid nitrogen. The polymer is produced on heating the tetramer to about 300°C.

(9.29)

Rather than the simple up-and-down alternating structure for polythiazyls (9.30), the actual structure is more crankshaft-like (9.31) caused by the uneven bond angles of SNS (120°) and NSN (103°).

(9.30)

(9.31)

Much of the interest in the polysilanes, polygermanes, and polystannanes involves their sigma delocalization and their sigma-pi delocalization when coupled with arenes or acetylenes. This is not unexpected since silicon exists as a covalent network similar to diamond. In exhibiting electrical conductivity, germanium and tin show more typical "metallic" bonding. Some polystannanes have been referred to as "molecular metals."

Because of the interesting electronic and physical properties of polysilanes, a number of potential uses have been suggested including precursors of β-SiC fibers, impregnation of ceramics, polymerization initiators, photoconductors for electrophotography, contrast enhancement layers in photolithography, deep UV-sensitive photoresists, nonlinear optical materials, and self-developing by excimer laser for deep UV exposure. The unusual absorption spectra of polysilanes have indicated potential use in a number of conducting areas.

One area of active interest in ceramics is the formation of ceramics that may contain some fiber structure. Currently, ceramics, while very strong, are very brittle. The introduction of thermally stable fiber-like materials might allow the ceramics some flexibility before cleavage. Such materials might be considered as ceramic composites, where the matrix is the ceramic portion and the fibers are the thermally stable fibers. The introduction of the fibers during the ceramic-forming step is a major obstacle that must be overcome. Carbon fibers have been investigated similar to other high temperature materials such as the polysilanes. Polysilanes are formed from the six-membered ring through extended heating at 400°C.

$$(9.32)$$

The above compound on further heating gives silicon carbide.

Table 9.4 lists a number of nonoxide ceramics that have been produced due to the pyrolysis of polymers.

Boron carbonitride ceramics are formed by heating borazine with borazine derivatives (9.33).

Heating borazine yields a polymer connected by B–N bonds (9.34).

TABLE 9.4
Nonoxide Ceramics Produced due to the Pyrolysis of Polymeric Materials

Polymer(s)	Resultant Ceramic
Poly(phosphonitric chlorides)	PN
Polysilanes, polycarbosilanes	SiC
Polyphenylborazole	BN
Polytitanocarbosilanes	Si–Ti–C
Polysilazanes	Si_3N_4, Si–C–N

Borazine

$$+$$

Boron carbonitride   (9.33)

(9.34)

Polyborazylene

9.6 SUMMARY

1. There is a wide variety of inorganic and metal-containing polymers. The potential uses are many and include the broad areas of biomedical, electrical, optical, analytical, catalytic, building, and photochemical applications.

2. The bond strength from any combination is higher than many traditional polymers with many having superior thermal stabilities.

3. Metal and inorganic polymers can be formed through a variety of reaction types including condensation, coordination, and addition reactions.

4. The majority of the condensation polymerizations can be considered to be extensions of typical Lewis acid–base reactions.

5. Polysiloxanes (silicons) offer a good combination of properties not found in organic polymers. Silicons are employed in a number of applications including antifoaming agents, lubricants, caulks, sealants, gaskets, and as biomaterials.

6. Polyphosphazenes offer unique thermal properties and have shown a number of uses in the field of electronics and medicine.

7. The number and variety of organometallic polymers, and the potential applications for organometallic polymers are great. Because of the high cost of production of many of these materials, uses will often be limited to applications employing minute quantities of the polymers. This is not true for many polymers containing silicon, tin, and main group materials since these are available in large quantities at a reasonable cost.

GLOSSARY

Borazoles: Molecules composed of boron and nitrogen atoms.
Capping: Protecting end groups.
Carboranes: Molecules composed of carbon and boron atoms.

Coordination polymers: Polymers based on coordination complexes.

Metallocenes: Sandwich or distorted sandwich-like molecules containing generally two cyclopen-
tadienes and a metal atom bonded to them.

Polyphosphonitrile: Polymer with a repeat unit of $-P=N-$

RTV: Room temperature vulcanization.

Siloxanes and Silicones: Polymers containing $-Si-O-$ backbones.

EXERCISES

1. What is meant by "lost loops" in the production of silicones?
2. How could you produce a silicone with a low $\overline{DP}$?
3. What would you estimate the solubility parameters of silicones to be?
4. Sodium silicate is water soluble (forming water glass), but silicones are water repellents. Explain the difference.
5. How might you polymerize an aqueous solution of sodium silicate?
6. How might you explain the good thermal stability of silicones?
7. Show the repeat unit for polydiethylsiloxane.
8. What are the reactants used to make phosphazenes?
9. Why would you predict that the chloro groups in phosphonitrilic polymers would be attacked by water?
10. Which of the phosphazenes would be more flexible: one made by the reaction of poly(phosphonitrilic chloride) with (a) sodium trifluoroethoxide or (b) sodium trifluorobutoxide?
11. Show the structure of borazole.
12. Since tin-containing organometallic polymers are used in marine antifouling coatings, what would you predict about their water resistance?
13. In addition to high cost, name another disadvantage of coordination polymers.
14. What is the ceiling temperature of sulfur nitride polymers?

FURTHER READINGS

Abd-El-Aziz, A., Carraher, C., Pittman, C., Sheats, J., and Zeldin, M. (2003): *A Half Century of Metal- and Metalloid-Containing Polymers*, Wiley, Hoboken, NJ.

Abd-El-Aziz, A., Carraher, C., Pittman, C., Sheats, J., and Zeldin, M. (2004a): *Ferrocene Polymers*, Wiley, Hoboken, NJ.

Abd-El-Aziz, A., Carraher, C., Pittman, C., Sheats, J., and Zeldin, M. (2004b): *Biomedical Applications of Metal-Containing Polymers*, Wiley, Hoboken, NJ.

Abd-El-Aziz, A., Carraher, C., Pittman, C., Zeldin, M. (2005a): *Group IVA-Containing Polymers*, Wiley, Hoboken, NJ.

Abd-El-Aziz, A., Carraher, C., Pittman, C., and Zeldin, M. (2005b): *Coordination Polymers*, Wiley, Hoboken, NJ.

Abd-El-Aziz, A., Carraher, C., Pittman, C., Sheats, J., and Zeldin, M. (2005c): *Transition Metal Polymers*, Wiley, Hoboken, NJ.

Abd-El-Aziz, A., Carraher, C., Pittman, C., Sheats, J., and Zeldin, M. (2006): *Nanoscale Interactions of Metal-Containing Polymers*, Wiley, Hoboken, NJ.

Abd-El-Aziz, A., Carraher, C., Pittman, C., Sheats, J., and Zeldin, M. (2007): *Boron-Containing Polymers*, Wiley, Hoboken, NJ.

Abd-El-Aziz, A., Carraher, C., Pittman, C., Sheats, J., and Zeldin, M. (2008): *Inorganic and Organometallic Macromolecules: Design and Application*, Springer, New York.

Abd-El-Aziz, A., Carraher, C., Pittman, C., Sheats, J., and Zeldin, M. (2009): *Inorganic Supramolecular Assemblies*, Wiley, Hoboken, NJ.

Abd-El-Aziz, A., Carraher, C., Pittman, C., Sheats, J., and Zeldin, M. (2010): *Photophysics and Photochemistry*, Wiley, Hoboken, NJ.

Allcock, H. R., (1972): *Phosphorus-Nitrogen Compounds*, Academic Press, New York.

Archer, R. (2004): *Inorganic and Organometallic Polymers*, Wiley, Hoboken, NJ.

Brinker, C. and Scherer, D. (1990): *The Physics and Chemistry of Sol-Gel Processing*, Academic, Orlando, FL.

Brook, M. (1999): *Silicon in Organic, Organometallic, and Polymer Chemistry*, Wiley, New York.

Chandrasekhar, V. (2005), *Inorganic and Organometallic Polymers*, Springer, New York.

Chen, L. and Hong, M. (2009): *Design and Construction of Coordination Polymers*, Wiley, Hoboken, NJ.

Dvornic, P. and Owen, M. (2009): *Silicon-Containing Dendrite Polymers*, Springer, New York.

Ganachaud, F., Boileau, S., and Boury, B. (2008): *Silicon Based Polymers: Advances in Synthesis and Supramolecular Organization*, Springer, New York.

Innocenzi, P., Zub, Y., and Kessler, V. (2008): *Sol-Gel for Materials Processing: Focusing on Materials for Pollution Control, Water Purification, and Soil Remediation*, Springer, New York.

Jones, R. (2001): *Silicon-Containing Polymers*, Kluwer, New York.

Jones, R., Andeo, W., and Chojnowski, J. (2000): *Silicon-Containing Polymers*, Kluwer, New York.

Kipping, F. S. (1927): Silicones, *J. Chem. Soc.*, 130:104.

Ladenburg, A. (1872): Silicones, *Ann. Chem.*, 164:300.

Manners, I. (2004): *Synthetic Metal-Containing Polymers*, Wiley, Hoboken, NJ.

Marklund, S. (2008): *Organophosphorus Flame Retarders and Plasticizers*, VDM Verlag, New York.

Neuse, E. W. and Rosenberg, H. (1970): *Metallocene Polymers*, Dekker, New York.

Pittman, C. Carraher, C., Zeldin, M., Culbertson, B., and Sheats, J. (1996): *Metal-Containing Polymeric Materials*, Plenum, New York.

Rochow, E. G. (1951): *An Introduction to the Chemistry of Silicones*, Wiley, New York.

Stepnicka, P. (2008): *Ferrocenes: Ligands, Materials and Biomolecules*, Wiley, Hoboken, NJ.

Tsuchida, E. (2000): *Macromolecular-Metal Complexes*, Wiley, New York.

10 Inorganic Polymers

10.1 INTRODUCTION

Just as polymers abound in the world of organics, so do they abound in the world of inorganics. Inorganic polymers are the major components of soil, mountains, and sand. Inorganic polymers are also extensively employed as abrasives and cutting materials [diamond, boron carbide, silicon carbide (carborundum), aluminum oxide], coatings, flame retardants, building and construction materials (window glass, stone, Portland cement, brick, tiles), and lubricants and catalysts (zinc oxide, nickel oxide, carbon black, graphite, silica gel, alumina, aluminum silicate, chromium oxides, clays).

The first somewhat man-made, synthetic polymer was probably inorganic in nature. Alkaline silicate glass was used in the Badarian period in Egypt (about 12,000 BC) as a glaze, which was applied to steatite after it had been carved into various animal, etc., shapes. Frience, a composite containing powered quartz or steatite core covered with a layer of opaque glass, was employed from about 9000 BC to make decorative objects. The earliest known piece of regular (modern day type) glass, dated to 3000 BC, is a lion's amulet found at Thebes and now housed in the British Museum. This is a blue opaque glass partially covered with a dark green glass. Transparent glass appeared about 1500 BC. Several fine pieces of glass jewelry were found in Tutankhamen's tomb (ca. 1300 BC) including two bird heads of light blue glass incorporated into the gold pectoral worn by the Pharaoh.

Because of the wide variety and great number of inorganic polymers, this chapter will focus on only a few of the more well-known inorganic polymers. Table 10.1 contains a partial listing of common inorganic polymers.

Along with the silicates that will be dealt with in other sections, there are many other inorganic polymers based on other units. One of these is the hydroxylapatite or hydroxyapatite materials that have a general formula of $Ca_5(PO_4)_3(OH)$, which is a member of the apatite group. Seventy percent of bone is composed of hydroxyapatite. Dental enamel is made from carbonated-calcium deficient hydroxylapatite nanotubes. It is believed that the initial step is the production of composite nanospheres made of nanocrystallite apatite and amelogenin. Thesc then aggregate forming nanorods about 50 nm in diameter and 250 nm in length. These nanorods further assemble forming elongated crystals that compose our dental enamel.

10.2 PORTLAND CEMENT

Portland cement is the least expensive, most widely used synthetic inorganic polymer. It is employed as the basic nonmetallic, nonwoody material of construction. Concrete highways and streets span our country side and concrete skyscrapers silhouette the urban skyline. Less spectacular uses are found in everyday life as sidewalks, fence posts, and parking bumpers.

The name "Portland" is derived from the cement having the same color as the natural stone quarried on the Isle of Portland, a peninsula on the south of Great Britain. The word cement comes from the Latin word *caementum*, which means "pieces of rough, uncut stone." Concrete comes from the Latin word *concretus*, which means, "to grow together."

Common cement consists of anhydrous crystalline calcium silicates (the major ones being tricalcium silicate, Ca_3SiO_5, and β-dicalcium silicate, Ca_2SiO_4), lime (CaO, 60%), and alumina

TABLE 10.1
Important Inorganic Polymers

Agate	Chabazite	Glasses (many kinds)	Silicon carbide
Alumina	Chett	Graphite	Spodumene
Aluminum oxide	Chrysotile	Imogolite	Stilbite
Amphiboles	Concrete	Jasperite	Stishorite
Anthophylite	Cristobalite	Kaolinite	Sulfur nitride
Arsenic selenide	Crocidolite	Mesolite	Talc
Arsenic sulfide	Diamond	Mica	Thomsonite
Asbestos	Dickite	Montmorillonite	Tremolite
Berlinite	Epistilbite	Muscovite	Tridymite
Beryllium oxide	Feldspars	Phosphorus oxynitride	Valentinite
Boron nitride	Flint	Polyphosphates (many)	Vermiculite
Boron oxides	Fuller's earth	Quartz	Wollastonite
Boron phosphate	Garnet	Rhodonite	Xonotlite
Calcite	Germanium selenide	Serpentine	Ziolites
Carbon black	Gibbsite	Silicon dioxides (many)	Zirconia

(a complex aluminum silicate, 5%). White cement is widely used and has been studied in good detail, yet its structure and the process whereby it is formed are not completely known. This is due to at least two factors. First, its three-dimensional arrangement of various atoms has a somewhat ordered array when a small (molecular level) portion is studied, but as larger portions are viewed, less order is observed giving only an average overall structure. This arrangement is referred to as short-range order and long-range disorder and is a good description of many three-dimensional, somewhat amorphous inorganic and organic polymers. Thus, there exists only an average structure for the cement, which varies with the amount of water and other components added, time after application (i.e., age of the cement), and the source of concrete mix and location (surface or internal). Second, three-dimensional materials are insoluble in all liquids; therefore, the tools of characterization and identification that require materials to be in solution cannot be employed to assist in the structural identification of cement.

When anhydrous cement mix is added to water, the silicates react, forming hydrates and calcium hydroxide. Hardened Portland cement contains about 70% cross-linked calcium silicate hydrate and 20% crystalline calcium hydroxide.

$$2Ca_3SiO_5 + 6H_2O \rightarrow Ca_3Si_2O_7 \cdot 3H_2O + 3Ca(OH)_2 \tag{10.1}$$

$$2Ca_2SiO_4 + 4H_2O \rightarrow Ca_3Si_2O_7 \cdot 3H_2O + Ca(OH)_2 \tag{10.2}$$

A typical cement paste contains about 60%–75% water by volume and only about 40%–25% solids. The hardening occurs through at least two major steps (Figure 10.1). First, a gelatinous layer is formed on the surface of the calcium silicate particles. The layer consists mainly of water with some calcium hydroxide. After about 2 h, the gel layer sprouts fibrillar outgrowths that radiate from each calcium silicate particle. The fibrillar tentacles increase in number and length, becoming enmeshed and integrated. The lengthwise growth slows, with the fibrils now joining up sideways, forming striated sheets that contain tunnels and holes. During this time, calcium ions are washed away from the solid silicate polymeric structures by water molecules and react further, forming additional

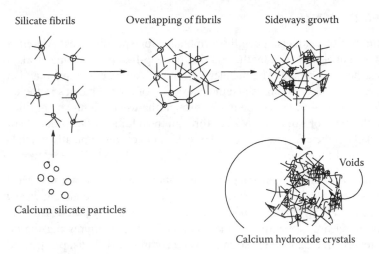

Silicate fibrils   Overlapping of fibrils   Sideways growth

Voids

Calcium silicate particles

Calcium hydroxide crystals

FIGURE 10.1   Steps in the hardening of Portland cement.

calcium hydroxide. As particular local sites become saturated with calcium hydroxide, calcium hydroxide itself begins to crystallize, occupying once vacant sites, and carrying on the process of interconnecting about and with the silicate "jungle."

In spite of attempts by the silicate and the calcium hydroxide to occupy all of the space, voids are formed, probably from the shrinkage of the calcium hydroxide as it forms a crystalline matrix. (Generally, crystalline materials have higher densities than amorphous materials; thus, a given amount will occupy less volume, leaving some unfilled sites.) Just as a chain is no stronger than its weakest link, so also is cement no stronger than its weakest sites, i.e., its voids. Much current research concerns attempts to generate stronger cement with the focus on filling these voids. Interestingly, two of the more successful cement-void-fillers are also polymers-dextran, a polysaccharide, and polymeric sulfur.

Table 10.2 shows a typical concrete mix. The exact amounts may vary by as much as 50% depending on the intended use and preference of the concrete maker.

The manufacture of Portland concrete consists of three basic steps—crushing, burning, and finish grinding. As noted before, Portland cement contains about 60% lime, 25% silicates, and 5% alumina with the remainder being iron oxides and gypsum. Most cement plants are located near limestone ($CaCO_3$) quarries, since this is the major source of lime. Lime may also come from oyster shells, chalk, and a type of clay called marl. The silicates and alumina are derived from clay, silicon sand, shale, and blast-furnace slag.

TABLE 10.2
Sample Concrete Mix

Material	Amount	
	By Volume	By Weight
Portland cement	90–100 lb (1 ft³)	90–100 lb (40–50 kg)
Water	5.5 gal	45 lb (20 kg)
Sand	2 ft³	200 lb (90 kg)
Gravel (small rocks)	3 ft³	250 lb (120 kg)

10.3   OTHER CEMENTS

There are a number of cements specially formulated for specific uses. *Air-entrained concrete* contains small air bubbles formed by the addition of soap-like resinous materials to the cement or to the concrete when it is mixed. The bubbles permit the concrete to expand and contract (as temperature changes) without breaking (since the resistance of air to changes in the concrete volumes is small). *Lightweight concrete* may be made through the use of light-weight fillers, such as clays and pumice, in the place of sand and rocks or through the addition of chemical foaming agents that produce air pockets as the concrete hardens. These air pockets are typically much larger than those found in air-entrained concrete.

Reinforced concrete is made by casting concrete about steel bars or rods. Most large cement-intense structures, such as bridges and skyscrapers, employ reinforced concrete. *Prestressed concrete* is typically made by casting concrete about steel cables stretched by jacks. After the concrete hardens, the tension is released, resulting in the entrapped cables compressing the concrete. Steel is stronger when tensed, and concrete is stronger when compressed. Thus, prestressed concrete takes the advantage of both of these factors. Archways and bridge connections are often made from prestressed concrete.

Concrete masonry is simply the name given to the cement building blocks employed in the construction of many homes, and it is simply a precast block of cement, usually with lots of voids. *Precast concrete* is concrete that is cast and hardened before it is taken to the site of construction. Concrete sewer piped, wall panels, beams, grinders, and spillways are all examples of precast cements.

The cements cited above are all typically derived from Portland cement. Following are non-Portland cements.

Calcium aluminate cement has a much higher percentage of alumina than does Portland cement. Furthermore, the active ingredients are lime, CaO, and alumina. In Europe, it is called melted or fused cement. In the United States, it is manufactured under the trade name Lumnite. Its major advantage is its rapidity of hardening, developing high strength within a day or two.

Magnesia cement is largely composed of magnesium oxide (MgO). In practice, the magnesium oxide is mixed with fillers, rocks, and an aqueous solution of magnesium chloride. This cement sets up (hardens) within 2–8 h and is employed for flooring in special circumstances.

Gypsum, or hydrated calcium sulfate ($CaSO_4 \cdot 2H_2O$), serves as the basis for a number of products including plaster of Paris (also known as molding plaster, wall plaster, and finishing plaster). The ease with which plaster of Paris and other gypsum cements can be mixed and cast (applied), and the rapidity with which they harden contribute to their importance in the construction field as a major component for plaster wall boards. Plaster of Paris' lack of shrinkage in hardening accounts for its use in casts. Plaster of Paris is also employed as a dental plaster, pottery plaster, and as molds for decorative figures. Unlike Portland cement, plaster of Paris requires only about 20% water and dries to the touch in 30–60 min giving maximum strength after 2–3 days. Portland cement requires several weeks to reach maximum strength.

10.4   SILICATES

Silicon is the most abundant metal-like element in the earth's crust. It is seldom present in pure elemental form, but rather is present in a large number of polymers largely based on the polycondensation of the orthosilicate anion, SiO_4^{-4} as illustrated in the following:

$$SiO_4^{-4} \rightleftharpoons Si_2O_7^{-6} + O^{-2} \qquad (10.3)$$
$$\quad (1) \qquad\quad (2)$$

$$Si_2O_7^{-6} + SiO_4^{-4} \rightleftharpoons Si_3O_9^{-6} + 2O^{-2} \qquad (10.4)$$
$$(3)$$

$$2Si_2O_7^{-6} \rightleftharpoons Si_4O_{10}^{-4} + 4O^{-2} \qquad (10.5)$$
$$(4)$$

$$2Si_2O_7^{-6} \rightleftharpoons Si_4O_{11}^{-6} + 3O^{-2} \qquad (10.6)$$
$$(5)$$

$$2Si_2O_7^{-6} \rightleftharpoons Si_4O_{12}^{-8} + 2O^{-2} \qquad (10.7)$$
$$(6)$$

$$3Si_2O_7^{-6} \rightleftharpoons Si_6O_{18}^{-12} + 3O^{-2} \qquad (10.8)$$
$$(7)$$

$$SiO_4^{-4} \rightarrow SiO_2 \qquad (10.9)$$
$$(8)$$

The number listed with each product corresponds with the "Geometric ID Number" given in Table 10.3 and the "Structural Geometry" given in Table 10.3, and is depicted in Figure 10.2.

TABLE 10.3
Inorganic Polymeric Silicates as a Function of Common Geometry

Geometric ID Number (Text)	Basic Geometric Unit	Structural Geometry (Figure 10.2)	General Silicate Formula*	Examples[a]
1	Tetrahedron	A	SiO_4^{-4}	Granite olivine—$(Mg, Fe)_2SiO_4$
				Fosterite—Mg_2SiO_4
				Topez
2	Double tetrahedron	B	$Si_2O_7^{-6}$	Akermanite—$Ca_2MgSi_2O_7$
3	Triple ring	C	$Si_3O_9^{-6}$	Wollastonite
4	Tetra ring	D	$Si_4O_{12}^{-8}$	Neptunite
5	Six ring	E	$Si_6O_{18}^{-12}$	Beryl—$Al_2Be_3Si_6O_{18}$
6a	Linear chain	F	$Si_4O_{12}^{-8}$	Augite, enstatite—$MgSiO_3$
				Diopside—$CaMg(SiO_3)_2$
				Chrysotile—$Mg_6Si_4O_{11}(OH)_6$
6b	Double-stranded ladder	G	$Si_4O_{11}^{-6}$	Hornblende
7	Parquet (layered)	H	$Si_4O_{10}^{-4}$	Talc—$Mg_3Si_4O_{10}(OH)_2$
				Mica—$KAl_3Si_3O_{10}(OH)_2$
				Kaolinite—$Al_2Si_2O_5(OH)_4$
				(Condensed) silicic acid—$H_2Si_2O_5$
8	Network	I	SiO_2	Quartz, feldspar (orthoclase)—$KAlSi_3O_8$

[a] The formulas given are for the most part simplified.

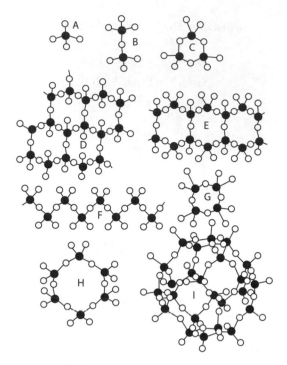

FIGURE 10.2   General silicate structures.

Each of these steps is based on a tetrahedral silicon atom attached to four oxygen atoms. The complexity and variety of naturally occurring silicates is due to two major factors. First, the ability of the tetrahedral SiO_4^{-4} unit to be linked together often gives polymeric structures. Second, the substitution of different metal atoms of the same approximate size as that of Si often occurs giving many different materials.

In the structures cited in Table 10.3, except for pure SiO_2, metal ions are required for overall electrical neutrality. These metal ions are positioned in tetrahedral, octahedral, etc., positions in the silicate-like lattice. Sometimes they replace the silicon atom. Kaolinite asbestos (Figure 10.3) has aluminum substituted for silicon in the Gibbosite sheet.

FIGURE 10.3   Two-dimensional representation of a multicomponent silicon dioxide intense glass where the black spheres are oxygen, white spheres are silicon, and the off-color spheres are an impurity cation.

Further, sites for additional anions, such as the hydroxyl anion, are available. In ring, chain, and sheet structures neighboring rings, chains, and sheets are often bonded together by metal ions held between the rings. In vermiculite asbestos, the silicate sheets are held together by nonsilicon cations.

For sheet-layered compounds, the forces holding together the various sheets can be substantially less than the forces within the individual sheets. Similar to graphite, such structures may be easily cleaved parallel to the sheets. Examples of such materials are mica, kaolinite, and talc.

Bonding occurs through a combination of ionic and covalent contributions just as they are present in organic polymers except that the ionic character is a little higher. "Back-bonding" from electrons associated with the oxygen to vacant orbitals in the silicon (or other tetrahedral metal atom) occurs giving the silicon–oxygen linkages some double- or pi-bond character.

As noted before, cations other than silicon may occupy the tetrahedral centers. A major factor in predicting which cations will be found to substitute for silicon is ionic size. In general, cations whose size is about 0.03–0.1 nm are the best candidates. Si^{+4} has an ionic radius of about 0.041 nm. Cations such as Fe^{+2} (ionic radius = 0.07 nm), Al^{+3} (0.05 nm), Ca^{+2} (0.1 nm), and Mg^{+2} (0.065 nm) are most often found in silicate-like structures and meet this requirement.

Most silicate-like polymers can be divided into three major classes—the network structures based on a three-dimensional tetrahedral geometry (such as quartz), layered geometries with stronger bonding occurring within the "two-dimensional" layer (such as talc), and linear structures.

10.4.1 NETWORK

Quartz is an important network silicate (Section 10.6). A number of additional tetrahedral silicate-like materials possess some AlO_4 tetrahedra substituted for the SiO_4 tetrahedra. Such structures offer a little larger "hole" in comparison to the entirely SiO_4 structures allowing alkali and alkaline-earth cations to be introduced. Feldspar (orthoclase) is such a mineral. The aluminosilicate networks are almost as hard as quartz. For feldspar and other tetrahedral networks, the number of oxygen atoms is twice the summation of silicon and other MO_4 cations.

The feldspars are widely distributed and comprise almost two-thirds of all igneous rocks. Orthoclase and albite, $NaAlSi_3O_8$, are feldspars where one-fourth of the silicon atoms are replaced by aluminum and anorthite, $CaAl_2Si_2O_8$, has one-half of the silicon atoms replaced by aluminum. Because the ionic radius of Na^+ (0.095 nm) and Ca^{+2} (0.1 nm) is about the same, solid solutions are often formed between albite and anorthite. Good stones of albite and orthoclase are known as moonstones.

Some of the network structures exhibit a framework sufficiently "open" to permit ions to move in and out. The zeolite minerals used for softening water are of this type.

Ultramarines are three-dimensional cage-like structures. They differ from feldspars and zeolites because of the large spaces within the structures that can contain cations and anions but no water illustrating a natural "buckeyball-like" structure and cavity, and a diversity of environment between the internal and external cages. Ultramarines can act as ion exchangers for both anions and cations. The blue color of ultramarines is due to the presence of the S_3^- ion, although a yellow ion S_2^- also exists in the same structure.

10.4.2 LAYER

Layered structures typically conform to the approximate composition $Si_4O_{10}^{-4}$ or $Si_2O_5^{-2}$. For most of these, three of the oxygen atoms of each tetrahedron are shared by other tetrahedra, and the fourth oxygen is present on one side of the sheet.

In talc and kaolinite, the layers are neutral. Thus, the layers slide over one another easily imparting to these minerals a softness and ease in cleavage. In other minerals, the layers are charged and are held together by cations. In mica, the aluminosilicate layers are negatively charged and cations, generally K^+, are present between the layers giving the entire system of layers an electronic neutrality. The ionic attractive forces between the layers result in mica being much harder than talc and kaolinite. Even so, these inter-sheet bonding forces are less than the "within-the-sheet" bonding forces permitting relatively easy and clean cleavage of mica. Mica is used as an insulator for furnaces and electric equipments. Montmorillonite is an important ingredient in soils and is employed industrially as a catalyst in the conversion of straight-chain hydrocarbons to more branched hydrocarbons, and more recently as a sheet or a clad material in the manufacture of tires.

Vermiculites are formed by the decomposition of mica. They contain layers of water and magnesium ions in the place of potassium ions. When heated from 800°C to 1100°C, vermiculite expands because of the conversion of the water to a gas. The expanded vermiculite has a low thermal conductivity and density, and is used as a thermal and sound barrier and also as an aggregate in lightweight concrete. It is also used as a moisture-retaining soil conditioner in planting.

A number of clays are layered silicate-like materials. Most clay contains finely divided quartz, micas, and feldspars. Iron oxide-rich clays are employed to make pottery and terra cotta articles. Clays containing iron oxide and sand are used to make bricks and tiles. Clays rich in calcium and magnesium carbonate are known as marls, and are used in the cement industry (Section 10.2).

Kaolinite is the main constituent in china clay that is used to make porcelain. The layers are largely held together by van der Waals' forces. Bentonite is used in cosmetics, as a filler for soaps, and as a plasticizer, and it is used in drilling muds as a suspension stabilizer. Bentonite and kaolinite clays are used, after treatment with sulfuric acid to create acidic surface sites, as petroleum-cracking catalysts. Asbestos also has a layered structure (Section 10.8).

10.4.3 CHAIN

Both single and double stranded chains are found. The most important members of single chains are the pyroxenes and includes diopside. The most important double-chained minerals are the amphiboles. Some of these contain hydroxyl and fluoride ions, bonded directly to the metal cation and not to the silicon atom.

Jade, which has been valued in carving by eastern Asians for centuries, is generally one of the two minerals—pyroxene or jadeite, $NaAl(SiO_3)_2$, and the amphibole nephrite, $Ca_2(Fe^{+2}$ and/or $Mg^{+2})_5$ $(Si_4O_{11})_2(OH)_2$. X-ray diffraction has shown the presence of triple chains in nephrite.

Because the interchain bonding is weaker than the Si–O backbone bonding, these chain structures can generally be easily cleaved between the chains.

Several amphiboles are fibrous, and fibers from then can be processed to give heat-insulating materials. Among these are tremolite and crocidolite. These minerals are also used as fibers in composites.

10.5 SILICON DIOXIDE (AMORPHOUS)

Silicon dioxide (SiO_2) is the major repeating general formula for the vast majority of rock, sand, and dirt about us and for the material we refer to as glass. The term glass can refer to many materials, but here we will use the ASTM definition that glass is an inorganic product of fusion that has been cooled to a rigid condition without crystallization. In this section, silicate glasses, the common glasses for electric light bulbs, window glass, drinking glasses, glass bottles, glass test tubes and beakers, and glass cookware, will be emphasized.

Glass has many useful properties, as listed in Table 10.4. It ages (changes chemical composition and physical property) slowly, typically retaining its fine optical and hardness-related properties for centuries. Glass is referred to as a supercooled liquid or a very viscous liquid. Indeed, it is a slow-moving liquid as attested to by sensitive measurements carried out in some laboratories. Concurrent with this is the observation that the old, stained glass windows adorning European cathedrals are

---

TABLE 10.4
General Properties of Silicate Glasses

High transparency to light	Permanent (long-term) transparency
Hard	Scratch-resistant
Chemically inert	Low thermal expansion coefficient
Good electrical insulator	High sparkle
Good luster	Low porosity
Good ease of reforming	Easily recyclable
(Relatively) inexpensive	Available in large amounts

---

a little thicker at the bottom of each small, individual piece than at the top of the piece. For most purposes though, glass can be treated as a brittle solid that shatters on sharp impact.

Glass is mainly silica sand (SiO_2) and is made by heating silica sand and powdered additives together in a specified manner and proportion much as a cake is baked, following a recipe that describes the items to be included, amounts, mixing procedure (including sequence), oven temperature, and heating time. The amounts, nature of additives, etc., all affect the physical properties of the final glass.

Typically, cullet, recycled, or waste glass (5%–40%), is added along with the principle raw materials (mostly SiO_2). The mixture is thoroughly mixed and then added to a furnace where the mixture is heated to near 1500°C to form a viscous, syrup-like liquid. The size and nature of the furnace corresponds to the glasses' intended uses. For small, individual items, the mixture may be heated in a small clay (refractory) pot.

Most glass is melted in large (continuous) tanks that can melt 400–600 metric ton a day for the production of different glass products. The process is continuous with the raw materials fed into one end as molten glass is removed from the other end. Once the process (called a campaign) is begun, it is continued indefinitely, night and day, often for several years until the demand is met or the furnace breaks down.

A typical window glass will contain 95%–99% silica sand with the remainder being soda ash (Na_2CO_3), limestone ($CaCO_3$), feldspar, and borax or boric acid along with the appropriate coloring and decolorizing agents, oxidizing agents, etc. As noted previously, 5%–40% by weight of crushed cullet is also added. The soda ash, limestone, feldspar, and borax or boric acid all form oxides as they are heated, which become integrated into the silicon structure:

$$Na_2CO_3 \quad \rightarrow Na_2O + CO_2 \tag{10.10}$$
$$\text{Soda ash}$$
$$\text{(sodium carbonate)}$$

$$CaCO_3 \quad \rightarrow CaO + CO_2 \tag{10.11}$$
$$\text{Limestone}$$
$$\text{(calcium carbonate)}$$

$$R_2OAl_2O_3 \cdot 6H_2O \rightarrow R_2O + \quad Al_2O_3 \quad + \quad 6SiO_2 \tag{10.12}$$
$$\text{Feldspar} \qquad\qquad \text{Alumina}$$
$$\text{(aluminum oxide)}$$

$$H_3BO_4 \rightarrow B_2O_3 \tag{10.13}$$
$$\text{Boric acid}$$

The exact structure varies according to the ingredients and actual processing conditions. As in the case of Portland cement, glass is a three-dimensional array that offers short-range order and long-range disorders—it is amorphous offering little or no areas of crystallinity. The structure is based on the silicon atoms existing in a tetrahedral geometry with each silicon atom attached to four oxygen atoms, generating a three-dimensional array of inexact tetrahedra. Thus, structural defects occur, due in part to the presence of impurities such as Al, B, and Ca, intentionally or unintentionally introduced. These impurities encourage the glass to cool to an amorphous structure since the different sized impurity metal ions, etc., disrupt the rigorous space requirement necessary to allow crystal formation. Figure 10.3 is an illustration of this situation and contains two atoms of impurity.

Processing includes shaping and pretreatments of the glass. Since shaping may create undue sites of amorphous structure, most glass objects are again heated to near their melting point. This process is called **annealing**. Since many materials tend to form more ordered structures when heated and recooled slowly, the effect of annealing is to "heal" the sites of major dissymmetry. It is important to heal these sites, since they represent preferential locations for chemical and physical attacks such as fracture.

Four main methods are employed for shaping glass. They are drawing, pressing, casting, and blowing. Drawing is employed for shaping flat glass, glass tubing, and for creating fibrous glass. Most flat glass is shaped by drawing a sheet of molten glass (heated so that it can be shaped, but not heated so much that it freely flows) onto a tank of molten tin. Since the glass literally floats on the tin, it is called "float glass." The temperature is carefully controlled. The glass from the float bath typically has both sides quite smooth with a brilliant finish that requires no polishing.

As noted above, annealing encourages the removal of the sites of stress and strain. Slow cooling results in a glass with more crystallinity, which is stronger but more brittle. **Tempering** is the name given when the glass is rapidly cooled, resulting in an amorphous glass that is weaker but less brittle. The correlation between crystallinity, rate of cooling, and brittleness is demonstrated by noting that an older window glass exposed to full sun for years is more brittle and more easily shattered since the sunlight raises the temperature sufficiently to permit small molecular movements (even though in full sunlight the glass is not near the temperature required for ready movement) and over the years gives a glass with small regions of greater order.

Silicon-based glasses account for almost all of the glasses manufactured. Silica is finely ground silica sand. Yet most sand is unsuitable for general glassmaking due to the presence of excessive impurities. Thus, while sand is plentiful, sand that is useful for the production of glass is much less common. In fact, the scarcity of large deposits of glass sand is one major reason for the need to recycle glass items. The second major reason is associated with the lowered energy requirements for glass to be made molten again for reshaping compared with a virgin glass mixture, i.e., culled glass becomes molten at lower temperatures than virgin glass.

10.5.1 KINDS OF GLASSES

The types and properties of glass can be readily varied by changing the relative amounts and nature of ingredients. **Soda-lime glass** is the most common of all glasses accounting for about 90% of glass made. Window glass, glass for bottles, etc., are all soda-lime glass. Soda-lime glass (75% silica, 15% soda (sodium oxide), 9% lime (calcium oxide), and the remaining 4% minor ingredients) has a relatively low softening temperature and low thermal shock resistance limiting its high-temperature applications.

Vycor, or 96% silicon glass, is made using silicon and boron oxide. Initially the alkali–borosilicate mixture is melted and shaped using conventional procedures. The article is then heat treated, resulting in the formation of two separate phases—one that is high in alkalis and boron oxide, and the other containing 96% silica and 3% boron oxide. The alkali-boron oxide phase is soluble in strong acids and is leached away by immersion in hot acid. The remaining silica-boron oxide phase is quite porous. The porous glass is again heated to about 1200°C, resulting in a 14% shrinkage due to the remaining portions filling the porous voids. The best variety is "crystal" clear and called *fused quartz*. The 96% silica glasses are more stable and exhibit higher melting points (1500°C) than soda-lime glass. Crucibles, ultraviolet filters, range burner plates, induction furnace linings, optically clear filters and cells, and super heat-resistant laboratory ware are often 96% silicon glass.

Borosilicate glass contains about 80% silica, 13% boric oxide, 4% alkali, and 2% alumina. It is more heat-shock-resistant than most glasses due to its unusually small coefficient of thermal expansion (typically between 2 and 5×10^{-6} cm/cm/°C; for soda-lime glass it is $8-9 \times 10^{-6}$ cm/cm/°C). It is better known by trade names such as Kimax and Pyrex. Bakeware and glass pipelines are often made of borosilicate glass.

Lead glasses (often called heavy glasses) are made by replacing some or all of the calcium oxide by lead oxide (PbO). Very high amounts of lead oxide can be incorporated—up to 80%. Lead glasses are more expensive than soda-lime glasses, and they are easier to melt and work with. They are more easily cut and engraved, giving a product with high sparkle and luster (due to the high refractive indexes). Fine glass and tableware are often made of lead glass.

Silicon glass is made by fusing pure quartz crystals or glass sand (impure crystals), and it is typically about 99.8% SiO_2. It has a high melting point, and this makes it difficult to fabricate.

Colored or *stained glass* has been made for thousands of years, first by Egyptians and later by Romans. Color is typically introduced by the addition of transition metals and oxides. Table 10.5 contains selected inorganic colorants and the resulting colors. Because of the high clarity of glass, a small amount of coloring agent goes a long way. One part of cobalt oxide in 10,000 parts of glass gives an intense blue glass. The most well-known use for colored glass is the construction of stain-glass windows. In truth, there are many other uses such as industrial color filters and lenses.

Glazes are thin, transparent coatings (colored or colorless) fused on ceramic materials. *Vitreous enamels* are thin, normally opaque or semi-opaque, colored coatings fused on metals, glasses, or ceramic materials. Both are special glasses but can contain little silica. They are typically low melting and often cannot be easily mixed with more traditional glasses.

Optical fibers can be glass fibers that are coated with a highly refractive polymer coating such that light entering through one end of the fiber is transmitted through the fiber (even around corners as within a person's stomach), emerging from the other end with little loss of energy. These optical fibers can also be made to transmit sound and serve as the basis for the transmission of television and telephone signals over great distances through cables.

There are many silica-intensive fibers lumped together as *fibrous glass* or *fiberglass*. A general-purpose fiberglass may contain silica (72%), calcium oxide (9.5%), magnesium oxide (3.5%), aluminum oxide (2%), and sodium oxide (13%). The fibers are produced by melting the "glass mixture" with the molten glass drawn through an orifice. The filaments are passed through a pan containing sizing solution onto a winding drum. The take-up rate of the filament is more rapid than the exit rate

TABLE 10.5
Colorants for Stained Glass

Colorant	Color	Colorant	Color
Nickel (II) oxide	Yellow to purple	Calcium fluoride	Milky white
Cobalt (II) oxide	Blue	Iron (II) compounds	Green
Iron (III) compounds	Yellow	Copper (I) oxide	Red, blue, or green
Tin (IV) oxide	Opaque	Manganese (IV) oxide	Violet
Gold (III) oxide	Red		

TABLE 10.6
Leading U.S. Glass Companies

Owen-Illinois, Inc.	PPG Industries, Inc.
Corning Glass Works	Owens-Corning Fiberglass Corporation
Libbey-Owens-Ford Company	

from the orifice that acts to align the molecules and draw the fibers into thinner filaments. Thus, a fiber forced through a 0.1 cm orifice may result in filaments of 0.0005 cm diameter. This drawing increases the strength and the flexibility of the fiberglass. The applications of fiberglass include insulation and use in composites.

Table 10.6 contains a listing of major glass-producing companies in the United States.

Optical glass for eyeglass lenses and camera lenses is typically soda-lime glass that is highly purified so that it is highly transmissive of light. Today, there exists many other special glasses that are important in today's society as laser glasses, photosensitive glasses, photochromic windows and eyeglass glass, invisible glasses, radiation absorbing glasses, etc.

10.5.2 Safety Glass

Safety glass is defined as "glass" that diminishes the threat of injuries and robberies because of impacts, distortion, or fire.

In 1905, British inventor John C. Wood was working with cellulose and developed a method to adhere glass panes using celluloid as the adhesive. Wood's version of shatter-resistant glass was produced under the brand name Triplex since it consisted of outer layers of glass with an inner layer of celluloid polymer.

About the same time, Edouard Benedictus, a French chemist, was climbing a ladder to get chemicals from a shelf and accidentally (another discovery due to an accident) knocked a glass flask onto the floor. He heard the flask shatter but when he looked at the broken flask, the broken pieces hung together instead of breaking into many pieces and scattering over the floor. Benedictus learned from his assistant that the flask had recently held a solution of cellulose nitrate. The solution of cellulose nitrate dried to give a thin film that was transparent, so the flask was set aside as a "cleaned" flask. It was this thin film that coated the inside of the flask that held together the broken pieces. The film was formed on the evaporation of cellulose nitrate prepared from cellulose and nitric acid.

Shortly after the laboratory accident, he read about a girl who had been badly cut from flying glass resulting from an automobile accident. Later he read about other people being cut by flying glass in automobile accidents. He remembered the flask that did not splinter into small pieces when broken because of the cellulose nitrate coating. He experimented with placing cellulose nitrate between sheets of glass, using pressure to help adhere the glass with the cellulose nitrate. By 1909, Benedictus had patented the material.

Prior to its use in windshields, safety glass found its initial major application as the lenses for gas masks during World War I. Manufacturers found it easy to work with, so that the technology and ability to manufacture safety glass windshields came easily.

As automobiles became more common, so did the hazards of mud, rocks, etc., so that by 1904 windshields were introduced. These first windshields could be folded moved if they became blocked by excessive mud. While the usefulness of the windshield was abundantly obvious, drivers found that they were dangerous during a wreck cutting passengers, drivers, and passer-byes alike. Because the drivers were believed to be primarily responsible for automobile safety, most manufactures were slow to adopt safety glass. In 1919, Henry Ford addressed the windshield problem by having safety glass windshields on his automobiles.

This safety glass turned yellow after several years of exposure to light. The bonding layer was replaced in 1933 by cellulose acetate, made from the reaction of cotton with acetic acid. By 1939, this was replaced by poly(vinyl butyral), PVB, still in use today as the adhesive placed between the sheets of glass to produce laminated safety glass. This is one of the very few modern-use materials that has retained the same basic materials for over 60 years.

$$(10.14)$$

Poly(vinyl alcohol)/butyraldehyde   Poly(vinyl butyral)

PVB is made from poly(vinyl alcohol), which itself is made from poly(vinyl acetate) because the monomer vinyl alcohol does not exist (10.14).

Today, safety glass is divided into three general categories, laminated safety glass, tempered safety glass, and armed glass. **Tempered safety glass** is made by heating the glass to its melting point, about 700°C, and then cooling it rapidly by blowing cold air onto its surfaces. The effect is similar to the production of stressed concrete where the concrete is allowed to harden under stress giving a stronger concrete. In the case of glass, when it is rapidly cooled, a structure is locked in that produces extra stress on the glass structure making it stronger. As the glass is cooled, the surfaces harden first locking in the overall glass volume. As the center cools, it forces the surfaces and edges into compression. With appropriate rapid cooling, the glass is not only stronger, but when shattered, produces granulates rather than sharp cutting shards. The typical force necessary to break tempered glass is about four times that required to shatter an ordinary glass of the same thickness.

While the front "windshield" is made of safety glass, the remainder of the automotive glass windows is generally made from tempered glass. Tempered glass is also used for commercial building doors and windows, sidelights, patio-door assemblies, storm doors, shower and tub enclosures, refrigerator, oven, and stove shelves, and fireplace screens.

Armed glass is most commonly used as roofing on factory buildings. It is glass that has a built-in metal grill that strengthens the glass. The metal grill is often like chicken wire in appearance. The glass breaks similar to regular glass, but the wire mesh helps hold it in place. The visibility is reduced because of the presence of the metal mesh.

Laminated glass is used in automobiles and often used for added protection in windows, balconies, or sloping glass roofs. The laminated glass resists breakage in comparison to ordinary glass because the PVB inner layer(s) helps dampen sharp blows. Even so, laminated glass is more easily cracked than tempered glass. But, it is harder to pierce than tempered glass because of the PVB inner layer(s). Another difference is that laminated glass can be cut, sawn, or drilled whereas tempered glass cannot. The PVB film also has ultraviolet-screening properties reducing the discoloration of objects placed behind the safety glass. It also acts to dampen sound for additional soundproofing.

Laminated safety glass is available in different thicknesses depending on the number of PVB-glass layers. One of the uses of **bullet-resistant glass** is in the manufacture of thick, multilayer laminated safety glass. The laminated construction allows the multilayer assembly to have some additional flexibility with the multiple layers allowing the PVB layers to absorb some of the energy of the bullet. The plastic layers help hold the shattered glass fragments together, aiding in retaining a restraining barrier. Such glass is used in bank teller windows, and in windshields for aircraft, tanks, and special automobiles and trucks.

The ability to resist bullets and blasts is increased by increasing the number of layers of laminated safety glass. Such increases give an increased glass thickness and weight. Therefore, there is a tradeoff between expected abuse and practicality.

10.6 SOL–GEL

In the sol–gel process, ceramic polymer precursors are formed in solution at ambient temperature; shaped by casting, film formation or fiber drawing; and then consolidated to furnish dense glasses or polycrystalline ceramics. The most common sol–gel procedures involve alkoxides of silicon, boron, titanium and aluminum. In alcohol–water solution, the alkoxide groups are removed stepwise by hydrolysis under acidic or basic catalysis and replaced by hydroxyl groups that subsequently form –M–O–M– linkages. Branched polymer chains grow and interconnect forming a network that can span the entire solution volume. At this point, the gel point, the viscosity, and the elastic modulus rapidly increase.

The gel is a viscoelastic material composed of interpenetrating liquid and solid phases. The network retards the escape of the liquid and prevents structural collapse. The shapes formed by casting, drawing of fibers, or film formation are locked in by the gel formation. Some gels are oriented by drawing or shearing. The gel is dried by evaporation forming a xerogel or by supercritical fluid extraction forming an aerogel. Consolidation to dense glasses or ceramics is carried out by thermal treatment and sintering.

Since both aerogels and xerogels have high surface areas and small pore diameters they are used as ultrafiltration media, antireflective coatings, and catalysts supports. Final densification is carried out by viscous sintering.

The rate of silicate sol and gel formation is pH sensitive and water-alcohol sensitive as is the solubility of the amorphous silica that is formed. Silica networks are based on $(SiO_4)^{-4}$ tetrahedra modified by $(O_3Si–O^-, M^+)$ units and often by the addition of boron oxide, aluminum oxide, titanium IV oxide or zirconium IV oxide.

The nature of the reactants can be varied giving various silicate-like products. Following is described as the formation of borosilicate glasses using the sol–gel approach:

$$NaOR + B(OR)_3 + Si(OR)_4 + H_2O \rightarrow NaOH + B(OH)_3 + Si(OH)_4 + ROH \qquad (10.15)$$

$$NaOH + B(OH)_3 + Si(OH)_4 \rightarrow Na_2O \cdot B_2O_3 \cdot SiO_2 \cdot H_2O \rightarrow Na_2O \cdot B_2O_3 \cdot SiO_2 + H_2O \qquad (10.16)$$

The use of organically modified silicates (ceramers) gives a wide variety of products with a variety of structures and properties. Such ceramers have been used as adhesives for glass surfaces, protective coating for medieval stained glass, and as scratch-resistant coatings for plastic eyeglass lenses. They have also been used in the reinforcement of plastics and elastomers, and their nanoscale pores allow their use as porous supports and as selective absorbents.

Sol–gel preparations of tetraethoxysilane can be spun into fibers once the appropriate viscosity has been achieved. These fibers are only slightly weaker than silica-glass fibers.

Hybrid materials have been made by incorporating end-capped poly(tetramethylene oxide) blocks to tetramethoxysilane sol–gel glasses. These materials have high extensibility with interdispersed organic and inorganic regions.

10.6.1 AEROGELS

Aerogels are highly porous materials where the pore sizes are truly on a molecular level, less than 50 nm in diameter. This gives a material with the highest known internal surface area per unit weight. One ounce can have a surface area equal to 10 football fields, over $1000 \, m^2$ in 1 g.

Porous materials can be either open pored such as a common sponge, or closed pored such as the bubble-wrap packaging. Aerogels are open-pored materials such that unbonded material can move from one pore to another.

While in the gel state, the preaerogel has some flexibility, but as a solid aerogel behaves as a fragile glass. It may be very strong in comparison to its weight, but remember it is very light. Aerogels are more durable when under compression. Compression can be simple such as sealing the aerogel sample in a typical food sealer packing. Aerogels are best cut using a diamond-coated saw similar to that used by rock cutters to slice rocks.

When handled, aerogel samples will initially appear to exhibit some flexibility but then burst into millions of pieces. For large arrays of atoms, such as solid metals and polymers below their glass transition temperature, energy can be absorbed through bond flexing or bending. For polymers between their glass transition and melting points, kinetic energy can also be absorbed through segmental movement. Silica aerogel, being an inorganic polymer in the glassy state at room temperature, is a brittle material. As force is applied, there is very little bond flexing, so that the applied kinetic energy results in the collapse of the network with the force of impact spread over a large part of the aerogel and over a time because of the time required to transfer this energy from one cell to another within the aerogel matrix. Because the aerogel is open pored, gas contained within the solid is forced outward as collapse occurs. The frictional force caused by the gas passing through a restricted opening is indirectly proportional to the square of the pore diameter. Because the pore sizes are so small, the rapidly moving gas also absorbs a lot of the energy. Thus, energy is absorbed by the aerogel through both collapse of the solid network structure and release of the gas within the aerogel.

Aerogels contain particles that are about 2–5 nm in diameter with large surface to volume ratios on the order of 10^9 m^{-1} and high specific surface areas approaching 1000 m^2/g. Such large surface-to-volume ratios make the surface particularly active and potential materials as catalysts, absorbents, and catalyst substrates.

The precise chemical makeup of the surface depends on the materials used to make the aerogel and the method of processing. Typical aerogel sequences produce products whose surfaces are rich in hydroxyl groups. Because of the high surface area, −Si−OH groups act as weak acids and are reactive in typical Lewis acid–base reactions. As noted before, aerogels have many hydrogen-bonding hydroxyls at their surface, making aerogels extremely hygroscopic. Dry aerogel materials will increase their weight by 20% through the uptake of moisture from the air. This absorption is reversible and appears to have little or no effect on the aerogel. Water is removed through heating from 100°C to 120°C.

While the adsorption of water vapor has little effect on aerogels, contact with liquid water has devastating effects on aerogels. When water enters the nanometer-sized pores, the surface tension of the water exerts capillary forces sufficient to fracture the silica backbone resulting in a collapse of the complex matrix structure. This tendency to be attacked by water is overcome through conversion of the surface polar −OH groups to nonpolar −OR groups. The "R" is typically a trimethylsilyl group, though any aliphatic group would work. Conversion can be accomplished within the wet stage (preaerogel) or after the supercritical drying. These treatments result in an aerogel that is called "hydrophobic" aerogel, which is stable in water.

The pore size of aerogels varies. The International Union of Pure and Applied Chemistry classifies materials with pore sizes of less than 2 nm as "micropores," 2–50 nm are called "mesopores," and those greater than 50 nm in diameter are called "macropores." While aerogels have some pores that fall within the micropore region, the majority of pores are in the mesopore region.

Most of the aerogels produced today are described as being transparent. While it might be assumed that since aerogels are made of the same material as window glass and quartz, SiO_2, they would be transparent, this is not necessarily the case. Transparency requires a number of factors. Thus, so-called smokey and white quarts are colored because of the presence of impurities. Mixtures of amorphous and crystalline silicon dioxide can be made that are not transparent.

The size and distribution of reflecting and refractive sites are important factors in determining if a material is transparent with the theme of "sameness" contributing to making a material transparent.

The majority of light we see is scattered light, i.e., light that reaches our eyes in an indirect manner. The scattering phenomenon is what gives us blue skies, white to gray clouds, and poor visibility in fog. This scattering is not simply reflecting but results from the interaction of light with an inhomogeneous site. Light scattering photometry is used to determine the size of polymers. Scattering is most effective when the scattering particle size is about that of the wavelength of the light. For visible light, this occurs with scattering sites that are about 400–700 nm. Scattering centers that are much less than the incoming light wavelength are much less effective at scattering the light. Since the particle sizes in an aerogel are much smaller, the individual sites are ineffective scattering sites. Similar to classical polymer chains, where the entire chain or segments of the entire chain act as a scattering site, the clusters of individual sites within the aerogel act as scattering sites. Most of these scattering sites are again smaller than the wavelength of visible light, but some are within the range to scatter visible light so that a soft reflected light results. The different-sized scattering sites and variable wavelengths present in light cause a reddening of the transmitted light (red light has a longer wavelength and is scattered less by small clusters present in the aerogel) resulting in the blue appearance of the reflected light from the aerogel.

The good visible light transmission and good insulting power make aerogel materials of interest in window manufacturing. The visible transmission spectra of light show little absorption in the range of about 300–2700 nm giving aerogels a good visible light "window" making aerogels attractive for day-lighting applications. Aerogels provide about 40-fold more insulation than does fiberglass insulation. While such aerogels may eventually be used as the entire window component, on a shorter range, currently they may be used as the material sandwiched between two panes of clear plastic or glass. *Thermoglass* is generally simply glass sheets that are separated by a vacuum. The seals on such thermoglass often spring small leaks causing greater diminished insulation properties. Aerogel inner cores will not suffer from this problem. Currently, about 40%–50% of a house's heating bill literally goes out of the window because of lost heat or cold through windows. A single 1 in. thick glass pane of aerogel offers the insulation equivalent to over 30 windowpanes of R-20 insulation rated glass.

Another commercial area that is being considered is the use of aerogels as nanocomposite materials. Approaches are varied. In one approach, material is added to the silica sol before gelation. The material can be inorganic, organic, polymeric, bulk fibers, woven cloths, etc. The additional material must be able to withstand the subsequent processing steps including carbon dioxide drying. The added material must be present in a somewhat homogeneous manner throughout the system. Gentle agitation appears to be sufficient to give a product with decent homogeneity. Aerogels may be good materials for optical sensors. They have good visible transparency, high surface area, good temperature and chemical stabilities, and facile transport of gases through their pores.

10.7   SILICON DIOXIDE (CRYSTALLINE FORMS)—QUARTZ FORMS

Just as silicon dioxide forms the basis of glass, so also does it form the basis of many of the rocks, grains of sand, and dirt particles that compose the Earth's crust. Most rocks are inorganic polymers, but here we will deal with only a few of these containing significant amounts of silicon.

Silicon oxide crystallizes in mainly three forms—quarts, tridymite, and cristobalite. After the feldspars, quart is the most abundant material in the Earth's crust, being a major component of igneous rocks and one of the most common sedimentary materials, in the form of sandstone and sand. Quartz can occur as large (several pounds) single crystals, but normally is present as granular materials. The structure of quarts (Figure 10.4) is a three-dimensional network of six-membered Si–O rings (three SiO_4 tetrahedra) connected such that every six rings enclose a 12-membered Si–O (six SiO_4 tetrahedra) ring.

FIGURE 10.4 Structure of crystalline SiO_2 tetrahedra found in quartz. (a) Structural formula and (b) ball and stick model of one unit cell.

Quartz is found in several forms in all three major kinds of rocks—igneous, metamorphic, and sedimentary. It is one of the hardest minerals known. Geologists often divide quartz into two main groupings—course crystalline and cryptocrystalline quartz. Course crystalline quarts include six-sided quartz crystals and massive granular clumps. Some colored varieties of coarse crystalline quartz crystals, amethyst and citrine, are cut into gemstones. Others include pink (rose), purple, and milky quartz, but most coarse crystalline quartz is colorless and transparent. Sandstone is a ready example of granular quartz. Color is a result of the presence of small amounts of metal cations such as calcium, iron, magnesium, and aluminum.

Cryptocrystalline forms contain microscopic quartz crystals and include the chalcedony grouping of rocks such as chert, agate, jastper, and flint.

Quartz exhibits an important property that allows the ***piezoelectric effect***. When pressure is applied to a slice of quartz, it develops a net positive charge on one side of the quartz slice and a negative charge on the other side. This phenomenon is the piezoelectric generation of a voltage difference across the two sides of the quartz crystal. Furthermore, the same effect is found when pressure is applied not mechanically, but through the application of an alternating electrical field with only certain frequencies allowed to pass through the crystal. The frequencies allowed to pass vary with the crystal shape and thickness. Such crystals are used in radios, televisions, and radars. This effect also forms the basis for quarts watches and clocks. Voltage applied to a quartz crystal causes the crystal to expand and contract at a set rate, producing vibrations. The vibrations are then translated into a uniform measure of time.

While quartz crystals are suitable for the production of optical lenses, most lenses are manufactured from synthetically produced quartz due to the scarcity of good-grade large quartz crystals.

The feldspars are the most abundant minerals in the Earth's crust, accounting for about 60% of all igneous rocks. They are derivatives of silica, where about one-half or one-quarter of the silicon atoms have been replaced by aluminum. Feldspar is used in the manufacture of certain types of glass and pottery. Some feldspar crystals—such as moonstone (white perthilte), Amazon stone (green microcline), and multicolored labradorite—are used as gemstones and in architectural decorations. Some feldspar is used as a coating and filler in the production of paper.

Granite is a hard crystalline rock chiefly composed of quartz and feldspar. It is used in building bridges and building where great strength is needed. It is also employed in the construction of monuments and gravestones, since it can be polished giving a lasting luster and because of its ability to withstand wear by the natural elements.

Sand is composed of loose grains of minerals or rocks, larger than silt but smaller than gravel. Soil contains mineral (often in the form of small sand granules) and organic matter.

Micas are also composed of silicon-oxygen tetrahedra. The anionic charge on the silicate sheet is the result of the replacement of silicon by aluminum. Cations, such as potassium, are interspaced between these negatively charged sheets. Some micas are used in construction and electrical engineering applications. Synthetic mica is manufactured on a large scale for industrial consumption in coatings, as fillers, etc. Micas are one of the many-layered silicon-oxygen-intense materials found in nature.

10.8   SILICON DIOXIDE IN ELECTRONIC CHIPS

Silicon dioxide plays a critical role in the electronics industry. The silicon used to produce silicon chips is derived from silicon dioxide. Semi-pure silicon dioxide (to about 99%) is prepared from the reaction of silicon dioxide with coke (a poor grade of graphite) using high temperature and an electronic arc.

$$SiO_2 + C \rightarrow Si + CO_2 \tag{10.17}$$

Even so, this level of purity falls far short of the purity needed to produce the chips used in computers. The purity required is about 99.9999996 or a level of impurity of about one part in a billion. This is achieved through multistep processes. One of these requires the silicon to be heated with HCl at high temperatures forming the desirable volatile trichlorosilane. The vapor is condensed, then purified using distillation and absorption columns. The trichlorosilane is reacted with hydrogen gas at about 1200°C depositing polycrystalline chip-grade silicon. The other product of this reaction is HCl, which can be again used to create more trichlorosilane, thus eliminating the production of unwanted by-products.

$$Si + HCl \rightarrow HSiCl_3 \tag{10.18}$$

$$HSiCl_3 + H_2 \rightarrow Si + HCl \tag{10.19}$$

Silicon dioxide is also used to insulate regions of the integrated circuit. Here silicon dioxide is grown on the silicon surface by heating the surface to about 1000°C in the presence of oxygen.

$$Si + O_2 \rightarrow SiO_2 \tag{10.20}$$

An alternate approach employs heating gaseous tetraethoxysilane to form layers of silicon dioxide.

$$Si(OC_2H_5)_4 \rightarrow SiO_2 + 2H_2O + 4C_2H_4 \tag{10.21}$$

10.9   ASBESTOS

Asbestos has been known and used for over 2000 years. Egyptians used asbestos cloth to prepare bodies for burial. The Romans called it *aminatus*, and used it as a cremation cloth and for lamp wicks. Marco Polo described its use in the preparation of fire-resistant textiles in the thirteenth century. Asbestos is not a single mineral but rather a grouping of materials that give soft, thread-like fibers. These materials will serve as an example of two-dimensional sheet polymers containing two-dimensional silicate $(Si_4O_{10})^{-4}$ anions bound on either or both sides by a layer of aluminum hydroxide ($Al(OH)_3$; gibbsite) or magnesium hydroxide ($Mg(OH)_2$; brucite). Aluminum and magnesium are present as positively charged ions. These cations can also contain varying amounts of water molecules associated with them as hydrates. The spacing between silicate layers varies with the nature of the cation and amount of its hydration.

Due to its fibrous nature and ability to resist elevated temperatures (compared with most organic-based fabrics) it is used to make fabrics for the production of fire-resistant fabric including laboratory glove-wear. Shorter fibers were used in electrical insulation, building insulation, and in automotive brake linings. Though asbestos has been known for thousands of years, it has only recently become known that asbestos can be dangerous. For instance, asbestos miners and manufacturing personnel who worked with it for 20 years or longer are 10 times more likely to contract asbestosis. Families of these workers and those living near the mines also have a greater than average chance of getting asbestosis. Asbestosis is a disease that blocks the lungs with thick fibrous tissue, causing shortness of breath, and swollen fingers and toes. Bronchogenic cancer (cancer of the bronchial tubes) is prevalent among asbestos workers who also smoke. Asbestos also causes mesothelioma, a fatal cancer of the lining of the abdomen or chest. These diseases may lay dormant for many years after exposure. It is believe that these diseases are caused by asbestos particles (whether asbestos or other sharp particles) about 5–20 μm in length corresponding to the approximate sizes of the mucous openings in the lungs. Thus, they become caught in the mucous openings. Because they are sharp, they cut the lining when people cough. Scar tissue and the repeated healing process cause scar tissue to build up and the opportunity for cancerous mutations to begin.

10.10 POLYMERIC CARBON—DIAMOND

Just as carbon serves as the basic building element for organic materials, so also does it form a building block in the world of inorganic materials. Elemental carbon exists in many different forms including the two longest known—diamond and graphite. Graphite is the more stable allotrope of carbon, with graphite readily formed by heating diamonds.

Natural diamonds (Figure 10.5) are believed to have been formed millions of years ago when concentrations of pure carbon were subjected by the Earth's mantle to great pressure and heat. They

FIGURE 10.5 Representation of diamond where each carbon is at the center of a tetrahedron composed of four other carbon atoms.

are the hardest known natural material. The majority of diamonds (nongem) are now man made. Most of the synthetic diamonds are no larger than a grain of common sand. The major use of synthetic diamonds is as industrial shaping and cutting agents to cut, grind, and bore (drill). By 1970, General Electric was manufacturing diamonds of gem quality and size by compressing pure carbon under extreme pressure and heat. It was found that addition of small amounts of boron to diamonds causes them to become semiconductors. Today, such doped diamonds are used to make transistors.

While diamonds can be cut, shaping is done by trained gem cutters striking the rough diamond on one of its cleavage plants. These cleavage plates are called faces, and represent sites of preferential cleavage and reflection of light. This balance between strength and flexibility, crystalline and amorphous regions is demonstrated to one extreme by diamonds that are very crystalline resulting in a strong, inflexible, and brittle material.

10.11   POLYMERIC CARBON—GRAPHITE

While diamond is the hardest naturally occurring material, the most common form of crystalline carbon is the much softer and flexible graphite. Graphite occurs as sheets of hexagonally fused benzene rings (Figure 10.6) or "hexachicken wire." The bonds holding the fused hexagons together are traditional covalent bonds. The bonds holding the sheets together are weaker than the bonding within the sheets consisting of a weak overlapping of pi-electron orbitals. Thus, graphite exhibits many properties that are dependent on the angle at which they are measured. They show some strength when measured along the sheet, but very little strength if the layers are allowed to slide past one another. This sliding allows the graphite its flexibility, much like the bending of bundles of proteins sliding past one another allowing our hair flexibility. The fused hexagons are situated such that the atoms in each layer lie opposite to the centers of the six-membered rings in the next layer. This arrangement further weakens the overlapping of the pi electrons between layers such that the magnitude of layer-to-layer attraction is on the order of ordinary secondary van der Waals forces. The "slipperiness" of the layers accounts for graphite's ability to be a good lubricant.

The variance of property with angle of applied force, light, magnetism, etc., is called anisotropic behavior. Calcite is anisotropic in its crystal structure, resulting in a dependency of its interaction with light with the angle of incidence of the light.

As with diamond, graphite's discovery and initial usage is lost in antiquity. It was long confused with other minerals such as molybdenite (MoS_2). At one time, it was known as plumbago (like lead),

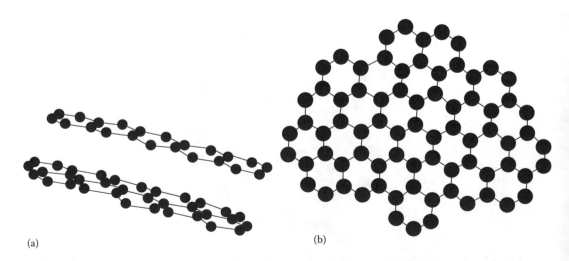

(a)   (b)

FIGURE 10.6   Representations of graphite emphasizing the layered (a) and sheet (b) nature of graphite.

crayon noir, silver lead, black lead, and *carbo mineralis*. Werner in 1789 first named it *graphit*, meaning (in Greek) "to write."

The Acheson process for graphite production begins by heating a mixture of charcoal, or coke, and sand. The silica is believed to be reduced to silicon that combines with carbon-forming silicon carbide, which subsequently dissociates into carbon and silicon. The silicon vaporizes and the carbon condenses forming graphite. Graphite is also produced using other techniques.

Today, graphite is mixed with clay to form the "lead" in pencils. Graphite conducts electricity and is not easily burned, so many industrial electrical contact points (electrodes) are made of graphite. Graphite is a good conductor of heat and is chemically inert, even at high temperatures. Thus, many crucibles for melting metals are graphite lined. Graphite has good stability to even strong acids, thus it is employed to coat acid tanks. It is also effective at slowing down neutrons, and thus composite bricks and rods (often called carbon rods) are used in some nuclear generators to regulate the progress of the nuclear reaction. Its slipperiness allows its use as a lubricant for clocks, door locks, and handheld tools. Graphite is also the major starting material for the synthesis of synthetic diamonds. Graphite is sometimes used as a component of industrial coatings. Dry cells and some types of alkali storage batteries also employ graphite. Graphite fibers are used for the reinforcement of certain composites.

10.12 INTERNAL CYCLIZATION—CARBON FIBERS AND RELATED MATERIALS

There are a number of important polymers that are formed through internal cyclization. In almost all cases, these are five- and six-membered rings with the vast majority being six membered. The tendency to form six-membered rings is related to a statistical feature. In studying the most probable distances from the beginning point using random statistics for units with a bond angle of about 109.5° (for a tetrahedral), the most probable distance for a chain of six units long is back at the starting point. The number of units required before the most probable distance is the starting point is dependent upon the bond angle and is called the *Kuhn element*. As note before, the Kuhn element for connected methylenes is six. The Kuhn element for sp^2 geometry is not six, but in this case, the driving force is the formation of six-membered rings with three alternating pi bonds, that is, the formation of the aromatic structure.

Often these internal cyclizations are incomplete giving products with mixed moieties. Even so, such internal cyclization is the source of a number of interesting and important polymers. A number of ladder-like structures have been synthesized from the internal cyclization of polymers. Following are several examples that illustrate this. The most important commercial products are those utilized to form the so-called carbon fibers.

Carbon fibers, and associated composite materials, are the result of internal cyclization. Polyacrylonitrile, when heated, undergoes internal addition forming a condensed polycyclic material called "black orlon." Further heating to about 1000°C removes the hydrogen atoms and most of the nitrogens giving a polyaromatic structure containing about 95% carbon (Figure 10.7). Further heating to about 2800°C gives a product with almost 99% carbon. These products can be forced through tiny holes to form tiny fiber-like materials, fibrals, that are combined to give fibrous-like materials that can finally be woven together to give fabrics. The fibrals, fibers, and fabrics act as the fibrous portion of many high strength composite materials. These fibers are lightweight, very strong, chemically inert, and can form semiconductor and conductor materials.

Diene polymers undergo cyclization in the presence of cationic initiators such as sulfuric acid. 1,2- and 3,4-diene polymers undergo this cyclization forming extensive fused ring groupings.

The polymerization of butadiene using certain catalytic systems such as butyl lithium and tetramethylenethylene diamine gives poly(1,2-polybutadiene). The polybutadiene, in turn, can undergo internal cyclization via cationic reactions forming a sort of linear saturated polycyclohexane. Further heating, resulting in dehydrogenation, with chloranil gives a fused ring product similar to that of carbon fibers when heated to about 1500°C (Figure 10.8). This material is sold under the trade name

FIGURE 10.7   Idealized structure of "black orlon."

FIGURE 10.8   Idealized structures for the synthesis of carbon-like materials from poly(1,2-butadiene).

of Pluton by 3 M. This sequence is described in Figure 10.8. While the final product is insoluble and infusible, prespinning is done on the soluble saturated intermediate.

10.13   CARBON NANOTUBES

There are several materials that form so-called nanotubes including boron–nitrogen compounds. Here we will focus on nanotubes derived from carbon. Carbon nanotubes, CNTs, have probably been made in small amounts since the first fires reduced trees and organic material to ashes. It has been found that certain ancient steel products may have possessed carbon nanotubes derived from the exposure of the processed steel to carbon sources. Thus, the Damascus steel used in making very strong weapons is believed to have profited from the presence of carbon nanotubes. It was not until recently, as part of the so-called nano revolution, that we first recognized the existence of these nanotubes. In 1952, Radushkevick and Lukyanovich published pictures of tubes of carbon. This discovery was largely unnoticed. Others contributed to the early history of carbon in the form of tubes. In 1991 Sumio Iijima, NEC Fundamental Research Laboratory in Japan, first observed carbon nanotubes as by-products of the arc-discharge synthesis of fullerenes, and it is often accepted that this observation prompted the current activities with carbon nanotubes.

Carbon nanotubes (CNTs) are carbon allotropes that have attracted much attention. They have a diameter of about 1/50,000 that of a human hair. Some have suggested that carbon nanotubes will be one of the most important twenty-first century materials because of the exceptional properties and ready abundance of the feedstock, carbon. CNTs are generally classified into two groups. Multi-walled carbon nanotubes, MWCNTs, are comprised of 2–30 concentric graphitic layers with diameters ranging from 10 to 50 nm with lengths that can exceed 10 μm. Single-walled carbon nanotubes, SWCNTs, have diameters ranging from 1.0 to 1.4 nm with lengths that can reach several micrometers.

An ideal CNT can be envisioned as a single sheet of fused hexagonal rings, that is graphite, that has been rolled up forming a seamless cylinder with each end "capped" with half of a fullerene molecule. Single-walled CNT can be thought of as the fundamental cylindrical structure, with multi-walled CNTs simply being concentric tubes. They can also be conceived of as being the fundamental building block of ordered arrays of single-walled nanotubes called ropes.

10.13.1 STRUCTURES

Carbon nanotubes are composed of carbon sp^2 bonded structures, similar to those of graphite. These bonds are stronger than the sp^3 bonds resulting in the strength of the carbon nanotubes. Unlike two dimension sheets of graphite, the carbon nanotubes align themselves into rope-like structures. Because of bonding similarity to graphite, these materials are often referred to as graphene.

Geometrically, carbon nanotubes can be described in terms of a two-dimensional graphene (graphite) sheet. A chiral vector is defined on the hexagonal lattice as

$$\mathbf{C_h} = n\mathbf{x} + m\mathbf{y} \tag{10.22}$$

where $\mathbf{x}$ and $\mathbf{y}$ are unit vectors, and n and m are integers, also tube indices. The chiral angle is measured relative to the direction defined by $n\mathbf{x}$.

When the graphene sheet is rolled up forming a nanotube, the two ends of the chiral vector meet one another. The chiral vector thus forms the circumference of the CNTs circular cross section. Different values of n and m give different nanotube structures with different diameters (Figure 10.9).

There are three general types of CNT structure (Figure 10.10). The zigzag nanotubes correspond to $(n, 0)$ or $(0, m)$ and have a chiral angle of 0°. The carbon–carbon position is parallel to the tube axis. Armchair nanotubes have (n, n) with a chiral angle of 30°. The carbon–carbon positions are perpendicular to the tube axis. Chiral nanotubes have general (n, m) values and a chiral angle of between 0° and 30°, and as the name implies, they are chiral.

In real life, nothing is perfect. As is the case with CNTs, the defects are mainly inclusion of wrong-membered rings. Pentagonal defects, i.e., the replacement of a hexagonal with a five-membered ring, result in a positive curvature causing the tube to curve inward like a horseshoe. The closure of an open cylindrical surface necessarily involves topological defects—often the formation of pentagons. Heptagonal defects result in a negative curvature with the lattice looking expanded around the defect.

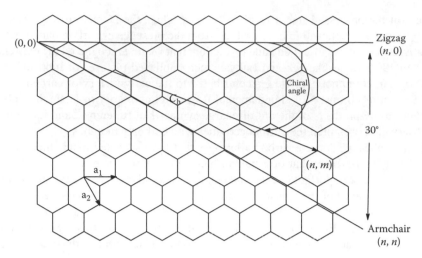

FIGURE 10.9 Graphite sheet representation showing the θ or chiral angle. For a chiral angle of 0°, the structure is a zigzag; for a chiral angle of 30° it is an armchair; and in between it is helical. The figure is drawn with a 30° angle between the zigzag and the armchair.

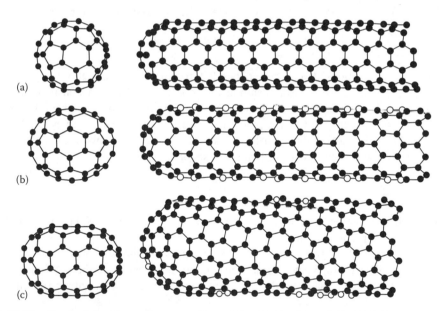

FIGURE 10.10 On the left side are representations of C_{60} (a), C_{70} (b), and C_{80} (c) fullerene structures and on the right representations of the three major structural forms of carbon nanotubes: armchair (a), zigzag (b), and helical (c).

The tendency to include pentagonal units can be seen by comparing the presence of pentagonal units in fullerene structures. The C_{60} structure contains twelve pentagons and twenty hexagons. The larger the fullerene, the smaller is the ratio of pentagons to hexagons. This is consistent with the use of pentagons to "cause" sharper bends and greater curvature in comparison to hexagons. Interestingly, fullerene C_{60} is one of the most strained molecules known but it exhibits good kinetic stability. It begins to decompose at about 750°C. There are a number of higher numbered (carbon number) fullerenes including C_{70}, C_{76}, C_{78} (two geometric isomers), C_{80}, ... Fullerenes can act as a source of the carbon nanotubes with the different-sized fullerenes producing different nanotubes. The three general structures of nanotubes can be produced using different fullerenes with C_{60} giving armchair nanotubes, C_{70} giving zigzag structures, and C_{80} giving helical forms of nanotubes. Figure 10.10 contains representations of these three fullerenes.

This difference in structure also influences the electrical conductivity with the armchair form being conductive or metal-like in its conductivity, and most of the other forms act as semiconductors.

One of the major reasons for the intense interest in CNT is their extreme and varied properties. Carbon nanotubes are among the strongest and stiffest materials known. Tensile strengths to about 65 GPa, have been found. This translates to a cable of 1 mm^2 cross section capable of holding about 3200 ton. A general density of carbon nanotubes is about 1.4 g/cc. Its specific strength is over 300 times as great as steel. Under stress, tubes can undergo permanent deformation, plastic deformation. Interestingly, because the tubes are hollow, they are not nearly as strong buckling under compression, torsion, and bending stress. Nanotubes conduct and transport along the lengthwise direction resulting in them often being referred to as being one dimensional.

Table 10.7 contains a comparison between single-walled CNTs and competitive materials/techniques. Another reason for the intense interest is that carbon nanotubes are produced from readily available inexpensive materials they are being considered for use as both bulk materials, such as in composites and clothing, and as components in computers, electrical devices, etc. Following is a discussion of some of the application areas for carbon nanotubes.

Electrical—Nanotubes can be metallic or semiconducting, depending on their diameters and helical arrangements. Armchair ($n = m$) tubes are metallic. For all other tubes (chiral and zigzag) when tube indices ($n + m$)/3 is a whole number integer, the tubes are metallic and otherwise they are semiconducting. Carbon nanotubes can, in principle, play the same role as silicon does in electronic circuits, but on a molecular scale where silicon and other standard semiconductors cease to work. Single carbon nanotube bundles have been used to construct elementary computing circuits known as logic gates. Metallic nanotubes can have electrical current densities on the order of one thousand times that of silver and copper.

Mechanical—CNTs have superior resilience and tensile strength. They can be bent and pressed over a large angle before they begin to ripple or buckle, finally developing kinks. Until the elastic

---

TABLE 10.7
Comparison between Selected Properties of Single-Walled CNTs and Competitive Materials/Techniques

Property	Single-Walled Carbon Nanotubes	Comparison
Size	0.6–1.8 nm in diameter	Electron beam lithography can create lines 50 nm wide and a few nm thick
Density	1.33–1.40 g/cc	Aluminum has a density of 2.7 g/cc, and titanium has a density of 4.5 g/cc
Tensile strength	ca. 45 billion pascal	High-strength steel alloys break at about 2 billion pascal
Resilience	Can be bent at large angles and re-straightened without damage	Metals and carbon fibers fracture at grain boundaries
Current carrying capacity	Estimated at 1 billion A/cc	Copper wires burn out at about 1 million A/cc
Heat transmission	Predicted to be as high as 6000 W/m K	Diamond transmits 3320 W/m K
Field emission	Can activate phosphors at 1–3 V if electrodes are spaced 1 micron apart	Molybdenum tips require fields of 50–100 V/μm and have limited lifetimes
Temperature stability	Stable up to 2800°C in vacuum, 750°C in air	Metal wires in microchips melt at 600°C–1000°C

limit is exceeded, the deformations are elastic with the deformation disappearing when the stress is removed. It is envisioned that buildings and bridges built from them may sway during an earthquake rather than fracturing and crumbling. Multi-walled nanotubes can easily slide past one another without friction, resulting in a rotating bearing used in rotational motors.

Field emission—When stood on end electrified carbon nanotubes act as a lightning rod concentrating the electrical field at their tips. While a lightning rod conducts an arc of electricity to a ground, a nanotube emits electrons from its tip at a rapid rate. Because the ends are so sharp, the nanotube emits electrons at lower voltages than do electrodes made from other materials, and their strength allows nanotubes to operate for longer periods without damage. Field emission is important in several industrial areas including lighting and displays. Commercial use of CNTs as field emitters has begun. Vacuum-tube lamps in six colors have been developed that are twice as bright as conventional light bulbs, longer lived, and at least 10 times more energy efficient as conventional light bulbs.

Hydrogen and ion storage—While we can picture CNTs as being composed of hexagonal carbon atoms with lots of empty space between the carbons, atoms "thrown" against them generally just bounce off. Even helium atoms at an energy up to 5 eV do not readily penetrate the nanotube. Thus, the graphene sheet and the CNTs are really membranes or fabrics that are one atom thick made of the strongest material that is also impenetrable (to a limit). Thus, CNTs can be used for hydrogen storage in their hollow centers with release being controlled, allowing them to act as inexpensive and effective fuel cells.

Chemical and genetic probes—Nanotube-tipped atomic force microscopes can trace a strand of DNA and identify chemical markers that reveal DNA fine structure. A miniaturized sensor has been constructed based on coupling the electronic properties of nanotubes with the specific recognition properties of immobilized biomolecules by attaching organic molecules-handles to these tubular nanostructures. In one study, the pi-electron network on the CNT is used to anchor a molecule that irreversibly adsorbs to the surface of the SWNT. The anchored molecules have a "tail" to which proteins, or a variety of other molecules, can be covalently attached. The result is that these molecules are immobilized on the sidewall of the nanotube with high specificity and efficiency. The molecule's tail is tipped with a succinimidyl ester group, which is readily displaced when an amine group attacks the ester function, forming an amide bond. Thus, the CNTs are used as both highly sensitive probes and highly selective immobilizing sites that allow specific reactions to occur.

Analytical tools—Single-walled CNTs are being used as the tips of scanning probe microscopes. Because of their strength, stability, and controllable and reproducible size, the tube probes allow better image fidelity and longer tip lifetimes.

Superconductors—Carbon nanotubes offer unique electronic properties due to quantum confinement. According to quantum confinement, electrons can only move along the nanotube axis. Metallic CNTs are found to be high temperature superconductors.

A major concern involves their possible toxicity. Research involving their potential and real toxic effects is just beginning. This is particularly important since these materials may become part of our everyday lives in many ways. Current results indicate that nanotubes can cross membrane barriers. If this is true, then access to our organs will occur. There is building biological evidence suggesting that under certain conditions nanotubes, of any origin, pose a real risk to our health.

Compounding the question of toxicity is the large variety of nanotubes potentially and really available each with its own toxicity profile. Further, the description of these various materials by name or other convention has yet to be settled.

10.14 BITUMENS

The petroleum industry, including the commercial bitumen industry, was born in the United States in August 27, 1859 when Colonel Drake drilled a hole about 70 ft deep near Titusville, Pennsylvania to "bring in" the first producing well. By 1908, Henry Ford began to mass-produce his Model "T" Ford creating an additional need for this petroleum in the form of gasoline. The distillation residue became more plentiful and a need for large-scale usage of bitumens increased.

Even so, the bitumens are a very old material. They were used in the waterproofing of the cradle that baby Moses was floated in. It was used by the ancient Egyptians in their mummification process. Bitumens were used in sand stabilization and for lighting the naval base by the Second Muslim Caliph, Omar ben Khattab, at Basra on Shattul-Arab on the West Coast of what is now Saudi Arabia around 640 AD.

Bitumens occur naturally or are formed as the residue in the distillation of coal tar, petroleum, etc. Industrially, the two most important bitumens are asphalt and coal tar. Asphalt is a brown to black tar-like variety of bitumen that again occurs naturally or is the residue of distillation. Coal tar is the black, thick liquid obtained as the residue from the distillation of bituminous coal.

Bitumens are examples of materials that have only an approximate structure. Bitumens are carbon-intense small polymers with molecular weights from about 200–1000 Da for coal tar, with a calculated average number of carbons of about 15–70 in a chain. Asphalt has a molecular weight averaging about 400–5000 Da, with a calculated average number of carbons in a chain of about 30–400. Thus, they are generally oligomers to short polymers. Asphalt has a C/H ratio of about 0.7, while coal tar has a C/H ratio of about 1.5 approaching that of a typical hydrocarbon where the C/H ratio is about 2.

As with most nonpolar hydrocarbon-intense polymers, bitumens exhibit good resistance to attack by inorganic salts and weak acids. They are dark, generally brown to black, with their color difficult to mask with pigments. They are thermoplastic materials with a narrow service temperature range unless modified with fibrous fillers and/or synthetic resins. They are abundant materials that are relatively inexpensive, thus they are used in many bulk applications.

At temperatures above the T_g, bitumens generally show Newtonian behavior. Below the T_g, bitumens have rheological properties similar to elastomers.

Bitumens are consumed at an annual rate in excess of 75 billion pounds in the United States. Bitumens are generally used in bulk such as in pavements (about 75%), and in coatings for roofs (15%), driveways, adhesive applications, construction, metal protection, etc. where the bitumen acts as a weather barrier. Bituminous coatings are generally applied either hot or cold. Hot-applied coatings are generally either filled or nonfilled. Cold-applied coatings are generally either nonwater or water containing. In the hot-applied coatings, the solid is obtained through a combination of cooling and liquid evaporation whereas in the cold-applied coatings, the solid material is arrived at through liquid evaporation. One often-used coating employs aluminum pigments compounded along with solvents. These coatings are heat reflective and decrease the energy needs of building using them. The aluminum-metallic appearance is generally more desirable than black, and the reflective nature of the aluminum reflects light that may damage the bitumen coating allowing the coating a longer useful life. Today, many of the bitumen coatings contain epoxy resins, various rubbers, and urethane polymers.

10.15 CARBON BLACK

Carbon black is another of the carbon intensive materials. It is formed from the burning of gaseous or liquid hydrocarbons under conditions where the amount of air is limited. Such burning favors "soot" formation, i.e., carbon black formation. It was produced by the Chinese over 1000 years ago. Today, it is produced in excess of 1.5 million ton annually in the United States. The most widely used carbon black is furnace carbon black. The particle size of this raw material is relatively large, about 0.08 mm. It is soft with a Mohs scale hardness of less than one.

FIGURE 10.11   Sheets of carbon black.

In addition to carbon, carbon black also contains varying amounts of oxygen, hydrogen, nitrogen, and sulfur. A typical carbon black contains about 98% carbon, 0.3%–0.8% hydrogen, 0.3%–1.2% oxygen, 0.0%–0.3% nitrogen, and 0.3%–1% sulfur. The impurities in the water employed to quench the burning carbon is mainly responsible for the noncarbon and hydrogen contents, and the sulfur comes mainly from the feedstock.

While some describe the structure of carbon black as being small chain-like structures, recent information shows it as being somewhat graphite-like in structure where the parallel planes are separated by about 0.35–0.38 nm, always greater than the inter-layer distance of 0.335 nm for graphite (Figure 10.11). The microstructure of carbon black has been studied extensively employing various spectroscopies including ATF, high-resolution transmission electron microscopy (TEM), and x-ray diffraction. Various models have been developed to describe the average structure in greater detail. These models are related to the structure of graphite.

As with other polymeric materials, the surface structure is different from the bulk and is related to the conditions under which the material was manufactured. Since the material is formed in air, the surface is rich in oxygen. These oxygen atoms play an important role in the resulting carbon black properties. The surface can be modified using a variety of treatments. Heat treatments above 800°C act to increase the amount of crystallinity in the overall structure. Under inert conditions, surface groups are modified and the amount of oxygen decreased with heating from 200°C to 1200°C. Plasma treatments are employed to modify the carbon black surface creating and destroying various functional groups in the presence of other reactants. Chemical oxidation is also employed to modify carbon black employing oxidizing agents such as air, nitric acid, and ozone. Surface grafting is also employed to provide desired surface functional groups.

The nature of the surface is especially important for carbon black since most applications employ the carbon black as an additive forming blends and alloys. These blends can be bound by simple physical contact or by chemical binding through the formation of chemical bonding between the various phases.

A major use of carbon black is in rubbers. Incorporation of carbon black into a rubber matrix can result in an increase in strength-related properties such as abrasion resistance, viscosity, and modulus. Thus, carbon black is a reinforcing agent rather than simply an inexpensive additive. The reinforcing effect is dependent on the particle size, amount, aggregate structure, surface area, surface activity, etc. The reinforcing effect mainly occurs because of the interaction between the interfacial surfaces of the rubber matrix and the carbon black. Carbon blacks can be divided roughly as to being reinforcing (<35 nm) or semi-reinforcing (>35 nm) depending on the particle size. The antiabrasion of carbon black increases with an increase in particle size.

Carbon black is the most important additive to rubber composing of between 30% and 70% of the bulk rubber product. Tire goods consume about 65% of the carbon black, mechanical goods another 25%, with only about 10% employed for nonrubber applications.

A typical tire rubber formulation for tire tread will contain various rubbers, mainly styrene-butadiene (50%) and *cis*-polybutadiene (12%), various processing aids (2%), softeners (3%), vulcanizing agent (mainly sulfur; 1%), accelerators, and reinforcing filler (namely carbon black; 30%) so that by bulk, carbon black is the second most used material.

Of the 10% of carbon black used for nonrubber applications, about 35% is used for plastics, 30% for printing inks, 10% for coating, and 5% for paper. In plastics, carbon black enhances a variety of properties including UV shielding, electrical conductance, pigment, opacity, and mechanical properties. Plastics that use carbon black filler may contain only several percent to having over half of the weight being carbon black.

In xerography, carbon black is the most important pigment in printing or duplicating toner applications. Here, blackness, good dispersion, and needed electrical properties are provided by carbon black. Similarly, these properties are useful for applications in coatings, inks, and printing. Here though, rheological properties in liquid media are also important.

10.16   POLYSULFUR

Sulfur is present in the petrochemicals derived from once-living matter because of its presence in certain amino acids. Because of its removal from industrial waste, sulfur has been stockpiled and is available at a low price in large amounts. While the stable form of sulfur at room temperature is cyclooctasulfur, S_8, linear polysulfur is formed on heating. Unfortunately, the thermodynamically stable form of sulfur is the cyclooctasulfur monomer, and the polymer undergoes depolymerization after awhile. Methods have been studied to inhibit this reversal process. Some have involved the addition of olefins, such as limonene, myrcene, and cyclopentadiene, to the ends to inhibit the depolymerization. Such stabilized polysulfur has been incorporated into concrete and asphalt mixes to strengthen them. Concrete blocks, posts, and parking tire restrainers containing polysulfur are now being produced.

10.17   CERAMICS

The term "ceramics" comes from the Greek word *keramos*, which means "potter's clay" or "burnt stuff." While traditional ceramics were often based on natural clays, today's ceramics are largely synthetic materials. Depending on which ceramic and which definition is to be applied, ceramics have been described as inorganic ionic materials and as inorganic covalent (polymeric) materials. In truth, many ceramics contain both covalent and ionic bonds, and thus can be considered "to be or not to be" (shades of Shakespeare) polymeric materials. Many of the new ceramics, such as the boron nitriles and the silicon carbides, are polymeric without containing any ionic bonds.

Ceramics are typically brittle, strong; resistant to chemicals such as acids, bases, salts, and reducing agents; and they are high melting. They are largely composed of carbon, oxygen, and nitrogen, and are made from silicates such as clay, feldspar, bauxite, silica. But now, ceramics contain other materials such as borides, carbides, silicides, and nitrides.

Ceramics are generally made by two processes—sintering and fusing. In sintering, the starting material is reduced to a powder or granular form by a series of crushing, powdering, ball-milling, etc. The ground preceramic material is then sized, separated according to particle size, using different-sized screens.

Ceramic material is generally shaped by pressing it into a form or through extruding, molding, jiggering, or slip-casting. Slip-casting uses a suspension of the preceramic material in water. The mixture must be dilute enough to allow it to be poured. Deflocculates are often added to assist in maintaining the suspension. The "slip" is poured into a plaster of Paris mold that absorbs water, leaving the finished shape. The preceramic material hardens next to the mold, and surplus "slip" material is poured off leaving a hollow item. At this point, the molded material is referred to as a "green body," which has little strength. Coffee pots and vases are formed using this technique.

In jiggering, machines press the preceramic material into a rotating mold of desired shape. Dinnerware products are often made using jiggering.

Abrasives and insulators are formed from simply pressing the preceramic material into a mold of desired shape. In extrusion, the preceramic material is forced through an opening in a "shaping" tool. Bricks and drainpipes are formed using extrusion.

After the product has dried, it is heated or fired in a furnace or kiln. Modern ceramics generally require certain heating schedules that include the rate and duration of heating, and under what conditions such as in the presence or absence of air. This is similar to procedures used to produce carbon fibers where the heating schedule is critical to the end product's properties.

In one approach, six-membered silicon-containing rings are pyrolyzed giving mixed carbosilane preceramic polymers through heating to 400°C, and subsequently forming silicon carbides or poly(carbosilanes) at 800°C (10.23).

$$
\begin{array}{c}
\text{Me} \quad \text{Me} \\
\backslash \quad / \\
\text{Si} \\
/ \quad \backslash \\
(\text{Me})_2\text{Si} \qquad \text{Si}(\text{Me})_2 \\
| \qquad\qquad | \\
(\text{Me})_2\text{Si} \qquad \text{Si}(\text{Me})_2 \\
\backslash \quad / \\
\text{Si} \\
/ \quad \backslash \\
\text{Me} \quad \text{Me}
\end{array}
\rightarrow
-(-\underset{\underset{\text{H}}{|}}{\overset{\overset{\text{Me}}{|}}{\text{Si}}}-\text{HCH}-)-\ \text{and}\ -(-\underset{\underset{\text{Me}}{|}}{\overset{\overset{\text{Me}}{|}}{\text{Si}}}-\text{HCH}-)-\rightarrow \text{SiC}
\qquad (10.23)
$$

SiC fibers can be formed using dimethyl dichlorosilicon and diphenyl dichlorosilicon heated together (10.24).

$$
\text{Cl}-\underset{\underset{\text{Me}}{|}}{\overset{\overset{\text{Me}}{|}}{\text{Si}}}-\text{Cl}+\text{Cl}-\underset{\underset{\text{Ph}}{|}}{\overset{\overset{\text{Ph}}{|}}{\text{Si}}}-\text{Cl}\rightarrow-(-\underset{\underset{\text{Me}}{|}}{\overset{\overset{\text{Me}}{|}}{\text{Si}}}-)-(-\underset{\underset{\text{Ph}}{|}}{\overset{\overset{\text{Ph}}{|}}{\text{Si}}}-)-\rightarrow \text{SiC}
\qquad (10.24)
$$

Such "ceramic" fibers offer uniquely strong and resistant inexpensive materials including new ceramic composites that have great fracture toughness.

There are a number of other "nonoxygen" or nonoxide ceramics including phosphonitric chlorides (PN backbone), boron nitriles (BN), aluminum nitriles (AlN), titanocarbosilanes (Si–Ti–C backbone), and silazanes (Si–C–N backbones).

Many ceramic products are coated with a glassy coating called a glaze. The glaze increases the resistance of the material to gas and solvent permeability and makes the surface smoother in art objects used for decoration.

One group of advanced material ceramics are the zirconia ceramics. Most of these are based on zirconia (zirconium (IV) oxide) that contains small amounts of magnesia (magnesium oxide). They have bending strengths two to three times that of corundum and Alumina; high fracture toughness (about five times that of corundum ceramics); high resistance to wear and corrosion; and they have high density (5.8 g/cc). They are used in the construction of shear blades in the textile industry, plungers in the food and drink industry, valves in the petroleum industry, and as milling balls in the materials industry.

Zirconia exists in three solid phases (10.25).

$$\text{Monoclinic} \leftarrow 1200°C \rightarrow \text{Tetragonal} \leftarrow 2400°C \rightarrow \text{Cubic} \leftarrow 2700°C \rightarrow \text{Liquid} \quad (10.25)$$

The transformations between the monoclinic and tetragonal phases involve large and abrupt volume changes introducing fractures in the material. This is minimized through the use of di- and tri-valent oxides of the cubic symmetry such as calcium oxide and magnesium oxide. This results in a lowering of the transition temperatures (M $\rightarrow$ T and T $\rightarrow$ C), and also lowers the expansion coefficient of the zirconia and the subsequent volume changes associated with the phase changes reducing ceramic fracturing. The addition of calcium oxide or magnesium oxide is said to "stabilize" the ceramic.

Strength, brittleness, and solvent permeability properties are limited because of lack of control of the ceramic composition on a macro- and micro-level. Even small particle sizes are large compared with the molecular level. There have been a number of attempts to produce uniform ceramic powders including the sol–gel synthesis where processing involves a stable liquid medium, co-precipitation where two or more ions are precipitated simultaneously. More recently, Carraher and Xu have used the thermal degradation of metal-containing polymers to deposit metal atoms and oxides on a molecular level.

10.18   HIGH-TEMPERATURE SUPERCONDUCTORS

10.18.1   Discovery of the 123-Compound

In early 1986, George Bedorz and K. Alex Muller reported a startling discovery—a ceramic material, La-Ba-Cu-O, lost its resistance to electrical current at about 30 K. This was the first report of a so-called high T_c superconductor. Intensive efforts were then concentrated on substituting the component ions with similar elements on both the La and Ba sites. The first success was reported by Kishio et al. with an $(La_{1-x}Sr_x)_2CuO_4$ system that exhibited a higher T_c to about 37 K. Then the substitution on the La sites led Wu and coworkers to find another superconductor, the Y-Ba-Cu-O system with a T_c of 93 K in February 1987. This finally broke the technological barrier that would allow superconductivity at temperatures above liquid nitrogen. The superconducting phase was identified as $Y_1Ba_2Cu_3O_7$ (commonly referred to as the 123-compound). The 123-compound was the first of the 90 K plus superconductors to be discovered, and it has been the most thoroughly studied.

10.18.2   Structure of the 123-Compound

The structure of the 123-compound is related to that of an important class of minerals called perovskites. These minerals contain three oxygen atoms for every two metal atoms. It has six metal atoms in its unit cell, and it would be expected to have nine oxygens if it were an ideal perovskite. In fact, it has, in most samples, between 6.5 and 7 oxygens. In other words, by comparison to an ideal perovskite, about one-quarter of the oxygens are missing. The unit cell can be thought of as a pile of three cubes. Each cube has a metal atom at its center: Barium in the bottom cube, yttrium in the middle one, and barium in the top one. At the corners of each cube, a copper would be surrounded by six oxygens in an octahedral arrangement linked to each oxygen atom in an ideal perovskite. Each barium and yttrium would then be surrounded by 12 oxygen atoms. But, x-ray and neutron diffraction studies have shown that the unit cell does not conform to this simple picture because certain oxygen positions are vacate. All oxygen positions in the horizontal plane containing yttrium are vacant. The other vacancies are located in the top and bottom Cu–O planes.

The two copper oxide layers can be considered as polymeric since the covalent character is in the same range as for the carbon fluoride bond in Teflon. Thus, the 123-superconductors consist of two types of polymeric copper oxide layers held together by ionic bonding metals such as barium and yttrium. This theme of polymeric layers held together by ionic bonding to metals is common in the silicates and other minerals.

10.19 ZEOLITES

At least three major themes are helping to drive polymer synthesis and use of polymers today. These involve synthesis and assembling on an individual scale (nano-level); synthesis in confined spaces (selected inorganic zeolites and biological syntheses); and single-site catalysis (both selected biological and synthetic polymer synthesis). Superimposed on this is the applications aspects including the human genome/biomedical, electronic/communications, etc.

Zeolites include a whole group of aluminosilicates with an approximate formula of $SiAlO_4$. With respect to the type of bonding, zeolites can be divided into three groups. The natrolite group (mesolite, thomsonite, edingtonite, natrolite) consists of structures built up from the rings of four $SiAlO_4$ tetrahedral groups linked together into chains with fewer linkages between the chains so that a cleavage along the chain direction is preferred. These materials generally have a fibrous character. In the heulandite group (stilbite, epistilbite, and heulandite), the $SiAlO_4$ tetrahedra form sheets of six-membered rings with few linkages between the sheets. These materials are mica-like in behavior. The third group, the so-called framework zeolites, has the density of bonding similar in all three directions. This group includes most of the zeolites mentioned below.

Framework zeolites can be described as aluminosilicates composed of tetrahedra linked by the sharing of oxygen atoms into rings and cages that can accommodate water molecules, metallic ions, as well as selected organic molecules. While there are a variety of structures, the framework zeolites can be briefly described as having an open arrangement of corner-sharing tetrahedra where the SiO_4 units are partially replaced by AlO_4 units in which there are enough cations present to be neutral. There are well over 100 different synthetic and natural framework zeolites known today with more being found. Magic-angle NMR indicates that there are five distinct zeolite groups where $n = 0$–4 for $Ai(AiO)_n(SiO)_{4-n}$. The open structures give materials with lower densities (on the order of 2.0–2.2 g/mL), as expected, than similar materials with closed structures such as feldspar (density about 2.6–2.7 g/mL).

Molecular sieves was the name first given to framework zeolites dehydrated by heating in vacuum to about 350°C because of their ability to capture and remove water and certain other species. Today, other materials, such as microporous silicas and aluminum phosphate, are also employed as molecular sieves.

Syntheses that occur within confined spaces typically give products with a specificity that is not available by other modes. The specificity may be general such as in the case of the synthesis of lignin in plants where the synthesis between plant layers produces a largely two-dimensional material with only an average structure. It may be highly selective to not totally selective as in the case of zeolites and hollow nanofibers. Or it may be essentially totally selective as in the case of many of the biologically important proteins and nucleic acids where both spacial and electronic interactions act to give a highly "preordained" structure.

Zeolites and related ordered clay-associated materials have been suggested to be involved in the initial primeval synthesis of basic elements of life. They are also being involved in the synthesis of a number of polymeric and nonpolymeric materials. Zeolites come in a variety of forms with differing shapes and sizes with researchers associating the particular size and shape with a particular desired synthesis. This is somewhat akin to the considerations that are made in effectively employing crown phase transfer agents and related materials.

Zeolites have been employed in the preferential synthesis of optically active sites and in determining the particular products formed from certain reactions. In looking at the products formed from the decomposition and reformations involving the unsymmetrical ketone, the major products are a combination of products listed below:

$$Ph_3C-\underset{\underset{A}{}}{\overset{\overset{O}{\|}}{C}}-CMe_3 \rightarrow Ph_3C-\underset{\underset{B}{}}{\overset{\overset{O}{\|}}{C}}-CPh_3 + Ph_3C-CPh_3 + Me_3C-\underset{\underset{C}{}}{\overset{\overset{O}{\|}}{C}}-C-Me_3 + Me_3C-CMe_3 \qquad (10.26)$$

Employing a single zeolite that selectively accommodates the Me_3C radical results in a preferential formation of D and little C because of preferential diffusion of the Me_3C radical as it is formed with the other radicals being "washed" past. The use of two zeolites in conjunction, one preferentially accepting the Me_3C radical and the second accepting the Ph_3C radical, results in the preferential formation of B and D.

An additional consideration involves matching the so-called hardness or flexibility of the confined space. Zeolites offer hard or inflexible confined spaces whereas liquid crystals and some other polymer media offer softer confined spaces. While the zeolites offer a "safe haven" for selected species enabling the hosted species a relatively long existence, they do not allow for ready movement and mixing. By comparison, a more flexible container such as especially designed polyethylenes allow a more flexible environment that allows for variations in the product sizes and shapes and some assembling to occur within the more flexible confines. Thus, the reaction of 1-naphthyl esters was carried out in an appropriate zeolite and appropriate polyethylene. In solution, eight products are formed. In the zeolite only one product, the result of a specific geminate-pair recombination is formed. In the polyethylene, several products are formed, primarily the isomeric products of geminate-pair recombinations.

10.20   SUMMARY

1. Inorganic polymers are widely employed in the construction and building businesses, as abrasives and cutting materials, as fibers in composites, as coatings and lubricants, and as catalysts. They also serve as the basis of rocks and soils.
2. Portland cement is the least expensive, most widely used synthetic polymer. It has a complex (short-range order and long-range disorder; average structure), three-dimensional structure.
3. There are many specialty cements including reinforced concrete, lightweight concrete, prestressed concrete, gypsum, and heavy glass.
4. Silicates are among the most widely found materials on the face of the earth. They form the basis for much of the soil, sand, and rocks with most of this being of the crystalline variety of silicates. As amorphous materials, they are found as the fiber material in fiberglass, as window glass, and a whole host of specialty glasses including safety glass, borosilicate glasses, lead (or heavy) glasses, colored glasses, and glazes. Glass can be shaped by drawing, pressing, casting, and blowing. Most of these glasses have the approximate structure of SiO_2 or silicon dioxide. Many of the rocks are present in crystalline silicate forms as quartz and feldspar. Today the sol–gel technique is important in making many materials including aerogels, among the least dense solids known today. In nature, silicates are found as polymeric sheets such as asbestos, three-dimensional materials such as the zeolites, and linear materials.
5. There are many widely used carbon polymers. These include the hardest known material diamond; graphite, a sheet material with little strength holding together the sheets; carbon fibers used in high strength composites; carbon nanotubes, one of the most important

new materials for the twenty-first century; bitumens used in asphalt; and carbon black widely used as a filler material. Diamonds structurally have a carbon at the center of a tetrahedron composed of four other carbon atoms. They can be industrially synthesized, and used for cutting and shaping. Graphite occurs as sheets of hexagonally fused benzene rings. The bonds holding the fused benzene rings together are covalent bonds, while the bonding between the sheets results from the weaker overlapping of pi-electron orbitals. Thus, many of the properties of graphite are anisotropic. The weak forces holding the sheets together are responsible for its "slipperiness." Graphite is commercially made from charcoal or coke.

6. Many ceramics are partially polymeric in structure. These include the new superconductive materials that exist as polymeric sheets connected by metal ions similar to many of the silicate sheets.

GLOSSARY

Aluminum oxide: Complex aluminum-containing silicate.

Annealing: Subjecting a material to near its melting point.

Asbestos: Group of silica-intensive materials containing aluminum and magnesium that gives soft, threadlike fibers.

Asbestosis: Disease that blocks the lungs with thick fibrous tissue.

Borosilicate glass: Relatively heat-shock-resistant glass with a small coefficient of thermal expansion, such as Kimax and Pyrex.

Calcium–aluminate cement: Contains more alumina than Portland cement.

Chrysotile: Most abundant type of asbestos.

Concrete: Combination of cement, water, and filler material such as rock and sand.

Diamonds: Polymeric carbon where the carbon atoms are at the centers of tetrahedra composed of four other carbon atoms; hardest known natural material.

Feldspars: Derivatives of silica where one-half to one-quarter of the silicon atoms are replaced by aluminum atoms.

Fibrous glass (fiberglass): Fibers of drawn glass.

Float glass: Glass made by cooling sheets of molten glass in a tank of molten time: most common window glass is of this type.

Glass: Inorganic product of fusion that has been cooled to a rigid condition without crystallization; most glasses are amorphous silicon dioxide.

Glass sand: Impure quartz crystals.

Glaze: Thin, transparent coatings fused on ceramic materials.

Granite: Hard crystalline rock containing mainly quartz and feldspar.

Graphite: Polymeric carbon consisting of sheets of hexagonally fused rings where the sheets are held together by weak overlapping pi-electron orbitals; anisotropic in behavior.

Gypsum: Serves as the basis of plaster of Paris, Martin's cement, etc.; shrinks very little on hardening and rapid drying.

High-temperature superconductors: Polymeric copper oxide layers containing metal atoms that hold them together, which are superconductors above the boiling point of liquid nitrogen.

Inorganic polymers: Polymers containing no organic moieties.

Kaolinite: Important type of asbestos.

Lead glass (heavy glass): Glass where some or all of the calcium oxide is replaced by lead oxide.

Lime: calcium carbonate from oyster shells, chalk, and marl.

Magnesia cement: Composed mainly of magnesium oxide; rapid hardening.

Optical fibers: Glass fibers coated with highly reflective polymeric coatings; allows light entering one end to pass through the fiber to the other end with little loss of energy.

Piezoelectric effect: Materials that develop net electronic charges when pressure is applied; sliced quart is piezoelectric.

Portland cement: Major three-dimensional inorganic construction polymer containing calcium silicates, lime, and alumina.

Precast concrete: Portland concrete cast and hardened prior to being taken to the site of use.

Prestressed concrete: Portland concrete cast about steel cables stretched by jacks.

Quartz: Crystalline forms of silicon dioxide: basic material of many sands, soils, and rocks.

Reinforced concrete: Portland concrete cast about steel rods or bars.

Safety glass: Laminated glass; sandwich containing alternate layers of soda-lime glass and poly(vinyl butyral).

Sand: Loose grains of minerals or rocks larger than silt but smaller than gravel.

Sandstone: Granular quartz.

Silica: Based on SiO_2; finely ground sand.

Silicon glass: Made by fusing pure quartz crystals or glass sand; high melting.

Soda: Na_2O.

Soda ash: Na_2CO_3.

Soda-lime glass: Most common glass; based on silica, soda, and lime.

Soil: Contains mineral and organic particles; majority by weight is sand.

Tempered safety glass: Single piece of specially treated glass.

Tempering: Process of rapidly cooling glass resulting in an amorphous glass that is weaker but less brittle.

Vitreous enamels: Thin, normally somewhat opaque-colored inorganic coatings fused on materials.

Vycor: 96% silicon glass; made form silicon and boron oxides; best variety is called fused quartz.

EXERCISES

1. What properties of glass correspond to those of organic polymers?
2. Why is Portland cement an attractive large-bulk use building material?
3. Name five important synthetic inorganic polymers. Name five important natural inorganic polymers.
4. Describe what is meant by three-dimensional polymers. Name five important three-dimensional polymers. Name two general properties typical of three-dimensional polymers.
5. Why are specialty cements and concretes necessary.
6. What is intended by the comment that "glass is a supercooled liquid"?
7. Briefly describe the techniques employed to shape glass.
8. Why are specialty glasses important in today's society?
9. Name two important inorganic fibers employed with resins to form useful materials.
10. Which would you predict to be more brittle—quartz, fibrous glass, or window glass?
11. Briefly describe the piezoelectric effect.
12. We do not live in a risk-free society. Discuss this statement in terms of asbestos.
13. What does anisotropic behavior mean? Why does graphite exhibit anisotropic behavior?
14. Briefly compare the structures of diamond and graphite.

FURTHER READINGS

Andrady, A. (2008): *Science and Technology of Polymer Nanofibers*, Wiley, Hoboken, NJ.

Bruce, D. and O'Hare, D., (1997): *Inorganic Materials*, 2nd ed., Wiley, New York.

Bunsell, A. and Berger, M.-H. (1999): *Fine Ceramic Fibers*, Dekker, New York.

Bye, G. (1999): *Portland Cement*, Telford, London, U.K.

Fennessey, S. (2008): *Continuous Carbon Nanofibers*, VDM Verlag, New York.

Jaeger, R. and Gleria, M. (2009): *Silicon-Based Inorganic Polymers*, Nova, Hauppauge, New York.

Mallada, R. and Menendez, M. (2008): *Inorganic Membranes: Synthesis, Characterization and Applications*, Elsevier, New York.

Mark, J. (2005), *Inorganic Polymers*, Oxford University Press, Cary, NC.

Ohama, T., Kawakami, M., and Fukuzawa, K. (1997): *Polymers in Concrete*, Routledge, New York.

Schubert, U. and Husing, N. (2000): *Inorganic Materials*, Wiley, New York.

Sirok, B., Blagoievic, B., and Bullen, P. (2008): *Mineral Wool: Production and Properties*, CRC, Boca Raton, FL.

Weller, M. (1995): *Inorganic Materials Chemistry*, Oxford University Press, New York.

Wesche, R. (1999): *High-Temperature Superconductors*, Kluwer, New York.

11 Reactions of Polymers

Reactions of polymers in nature and industry is widespread. In nature, reactions of polymers serve as the basis for information transfer, synthesis of needed biomaterials, degradation of biomaterials, and in fact, they are at the core of life itself. Nature also reacts with synthetic polymers as they age, degrade, cross-link, etc. Synthetic polymers also serve as the basic material for the production of many important fibers, elastomers, and plastics. The nature of these reactions is governed similarly irrespective of whether the source and site of polymer is natural or man-made. There are some differences especially with respect to the precision of the interaction and the predicted outcome. Some polymer interactions, such as those that transfer information, must occur with a very high degree of precision for each and every incident. By comparison, naturally induced degradation of natural and synthetic polymers through weathering occurs via general steps that can be described in some statistical manner.

Reactions of synthetic polymers often mimic similar reactions involving small molecules where size is the main difference. There are some exceptions, for example, nearby groups may hinder or assist in the reaction when size differences occur. Here, the main differences are often kinetic, though some geometrical differences are also found in specific cases. A reaction where the rate of reaction in polymers is enhanced by the presence of neighboring groups is called anchimeric assistance.

While there are many possible routes, most degradation can be described as occurring through two general routes—unzipping and random scission. Polymers, such as polysiloxanes and polysulfur, undergo unzipping reactions, forming preferred internal cyclic products unless the end group is capped in such a manner so as to discourage unzipping. In unzipping, one end begins to "unzip" and the process continues down the chain until it is completed. Random scission is the normal degradation pathway for most natural [such as polycarbonates (PCs)] and synthetic polymers. Here, a long chain is attacked at some site, normally one that is exposed, and of the exposed sites, one that is stressed. Thus, while random scission implies a random statistical manner of chain breakage, superimposed on this are more complex considerations of exposure and stress and the likelihood that a particular site is susceptible to that particular type of bond breakage.

This chapter describes many of the important reactions of polymers. Synthesis and curing (cross-linking) of polymers and telomerization are chemical reactions of polymers that have been discussed in previous chapters.

11.1 REACTIONS WITH POLYOLEFINES AND POLYENES

As with other carbon, hydrogen, and oxygen-containing materials, the main products of combustion in the presence of oxygen are water and carbon dioxide. Such polymers can be treated with various reactants, and the resulting products are analogous to those obtained from small alkanes and alkane-intense compounds. The moderating conditions between reactions of small molecules and polymers involve contact between the polymer segments and the reactants.

Polyolefines, like simple alkanes, can be chlorinated producing hydrogen chloride and other chlorinated products, such as Tyrin, which is used as plasticizers and flame retardants, and poly(vinyl dichloride), which has better heat resistance than poly(vinyl chloride) (PVC) and is used for hot-water piping.

Reactions with polyenes are similar to the reactions of alkenes. Thus, Hermann Staudinger found that polyenes such as *Hevea brasiliensis* could be hydrogenated, halogenated, hydrohalogenated, and cyclized. This classic work was done in the early 1900s. In fact, Berthelot hydrogenated *H. Brasiliensis* in 1869. Chlorinated rubber (Tornesit and Parlon) is produced by the chlorination of rubber resulting in products with varying amounts of chlorine and is used for the coating of concrete. The hydrohalogenation of *H. Brasiliensis* gives a product (such as Pliofilm), which is a packaging film.

The product obtained by the partial hydrogenation of polybutadiene (Hydropol) has been used as a wire coating, and a saturated ABA copolymer (Kraton) is produced by the hydrogenation of the ABA block copolymer of styrene and butadiene.

11.2   REACTIONS OF AROMATIC AND ALIPHATIC PENDANT GROUPS

Polymers with aromatic pendant groups, such as polystyrene (PS), undergo all of the characteristic reactions of benzene including alkylation, halogenation, nitration, and sulfonation. Polystyrene has been sulfonated by fuming sulfuric acid with cross-linked products used as ion exchange resins. Living polymers (cation propagation) are used to form a variety of products with hydroxyl and dihydroxyl products forming telechelic polymers and macromers. Polyaminostyrene is formed from the nitration of PS. These polyaminostyrene products can be diazotized giving polymeric dyes.

Esters, such as poly(vinyl acetate) (PVAc), may be hydrolyzed producing alcohols such as poly(vinyl alcohol) (PVA; Equation 11.1), which will have the same $\overline{DP}$ as the ester. In truth, PVAc does not totally hydrolyze, but with reasonable effort the extent of hydrolysis is greater than 90%. Since PVAc is not water soluble, but PVA is, the extent of water solubility is dependent on the extent of hydrolysis.

$$\text{(11.1)}$$

Esters of poly(carboxylic acids), nitrites, and amides may be hydrolyzed to produce poly(carboxylic acids). Thus, polyacrylonitrile, polyacrylamide, or poly(methyl acrylate) may be hydrolyzed producing poly(acrylic acid) (Equation 11.2)

$$\text{(11.2)}$$

Poly(acrylic acid) and partially hydrolyzed polyacrylamide are used for the prevention of scale in water used for boilers and as flocculating agents in water purification.

Neutralization of ionic polymers, such as poly(acrylic acid), causes the now fully negatively charged carboxylate groups to repel one another, resulting in the chain changing conformation from that of a free-draining random shape to that approaching a rigid rod. This increase in length results in an increase in polymer viscosity with the extent of viscosity increasing related to the extent of ionic groups present. Addition of salts, such as sodium chloride, allows the negatively charged chains to return to a more random conformation and subsequent decrease in viscosity.

Some of the largest polymers, with molecular weights of the order of 10–15 million dalton, contain sodium acrylate and acrylamide groups. Simple stirring can result in chain breakage for these extremely high molecular weight chains. Copolymers containing sodium acrylate and acrylamide groups are used in tertiary oil recovery and in water purification.

11.3 DEGRADATION

Here, the term degradation includes any change, decrease, in polymer property because of the impact of environmental factors, namely, light, heat, mechanical, and chemicals. Seven polymers represent the majority of the synthetic polymers. These are the various polyethylenes (PE), polypropylene (PP), nylons, poly(ethylene terephthalate) (PET), PS, PVC, and PC. Each of these has its own particular mode of degradation. Even so, there are some common generalities for the condensation (PET, PC, nylons) and vinyl (PE, PP, PS, PVC) polymers.

Some reactions on polymers are intended and give a material with different desired properties. The (positive) modification of polymers is an area of vigorous activity. Other reactions on polymers are unintended and generally result in a material with unfavorable properties. Included in the latter are a host of polymer degradation reactions. Some of these degradation reactions are covered elsewhere. Here we will focus on some general concepts. Degradation can be promoted by many means and any combination of means. The major means of polymer degradation are given in Table 11.1.

Backbone chain scission degradation can be divided as occurring via depolymerization, random chain breakage, weak-link or preferential site degradation, or some combination of these general routes. In depolymerization, a monomer is split off from an activated end group. This is the opposite of the addition polymerization, and it is often referred to as "unzipping."

$$P—MMMMMMMM \xrightarrow{-M} P—MMMMMMM \xrightarrow{-M} P—MMMMMMM \xrightarrow{-M} P—MMMMM \xrightarrow{-M} \text{Etc.}$$
(11.3)

Chain scission is similar to the opposite of stepwise polycondensation where units are split apart in a random manner.

$$P_m - P_n \rightarrow P_m + P_n$$
(11.4)

Depolymerization can result in backbone degradation and/or in the formation of cyclic (Section 10.12) or other products. The thermal degradation of PVA and PVC occurs with the splitting-out

TABLE 11.1
Major Synthetic Polymer Degradative Agents

Degradation Agent	(Most) Susceptible Polymer Types	Examples
Acids and bases	Heterochain polymers	Polyesters, polyurethanes
Moisture	Heterochain polymers	Polyesters, nylons, polyurethanes
High energy radiation	Aliphatic polymers with quaternary carbons	Polypropylene, LDPE, PMMA, poly(α-methylstyrene)
Ozone	Unsaturated polymers	Polybutadienes, polyisoprene
Organic liquids/vapors	Amorphous polymers	
Biodegradation	Heterochain polymers	Polyesters, nylons, Polyurethanes
Heat	Vinyl polymers	PVC, POLY(α-methylstyrene)
Mechanical (applied stresses)	Polymers below T_g	

of water or HCl followed by a combination of further chain degradation to finally give small products and formation of complex cyclic products. The elimination of HCl further accelerates additional HCl elimination and increased property loss. PVC degradation is decreased by the addition of agents that impede degradation such as those that neutralize HCl, trap free radicals, and/or that react with the double-bond formation to impede further depolymerization. Commercial PVC often contains organotin or antimony mercaptide compounds that act as stabilizers.

In general, for vinyl polymers, thermal degradation in air (combustion) produces the expected products of water, carbon dioxide, and char along with numerous hydrocarbon products. Thermally, simple combustion of polymeric materials gives a complex of compounds that varies according to the particular reaction conditions. Application of heat under controlled conditions can result in true depolymerization, generally occurring via an unzipping. Such depolymerization may be related to the ceiling temperature of the particular polymer. Polymers such as PMMA and poly(α-methylstyrene) depolymerize to give large amounts of monomer when heated under appropriate conditions. Thermal depolymerization generally results in some char and formation of smaller molecules including water, methanol, and carbon dioxide.

Most polymers are susceptible to degradation under natural radiation, sunlight, and high temperatures even in the presence of antioxidants. Thus, LDPE sheets, impregnated with carbon black, become brittle after exposure to one year's elements in South Florida. HDPE, while more costly, does stand up better to these elements, but again after several seasons, the elements win, and the HDPE sheets become brittle and break. Long-term degradation is often indicated in clear polymers by a yellowing and a decrease in mechanical properties.

Most polymers are subject to oxidative degradation, particularly in the presence of other "enticers" such as heat, a good supply of air, various catalysts, high-energy radiation including UV and higher energy visible light, and mechanical stressing that not only exposes additional polymer to the "elements" but also brings about the actual breakage of bonds subsequently leading to additional breakdown.

While polymers that contain sites of unsaturation, such as polyisoprene and the polybutadienes, are most susceptible to oxygen and ozone oxidation, most other polymers also show some susceptibility to such degradation including natural rubber, PS, PP, nylons, PE, and most natural and naturally derived polymers.

Because of the prevalence of degradation by oxidation, antioxidants are generally added. These antioxidants are generally compounds that readily react with free radicals or those that may act to lessen the effects of "enticers" such as UV radiation.

Mechanical degradation, while applied on a macro level, can result in not only chain rearrangement but also chain degradation. Such forces may be repetitive or abrupt and may act on the polymer while it is in solution, melt, elastic, or below its T_g. Passage of polymer melts through a tiny orifice for fabrication purposes can result in both chain alignment and chain breakage. In the case of rubber, mastication of the elastomer, breaking of the polymer chains, is intentional, allowing easier deformation and processability. While shearing itself can result in chain breakage, chain breakage is often associated with localized heat buildup, which is a consequence of chains "rubbing" together, etc. (molecular friction).

Most heterochained polymers, including condensation polymers, are susceptible to aqueous associated acid or base degradation. This mode of degradation is referred to as hydrolysis. This susceptibility is due to a combination of the chemical reactivity of heteroatom sites and to the materials being at least wetted by the aqueous solution allowing contact between the proton and the hydroxide ion to occur. Both of these factors are related to the difference in the electronegativity of the two different atoms, resulting in the formation of a dipole that acts as a site for nucleophilic/electrophilic chemical attack and that allows polar materials to come into contact with it. Such polymers can be partially protected by application of a thin film of hydrocarbon polymer that acts to repeal the aqueous solutions.

Enzymatic degradation is complex and not totally agreed upon. Microbes have enzymes, some of which are capable of breaking selected bonds, such as those that appear naturally—including amide, ester, and ether linkages—and including natural, naturally derived, and synthetic materials. While a purpose of these enzymes is to digest nutrients for the host, when polymers with susceptible linkages come into contact with a microbe that contains appropriate enzymes, polymer degradation can occur. Although they are often similar to the acid and base associated degradations, enzymatic degradations are more specific, bringing about only specific reactions. Even so, it is often difficult to differentiate between the two, and both may occur together.

Some polymer deterioration reactions occur without loss in molecular weight. These include a wide variety of reactions where free radicals (most typical) or ions are formed, and cross-linking or other nonchain scission reaction occurs. Cross-linking discourages chain and segmental chain movement. At times, this cross-link is desired such as in permanent press fabric and in elastomeric materials. Often the cross-links bring about an increased brittleness beyond that desired.

Some degradation reactions occur without an increase in cross-linking or a lessening in chain length. Thus, in minute amounts of HCl, water, ester, etc. elimination can occur with vinyl polymers giving localized sites of double-bond formation. Because such sites are less flexible and more susceptible to further degradation, these reactions are generally considered unwanted.

Cross-linking reactions can give products with desired increased strength, memory retention, etc. However, accompanying such cross-linking can be unwanted increases in brittleness. Throughout the text, cross-linking is an important reaction that allows the introduction of desirable properties. For instance, cross-linking is the basis for many of the elastomeric materials.

Stabilizers are often added to prevent some forms of degradation. Hindered-amine light stabilizers are added to scavenge free radicals that are light produced in many polymers. UV stabilizers are added to absorb UV radiation converting some of it into heat. Antioxidants are added to stop free radical reactions that are caused by UV radiation.

11.4 CROSS-LINKING

Cross-linking reactions are common for both natural and synthetic and vinyl and condensation polymers. These cross-links can act to lock in "memory" preventing free-chain movement. Cross-linking can be chemical or physical. Physical cross-linking occurs in two major modes. First, chain entanglement acts to cause the tangled chains to act as a whole. Second, crystalline formation, large scale or small scale, act to lock in particular structures. This crystalline formation typically increases the strength of a material as well as reduces wholesale chain movement.

Chemical cross-linking often occurs through the use of double bonds that are exploited to be the sites of cross-linking. Cross-linking can be effected either through the use of such preferential sites as double bonds, or through the use of other especially susceptible sites such as tertiary hydrogens. It can occur without the addition of an external chemical agent or, as in the case of vulcanization, an external agent, a cross-linking agent such as sulfur, is added. Cross-linking can be effected through the application of heat, mechanically, through exposure to ionizing radiation and nonionizing (such as microwave) radiation, through exposure to active chemical agents, or through any combination of these.

Cross-linking can be positive or negative depending on the extent and the intended result. Chemical cross-linking generally renders the material insoluble. It often increases the strength of the cross-linked material but decreases its flexibility and increases its brittleness. Most chemical cross-linking is not easily reversible. The progress of formation of a network polymer has been described in a variety of ways. As the extent of cross-linking increases, there is a steady increase in the viscosity of the melt. At some point, there is a rapid increase in viscosity and the mixture becomes elastic and begins to feel like a rubber. At this point, the mixture is said to be "gelled." Beyond this point, the polymer is insoluble. The extent of cross-linking can continue beyond this gel point.

11.5 REACTIVITIES OF END GROUPS

Polymers containing one reactive end group are often referred to as macromolecular monomers or *macromers*. They can be represented as

$$\text{M-M-M-M-M-M-M-M-M-M-M-M-M-M-M-M-M-M-M-M-M-M-M-Reactive Group} \quad (11.5)$$

They are used in the formation of a number of products including star polymers, and some graft polymers.

Polymers that contain two active end groups are referred to as *telechelic* polymers. These are then of the form

$$\text{Reactive Group-M-M-M-M-M-M-M-M-M-M-M-M-M-M-M-M-M-M-Reactive Group} \quad (11.6)$$

Telechelic polymers are also used in the synthesis of many products including segmented polyurethane and polyester products.

Macromers are then "short" polymers that contain an active end group. This end group can be a site of unsaturation, heterocycle, or other group that can further react. Macromers are usually designed as intermediates in the complete synthesis of a polymeric material. These macromers can be introduced as side-chains (grafts) or they may serve as the backbones (comonomer) of polymers. The macromers can also act as separate phases.

Macromers have been used to produce thermoplastic elastomers. Generally, the backbone serves as the elastomeric phase and the branches as the hard phases. These structures are often referred to as "comb"-shaped because of the similarity between the rigid part of the comb and its teeth and the structure of these graft polymers.

```
-M-M-M M-M-M-M-M-M-M-M-M-M-M-M-M-M-M-M-M-M-M-M-M-M-M-M-M-M-M-
   |  |    |       |           |           |           |
   N  N    N       N           N           N           N
   N  N    N       N           N           N           N
   N  N    N       N           N           N           N
   N  N    N       N                       N           N      (11.7)
   N     N       N                       N           N
   N     N       N                       N
   N           N                       N
   N           N                       N
             N                       N
```

Block copolymers are also produced using such macromers. Again, the various units of the polymer can be designed to act as needed with hard/soft or hydrophilic/hydrophobic or other combinations. Also, these macromers generally have two active functional groups that allow polymerization to occur.

Telechelic polymers can also be used as cross-linkers between already formed polymers because both ends are active. These functional groups could be either two vinyl ends or two Lewis acids or bases such as two hydroxyl or amine groups. Interpenetrating polymer networks, IPNs, and related structures such as dendrites and stars can also be formed using macromers.

The use of placing specific end groups onto polymers and oligomers is widespread, and this will be illustrated here for dimethylsiloxanes, the most widely used form of the siloxanes. These end groups are given various names depending on their intended use. Some names for these end groups are capping agents, blocking agents, chain length modifiers, and coupling agents. Typical siloxane end groups include trimethylsiloxyls, acyloxyls, amines, oximes, alcohols, and alkoxyls. These reactive end groups are reacted forming a wide variety of useful materials. Most silicone

RTV adhesives, sealants, and caulks are moisture curing, i.e., they contain hydroxyl-capped silanes that are reacted with acyloxy, amine, oxime, or alkoxyl moisture-sensitive compounds.

11.6 SUPRAMOLECULES AND SELF-ASSEMBLY

The terms self-assembly, self-organization, and self-synthesis are closely related and sometimes used interchangeably. *Self-assembly* involves the aggregation of molecules, including macromolecules, into thermodynamically stable structures that are held together often using secondary bonding including hydrogen bonding, van der Waals and electrostatic forces, pi–pi interactions, and hydrophobic and hydrophilic interactions. The term *self-organization* is used for situations where the secondary bonding interactions are more specific and directional giving a higher degree of order to the self-assembled molecules. Finally, the term *self-synthesis* includes self-assembly and self-organization, but also includes situations where self-replication and template-type synthesis occurs.

Self-assembly is the spontaneous organization of molecules into stable, well-defined structures with the driving forces being noncovalent associations. The final structure is normally near or at the thermodynamic equilibrium arrangement allowing it to readily spontaneously form. Such formations can be done under conditions where defects are either minimized or eliminated. In nature, self-assembly is common such as the folding of proteins, formation of the DNA double helix, etc.

Self-assembled monolayers, SAMs, are the most widely studied nonnatural self-assembly systems. They are generally spontaneously formed from chemisorption and self-organization of organic molecules onto appropriate surfaces.

Natural polymers utilize a combination of primary and secondary forces and bond angles and distances to form polymers with both long-range (multi-macromolecular) and short-range structures with both structures essential for the "proper" functioning of the macromolecular structure. While most synthetic polymer chemists have focused on what is referred to as primary and secondary structures (short-order structure control), work is just beginning on developing the appropriate structure control to allow tertiary and quaternary structural controls (long-range control). While the "backbones" of these structures are held together by primary bonds, the secondary, tertiary, and quaternary structures are generally "driven" by secondary forces with the resulting tertiary and quaternary structures fixed in place through a combination of these secondary forces and small amounts of ionic and covalent cross-linking.

As noted above, self-assembled structures normally are the result of what are often referred to as weaker force interactions that include van der Waals forces, pi–pi interactions, capillary interactions, etc. Structures that successfully form ordered structures often take advantage of these short-range attractive forces but also of longer-range repulsive forces. Almost all of these structures have some thermodynamic stability with the overall structure more stable than the unassembled parts. These driving forces make the structures especially susceptible to small variations in the conditions leading to ready reversibility.

Several strategies are being developed that allow this long-range control including the use of secondary forces to "hold" in place monomers that subsequently will be polymerized "in place." In another approach, already molecular "architectured" templates are employed to hold the polymer, pre-polymer, or monomers in the desired shape with subsequent reactions and interactions enacted producing a material with somewhat "robust" tertiary and quaternary structures. Some of these "molecular molds" are being produced using nanotechnology.

Self-assembly tendencies are apparent in the simple crystallization of inorganics and organics. Structure, size, and chemical tendencies (such as "like-liking-like" and "unlikes" repelling, secondary and primary bonding tendencies) are all involved. Proteins "self-assemble" in a much more diverse manner than so-called "simple" crystallization of common organics and inorganics. As noted above, we are just beginning to understand the nuances involved with the self-assembling, formation of giant macromolecules including organizations such as those present in the cells of our bodies. We are beginning to understand the major factors involved in making the cell membranes and are starting to mimic these features to form synthetic biological-like membranes. We are using self-assembling concepts and approaches to develop a large number of interesting and potentially useful macromolecular materials.

One of the applications of molecular self-assembling is the formation of ultrathin films using both synthetic and natural surfaces as two-dimensional templates. As noted above, the same chemical and physical factors that we recognize in other areas are at work here. We will begin considering the formation of a simple bilayer membrane such as that present in natural cell membranes. Using the concept of like-liking-like and unlikes rejecting one another, the orientation of molecules with two different polar environments will vary depending upon the particular environment in which they are placed. For a common soap molecule with hydrophilic and hydrophobic ends, the like ends will congregate together and will reside either internally together or externally together. This is exactly the same concept as given in most general chemistry texts when considering the formation of micelles in commercial detergents. In the presence of water, the hydrophilic ends will face outward and in a nonpolar organic solvent the polar ends will face inward. Researchers have extended these simple concepts to include specially designed molecules that contain not only the heads and tails, but also spacers, conductors, and to vary the flexibility of the various parts of the molecule, spacing, and number of heads and tails, etc.

An important concept in the creation of some of these structures is that a primary driving force is the solute–solvent immiscibilities (energy; enthalpic). Thus, the magnitude of the cohesive energy of the solute may be a secondary factor in determining these supermolecular or supramolecular structures for such systems.

The self-assembling character of bilayer membranes is demonstrated by the formation of free-standing cast films from aqueous dispersions of synthetic bilayer membranes. The tendencies for association are sufficiently strong so as to allow the addition of "guest" molecules (nanoparticles, proteins, and various small molecules) to these films where the connective forces are secondary in nature and not primary. Synthetic polymer chemists have made use of these self-assembling tendencies to synthesize monolayer films. Essentially a monomer that contains both reactive groups and hydrophobic and hydrophilic areas is "cast" onto an appropriate template that "self-assembles" the monomer, holding it for subsequent polymerization. Thus, a bilayered structure is formed by

$$
\begin{array}{c}
\quad\quad\quad O \quad\quad\quad O \quad\quad Me \\
\quad\quad\quad \| \quad\quad\quad \| \quad\quad | \\
H_3C\text{-}(\text{-}CH_2\text{-})_{15}\text{-}O\text{-}C\text{-}CH\text{-}NH\text{-}C\text{-}(\text{-}CH_2\text{-})_{10}\text{-}N\text{-}Me, Br \\
\quad\quad\quad\quad\quad\quad | \quad\quad\quad\quad\quad\quad\quad | \\
H_3C\text{-}(\text{-}CH_2\text{-})_{15}\text{-}O\text{-}C\text{-}CH_2 \quad\quad\quad Me \\
\quad\quad\quad\quad\quad \| \\
\quad\quad\quad\quad\quad O
\end{array}
\tag{11.8}
$$

The bis-acrylate monomer (11.8)

$$
\begin{array}{c}
\quad\quad\quad O \quad\quad\quad\quad\quad\quad\quad\quad\quad\quad O \\
\quad\quad\quad \| \quad\quad\quad\quad\quad\quad\quad\quad\quad\quad \| \\
H_2C\text{=}CH\text{—}C\text{—}O\text{—}(\text{—}CH_2\text{—}CH_2\text{—}O\text{—})_{14}\text{—}O\text{—}C\text{—}CH\text{=}CH_2
\end{array}
\tag{11.9}
$$

is accommodated by this bilayer. The bis-acrylate is photopolymerized. While there is some change in the particular bond lengths, the bilayer still holds on and a coherent film is formed. The bilayer template is subsequently washed away by the addition of methanol leaving a flexible, self-supporting film. Polymerization can also result in the creation of different bond lengths that can act to "release" the newly formed monolayer. Multilayered structures can be made by simply allowing the outer surface of one monolayer to act as the template for the next monolayer.

Inorganics can also be synthesized and used as templates. Thus, controlled siloxane networks were formed when dispersions of alkoxysilanes (such as $(MeO)_3SiMe$) were mixed with the suitable template matrixes. Ultrafine particles of metal oxides can be used as starting materials for the formation of metal oxide films. For instance, a mixture of a double-chained ammonium amphiphile and an aqueous solution of aluminum oxide particles (diameter about 10–100 nm) gives a multilayered aluminum oxide film when calcinated at over 300°C.

Metal-ligand structures, such as porphyrin-based structures, are able to "control" the geometry of forming superstructures. The interactions between the various internally chelated porphyrin metal and a combination of the planarity and pi–pi interactions between the porphyrin rings drive the resulting structures. Connecting groups can act as spacers or act as additional geometry-determining features. Smaller metal-based groups have been used to self-assemble structures such as nano-cages that can have within them molecules that give the combination cage and captive molecule unique chemical and electrical environments. A nano-cage has been formed using six cis-protected palladinum II nitrate molecules and four tridendate tripyridyl molecules. These structures, unlike other self-assembly molecules considered here, have primary bonds as major contributors to the self-assembly process. Many of these excursions are the exercise of combining organometallic chemistry with structural knowledge and purpose.

Graphite and carbon fibers have been used as templates. Thus, nylon 6 has been polymerized on a graphite matrix. Such syntheses of polymers in the presence of a solid template where the solid acts as a template have been described as polymerized-induced epitaxy, PIE. The monomer and resulting film is adsorbed on the template surface through only van der Waals forces. After polymerization, the polymer is washed from the template. The recovered polymer retains "special" structural features introduced by the template.

Block copolymers with amphiphilic groups have been used to give molecules with several "levels" of molecular architecture. Thus, a block copolymer of styrene and 4-vinylpyridine and long-tailed alkylphenols contains polar groups in the backbone that associate with the alkylphenol chains forming hydrogen bonds resulting in a bottle-brush-like structure. If the alkyl tails on the polymer are strongly repulsed by the amphiphilic portions, microphase separation between the tails and the rest of the copolymer results. Here, the alkylphenol portion, which is hydrogen bonded to the vinylpyridine blocks, separates as a microphase inside the poly(4-vinylpyridine) block domains. If the copolymer is heated, the bottle-brush structure undergoes an order–disorder transition around 100°C and a second order–order transition around 150°C as the amphiphilic alkylphenol chains diffuse into the PS-rich domains and cylindrical structures are formed.

While much of the emphasis on inter- and intra-molecular interactions, secondary bonding, and forces associated with association and dissociation involves attractive forces, we are finding that phobic effects are also important and for some systems are actually the major factors. Briefly, this can be described by such sayings as "the enemy of my enemy is my friend," or "given the choice between bad and worse, bad wins out." Formation of many self-assemblies is due in large measure to such phobic factors.

As we learn more about what drives molecular shapes we are finding more applications utilizing this information in designing, on a molecular level, oligomeric and polymeric chains. It is also becoming more apparent to us that we have been forming organized structures without knowing it. While the simple layered and linear structures are generally employed to describe the concept, the self-assembling approach holds for any two- and three-dimensional structure under the appropriate conditions. As in many of the areas of research, the potential is only limited by our imagination.

Considering only Lewis acid–base or donor–acceptor interactions we can envision a hydrogen bond donor site such as an alcohol, acid, thiol, or amine and an acceptor site such as a carbonyl oxygen on another molecule or part of the molecule. These components will bind with one another, acting to either bind the molecules containing the two differing bonding sites or if the two sites are on the same molecule attempt to contort, twist the molecule allowing the preferred bonding to occur.

Synthetic shapes are generally limited to sheets and polyhedral structures. Yet nature produces a much wider variety of shapes including curves, spirals, ripples, bowls, pores, tunnels, spheres, and circles. We are beginning to master such shapes. We are beginning to make these shapes based on especially "grown" shapes that act as templates for further growth. For instance, Geoffrey Ozin and coworkers mixed together alumina, phosphoric, and decylamine in an aqueous solution of tetraethylene glocol. After a few days, millimeter-sized aluminophosphate solid spheres and hollow shells were formed with the surfaces sculpted into patterns of pores, meshes, ripples, bowls, etc. A decylammonium dihydrogenphosphate liquid-crystal phase was formed and this surfactant, along with the glycol, was forming bilayer vesicles. The vesicles acted in different ways with some fusing to one another, others splitting apart or collapsing giving a variety of structures. Thus, appropriate conditions can be selected that favor certain template structures producing an array of geometric structures. Further, the templates themselves can be used to make selective separations. In a related study, the group employed a silica precursor, tetraethyl orthosilicate. Here the orthosilicate units assembled together forming micelles that in turn acted as liquid-seed crystals growing other assemblies with varying shapes. Rapid growth in the axial direction produces ropelike structures that can be made to form circles and loops through the application of external forces. Other structures included egg shapes, disks, spirals, knots, and spheres.

Metal coordination is another important bonding opportunity with respect to self-assembly. This is important in many natural molecules, such as hemoglobin and chlorophyll, where the metal atom acts as both the site of activity and as a "centralizing" agent with respect to shape, and thus acts as a nucleating agent for self-assembly.

There are many potential and real applications of self-assembly.

Pharmaceutical chiral drug sales top $100 billion yearly worldwide. More than one-half of the drugs on the market are asymmetric molecules with about 90% administered as racemates. Generally, one optical center of a drug will have the desired activity while the other often produces negative side effects. Because of this the FDA, in 1992, issued a statement saying that for every new racemic drug, the two enantiomers must be treated as separate substances that are required to undergo pharmocokinetic and topological studies.

One direct approach to the separation of chiral compounds is called molecular imprint polymers, MIPs, which involves the formation of a three-dimensional cavity with the shape and electronic features that are complementary to the imprinted or target molecule.

11.7 TRANSFER AND RETENTION OF OXYGEN

Today, the polymer scientist should be aware of synthetic, inorganic, and biological macromolecules. The field of biological macromolecules is large and is one of the most rapidly expanding areas of knowledge. It involves gene splicing and other related biological aspects including biological engineering, neurobiology, medicine, drugs, etc.; the very elements of life and death, of thought and caution, pain and health, of biological transference, of energy and biological matter. The polymer scientist can learn from these advances but must also contribute to their understanding on a molecular and chain aggregate level. The investigation of these biological macromolecules is happening using state-of-the-art instrumentation and techniques and the use of scientific intuition. The world of the natural macromolecule is yielding information allowing an understanding on a molecular level. A striking example involves oxygen transfer and retention in mammals.

Oxygen retention and transfer involves the iron-containing organometallic planer porphyrin-containing structure called *heme* (Figure 11.1). The iron is bonded through classical coordination. The ferrous or iron (II) has six coordination sites. Four of these coordination sites are occupied by the nitrogen atoms of the four pyrrole-related rings of the porphyrin. A fifth site is occupied by one of the nitrogens of an imidazole side chain found as a part of the protein structure and located just opposite the planer porphyrin moiety. The sixth site acts to bind oxygen. The ion remains in the +2 oxidation state, whether oxygen is being bound or the site is vacant. An additional histidine is

FIGURE 11.1 Porphyrin structures that serve as the basis of heme (left). Upon addition of iron, this porphyrin, which is called protoporphyrin IX, forms the heme group (right).

present, residing in the protein chain opposite the sixth site of the iron atom. This second histidine does not bind iron but serves to stabilize the binding site for oxygen. Experimentally, heme does not bind oxygen. Instead, a complex protein wrapping is necessary to both assist binding and protect the binding site from foreign competitor molecules that could render the heme site inactive, through structural change, change in iron oxidation state, or the occupation of this site, thus preventing oxygen access to the active binding site.

The precise electronic environment of iron deserves a special comment. In deoxyhemoglobin, the iron atom has four unpaired electrons, but in oxyhemoglobin iron has no unpaired electrons. The iron in the oxygen-free deoxyhemoglobin is referred to as "high-spin" iron, whereas iron in oxyhemoglobin is called "low-spin" iron. Hund's rule of maximum multiplicity calls for the most energy-favored, lowest energy form to be the structure containing the highest number of unpaired electrons. The binding of oxygen, itself with two unpaired electrons, is probably the result of a favored energy of binding brought about through the coupling of the two sets of unpaired electrons—the favorable energy allowing the violation of Hund's rule.

There are two major protein-heme-binding macromolecules. These are myoglobin (Figure 11.2), which is used as an oxygen storage molecule in mammalian muscle and hemoglobin, which is used in oxygen transport. Myoglobin is single stranded with one heme site per chain, whereas hemoglobin is composed of four protein chains, each one containing a single heme site (Figure 11.4). In hemoglobin, there are two sets of equivalent chains composing its quaternary structure. These two chain types are referred to as α and β chains. The α chains contain 141 amino acid units and the β chains contain 146 units. The myoglobin contains 153 amino acid units. Each of these chains is similar, and each forms the necessary environment to allow the heme site to bind oxygen in a reversible manner. The protein segments are described as being loosely helical with about 60%–80% of the structure being helical.

While the chains are similar in overall structure, there exist somewhat subtle differences. For instance, the quaternary structure of hemoglobin permits interaction between the four chains. Structural movement brought about through the binding of oxygen at one of the four heme sites acts to make it easier for subsequent oxygen addition at the other heme sites. Such cooperative binding of oxygen is not possible in the single-chained myoglobin. The consequence of this cooperative binding is clearly seen in a comparison of oxygen binding by both myoglobin and hemoglobin as a function of oxygen pressure (Figure 11.3). Oxygen binding occurs by myoglobin at even low oxygen pressures. This is referred to as hyperbolic behavior. By comparison, hemoglobin binding increases

FIGURE 11.2 Generalized myoglobin structure showing some amino acid units as open circles to illustrate the "folded" tertiary structure.

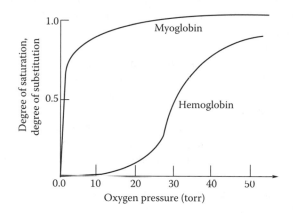

FIGURE 11.3 Degree of saturation as a function of oxygen pressure.

more slowly as the oxygen pressure increases, occurring in a more sigmoidal fashion. Thus, initial binding of oxygen at a heme site is relatively difficult, becoming increasingly easier as the number of heme sites that are bonded increases.

The differences in oxygen binding characteristics are related to the differing roles of hemoglobin and myoglobin. Thus, myoglobin is employed for the storage of oxygen in muscle. Binding must occur even at low oxygen contents. Hemoglobin is used for the transport of oxygen, and it becomes saturated only at higher oxygen concentrations. The oxygen content in the alveoli portion of our lungs is on the order of 100 torr (1 atm of pressure = 760 torr). Here, almost total saturation of the heme binding sites in hemoglobin occurs. By comparison, the oxygen level in the capillaries of active muscles is on the order of only 20 torr, allowing the hemoglobin to deliver about 75% of its oxygen and for myoglobin to almost reach saturation in oxygen binding.

Conformational changes accompany the binding and release of oxygen. These changes are clearly seen by superimposing the oxygen-containing form of hemoglobin–oxyhemoglobin, over

FIGURE 11.4 Space-filling models of deoxhemoglobin (a), and oxyhemoglobin (b). Notice the small shifts in the overall geometry of the various protein chains and the decreased size of the inner core.

the nonoxygen-containing form of hemoglobin–deoxyhemoglobin (Figure 11.4). It is interesting to note that most enzymes have their active sites on the surface but enzymes, such as myoglobin, that are dealing with "small molecules" such as oxygen, have their active sites "hidden away" in the internal portion so that other unwanted molecules cannot easily access them with the protein "tunnel" acting to restrain unwanted suitors.

$$(11.10)$$

Another metal-containing important enzyme, chlorophyll (11.10) is structurally related to myoglobin, and it is responsible for photosynthesis, the conversion of water and carbon dioxide, with the aid of solar energy, to carbohydrates.

11.8 NATURE'S MACROMOLECULAR CATALYSTS

Probably the most important reactions occurring on polymers involve the catalytic activity of a class of proteins called enzymes. The catalytic action is a result of a lowering of the activation energy for the rate-determining step in the reaction. In general terms, the catalytic action results from the formation of a complex between the enzyme and the molecule undergoing reaction. The decreased activation energy is a result of the reacting molecules being held by the enzyme in such a manner

FIGURE 11.5 There are two major models for the binding of reactant molecules to the active sites of nonallosteric enzymes. The sequence describes the essential steps in the lock-and-key model, where the reactant(s) is attracted to the active site on the enzyme where the active site is a cavity of the same general size, shape, and (complementary) electronic features. Binding occurs and the appropriate reaction(s) occurs resulting in a change in the geometry and the electronic configuration of the product, causing its release. The second model (not shown) is the induced-fit model where the individual steps are similar to the lock-and-key except that the reactants "induce" a change in the conformation of the active site on the enzyme, allowing it to accept the reactant(s).

so as to favor the appropriate reaction occurring. The two classical models employed to describe the formation of the complex between the reacting molecules(s) and the enzyme are the lock-and-key model and the induced-fit model. Briefly, the *lock-and-key model* calls for an exact or highly similar complementary fit between the enzyme and the reacting molecule(s) (Figure 11.5). Geometry plays an essential role in permitting the electronic (polar, electrostatic, etc.; attractions/repulsions) interactions to form the necessary complex with the correct geometry. Release is encouraged by the new geometry and electronic distribution of the resulting products of the reaction being sufficiently dissimilar. The *induced-fit model* is similar except that the enzyme originally does not fit the required shape (geometrical and electronic cavity). The required shape is achieved upon binding—the binding causing needed "assisting factors" or proximity and orientation to effect a decrease in the energy of the transition state. Figure 11.5 shows how the enzyme sucrase breaks sucrose into its basic units of glucose and fructose. The sucrose is held by the enzyme in such a fashion so that water easily and efficiency breaks the ether bond forming two hydroxyl units that are released because they no longer have the required shape/size/electronic features of the enzyme cavity.

Enzyme reactions generally follow one of two kinetic behaviors. Briefly, the oxygen-binding curve for myoglobin is hyperbolic, whereas that for hemoglobin it is sigmoidal (Figure 11.6). In general, it is found that similar enzymes, such as myoglobin, follow a similar hyperbolic relationship between reaction extent and reaction time. More complex enzymes, such as hemoglobin, follow a sigmoidal relationship between reaction extent and reaction time. The primary difference involves the ability of different portions of the overall hemoglobin structure, removed reaction sites, to affect each other. Molecules in which various removed sites affect the reactivity of other removed sites are called *allosteric enzymes*.

The Michaelis–Menten model is commonly employed in describing nonallosteric enzyme reactions. The overall model can be pictured as follows (11.11) where E represents the enzyme, M the reacting molecule(s), E + M → EM is associated with k_1, and the reverse reaction is associated with k_{-1} and EM → E + P is associated with k_2.

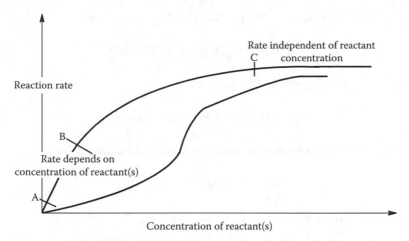

FIGURE 11.6 Dependence of reaction showing sigmoidal (bottom) and hyperbolic (top behavior). The top plot also shows the initial rate of reaction as a function of reactant concentration when the concentration of enzyme remains constant. The initial reaction rate, A to B, varies until the number of reactants clearly outnumbers the number of reaction sites on the enzyme, at which time, C, the rate becomes zero order, independent of the reactant concentration. Often it is difficult to directly measure the concentration of E as the reaction progresses.

Thus, the concentration of E is generally dealt with using an approach as shown in Equations 11.13 to 11.24.

$$E + M \underset{k_{-1}}{\overset{k_1}{\rightleftharpoons}} EM \overset{k_2}{\rightleftharpoons} E + P \tag{11.11}$$

Here, EM represents the enzyme complex and P the product(s). The rate of complex formation is described as

$$\text{Rate of complex formation} = \frac{d[EM]}{dt} = k_1[E][M] \tag{11.12}$$

The complex then either returns to form the initial reactants or forms the product(s) and the free enzyme, E. In kinetic terms, the change, or rate of breakdown of the complex is described as

$$\text{Rate of complex change} = -\frac{d[EM]}{dt} = k_{-1}[EM] + k_2[EM] \tag{11.13}$$

The negative sign associated with the equation means that the terms are describing the rate of decrease in complex concentration. The rate of complex formation is rapid, and fairly soon the rate at which the complex is formed is equal to the rate at which it breaks down. This situation is called a steady state. Mathematically, this is described by

$$\frac{d[EM]}{dt} = -\frac{d[EM]}{dt} \tag{11.14}$$

and

$$k_1[E][M] = k_{-1}[EM] + k_2[EM] \tag{11.15}$$

$$[E] = [E_0] - [EM] \tag{11.16a}$$

where $[E_0]$ is the initial enzyme concentration.

Substitution of this description for [E] into Equation 11.15 gives

$$k_1([E_0][M] - [EM][M]) = k_{-1}[EM] + k_2[EM] \tag{11.16b}$$

Separating out [EM] on the right-hand side gives

$$k_1([E_0][M] - [EM][M]) = (k_{-1} + k_2)[EM] \tag{11.17}$$

Bringing together all of the rate constants from both sides of the equation gives

$$[E_0][M] - [EM][M] = k'[EM] \tag{11.18}$$

Moving the [EM] containing terms to the right side gives

$$[E_0][M] = k'[EM] + [EM][M] \tag{11.19}$$

and separating out [EM] from the right side gives

$$[E_0][M] = [EM](k' + [M]) \tag{11.20}$$

Now, division of both sides by $k' + [M]$ gives

$$[EM] = \frac{[E_0][M]}{k' + [M]} \tag{11.21}$$

The initial rate of product formation, R_i, for the Michaelis–Menten model depends only on the rate of complex breakdown, i.e.,

$$R_i = k_2[EM] \tag{11.22}$$

Substitution from Equation 11.21 into Equation 11.22 gives

$$R_i = \frac{k_2[E_0][M]}{k' + [M]} \tag{11.23}$$

This expression is dependent on the concentration of M and describes the initial part of the plot given in Figure 11.6.

Generally, the concentration of M far exceeds that of the enzyme sites such that essentially all of the enzyme sites are complexed, i.e., [EM] = [E_0]. (This is similar to a situation that occurs regularly in south Florida where four- and six-lane roads are funneled into a two-lane section of road because of a wreck or road construction.) Thus, the rate of product formation is maximized under these conditions. This maximum rate, R_m, allows us to substitute [E_0] for [EM] in Equation 11.23 to give

$$R_m = k_2[E_0] \tag{11.24}$$

Since the enzyme concentration is constant, the rate of product formation under these conditions is independent of [M] and is said to be zero order (Figure 11.6).

The maximum rate is directly related to the rate at which the enzyme "processes" or permits conversion of the reactant molecule(s). The number of moles of reactants processed per mole of enzyme per second is called the ***turnover number***. Turnover numbers vary widely. Some are high, such as for the scavenging of harmful free radicals by catalase, with a turnover number of about 40 million. Others are small such as the hydrolysis of bacterial cell walls by the enzyme lysozyme, with a turnover number of about one-half.

The Michaelis–Menten approach does not describe the behavior of allosteric enzymes, such as hemoglobin, where rate curves are sigmoidal rather than hyperbolic. A more complex mode is called for to account for the biofeedback that occurs with allosteric enzymes. Such effects may be positive, such as those associated with hemoglobin, where binding by one site changes the geometry and electronic environment of the other remaining sites, allowing these addition sites to bind oxygen under more favorable conditions. The affects may also be negative, such as that of cytidine triphosphate, which inhibits ATCase and catalyzes the condensation of aspartate and carbamoly phosphate—forming carbamoyl aspartate.

Two major models are typically used to describe these situations: the concerted model and the sequential model. In the ***concerted model***, the enzyme has two major conformations—a relaxed form that can bind to the appropriate reactant molecule(s) and a tight form that is unable to tightly bind the reactant molecule(s). In this model, all subunits containing reactive sites change at the same time (Figure 11.7). An equilibrium exists between the active and inactive structures. Binding at one of the sites shifts the equilibrium to favor the active relaxed form.

The major feature in the ***sequential model*** is the induction of a conformational change from the inactive tight form to the active relaxed form as the reacting molecule(s) is bound at one of the sites. This change from an unfavorable to a favorable structure is signaled to other potentially reactive sites bringing about a change to the more favored structural arrangement in these other sites (Figure 11.8).

Structural changes can be brought about through simple electrostatic and steric events caused by the presence of the reacting molecule(s). Structural changes also result as cross-linking and other primary bonding changes occur.

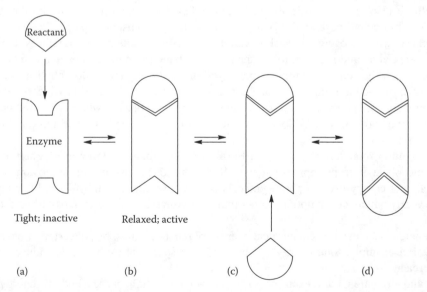

FIGURE 11.7 Concerted model for allosteric enzymes. The major steps are (a) and (b). An equilibrium exists between the tight (a) and relaxed (b) forms of the allosteric enzyme. The reactant molecule(s) approaches the reactive site of one of the enzyme sites present in the relaxed form (c). Binding occurs, shifting the equilibrium to the relaxed form(s). The second site is bound (d).

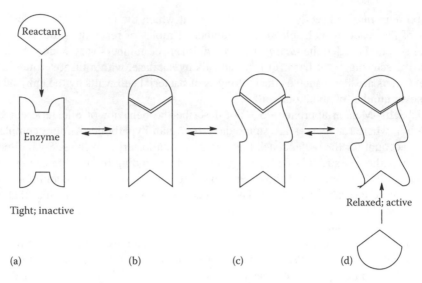

FIGURE 11.8 Sequential model for allosteric enzymes. The individual steps are as follows: the reactive molecule(s) approaches the reactive site (a), which typically has a cavity similar to, but not the same as, the reactant molecule(s); a conformational change is effected so that the reactive molecule(s) can be bound (b); the bound portion of the enzyme changes shape; this shape change is transmitted to the other unit(s) containing active sites (c); the remaining enzyme portion containing active sites undergoes a conformational change that makes binding easier (d).

11.9 PHOTOSYNTHESIS

The recent environmental issues related to the green house effect and atmospheric contaminations heighten the importance of obtaining energy from clean sources such as photosynthesis. Photosynthesis also acts as a model for the creation of synthetic light-harvesting systems that somewhat mimic chlorophyll in their ability to convert sunlight into usable energy. The basis of natural photosynthesis was discovered by Melvin Calvin, one of my academic grandfathers. Using carbon 14 as a tracer, Calvin and his team found the pathway that carbon undergoes in a plant during photosynthesis. They showed that sunlight supplies the energy through the chlorophyll site allowing the synthesis of carbon-containing units, saccharides or carbohydrates. Chlorophyll (Structure 11.10) is a metal embedded in a polymer matrix and illustrates the importance of metals in the field of photochemistry and photophysics. Following is a description of that activity of chlorophyll in creating energy from the sun.

The maximum solar power density on Earth is approximately $1350\,W/m^2$. When it enters the Earth's atmosphere, it drops approximately to $1000\,W/m^2$ due to absorption in the atmosphere. The amount of light energy used in photosynthesis is immense, about 100 TW yearly or about seven times the yearly human consumption. Here organisms convert about 10^{11} ton of carbon into biomass yearly.

Solar energy is a clean and economical energy source, but it must be converted into useful forms of energy. For example, solar energy can be used as a source of excitation to induce a variety of chemical reactions.

Plants and algae are natural examples of the conversion of light energy, and this energy is used to synthesize organic sugar-type compounds through photosynthesis. This process has a great importance for the survival of all life on our planet because it provides the energy and reduces the carbon dioxide as well as produces molecular oxygen necessary for oxygen-consuming organisms. In fact, this process can be considered as a vital link between material and energy cycling in the biosphere.

Light energy $\Longrightarrow$ (A) $\longrightarrow\longrightarrow\longrightarrow$ (R) $\longrightarrow\longrightarrow$ (C)

FIGURE 11.9 Light is absorbed by the antenna pigments, A, transferring the energy until it reaches the RC, where it is used as the driving force for electron transfer reactions where specialized chlorophylls use it to form carbohydrates.

In photosynthesis, green plants and some bacteria harvest the light coming from the sun by means of their photosynthetic antenna systems. The light harvesting starts with light gathering by antenna systems, which consist of pigment molecules including chlorophylls, carotenoids, and their derivatives. The absorbed photons are used to generate excitons, which then travels via energy transfers toward the reaction centers (RCs). This overall series of processes is represented in Figure 11.9. The series can be remembered by the initials ARC where A represents antenna pigments; R represents the RCs; and C represents chlorophylls and carbohydrates.

In RCs, the energy is used as a driving force for an electron transfer reaction, which in turn initiates a series of slower chemical reactions and the energy is saved as redox energy inducing a charge separation in a chlorophyll dimer called the special pair (chlorophyll)$_2$. Charge separation, which forms the basis for photosynthetic energy transfer, is achieved inside the RCs (11.25).

$$(\text{Chlorophyll})_2 + \text{energy} \rightarrow (\text{Chlorophyll})_2^+ + e^- \qquad (11.25)$$

Specialized reaction center proteins are the final destination for the energy where it is converted into chemical energy by means of electron-transfer reactions. These proteins consist of a mixture of polypeptides, chlorophylls (plus the special pair), and other redox-active cofactors. In the RCs, a series of downhill electron transfers occur resulting in the formation of a charge separated state. Based on the nature of the electron acceptors, two types of RCs can be described. For the first type (photosystem I), the RC contains iron–sulfur clusters (Fe_4S_4) as their electron acceptors and relays whereas for the second type (photosystem II), the RCs have quinones as their electron acceptors. Both types of RCs are present in plants, algae, and cyanobacteria, whereas the purple photosynthetic bacteria have only a photosystem II and the green sulfur bacteria contain a photosystem I. To gain a better understanding of these two types of RCs, each will be further discussed.

11.9.1 PURPLE PHOTOSYNTHETIC BACTERIA

In the mid-1980s, Deisenhofer reported his model for the structure of photosystem II for two species of purple photosynthetic bacteria (*Rhodopseudomonas viridis and Rhodobacter*) based on x-ray crystallography of the light-harvesting device II. Photosynthetic centers in purple bacteria are similar, but not exactly, examples for the green plants but they are simpler and better understood so will be described here. The photosynthetic membrane of purple photosynthetic bacteria is composed of many phospholipid-filled ring systems (LH II) and several larger dissymmetric rings (LH I) stacked almost like a honeycomb. Inside the LH I, is a protein called RC as illustrated in Figure 11.10.

The light-harvesting antenna complex LH II is composed of two rings of BChl *a* molecules that can be classified into two categories. The first one is a set of 18 molecules arranged in a slipped face-to-face arrangement, and it is located close to the membrane surface perpendicularly to these molecules. The second ring is composed of 9 BChl in the middle of the bilayer. The first 18 BChl has an absorption maximum at 850 nm and are called B850, while the second (9 BChl) have an absorption maximum at 800 nm and are called B800. These structures are contained within the walls of protein cylinders with radii of 1.8 and 3.4 nm. Once the LH II complex antenna absorbs light a series of very complex nonradiative photophysical processes are triggered. First, the excitation energy migrates via energy transfers involving the hopping of excitation energy within almost isoenergetic subunits

FIGURE 11.10 Simplified drawing showing two light-harvesting II assemblies next to one light-harvesting I unit. The circles are polypeptides and the lines represent rings of interacting bacteriochlorophylls (BChl) *a* (called B850). In the middle of LH I, there is a protein called the RC where the primary photoinduced electron transfer takes place from the special pair of BChl *b*.

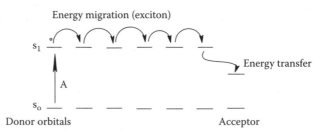

FIGURE 11.11 The exciton and energy transfer processes.

of a single complex, which is then followed by fast energy transfer to a lower energy complex with minimal losses (Figure 11.11). These ultrafast events occur in the singlet state (S_1) of the BChl pigments and, these are believed to occur over relatively long distances (30–100 Å).

The energy collected by the LH II antenna is transferred to another antenna complex known as LH I, which surrounds the RC. The photosynthetic RCs of bacteria consist mainly of a protein, which is embedded in and spans a lipid bilayer membrane. In the RC, a series of electron transfer reactions are driven by the captured solar energy. As a result of these electron transfer reactions, the captured solar energy is converted to chemical energy in the form of a charge separation process across the bilayer. The photosynthetic reaction center is where the harvested solar energy is converted to chemical energy via charge separation across the bilayer by means of an electron transfer reaction.

A special pair of BChl (P870) is excited from either the absorption of a photon or by acquiring this excitation energy from an energy transfer from the peripheral antenna BChl (not shown for simplicity) triggering a photoinduced electron transfer inside the RC. Two photoinduced electrons are transferred to a plastoquinone placed inside the photosynthetic membrane acting as an electron acceptor, and plastoquinone is consequently reduced to a semiquinone and finally to a hydroquinone. This reduction involves the uptake of two protons from water on the internal cytoplasmic side of the membrane. The formed hydroquinone then diffuses to the next component of the apparatus, a proton pump called the cytochrome bc1 complex.

The next step involves the oxidation of the hydroquinone back to a quinone, and the energy released is used for the translocation of the protons across the membrane, thus, establishing a proton concentration and a charge imbalance (proton motive force, pmf). Thus, the oxidation process takes place via a series of redox reactions triggered by the oxidized special pair, which at the end is reduced to its initial state. The oxidation process is ultimately driven, via various cytochrome redox relays, by the oxidized P870, which becomes reduced to its initial state. Finally, the enzyme

ATP synthase allows protons to flow back down across the membrane driven by the thermodynamic gradient, leading to the release of ATP formed from adenosine diphosphate and inorganic phosphate (Pi). The ATP fills the majority of the energy needs of the bacterium.

11.9.2 GREEN SULFUR BACTERIA

The observation of a photosynthetic reaction center in green sulfur bacteria dates back to 1963. Green sulfur bacteria RCs are of the type I or the Fe-S type (photosystem I) where the electron acceptor is not the quinine, but instead chlorophyll molecules (BChl 663, 8^1-OH-Chl a, or Chl a) serving as primary electron acceptors and three Fe_4-S_4 centers (ferredoxins) as secondary acceptors. A quinone molecule may or may not serve as an intermediate carrier between primary electron acceptor (Chl) and secondary acceptor (Fe-S centers).

A large number of chlorophyll antennas are used to harvest the solar energy, which in turn are used to excite the special pair P_{700}. The P_{700} donor will in turn transfer an electron to a primary acceptor (A_0, phyophytin) and in less than 100 ps to a secondary acceptor (A_1, a phylloquinone). The electron received by A_1 is in turn transferred to an iron-sulfur cluster and then to the terminal iron-sulfur acceptor.

11.10 MECHANISMS OF ENERGY ABSORPTION

Let us consider a force, stress, acting on a material producing a deformation. The action of this force can be described in terms of two models—a Hookean spring and a Newtonian dashpot (Section 13.1). In the Hookean spring, the energy of deformation is stored in the spring and may be recovered by allowing the spring to act on another load or through the release of the stress; in either case, the site is returned to zero strain. A Newtonian dashpot is pictorially a frictionless piston, and it is used to describe chains flowing past one another. The energy of deformation is converted to heat. In actuality, the deformation of most plastics results in some combination of the Hookean and Newtonian behaviors. The Newtonian behavior results in net energy adsorption by the stressed material, some of this energy producing the work of moving chains in an irreversible manner while some of the energy is converted to heat.

There are three major mechanisms of energy absorption: shear yielding, crazing, and cracking. The latter two are often dealt with together and called *normal stress yielding*.

We can distinguish between a crack and a craze. When stresses are applied to polymeric materials, the initial deformation involves shear flow of the macromolecules past one another if it is above T_g, or bond bending, stretching, or breaking for glassy polymers. Eventually a crack will begin to form, presumably at a microscopic flaw. This crack can then propagate at high speed, often causing catastrophic failure. The applied stress results in a realigning of the polymer chains. This results in greater order, but decreased volume occupied by the polymer chains, i.e., an increase in free volume. This unoccupied volume often acts as the site for opportunistic smaller molecules to attack, leading to cracking and crazing and eventually property failure.

A *crack* is an open fissure, whereas a *craze* is spanned top to bottom by fibrils that act to resist the entrance of opportunistic molecules such as water vapor. Even here, some smaller molecular interactions can occur within the void space, and eventually the specimen is weakened.

Crazing and cracking can be induced by stress or combined stress and solvent action. Most typical polymers show similar features. To the naked eye, crazing and cracking appear to be a fine, microscopic network of cracks generally advancing in a direction at right angles to the maximum principle stress. Such stress yielding can occur at low stress levels under long-term loading. Suppression of stress yielding has been observed for some polymers by the imposition of high pressure.

In shear yielding, oriented regions are formed at 45° angles to the stress. No void space is produced in shear yielding. Crazing often occurs prior to and in front of a crack tip. As noted before, the craze portion contains both fibrils and small voids that can be exploited after the stress is released or if the stress is maintained. Materials that are somewhat elastic are better at preventing small stress related crazing and cracks. Most plastics are not ideal elastomers and additional microscopic voids occur each time a material is stressed.

All three mechanisms result in a difference in the optical properties of the polymeric material because of the preferential reorientation, with realignment of the polymer chains resulting in a change in optical properties such as refractive index, allowing detection through various optical methods including visual examination, microscopy, and infrared spectroscopy of films and sheets. Crazed and cracked sites of optically clear materials appear opaque, whereas shear-yielded sites may appear to be "wavy" when properly viewed by the naked eye employing refracted light.

It is important to emphasize that the surface layers of most polymeric materials are different from that of the bulk material and are often more susceptible to environmental attack. Thus, special surface treatments are often employed in an attempt to protect the surface molecules.

Directly related to energy absorption is energy dissipation. Generally, the better a material can dissipate or share applied energy the more apt it is to retain its needed properties subsequent to the applied energy. Polymers dissipate applied energies through a variety of mechanisms including rotational, vibrational, electronic, and translational modes.

One area that illustrates aspects important to energy dissipation is the fabrication of protective armor. Such armor includes helmets, vests, vehicle exteriors and interiors, riot shields, bomb blankets, explosive containment boxes (aircraft cargo), and bus and taxi shields. In each case, energy dissipation is a critical element in the desired behavior of the device. To illustrate this let us look at body armor.

Most of the so-called bulletproof vests were made of PCs. More recently, layers of polyethylene were found to have similar "stopping power" for a lesser weight of material.

Today, most body armor is a complex of polymeric materials. Rapid dissipation of energy is critical, allowing the impact energy to be spread into a wide area. Materials should be strong enough so as not to immediately break when impacted and that they need to have enough contact with other parts of the body armor to allow ready transfer of some of the impact energy. If the material can adsorb some of the energy through bond breakage or heating, then additional energy can be absorbed at the site of impact. Along with high strength, the material should have some ability to stretch, to move allowing the material to transfer some of the energy to surrounding material. If the connective forces between the components are too strong, total energy dissipation is reduced because a strong bond discourages another way of reducing the impact energy, i.e., allowing the various materials to slide past one another. Thus, a balance is needed between the material strength, the strength of bonding holding the components together, and the ability to readily dissipate the impact energy.

Recently, it was found that some sequences of layered materials are more effective at energy dissipation than others. One of the better combinations is obtained when aramid layers are adjacent to UHMWPE

Another factor is breaking up the projectile. This is again done using polymeric materials—here composites such as boron carbide ceramics in combination with aramids, UHMWPE, or fibrous glass.

One of the most recent approaches employed to increase the effectiveness of body armor involves the use of *shear thickening fluids*, STFs. These combinations are generally referred to as "liquid armor." As force is inflected to a STF, it resists flow different to many liquids such as water. STF material is placed between the Kevlar sheets that perform several functions. Because it is a liquid under normal conditions, it adds to the flexibility of the body armor. Second, it is more effective at preventing penetration of a bullet, so allows less material to be used to achieve the same "stopping power" allowing the armor to weight less. Finally, it helps blunt the projectile through its action as a STF. Most STFs are polymeric. The liquid is generally low molecular weight poly(ethylene glycol), PEG, with silica particles contained dispersed in it. The PEG wets the Kevlar sheets assisting the Kevlar-STF combination to spread the impact dissipating the effect of the projectile.

11.11   BREAKAGE OF POLYMERIC MATERIALS

When a plastic is broken by a sharp blow or cut, are polymer chains broken? The important factors include the nature of the polymer, chain length, and arrangement of the chains.

Sperling and coworkers looked at the question of how many chains are broken and the defining factors related to this breakage when a polymeric material is cut or broken. They used various chain lengths of PS, and employed a dental drill as the cutting implement.

Pictorially, the problem can be described as looking at a robin pulling a worm out of a hole. Does the robin get the entire worm or some fraction of the worm? The factors are similar and deal with the length of the worm and how far into the hole, it is. If it is largely within the hole then it can grasp the dirt, roots, etc. about it to "hold on for dear life." If not, then the entire worm is a meal for the "early bird."

It turns out that the question dealt with here is related to determining the critical length of fibers that are to be used in a composite. When determining the optimum fiber length of a fiber in a matrix, measurements are made using fibers of differing lengths. If a fiber can be removed from the matrix unbroken, then it is too short and if the fiber breaks before it can be removed, then it is too long. Thus, fiber lengths should be such that the fiber just begins to be broken rather than allowing it to be removed intact. In a composite, the worm is the fiber and the soil is the matrix. For the plastic, the worm is the individual chain and the soil is the remainder of the plastic. For the composite, the fiber contains many individual polymer chains, while for the situation dealt with here, individual polymer chains will be examined.

The length of fiber or chain that can be removed without breaking is related to the frictional and attractive energies between the fiber and the matrix or other polymer chains holding onto the chain. Thus, if the strength holding together the polymer backbone is greater than the frictional energy holding the chain in place, then the polymer chain will be removed unbroken. In general, what was found through calculations was that PS chains up to 300 units in length are capable of being removed intact without breakage. This is in rough agreement with what Sperling found experimentally. Thus, individual PS chains up to about 300 units in length are removed from the plastic without chain breakage.

The relationship between chain length and chain breakage was found to be directly related to the typical length of chain necessary to produce physical cross-links, i.e., chain entanglements (Figure 11.12). (This is probably due to the fact that chain entanglements greatly increase the "apparent" chain length and frictional energy needed to be overcome to move a chain.) Typically, at least one chain entanglement is needed to guarantee some chain breakage. For many vinyl polymers, including PS, one chain entanglement occurs for every 300 units. Experimentally, it was found that as the length of the PS chain increases so does the number of chain entanglements so that with a chain length of about 2000 (or an average of 7 chain entanglements), 50% of the chains are broken

FIGURE 11.12   Idealized relationship between the chain length and the number of entanglements and polymer strength.

and when the chain length is about 4000 (or an average of 13 entanglements), approximately 100% of the chains break.

The production of chain entanglements is statically directly related to polymer length for linear chains, and is almost independent of the nature of the vinyl unit for many polymers.

Chain length and entanglement are also related to the strength of the polymeric material. As chain length increases, the number of entanglements increases as does strength. At about 8 entanglements, the relationship between number of entanglements (and chain length) and polymer strength levels off with only small changes in polymer strength occurring as the chain length and number of entanglements further increases as shown below.

Because the distance between each entanglement is about 300 units and to achieve nearly maximum strength 8 entanglements are needed a chain of about 2400 (300 units/entanglement times 8 entanglements) units is needed. Calculations relating chain strength (related to the energy necessary to break a chain) and frictional force to hold a chain of varying length in place, that is

$$\text{Chain strength} \, \alpha \, \text{Frictional force as a function of chain length} \tag{11.26}$$

have been carried out with chain lengths of about 2000–3000 required before the frictional force necessary to break the chain occurs. This is in rough agreement with the chain length of 2400 calculated above using chain entanglement as the limiting factor.

As the polymer chain length increases, so does the viscosity of the melted polymer requiring more energy to process the polymers using any technique that requires the polymer to flow. This includes most of the molding processing typically used to process vinyl polymers. For many applications, the maximum strength is not needed so that industry looks to a balance between chain length and necessary strength. This "balance" is often chosen such that the chain length is sufficient to give seven entanglements rather than the eight required to insure about 100% chain breakage and maximum polymer strength, as noted above.

There are exceptions to this. One exception involves ultrahigh molecular weight linear polyethylene, UHMWPE, which has few chain entanglements, so it is easier to pull long chains from a PE matrix without chain breakage. The equivalent energy necessary to pull a UHMWPE chain from a PE matrix occurs at about a chain length of 100,000 (or a $\overline{DP}$ of about 3300), much larger than that for PS.

For condensation polymers, the attractive forces between chains and chain units are greater, so that physical chain entanglement is not necessarily the limiting factor, but rather other factors including localized crystallization become important.

11.12 SUMMARY

1. Polymers undergo reactions that are analogous to smaller molecules. Variation generally involves the need for reactants to have contact with the active site. This is more difficult for polymers.
2. Polymer degradation typically occurs via random chain scission, depolymerization, or both, resulting in a loss of chain length and properties associated with polymer length.
3. Among the most important reactions on polymers are those that occur with biomacromolecules such as involved in the transfer of oxygen and the activity of enzymes. Shape, size, and electronic configuration are all essential factors in the transfer of oxygen and activity of enzymes. The transfer of oxygen can be mathematically described using the classical Michaelis–Menten approach. The two major models describing the activity of nonallosteric enzymes, such as myoglobin, are referred to as the lock-and-key model and the induced-fit model. Hemoglobin is an example of an allosteric enzyme where the two most popular models describing this behavior are the concerted and sequential models.

GLOSSARY

Anchimeric reactions: Reactions enhanced by the presence of a neighboring group.

Chain scission: Breaking of a polymer chain.

Curing: Cross-linking producing a polymer network.

Cyclized rubber: Isomerized rubber containing cyclohexane rings.

Heme: Iron-containing active site for hemoglobin and myoglobin.

Hemoglobin: Allosteric enzyme responsible for the transport of oxygen in our bodies; contains four myoglobin-like units.

Hydrogenation: Addition of hydrogen to an unsaturated compound.

Induced-fit model: One of the two basic models employed to describe enzymatic behavior of non-allosteric molecules. Here, the steps are similar to the lock-and-key model except that the reactants "induce" a change in the conformation of the active site allowing the active site to bind with the reactant.

Kraton: Tradename for ABA block copolymer of styrene (A) and butadiene (B).

Kuhn element: Number of repeat units needed so that the most probable distance between the first and the growing end is back at the first unit again.

Ladder polymer: Double stranded polymer for added stability.

Lock-and key model: One of the two basic models to describe the selectivity and the catalytic nature of nonallosteric enzymes. In this model, the reactant(s) is attracted to the active site on the enzyme, which is of the same general size, shape, and complementary electronic nature as the reactant.

Macromolecular monomers (macromers): Polymers containing one reactive end group.

Myoglobin: Nonallosteric enzyme responsible for the storage of oxygen in our bodies. The activity follows the kinetic scheme described by the Michaelis–Menten model.

Ozonolysis: Reaction of an unsaturated organic compound with ozone.

Photosynthesis: Metabolic pathway for the conversion of carbon dioxide into organic compounds, mainly carbohydrates, using sunlight as the energy source.

RIM: Reaction injection molding.

Telechelic polymers: Polymers containing two active ends.

Telomerization: Abstraction of an atom by a macroradical.

Topochemical reactions: Reactions on the surface.

Williamson reaction: Reaction of an alkoxide and an alkyl chloride.

EXERCISES

1. In general, which metal ions are better chelated by polymers such as the salt of poly(acrylic acid)?
2. What is the general mechanism for the curing of step-reaction polymers?
3. Write the formula for perdeuterated polyethylene.
4. What is the major difference between reactions occurring on benzene and PS.
5. When would you expect a polymer to undergo degradation via random scission?
6. How might you cross-link a polyethylene coating after it is applied to a wire?
7. How might you prepare a block copolymer of styrene and an alternating copolymer of ethylene and propylene?
8. What is the similarity between completely hydrogenated *Heveal brasiliensis* and completely hydrogenated gutta percha?
9. What part does the porphyrin play in transmitting oxygen?
10. Would you expect PP chains with a $\overline{DP}$ of 200 to break when a PP plastic is broken into two pieces?
11. What are the most important structural factors involved in natural catalysis?

12. What product would be produced from the ozonolysis of polybutadiene?
13. Write the structural formula for the polymeric hydrolytic products from (a) PVAc and (b) poly(methyl methacrylate).
14. Why is commercial methylcellulose more soluble in water than native cellulose?
15. Why is CMC used in detergent formulations?
16. Would you expect the addition of chlorine or hydrogen to a double bond in a polymer to occur 100%?
17. What is the DS of cellulose nitrate when it is used as an explosive?
18. Why is the DS of cellulose triacetate only 2.8 and not 3.0?
19. Which is more polar: (a) cellulose triacetate or (b) cellulose diacetate?
20. Why is poly(vinyl butyral) not made of 100% vinyl butyral units?
21. If you have PVA and you want to make it less hydrophilic using a condensation reaction, what might you use?
22. What products would you expect from the combustion of most vinyl polymers.
23. What monomer would be obtained by the decomposition of PVA?
24. Which would be more resistant to nitric acid: (a) PS or (b) perfluoropolystyrene?
25. What ions would be removed form water by sulfonated polystyrene: (a) cations or (b) anions?
26. What reaction occurs when tannic acid is added to proteins such as those present in cowhide?
27. Which of the following should be most susceptible to degradation by acids and bases: (a) PS, (b) PVC, (c) PET, (d) Nylon 6,6, (e) proteins, (f) DNA.
28. Propose a procedure for recovering monomeric methyl methacrylate from scrap PMMA.
29. Why is PVC so thermally unstable?
30. Outline the three general steps in operation for photosynthesis.

FURTHER READINGS

Cheng, S. and Gross, R. (2008): *Polymer Biocatalysis and Biomaterials*, Oxford University Press, Ithaca, NY.
Chiellini, E. (2003): *Biodegrable Polymers and Plastics*, Kluwer, New York.
Kharitonov, A. (2008): *Direct Fluorination of Polymers*, Nova, Hauppauge, NY.
Lee, Y. S. (2008): *Self-Assembly and Nanotechnology: A Force Balance Approach*, Wiley, Hoboken, NJ.
van Leeuwen, P. (2008): *Supramolecular Catalysis*, Wiley, Hoboken, NJ.
Lyland, R. and Browning, I. (2008): *RNA Interference Research Progress*, Nova, Hauppauge, NY.
Moeller, H. (2007): *Progress in Polymer Degradation and Stability Research*, Nova, Hauppauge, NY.
Rotello, V. M. and Thayumanavan, S. (2008): *Molecular Recognition and Polymers: Control of Polymer Structure and Self-Assembly*, Wiley, Hoboken, NJ.
Rudnik, E. (2008): *Compostable Polymer Materials*, Elsevier, New York.
Scott, G. (2003): *Degradable Polymers*, Kluwer, New York.

12 Testing and Spectrometric Characterization of Polymers

Public acceptance of polymers is usually associated with an assurance of quality based on the knowledge of successful long-term and reliable testing. In contrast, much of the dissatisfaction with synthetic polymers is related to failures that might have been prevented by proper testing, design, and quality control. The American Society for Testing and Materials (ASTM), through its various committees, has developed many standard tests, which may be referred to by all producers and consumers. There are also cooperating groups in many other technical societies: the American National Standards Institute (ANSDI), the International Standards Organization (ISO), and standards societies such as the British Standards Institution (BSI) in England, and Deutsche Normenausschuss (DNA) in Germany, and comparable groups in every nation with developed polymer technology throughout the world.

Testing is done by industry to satisfy product specifications and for public protection using standardized tests for stress–strain relationships, flex life, tensile strength, etc. The U.S. tests are overseen by the ASTM through a committee arrangement. For instance, Committee D-1 oversees tests related to coatings, while Committee D-20 oversees tests on plastics. New tests are continuously being developed, submitted to the appropriate ASTM committee, and after adequate verification through "round robin" testing, finally accepted as standard tests. These tests are published by the ASTM. Each ASTM test is specified by a unique combination of letters and numbers, along with exact specifications regarding data gathering, instrument design, and test conditions making it possible for laboratories throughout the world to reproduce the test and hopefully the test results if requested to do so. The Izod test, a common impact test, has the ASTM number D256–56(1961). The latter number, 1961, is the year the test was first accepted. The ASTM publication gives instructions for the Izod test specifying test material shape and size, exact specifications for the test equipment, detailed description of the test procedure, and how results should be reported. Most tests developed by one testing society have analogous tests or more often utilize the same tests so that they may have both ASTM, ISO, and other standardized society identification symbols.

A number of physical tests emphasizing stress–strain behavior will be covered in Chapter 13. Here we will concentrate on other areas of testing, emphasizing thermal and electrical properties, and on the characterization of polymers by spectral means. Spectroscopic characterization generally concentrates on the structural identification of materials. Most of these techniques, and those given in Chapter 13, can also be directly applied to nonpolymeric materials such as small organic molecules, inorganic compounds, and metals.

The testing of materials can be based on whether the tested material is chemically changed or is left unchanged. Nondestructive tests are those that result in no chemical change in the material, which may include the determination of many electrical properties, most spectroanalyzes, simple phase change tests (T_g and T_m), density, color, and the determination of most mechanical properties. Destructive tests result in a change in the chemical structure of at least a portion of the tested material. Examples include the flammability and chemical resistance tests when the material is not resistant to the tested material.

12.1 SPECTRONIC CHARACTERIZATION OF POLYMERS

12.1.1 INFRARED SPECTROSCOPY

The infrared spectral (IR) range spans the region bound by the red end of the visible region to the microwave region at the lower frequencies. Molecular interactions that involve vibrational modes correspond to this energy region. Infrared spectroscopy (IR) is one of the most common spectronic techniques used today to identify polymer structure. Briefly, when the frequency of incident radiation of a specific vibration is equal to the frequency of a specific molecular vibration the molecule absorbs the radiation. Today, most IR machines are rapid scan where the spectra are Fourier transformed. For the most part, IR band assignments for polymers are analogous to those made for small molecules.

In Fourier transform infrared spectroscopy (FTIR), the light is guided through an interferometer where the signal undergoes a mathematical Fourier transform giving a spectrum identical to the conventional dispersive IR.

With the advent of femtosecond infrared laser pulses, two-dimensional infrared correlative spectroscopy has become a new tool. Here, pump pulses are applied to the sample. After some time, which can be from zero to several picoseconds to allow the sample to relax, a second pulse is applied. The result is a 2D plot of the frequency that resulted from the initial pump pulse and a second plot resulting from the relaxed state spectrum. This allows the coupling of various vibrational modes. In some ways, this is similar to two-dimensional NMR spectroscopy in that the spectrum is spread out in two dimensions allowing certain "cross-peaks" to be observed.

Following are brief discussions of some of the more important techniques used in polymer analysis.

Attenuated total reflectance IR (ATR-IR) is used to study films, coatings, threads, powders, interfaces, and solutions. (It also serves as the basis of much of the communications systems based on fiber optics.) ATU occurs when radiation enters from a more-dense material (i.e., with a higher refractive index) into a less-dense material (i.e., with a lower refractive index). The fraction of the incident radiation reflected increases when the angle of incidence increases. The incident radiation is reflected at the interface when the angle of incidence is greater than the critical angle. The radiation penetrates a short depth into the interface before complete reflection occurs. This penetration is called the evanescent wave. Its intensity is reduced by the sample where the sample absorbs radiation.

Specular reflectance IR involves a mirrorlike reflection producing reflection measurements of a reflective material or a reflection–absorption spectrum of a film on a reflective surface. This technique is used to look at thin (from nanometers to micrometers thick) films.

Diffuse reflectance infrared Fourier transform spectroscopy (DRIFTS) is used to obtain spectra of powders and rough polymeric surfaces such as textiles and papers. IR radiation is focused onto the surface of the sample in a cup resulting in both specular reflectance (which directly reflects off the surface having equal angles of incidence and reflectance) and diffuse reflectance (which penetrates into the sample subsequently scattering in all angles). Special mirrors allow the specular reflectance to be minimized.

Photoacoustic spectroscopy IR (PAS) is used for highly absorbing materials. Generally, modulated IR radiation is focused onto a sample in a cup inside a chamber containing an IR-transparent gas such as nitrogen or helium. The IR radiation absorbed by the sample is converted into the heat inside the sample. The heat travels to the sample surface and then into the surrounding gas causing expansion of the boundary layer of gas next to the sample surface. The modulated IR radiation thus produces intermittent thermal expansion of the boundary layer creating pressure waves that are detected as photoacoustic signals.

PAS spectra are similar to those obtained using ordinary FTIR except that the truncation of strong absorption bands occurs because photoacoustic signal saturation often occurs. PAS allows the structure to be studied at different thicknesses because the slower the frequency of modulation, the deeper the penetration of IR radiation.

Emission infrared spectroscopy is used for thin films and opaque polymers. The sample is heated so that energy is emitted. The sample acts as the radiation source, and the emitted radiation is recorded giving spectra similar to those of classical FTIR. In some cases, IR frequencies vary because of differences in the structures at different depths and interactions between surface and interior emissions.

Infrared microscopy allows the characterization of minute amounts of a material or trace contaminants or additives. Samples as small as 10 microns can be studied. The microscope, often using fiber optics, allows IR radiation to be pinpointed.

Today, there are many so-called hyphenated methods with IR. Hyphenated methods involving IR include GC–IR where the IR spectra of materials are taken as they are evolved through the column. Related to this are HPLC–IR, TG–IR, and MS–IR.

12.1.2 Raman Spectroscopy

Raman spectroscopy is similar to IR spectroscopy in that it investigates polymer structure focusing on the vibrational modes. Whereas IR is a result of energy being absorbed by a molecule from the ground state to an excited state, Raman spectroscopy is a scattering phenomenon where the energy of photons is much greater than the vibrational transition energies. Most of these photons are scattered without change (so-called Rayleigh scattering). Even so, some are scattered from molecular sites with less energy than they had before the interaction resulting in Raman–Stokes lines. Another small fraction of photons have energies that are now greater than they originally had leading to the formation of anti-Stokes lines. Only the Raman–Stokes photons are important in Raman spectroscopy. While many chemical sites on a polymer are both IR active and Raman active, i.e., they both give rise to bands, some are less active or even nonactive because of the difference between groups that can absorb and those that scatter. These differences are generally described in terms of the symmetry of vibration. Briefly, an IR absorption occurs only if there is a change in the dipole moment during the vibration whereas a change in polarizability is required for Raman scattering to occur. Even so, the spectra are generally similar and a comparison of the two allows for additional structural characterization beyond that obtained from either of the techniques alone.

Carraher and Williams showed that for many polymers, differences in symmetry and band production were similar for small molecules as they were for the same groups found in polymers. Thus, observations from the literature and in model-compound studies with small molecules are generally applicable to similar moieties present in polymeric systems for both Raman and IR spectral analyzes.

As in the case with IR spectrometers, there exists a wide variety of specialty techniques especially applicable to polymer analysis.

In *surface-enhanced Raman spectroscopy* (SERS), samples are adsorbed onto microscopically roughened metal surfaces. Spectra are the intensities and frequencies of scattered radiation originating from a sample that has been irradiated with a monochromatic source such as a laser. SERS spectra are of molecules that are less than 50 Å from the surface.

12.1.3 Nuclear Magnetic Resonance Spectroscopy

Nuclear magnetic resonance spectroscopy (NMR) is a powerful tool for polymer structure characterization. Certain isotopes have two or more energy states available when exposed to a magnetic field. The transitions between these energy states are the basis for NMR. These magnetically active nuclei have a property called spin. As a consequence of this spin, these nuclei have an angular momentum and a magnetic moment. The ratio of these two properties is called the magnetogyric ratio. Each isotope has a distinct magnetogyric ratio that varies a little with the particular chemical environment in which it is placed.

While NMR has been a strong characterization tool for polymers for many years, it has increased in its usefulness because of continually improved instrumentation and techniques. When a nucleus is subjected to a magnetic field two phenomena are observed—Zeeman splitting and nuclear precession. Zeeman splitting creates $2I + 1$ magnetic energy states, where I is the spin quantum number. When the atomic mass and atomic number are even numbers, $I = 0$ so that these nuclei are unable to have multiple energy levels when exposed to a magnetic field. Thus ^{12}C, which has both an even atomic number and atomic mass is NMR inactive, whereas ^{13}C, which has an uneven atomic mass, is NMR active. Nuclear precession is the motion of a spinning body whose axis of rotation changes orientation. The precessional frequency is equal to the magnetic field strength times the magnetogyric ratio.

In a magnetic field, NMR active nuclei can be aligned with the magnetic field (low-energy state) or aligned against the field (high-energy state). At room temperature, there are slightly more nuclei in the lower energy state than in the higher energy state. As magnetic energy is supplied that corresponds to the energy gap, quantum level, between the low and high-energy states, some nuclei in the low-energy state move to the high-energy state resulting in an absorption of energy which is recorded as a NMR spectra. The difference between the two energy states is related to the strength of the external magnet. Better spectra are obtained when instruments with larger magnetic fields are employed.

Because of the small but consistent concentrations of carbon 13 present in all organic compounds, it is necessary to use more sophisticated NMR spectroscopy for determining the effect of neighboring electrons on these nuclei. However, ^{13}C-NMR spectroscopy is a valuable tool for investigating polymer structure.

Following is a short description of some of the newer advances.

Nuclear overhauser effect. The nuclear overhauser effect (NOE) only occurs between nuclei that share a dipole coupling, i.e., their nuclei are so close that their magnetic dipoles interact. Techniques that use NOE enhance ^{13}C spectra and allow spacial relationships of protons to be determined.

Two-dimensional NMR. Basically, the two-dimensional NMR techniques of nuclear overhauser effect spectroscopy (NOESY) and correlation spectroscopy (COSY) depend on the observation that spins on different protons interact with one another. Protons that are attached to adjacent atoms can be directly spin coupled, and thus can be studied using the COSY method. This technique allows the assignment of certain NMR frequencies by tracking from one atom to another. The NOESY approach is based on the observation that two protons closer than about 0.5 nm perturb one another's spins even if they are not closely coupled in the primary structure. This allows spacial geometry to be determined for certain molecules.

The use of actively shielded magnetic field gradients has made the use of pulsed field gradients possible. The use of pulsed field gradients reduces experiment time, minimizes artifacts, and allows for further solvent suppression.

In ***pulsed NMR,*** the magnetic field is turned on for the time necessary to rotate the magnetization vector into a plane called the 90° rotation or 90° pulse. The field is turned off and the magnetization vector rotates at a nuclear precession frequency relative to the coil. This induces a NMR signal that decays with time as the system returns to equilibrium. This signal is called the ***free induction decay (FID)***.

After a sample is excited, the spin loses excess energy through interactions with the surrounding eventually returning to its equilibrium state. This process is exponential and is called ***spin-lattice relaxation***. The decay is characterized by an exponential time constant.

Two-dimensional experiments allow the more precise determination of coupling relationships. Such experiments are carried out by collecting a series of FID spectra. The time between the pulses is called the ***evolution time***. The evolution time is systematically increased as each successive FID is obtained. Each new FID shows a continued change in the couplings in the polymer. The FID spectra are treated using Fourier transformation. A new series of FID spectra are now created by connecting points for each spectra, and these new FIDs are again treated by Fourier transformation producing two-dimensional spectra that are often presented as contour plots. Nuclei that share J-coupling produce a correlation peak. Such approaches allow a better interpretation of dipole couplings, molecular diffusion, J-coupling, and chemical exchange.

Solids. Many polymers are either difficultly soluble or insoluble. The NMR of solids generally gave broad lines because of the effects of dipolar coupling between nuclei and the effect of *chemical shift anisotropy (CSA)*. Both these effects are greatly reduced for polymers in solution and allow for the decent spectra of soluble polymers in solution.

CSA effects are large for solids and are the result of the directional dependence of electronic shielding. CSA effects are overcome through rapidly spinning the sample at an angle to the magnetic field known as the *magic angle*. Solid probes use spinning rotors to hold and spin the sample. The sample is rotated at the magic angle and spun fast enough to remove CSA effects. High-power decouplers or multiple pulse line narrowing allow the decoupling between protons and carbon by using a series of pulses to average the dipolar interactions through spin reorienting. Cross-polarization uses dipolar coupling to increase the sensitivity of less sensitive nuclei. The combination of cross-polarization and magic angle spinning (CPMAS) allows good spectra to be obtained for solid polymers.

12.1.4 NMR Applications

The various types of spectroscopes complement one another, sometimes giving "new" information and other times giving similar information. The amount of crystallinity and the preference for syndiotactic, isotactic, and atactic structures have been determined employing a number of techniques including x-ray. This information can be correlated with information that can be readily obtained from IR and NMR so that product control can be easily monitored. For instance, the NMR spectra of poly(methyl methacrylate) (PMMA) made by different synthetic routes give different proton shifts that allow the determination of the amounts of isotactic, syndiotactic, and atactic materials. Briefly, the production of PMMA via so-called high temperature free radical polymerization gives a largely atactic material. The proton chemical shifts for the α-methyl appear at about 8.8, 8.9, and 9.1 (chemical shifts are based on the tetramethylsilane peak having a value of 10.00 ppm). Largely isotactic PMMA produced using anionic polymerization shows an enhanced peak at about 8.8 that is assigned to the configuration where the α-methyl group in the PMMA repeat unit is flanked on both sides by the units of the same configuration. It is the major peak in the 8.8–9.1 triad. PMMA produced by employing low temperature free radical polymerization gives largely syndiotactic product. The peak at 8.9 is assigned to the α-methyl present in a heteroatctic configuration so that the units on either side of it are unlike the central mer since it now becomes the major peak. The 9.1 peak is the largest for the higher temperature product produced via free radical polymerization and is due to the triad central unit being flanked by a like and unlike mer.

Today's NMR capability allows the determination of additional structural features in solution and solid state including the identification of end groups, branches, and defects.

The combinations of solid state NMR and IR allow the molecular description of the effect of stress–strain, exposure to various conditions including chemical treatments, radiation, and heat on materials.

12.1.5 Electron Paramagnetic Resonance Spectroscopy

Electron paramagnetic resonance spectroscopy (EPR), or electron spin resonance (ESR) spectroscopy, is a valuable tool for measuring the relative abundance of unpaired electrons present in macromolecules. For example, macroradicals are formed by the homogeneous cleavage of nylon-66 chains when these filaments are broken, with the concentration of macroradicals increasing as the stress is increased.

12.1.6 X-Ray Spectroscopy

X-ray diffraction is a widely used tool for structural identification for almost all solids under the right conditions. X-ray diffractometers are generally either single crystal or powder.

Single-crystal studies allow the absolute configurational determination of polymeric materials that have high degrees of crystallinity. Such determination is costly with respect to time because of the complexity of polymeric materials.

Powder x-ray spectroscopy can employ smaller crystalline samples from one to several hundred nanometers. These crystallites have broadened peak profiles as a result of incomplete destructive interference at angles near the Bragg angle defined as

$$n\lambda = 2d \sin \theta \tag{12.1}$$

where
   n is the order of a reflection
   λ is the wavelength
   d is the distance between parallel lattice planes
   θ is the angle between the incident beam and a lattice plane known as the Bragg angle

This broadening allows the determination of crystallite size and size distribution. (Note that this is not particle size.)

The x-ray analysis of proteins and nucleic acids is especially important, as the absolute structure is needed for many advances in the fields of medicine and biochemistry.

12.2 SURFACE CHARACTERIZATION

Everything has a surface or an interface. These surfaces have their own kinetic and thermodynamic features that affect their formation and behavior. Sperling notes that for most polymers, the end groups reside perpendicular to the bulk of the polymers probably because the end is less hydrophobic to the bulk, and the polymer surfaces generally are "faced" with an air atmosphere that is more hydrophilic. When a polymer solution is deposited onto a surface to "dry," the concentration has an influence on the orientation of the polymer chains at the surface in the dried solid. Thus, when the amount of polymer is small, the polymer chain lays parallel to the surface in a so-called pancake form (Figure 12.1). As the concentration increases, the surface is not able to accommodate the entire polymer chain, and it begins to form an inner tangled chain with only the end and some of the chain segments facing the surface forming a "mushroom" shape. Finally, as the concentration of polymer increases, only the ends of the polymer chains occupy the surface with the polymer ends forming "brushes."

There is no exact, universally accepted definition of a surface. Here, the surface will be defined as the outermost atomic layers, including absorbed foreign atoms. The chemical and physical compositions, orientations, and properties of surfaces typically differ from those of the bulk material.

Current surface characterization techniques fall into two broad categories—those that focus on the outermost few layers (to within the 10–20 layer boundary) and those that focus on components present to several thousand angstroms into the solid (hundred to several hundred layers).

Pancake   Mushroom   Brushes

Increasing polymer concentration

FIGURE 12.1   General surface structures as a function of polymer concentration.

Attenuated total reflectance (ATF) typically employs special cells fitted onto traditional IR, FTIR, or UV instruments. While some outer surface aspects are gleaned from such techniques, information up to several thousand angstroms is also present in the spectra from ATF.

Techniques that analyze the first few atomic layers generally involve low-energy electrons or ions, since the incident radiation should penetrate only the top few layers. Normally, a combination of techniques is employed to clearly define the composition of the outer layers. Special precautions are employed to minimize sample surface contamination.

Following is a brief presentation of some of these modern surface characterization techniques.

12.2.1 Auger Electron Spectroscopy and X-Ray Photoelectron Spectroscopy

Auger electron spectroscopy (AES) and x-ray photoelectron spectroscopy (XPS) are two principle surface analysis techniques. They are used to identify the elemental composition, i.e., the amount and nature of species present at the surface to a depth of about 1 nm.

In Auger transitions, incident electrons interact with the inner shell electrons (E_i) of the sample. The vacancy created by an ejected inner shell electron is filled by an outer shell electron (E_1), and a second outer shell electron (E_2) is ejected leaving the atom in a doubly ionized state. The electrons ejected from the outer shells are called Auger electrons, named after Frenchman, Pierre Auger, who discovered the effect. Thus, AES measures the energies of the Auger electrons (E_a) emitted from the first 10 Å of a sample surface. The energy equation is expressed as

$$E_a = E_1 - E_i + E_2 \tag{12.2}$$

The kinetic energies of these ejected electrons originating within the first 30 Å of the sample surface are measured by XPS. In XPS, a sample is bombarded by a beam of x-rays with energy hv, and core electrons are ejected with a kinetic energy E_k that overcomes the binding energy E_b and the work function (ϕ). These core electrons are called the x-ray photoelectrons. The energy equation is expressed as follows:

$$E_k = hv - E_b - \phi \tag{12.3}$$

The energies of these photoelectrons are reflected in the type and environment of the atoms present at the surface.

12.2.2 Near-Field Scanning Optical Microscopy

Optical microscopes have one serious drawback, their resolution, resulting from the fundamental physics of lenses. Lord Rayleigh, over 100 years ago, defined the presently accepted maximum optical lens resolution to be one-half the wavelength of the imaging radiation. In truth, conventional optical microscopy did not achieve this level of definition mainly because of out-of-focus light. This prevented the observation of atoms and single molecules.

Near-field scanning optical microscopy (NSOM) allows an extension of optical microscopy to near that of electron microscopy. The central feature is the optical element that is similar, and sometimes the same, to that employed in atomic force microscopy (AFM). Essentially, light is directed through the probe tip onto the sample from just immediately above the sample surface. The light emanating from the probe tip is smaller than the light's wavelength and spreads out over the surface. This results in the maximum influence occurring at the surface with little contribution from regions nearby (such as within 30 nm) resulting in little out-of-focus light. Depending upon the surface and the sample thickness, the light is measured as absorption or fluorescence, and is collected and recorded electronically. NSOM can be fitted onto a regular optical microscope or coupled with scanning probe microscopy (SPM).

12.2.3 ELECTRON MICROSCOPY

The upper limit of magnification for optical microscopes is about 2000 times. Thus, additional forms of microscopy have been developed that allow near to actual atomic observation of polymer surfaces. Electron microscopy utilizes an electron beam to act as the sensing radiation in place of light. High-energy electrons take on wave character as they approach the speed of light. The wavelength is inversely proportional to the electron speed or velocity. When accelerated over large voltages, electrons can be made to travel at speeds to permit wavelengths on the order of 0.003 nm. The electron beam is focused, and the image is formed using magnetic lenses. The two most common forms of electron microscopy are transmission electron microscopy and scanning electron microscopy.

In *scanning electron microscopy* (SEM), the surface of the polymeric surface is scanned using an electron beam with the reflected or back-scattered beam of electrons collected and displayed on a cathode ray tube screen. The image represents the surface contour of the scanned material. Because the surface must be conductive, most polymer surfaces must be overlaid with a conductive coating. Magnifications up to about 50,000 are carried out using SEM.

Transmission electron microscopy (TEM) utilizes an image formed by an electron beam that passes through the sample. This allows internal microstructures to be determined. The structural details of materials can be observed on an atomic level by looking at contrasts in the image caused by the various concentrations of different elements. Very thin films are employed. Under good conditions, magnifications up to one million are possible by employing TEM.

12.2.4 SCANNING PROBE MICROSCOPY

Scanning probe microscopy (SPM) encompasses a group of surface-detection techniques that include AFM, and *scanning tunneling microscopy* (STM) that allow the topographic profiling of surfaces. SPM techniques investigate only the outermost few atomic layers of the surface with a nanometer resolution and at times atomic level resolution.

STM is generally used with electrically conductive materials applied to polymeric materials giving overlays consisting of conducting material layered over the surface of the sample. STM experiments typically require extremely low pressures less than 1×10^{-10} mbar. By comparison, AFM can be run under room conditions and does not require the use of electrically conductive material. In STM, the metallic tip is held close (about 0.5–1 nm) to the surface. A voltage, applied between the tip and the sample surface, drives a tunneling current. The conductive surface reconstructs the atomic positions via minimizing the surface free energy. This gives topographic superstructures with specific electronic states, which are recorded as surface contours or images.

AFM can be run under room conditions. AFM can be performed in either of the two forms—a contact mode and a noncontact mode. It does not require the use of electrically conductive material since (in the contact mode) the tip actually "touches" the surface rather than residing immediately above it as is the case in STM. In both the contact mode and the noncontact mode, light is used as the sensing source rather than an applied voltage. In contact-AFM, a cantilever with as sharp a point as possible is laid onto the sample surface with a small loading force in the range of 10^{-7}–10^{-10} N. Tips of differing sizes and shapes are tailor made. Data are obtained optically by bouncing an incident laser beam onto the cantilever toward a quadrant detector on or in an interferometer. The AFM can work in two modes—either a contact mode or in a noncontact mode. In the noncontact mode, the attractive force is important, and the experiment must be carried out under low pressures similar to those employed in STM.

In the contact mode, the tip acts as a low-load, high-resolution profiler. Along with structure determination, the AFM is also used to "move" atoms about allowing the construction of images at the atomic level. The AFM is also an important tool in the nanotechnology revolution.

Nanotubes are being used as points in some scanning probe microscopy units. The ends of these nanotubes can be closed or functionalized offering even "finer" tips and tips that interact with specific sites allowing manipulation on an atom-by-atom basis. These nanotubes are typically smaller than silicon tips and are generally more robust.

AFM is useful in identifying the nature and amount of surface objects. AFM, or any of its variations, also allow the studies of polymer phase changes, especially thermal phase changes, and the results of stress–strain experiments. In fact, any physical or chemical change that brings about a variation in the surface structure can, in theory, be examined and identified using AFM.

Today there exist a wide variety of AFMs that are the modifications or extensions of traditional AFM. Following is a brief summary of some of these techniques.

Contact mode AFM is the so-called traditional mode of AFM. Topography contours of solids can be obtained in air and fluids.

Tapping mode AFM measures contours by "tapping" the surface with an oscillating probe tip thereby minimizing shear forces that may damage soft surfaces. This allows an increased surface resolution. This is currently the most widely employed AFM mode.

There are several modes that employ an expected difference in the adhesion and the physical property (such as flexibility) as the chemical nature is varied. ***Phase imaging*** experiments can be carried out that rely on differences in surface adhesion and viscoelasticity. ***Lateral force microscopy*** (LFM) measures frictional forces between the probe tip and the sample surface. ***Force modulation*** measures differences between the stiffness and/or the elasticity of surface features. ***Nanoindenting/ scratching*** measures mechanical properties by "nanoindenting" to study hardness, scratching, or wear including film adhesion and coating durability. LEM identifies and maps relative differences in surface frictional characteristics. Polymer applications include identifying transitions between different components in polymer blends, composites, and other mixtures, identifying contaminants on surfaces, and looking as surface coatings.

Noncontact AFM measures the contour through sensing van der Waals attractive forces between the surface and the probe tip held above the sample. It provides less resolution than the tapping mode AFM and the contact mode AFM.

There are several modes that employ differences in a materials' surface electronic character and/or magnetic character as the chemical nature of the surface varies. ***Magnetic force microscopy*** (MFM) measures the force gradient distribution above the sample. ***Electric force microscopy*** (EFM) measures the electric field gradient distribution above a sample surface. EFM maps the gradient of the electric field between the tip and the sample surface. The field due to trapped charges, on or beneath the surface, is often sufficient to generate contrast in an EFM image. The voltage can be induced by applying a voltage between the tip and the surface. The voltage can be applied from the microscopes electronics under AFM control or from an external power supply. EFM is performed in one of the three modes—phase detection, frequency modulation, FM, or amplitude detection. Three-dimensional plots are formed by plotting the cantilever's phase or amplitude as a function of the surface location. ***Surface potential microscopy*** (SP) measures differences in the local surface potential across the sample surface. SP imaging is a nulling technique. As the tip travels above the surface, the tip and the cantilever experience a force whenever the surface potential differs from that of the tip. The force is nullified by varying the voltage of the tip so that the tip remains at the same potential as that of the immediate surface. The voltage is applied to the tip to maintain this constant potential as the tip surveys the surface with the results being plotted as a function of the surface coordinates creating a surface potential image. For best results, SP and EFM do the best job with conductive materials. ***Force volume*** measurements involve the production of two-dimensional arrays of force-distance values allowing a mapping of the force variation and the surface topology with individual force curves to be constructed.

Force–distance microscopy measures repulsive, attractive, and adhesion forces between the time and the surface during approach, contact, and separation. This technique combines electrical property with adhesion/physical property as a means to study sample surfaces.

Scanning thermal microscopy (SThM) measures two-dimensional temperature distributions across a sample surface. This is a special thermal technique.

In *electrochemical microscopy*, ECSTM and ECAFM, the material is immersed in electrolyte solution, and the surface and properties of conductive materials are studied.

Information derived from several of these techniques go together to give a clearer idea of the nature of the surface.

AFM results can be utilized in conjunction with other techniques. While some techniques, such as SAXS and SANS, allow structural information to be inferred, AFM gives real space results. While some of the polymeric structural designs may not be unambiguously determined, many can be determined employing ATM. The major limitation concerns whether the structures observed at or near the surface are similar to those in the interior. We are well aware that the surface composition differs from the interior composition. For instance, surfaces may be less organized being enriched in chain ends, loops, and switchboard chain segments. Further, for "sliced" samples the "slicing" disturbs the fine structure along the "cut" surface. For instance, the structure of linear polymers such as polyethylene has been suggested to consist of ordered or sharp folds, switchboard-like, loops with loose folds, buttressed loops, and combinations of these features. Magonov and Godovsky and others recently investigated the surface structures of a number of polymers employing ATM. For single crystals of PE, ordered grains 10–12 nm in size are found. For melt-crystallized LLDPE, spherulites of several microns are the major morphological features. Edge-on standing lamellae and lamellar sheets are found. Dark areas are assigned as amorphous regions. The lamellar edges are on the order of 25–40 nm, while the strands are several microns in length. By comparison, melt-crystallized LDPE, which is only about 30% crystalline, shows only spherulitic patterns with ill-defined ring patterns. The grain sizes are about 15–25 nm with fibrillar structures visible. A sample of melt-crystallized ULDPE with low crystallinity (about 15%) gives largely an ill-defined surface consistent with the surface being largely amorphous. As higher force is applied to press through the surface layer, grains of 0–10 nm and finally 9–11 nm become visible with some grains up to about 100–150 nm visible.

Other polymers have been studied. For instance, isotactic-polypropylene, i-PP, shows well-defined spherulites with grains (15–20 nm) embedded in an amorphous material. The grains are assembled in circles, and in some cases, along the radial direction, an ordered texture exists. PVDF shows numerous spherulites with fibrils 12–15 nm in width. The granular nanostructure of spherulites has also been found for polyesters and polyurethanes. ATM and other studies (including WAXS and SAXS) suggest that the nanoscale grains are elementary building blocks of the crystalline architecture in most polymers. These grains or blocks can have more or less structure within them. The overall crystalline structure may be developed as a one-dimensional assembling of grains into fibrils and as a two-dimensional assembling of grains into lamella. A correlation between the grain size and the size of molecular coils has not yet been answered using ATM.

Spin-cast films of poly(ethylene oxide) (PEO) show a flat crystalline morphology with lamellar sheets of different shapes. When melted and then cooled, PEO crystallizes with a similar morphology except that the lamellar sheets are smaller. When it is again melted and cooled, crystallization proceeds more slowly and the PEO morphology is dominated by spiral crystallites formed via a screw dislocation mechanism. In all cases, the thickness of the lamellar sheets is about 12 nm indicating multiple folding of the PEO chains. The lamellar sheets disappear at about 60°C though the melting point is listed to be 70°C. On cooling, the lamellar structures reappear at about 50°C.

The morphology of spin-cast film, thickness of 180 nm, from polycaprolactone shows many spherulitic structures with fibrillar nanostructures formed of lamellae lying edge on (about 10 nm thick) and areas with lamellar sheets lying flat on. Different crystalline structures are found when the sample is melted and crystallized as a function of temperature. These two studies reinforce the complex inner relationship between physical treatment and nanostructure.

While some structures show seemingly independent spherulitic structures on the surface, we know from other studies that these structures are connected to one another and to the more amorphous regions overall giving a material with a characteristic flexibility and strength. In general, chains are shared with adjacent areas allowing a sharing of stress–strain factors.

AFM is important for biological as well as synthetic macromolecules. Several examples are given to illustrate applications. Collagen is an important natural protein that is present in many tissues including bones, skin, tendons, and cornea. It is also employed in medical devices such as artificial skins, tendons, cardiac valves, ligaments, hemostatic sponges, and blood vessels. There are at least 13 different types of collagen. ATF can image collagen molecules and fibers, and their organization allows the identification of different kinds of collagen and at least surface interactions.

AFM allows the study of cell membranes. The precise organization of such cell membranes is important since they play a role in cell communication, replication, and regulation. It is possible to study real time interactions of such biologically important surfaces. Further, bilayers, modeled or containing naturally produced bilayers are used as biosensors. Again, interactions of these biomembranes can be studied employing AFM. For instance, the degradation of bilayers by phospholipases, the attachment of DNA, etc. can be studied on a molecular level. In another application, antibody–antigen interactions have been studied by employing AFM. One application of this is the creation of biosensors to detect specific interactions between antigens and antibodies.

12.2.5 SECONDARY ION MASS SPECTROSCOPY

Secondary ion mass spectroscopy (SIMS) is a sensitive surface analysis tool. Here, the mass analysis of negative and positive ions sputtered from the polymer surface through ion bombardment is analyzed. The sputtering ion beam is called the primary ion beam. This beam causes the erosion of the polymer surface removing the atomic and molecular ions. Then these newly created ions, composed of what is called the secondary ion beam, are analyzed as a function of mass and intensity. The depth of detection for SIMS is of the order of 20–50 Å. Because it is the ions in the secondary ion beam that are detected, the mass spectra obtained from SIMS are different from those obtained using simple electron impact methods. The extent of particular ion fragments observed is dependent on a number of factors including the ionization efficiency of the particular atoms and molecules composing the polymer surface.

SIMS can detect species that are present on surfaces of the order of parts per million to parts per billion

12.3 AMORPHOUS REGION DETERMINATIONS

Experimental tools that have been employed in an attempt to characterize amorphous regions are given in Table 12.1. Techniques, such as birefringence and Raman scattering, give information related to the short-range (less than 20 Å) nature of the amorphous domains while techniques, such

TABLE 12.1
Techniques Employed to Study the Amorphous Regions of Polymers

Short-Range Interactions	Long-Range Interactions
Magnetic birefringence	Electron diffraction
Raman scattering	Wide-angle x-ray scattering
Depolarized light scattering	Electron microscopy
Rayleigh scattering	Density
Brillouin scattering	Small-angle neutron scattering
NMR relaxation	
Small-angle x-ray scattering	

as neutron scattering, electron diffraction, and electron microscopy, give information concerning the longer range nature of these regions.

Birefringence measures order in the axial, backbone direction. The birefringence of a sample can be defined as the difference between the refractive indices for light polarized in two directions 90° apart. Thus, a polymer sample containing polymer chains oriented in a preferential direction by stretching or some other method will exhibit a different refractive index along the direction of preferred chain alignment compared to that obtained at right angles. This change in birefringence gives information concerning the amount of order, thus, information about disorder.

Small-angle neutron scattering (SANS) results indicate that vinyl polymers exist in random coils in the amorphous state. Results from the electron and x-ray diffraction studies show diffuse halos consistent with the nearest-neighbor spacings being somewhat irregular. It is possible that short-range order and long-range disorder exists within these amorphous regions.

12.4 THERMAL ANALYSIS

Because polymeric materials are expected to perform under a variety of temperature conditions, thermal properties are important. Thermal property investigations can also allow the better design of materials that meet the thermal requirements and may also give added structural data.

Major instrumentation involved with the generation of thermal property behavior of materials includes thermogravimetric analysis (TG, TGA), differential scanning calorimetry (DSC), differential thermal analysis (DTA), torsional braid analysis (TBA), thermomechanical analysis (TMA), thermogravimetric-mass spectrometry analysis (TG-MS), and pyrolysis gas chromatography (PGC). Most of these analysis techniques measure polymer response as a function of time, atmosphere, and temperature.

One of the simplest techniques is PGC in which the gases, resulting from the pyrolysis of a polymer, are analyzed by gas chromatography. This technique may be used for qualitative and quantitative analyzes. The latter requires calibration with known amounts of a standard polymer pyrolyzed under the same conditions as the unknown.

There are several different modes of thermal analysis described as DSC. DSC is a technique of nonequilibrium calorimetry in which the heat flow into or away from the polymer is measured as a function of temperature, atmosphere, and/or time. This differs from DTA where the temperature difference between a reference and a sample is measured. Even so, the distinction is not always clear. DSC equipment measures the heat flow by maintaining a thermal balance between the reference and the sample by changing current passing through the heaters under the two chambers. For instance, the heating of a sample and a reference proceeds at a predetermined rate until heat is emitted or consumed by the sample. The circuitry is programmed to give a constant temperature between the reference and the sample compartments. If an endothermic occurrence takes place, such as the melting of the sample, the temperature of the sample will be less than that of the reference. Less current is given to the sample, and the amount of the decrease reflects the amount of energy needed to melt the sample. The amount of energy required to melt the sample is generally given as the area under the resulting curve. The advantages of DSC and DTA over a good adiabatic calorimeter include speed, low cost, and the ability to use small samples.

The resultant plot of ΔT as a function of temperature is known as a thermogram. Figure 12.2 contains such a thermogram emphasizing the T_g and T_m endothermic regions for a polymer sample that is approximate 50% crystalline. Several things should be noted. First, the magnitude of the T_m is greater than the T_g because it requires more energy to gain entire chain mobility than to achieve segmental mobility. There is a small exothermic region after the T_g that reflects a tendency to form crystalline micelles when there is sufficient mobility to allow their formation.

Possible determinations from DSC/DTA measurements include (1) heat of transition, (2) heat of reaction, (3) sample purity, (4) phase diagram, (5) specific heat, (6) sample identification, (7) percentage incorporation of a substance, (8) reaction rate, (9) rate of crystallization or melting,

FIGURE 12.2   Representative DCS thermogram.

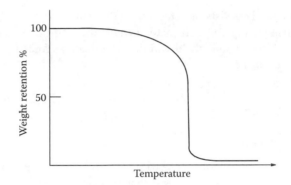

FIGURE 12.3   TGA thermogram for a typical vinyl polymer.

(10) solvent retention, and (11) activation energy. Thus, thermocalorimetric analysis can be a useful tool in describing the chemical and physical relationships of a polymer with respect to temperature.

In TGA, a sensitive balance is used to follow the weight change of a polymer as a function of time, atmosphere, or temperature. Figure 12.3 shows a typical TGA thermogram for a vinyl polymer such as PE. For such polymers where there is a similar thermal energy stability of the backbone, the TGA decrease generally occurs with one downward sweep beginning at about 200°C and being complete before 300°C. There is normally some residue that is high in carbon content remaining to over 1000°C.

Along with a determination of the thermal stability of the polymer, TGA can be used to determine the following: (1) sample purity, (2) identification, (3) solvent retention, (4) reaction rate, (5) activation energy, and (6) heat of reaction.

TMA measures the mechanical response of a polymer looking at (1) expansion properties including the coefficient of linear expansion, (2) tension properties such as measurement of shrinkage and expansion under tensile stress, i.e., elastic modulus, (3) volumetric expansion, i.e., specific volume, (4) single-fiber properties, and (5) compression properties such as measuring the softening or penetration under load.

In TBA, changes in structure are measured. The name TBA is derived from the fact that measurements were initially made of fibers that were "braided" together to give the test samples connected between or onto vice-like attachments or hooks.

Reading and coworkers have pioneered in the adaptation of nano-associated AFM for thermal analyzes. The tip of the AFM is replaced with a miniature resistive heater that is used to heat or measure temperature differences on a nanoscale. This allows differentiation between phases and (at times) individual polymer chains and segments (such as block and graft segments of copolymers).

T_g and T_m and melting are based on differences in thermal conductivity and diffusivity. This technique of using the micro-thermal sensor (MTDSC) is called calorimetric analysis with scanning microscopy (CASM). The probe can also be used to measure certain mechanical properties performing a micro-thermal mechanical analysis with scanning microscopy (MASH).

Newer techniques are being developed, and modifications of existing techniques continue.

Softening ranges are dependent on the technique and the procedure used to determine them. Thus, the listings of softening ranges should be accompanied with how they were determined. Following are some of the techniques used to determine softening ranges. While DSC gives a more precise measurement of not only the T_g, it often approximates the softening range. The capillary method of determining the melting point of small molecules can also be used to gain some idea as to the softening point. Another related technique is to simply use a Fisher–John melting point apparatus and apply some pressure to a sample contained between two glass slides.

The viscat needle method consists of determining the temperature at which a standard needle penetrates a sample. In the ring-and-ball method, the softening range is determined by noting the temperature at which the sample, held within a ring, can be forced through the ring by the application of a standard force.

12.5 THERMAL PROPERTY TESTS

12.5.1 THERMAL CONDUCTIVITY

As energy in the form of heat, magnetic, or electric, is applied to one side of a material, the energy is transmitted to the other areas of the sample. Heat energy is largely transmitted through the increased amplitude of molecular vibrations. The heat flow Q from any point in a solid is related to the temperature gradient, dt/dl, through the thermal conductivity λ as follows:

$$Q = -\lambda \left(\frac{dt}{dl} \right) \tag{12.4}$$

The transmission of heat is favored by the presence of ordered lattices and covalently bonded atoms for organic compounds. Thus, graphite, quartz, and diamond are good thermal conductors, while less ordered forms such as glass have lower thermal conductivities.

12.5.2 THERMAL EXPANSION

The coefficients of thermal expansion generally refer to differences in length, area, or volume as a function of a temperature unit. Relative to metals such as steel, polymers have large coefficients of thermal expansions. Polymers also have quite varied coefficients of thermal expansion. Both of these factors are troublesome when different materials are bound together, including composites, and exposed to wide temperature ranges. Such wide temperature ranges regularly occur in the aerospace industry (aircraft), within computer chips (and many other electrical devices), engines, motors, etc. Thus, it is critical to match the coefficients of thermal expansions of materials that are to be bound through mechanical (such as screws and bolts) and chemical (polymer blends, alloys, adhered through use of an adhesive) means or stress will develop between the various components resulting in fracture or separation.

For polymeric materials, factors that restrict gross movement, such as cross-linking, typically result in lowered coefficients of expansion. Thus, the typical range for coefficients of expansion for cross-linked thermosets is lower than the typical range found for thermoplastics. Further, materials, such as glass, graphite, and concrete, also exhibit low coefficients of expansion for the same reason.

12.5.3 GLASS TRANSITION TEMPERATURES

Qualitatively, the T_g corresponds to the onset of short-range (typically 1–5 chain atoms) coordinated motion. Actually, many more (often 10–100) chain atoms may attain sufficient thermal energy sufficient to move in a coordinated manner at T_g. The T_g (ASTM D-3418) is the temperature at which there is an absorption or a release of energy as the temperature is raised or lowered, respectively. The T_g can be measured by any means that allows changes in the conformation of the chains to be detected. Thus, most of the thermal techniques allow T_g to be determined. Also, most spectral techniques also allow T_g to be determined as well as techniques that measure volume changes. Dynamic mechanical spectroscopy (DMS) is also used to measure chain changes by subjecting the sample to repeated small-amplitude stains in a cyclic fashion as a function of temperature. The polymer molecules store some of the imparted energy and dissipate a portion in the form of heat. Since the amount of energy stored and converted to heat is related to molecular motion, changes in the ratios of energy stored to energy converted to heat are used to measure T_g. Sperling compared literature reports of T_g values for some common polymers and found differences of several hundred degrees sometimes occurs.

12.6 FLAMMABILITY

Since many polymeric materials are used as clothing, household items, components of automobiles and aircraft, etc. flammability is an important consideration. Some polymers such as polytetrafluoroethylene and PVC are "naturally" flame resistant, but most common polymers such as PE and PP are not. Small-scale horizontal flame tests have been used to estimate the flammability of solid (ASTM D-635), cellular (ASTM D-1692–74), and foamed (ASTM D-1992) polymers, but these tests are useful for comparative purposes only. Large-scale tunnel tests (ASTM E-84) are more accurate, but they are also more expensive to run than ordinary laboratory tests cited before.

One of the most useful laboratory flammability tests is the oxygen index (OI) test (ASTM D-2043 and ASTM D-2863). In this test, the polymer is burned by a candle in the controlled mixtures of oxygen and nitrogen. The minimum oxygen concentration that produces downward flame propagation is considered the OI or ignitability of the polymer.

12.7 ELECTRICAL PROPERTIES: THEORY

Polymers have served as important materials in the electronics industry. Generally, they have served as coating and containers because of their lack of conductivity; i.e., they are nonconductors. More recently, polymers have become major materials as conductors.

Some important dielectric behavior properties are dielectric loss, loss factor, dielectric constant, direct current (DC) conductivity, alternating current (AC) conductivity, and electric breakdown strength. The term "dielectric behavior" usually refers to the variation in these properties as a function of frequency, composition, voltage, pressure, and temperature.

The dielectric behavior is often studied by employing charging or polarization currents. Since polarization currents depend on the applied voltage and the dimensions of the condenser, it is customary to eliminate this dependence by dividing the charge, Q, by the voltage, V, to get a parameter called the capacitance (capacity; C)

$$C = \frac{Q}{V} \tag{12.5}$$

and then using the dielectric constant ε, which is defined as

$$\varepsilon = \frac{C}{C_0} \tag{12.6}$$

where

C is the capacity of the condenser when the dielectric material is placed between its plates in a vacuum

C_0 is the empty condenser capacity

Dielectric polarization is the polarized condition in a dielectric resulting from an applied AC or DC field. The polarizability is the electric dipole moment per unit volume induced by an applied field or unit effective intensity. The molar polarizability is a measure of the polarizability per molar volume; thus, it is related to the polarizability of the individual molecules or polymer repeat units.

Conductivity is a measure of the number of ions per unit volume and their average velocity in the direction of the applied field. Polarizability is a measure of the number of bound charged particles per cubic unit and their average displacement in the direction of the applied field.

There are two types of charging currents and condenser charges, which may be described as rapidly forming or instantaneous polarizations, and slowly forming or absorptive polarizations. The total polarizability of the dielectric is the sum of contributions due to several types of charge displacement in the materials caused by the applied field. The relaxation time is the time required for polarization to form or disappear. The magnitude of the polarizability, k, of a dielectric is related to the dielectric constant, ε, as follows:

$$k = \frac{3(\varepsilon - 1)}{4\pi(\varepsilon + 2)} \tag{12.7}$$

The terms "polarizability constant" and "dielectric constant" are often used interchangeably in a qualitative discussion of the magnitude of the dielectric constant. The k values obtained utilizing DC and low-frequency measurements are a summation of electronic (E), atomic (A), dipole (D), and interfacial (I) polarizations as shown in Figure 12.4. Only the contribution by electronic polarizations is evident at high frequencies. The contributions to dielectric constant at low frequencies are additive as shown in Figure 12.4.

Instantaneous polarization occurs when rapid (less than 10^{-10} s) transitions occur, i.e., at frequencies greater than 10^{10} Hz or at wavelengths less than 1 cm. Electronic polarization falls within this category and is due to the displacement of charges within the atoms. **Electronic polarization** is directly proportional to the number of bound electrons in a unit volume and inversely proportional to the forces binding these electrons to the nuclei of the atoms.

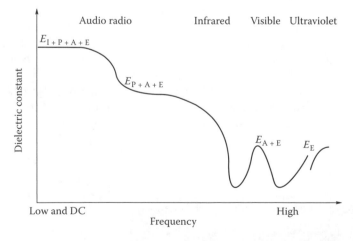

FIGURE 12.4 Relationship of dielectric constant with frequency emphasizing interfacial (I), dipole (P), atomic (A), and electronic (E) polarization contributions.

Electronic polarization occurs so rapidly that there is no observable effect of time or frequency on the dielectric constant until frequencies are reached that correspond to the visible and UV spectra. For convenience, the frequency range of the infrared region through the UV region is called the optical frequency range, and that of the radio range and the audio range is called the electric frequency range. Electronic polarization is an additive property dependent on the atomic bonds. Thus, the electronic polarizations and related properties are similar for both small molecules and polymers. Accordingly, values obtained for small molecules can be applied to analogous polymeric materials. This does not apply when the polymeric nature of the material plays an additional role in the conductance of electric charges, as in the case for whole-chain resonance or whole-chain delocalization of electrons.

Atomic polarization contributes to the relative motion of atoms in the molecule affected by perturbation by the applied field of the vibrations of atoms and ions having a characteristic resonance frequency in the infrared region. The atomic polarization is large in inorganic materials, which contain low-energy conductive bonds and approaches zero for nonconductive polymers. The atomic polarization is rapid, and this, as well as the electronic polarization, constitutes the instantaneous polarization components.

The remaining types of polarization are absorptive types with characteristic relaxation times corresponding to relaxation frequencies. Debye, in 1912, suggested that the high dielectric constants of water, ethanol, and other highly polar molecules were due to the presence of permanent dipoles within each individual molecule, and that there is a tendency for the molecules to align themselves with their dipole axes in the direction of the applied field. The major contributions to *dipole polarizations* are additive and are similar whether the moiety is within a small or large (macromolecule) molecule. Even so, the secondary contributions to the overall dipole polarization of a sample are dependent on both the chemical and physical environments of the specific dipole unit, its size, and its mobility. Thus, dipole contributions can be used to measure the T_g and the T_m.

The polarizations noted above are the major types found in homogeneous materials. Other types of polarization, called **interfacial polarizations**, are the result of heterogeneity. Ceramics, polymers with additives, and paper are considered to be electrically heterogeneous.

12.8 ELECTRIC MEASUREMENTS

Material response is typically studied using either direct (constant) applied voltage (DC) or alternating applied voltage (AC). The AC response as a function of frequency is characteristic of a material. In the future, such electric spectra may be used as a product identification tool, much like infrared spectroscopy. Factors such as current strength, duration of measurement, specimen shape, temperature, and applied pressure affect the electric responses of materials. The response may be delayed because of a number of factors including the interaction between polymer chains, the presence of specific molecular groupings within the chain, and effects related to interactions in the specific atoms themselves. A number of properties, such as relaxation time, power loss, dissipation factor, and power factor, are measures of this lag. The movement of dipoles (related to the dipole polarization, P) within a polymer can be divided into two types: an orientation polarization (P') and a dislocation or induced polarization.

The relaxation time required for the charge movement of electronic polarization E to reach equilibrium is extremely short (about 10^{-15} s), and this type of polarization is related to the square of the index of refraction. The relaxation time for atomic polarization A is about 10^{-3} s. The relaxation time for induced orientation polarization P' is dependent on molecular structure, and it is temperature dependent.

The electric properties of polymers are also related to their mechanical behaviors. The dielectric constant and the dielectric loss factor are analogous to the elastic compliance and the mechanical loss factor. Electric resistivity is analogous to viscosity. Polar polymers, such as ionomers, possess permanent dipole moments. These polar materials are capable of storing more electric energy than nonpolar materials. Nonpolar polymers are dependent almost entirely on induced dipoles for electric energy storage. Thus, orientation polarization is produced in addition to induced polarization when the polar polymers are placed in an electric field. The induced dipole moment of a polymer in an electric field is proportional to the field strength, and the proportionality constant is related to the polarizability of the atoms in the polymer. The dielectric properties of polymers are affected adversely by the presence of moisture, and this effect is greater in hydrophilic polymer than in hydrophobic polymer.

12.9 WEATHERABILITY

Polymers are used in almost every conceivable environment. They are tested for their ability to interact with radiation, weather, and microorganism. Weathering includes the ability to resist attacks by the freezing and heating cycles, resistance to frictional damage caused by rain and air, and influence of low and high temperatures as the polymeric materials are used.

Moisture is an important factor for some polymers, especially those with noncarbon backbones, where hydrolysis, and subsequent degradation, can bring about drastic changes in chain length and consequently, polymer properties. Such attack can occur on the surface and is indicated, as can many chemical attacks, by a discoloration generally followed by crazing and cracking. Other attacks can occur within the matrix with the polymer absorbing moisture.

The resistance to biological attack is important for many polymer applications including almost all of the biomedical applications, food storage and protection, and coatings where microorganism destruction is important. Most synthetic polymers are "naturally" resistant to destruction by microorganisms. This is particularly true for nonpolar polymers, but less so for condensation polymers such as nylons and polyesters where microorganisms may recognize similarities to bonds they ordinarily hydrolyze. Various preservatives and antimicroorganism additives are added, when appropriate, to protect the material against microbial attack. Tests include destructive degradation and simple growth of the microorganism on the material.

12.10 OPTICAL PROPERTIES—TESTS

Since polymers are often used as clear plastics or coatings and have many applications where transparency is an important property, knowledge of the optical properties of specific polymers is essential. The radiation scale, of course, includes microwave, infrared, ultraviolet, and visible regions. It is important to recognize the difference between refraction (associated with properties such as refractive index) and reflection (associated with properties such as haze). This difference is illustrated in Figure 12.5.

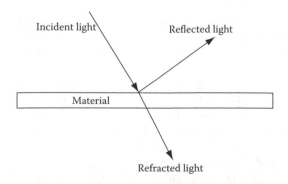

FIGURE 12.5   Refraction and reflection of incident light at the surface of a solid.

12.10.1 INDEX OF REFRACTION

Optical properties are related to both the degree of crystallinity and the actual polymer structure. Most polymers do not possess color site units, so they are colorless and transparent. But, some phenolic resins and polyacetylenes are colored, translucent, or opaque. Polymers that are transparent to visible light may be colored by the addition of colorants, and some become opaque as the result of the presence of additives such as fillers, stabilizers, moisture, gases, etc.

Many of the optical properties of a polymer are related to the refractive index, n, which is a measure of the ability of the polymer to refract or bend light as it passes through the polymer. The refractive index n is equal to the ratio of the sine of the angles of incidence, i, and refraction, r, of light passing through the polymer:

$$n = \frac{\sin i}{\sin r} \tag{12.8}$$

The magnitude of n is related to the density of the substance and varies from 1.000 for vacuum and 1.3333 for water, to about 1.5 for many polymers and 2.5 for white pigment titanium (IV) oxide (titanium dioxide). The value of n is often high for crystals, and it is dependent on the wavelength of the incident light and on the temperature. It is usually reported for the wavelength of the transparent sodium D line at 298 K. Typical refractive indices for polymers range from 1.35 for polytetrafluoroethylene to 1.67 for polyarylsulfone.

12.10.2 OPTICAL CLARITY

Optical clarity or the fraction of illumination transmitted through a material is related by the Beer–Lambert relationship:

$$\log \frac{I}{I_0} = -AL \quad \text{and} \quad \frac{I}{I_0} = e^{-AL} \tag{12.9}$$

where the fraction of illumination transmitted through a polymer (I/I_0) is dependent on the path length of the light (L) and the absorptivity of the polymer at that wavelength of light (A).

Clarity is typical for light passing through a homogeneous material, such as a crystalline ordered polymer or a completely amorphous polymer. Interference occurs when the light beam passes through a heterogeneous material in which the polarizability of the individual units varies causing interference that disrupts optical clarity.

12.10.3 ABSORPTION AND REFLECTANCE

Colorless materials range from being almost totally transparent to opaque. The opacity is related to the light scattering process occurring within the material. Incident radiation passes through nonabsorbing, isotropic, optically homogeneous samples with essentially little loss in radiation intensity. Actually, all materials scatter some light. The angular distribution of the scattered light is complex because of the scattering due to micromolecular differences.

Transparency is defined as the state permitting perception of objects through a sample. *Transmission* is the light transmitted. In more specific terms, transparency is the amount of undeviated light, i.e., the original intensity minus all light absorbed, scattered, or lost through other means. The ratio of the reflected light intensity to the incident light intensity is called the ***absorption coefficient***.

12.11 CHEMICAL RESISTANCE

The classic test for chemical resistance (ASTM D-543) measures the percentage weight change of test samples after immersion in different liquid systems. Tests for chemical resistance have been extended to include changes in mechanical properties after immersion. Since chemical attack involves changes in chemical structure, it can be readily observed by many instrumental methods that measure chemical structure, in particular, surface structure.

Tables 12.2 and 12.3 contain a summary of typical stability values for a number of polymers and elastomers against typical chemical agents. As expected, condensation polymers generally exhibit

TABLE 12.2
Stability of Various Polymers to Various Conditions

Polymer	Nonoxidizing Acid 20% Sulfuric Acid	Oxidizing Acid 10% Nitric Acid	Aqueous Salt Sol. NaCl	Aqueous Base NaOH	Polar Liquids Ethanol	Nonpolar Liquids Benzene	Water
Nylon 6,6	U	U	S	S	M	S	S
Polytetrafluorocthylene	S	S	S	S	S	S	S
Polycarbonate	M	U	S	M	S	U	S
Polyester	M	M	S	M	M	U	S
Polyetheretherketone	S	S	S	S	S	S	S
LDPE	S	M	S	—	S	M	S
HDPE	S	S	S	—	S	S	S
Poly(phenylene oxide)	S	M	S	S	S	U	S
Polypropylene	S	M	S	S	S	M	S
Polystyrene	S	M	S	S	S	U	S
Polyurethane	M	U	S	M	U	M	S
Epoxy	S	U	S	S	S	S	S
Silicone	M	U	S	S	S	M	S

Note: S, satisfactory; M, moderately to poor; U, unsatisfactory.

TABLE 12.3
Stability of Selected Elastomeric Materials to Various Conditions

Polymers	Weather— Sunlight Aging	Oxidation	Ozone Cracking	NaOH—Dil/ Con	Acid—Dil/ Con	Degreasers Chlorinated Hydrocarbons	Aliphatic Hydrocarbons
Butadiene	P	G	B	F/F	F/F	P	P
Neoprene	G	G	G	G/G	G/G	P	F
Nitrile	P	G	F	G/G	G/G	G	G
Polyisoprene (natural)	P	G	B	G/F	G/F	B	B
Polyisoprene (synthetic)	B	G	B	F/F	F/F	B	B
Styrene–butadiene	P	G	B	F/F	F/F	B	B
Silicone	G	G	G	G/G	G/F	B	F–P

Note: G, good; F, fair; P, poor; B, bad.

good stability to nonpolar liquids while they are generally only (relatively) moderately or unstable toward polar agents and acids and bases. This is because of the polarity of the connective "condensation" linkages within the polymer backbone. By comparison, vinyl type of polymers exhibit moderate to good stability toward both polar and nonpolar liquids and acids and bases. This is because the carbon–carbon backbone is not particularly susceptible to attack by polar agents and nonpolar liquids, at best, will simply solubilize the polymer. All of the materials show good stability to water alone because all of the polymers have sufficient hydrophobic character to repel the water.

12.12   SUMMARY

1. The ASTM and comparable organizations throughout the world have established meaningful standards for the testing of polymers.
2. Spectroscopic techniques that are useful for small molecules are equally as important with macromolecules. These techniques give both structural data and data related to the morphology of the polymers.
3. Surface properties are important to the physical and chemical behaviors of polymers. Similar to smaller molecules, polymer surface structures can be determined using a variety of techniques including AES, near-field optical microscopy, electron microcopy, scanning probe microscopy, secondary ion mass spectroscopy, and certain IR and MS procedures.
4. Thermal analysis measurements allow the measure of polymer behavior as a function of temperature, time, and atmosphere. DSC/DTA measures changes in energy as temperature is changed and allows the determination of many valuable parameters including T_g and T_m. TGA measures weight changes as a function of temperature.
5. The electrical properties of materials are important for many of the higher technology applications. Measurements can be made using alternating current and/or direct current. The electrical properties are dependent on voltage and frequency. Important electrical properties include dielectric loss, loss factor, dielectric constant, conductivity, relaxation time, induced dipole moment, electrical resistance, power loss, dissipation factor, and electrical breakdown. Electrical properties are related to polymer structure. Most organic polymers are nonconductors, but some are conductors.
6. Important physical properties of polymers include weatherability, chemical resistance, and optical properties. Polymers generally show good to moderate chemical resistance when compared to metals and nonpolymers.

GLOSSARY

ASTM: American Society for Testing and Materials.
BSI: British Standards Institution.
Chemical shifts: Peaks in NMR spectroscopy.
Dielectric constant: Ratio of the capacitance of a polymer to that in a vacuum.
Dielectric strength: Maximum applied voltage that a polymer can withstand.
Differential scanning calorimetry (DSC): Measurement of the difference in changes in the enthalpy of a heated polymer and a reference standard based on power input.
Differential thermal analysis (DTA, DT): Measurement of the difference in the temperature of a polymer and a reference standard when heated.
Environmental stress cracking: Cracking of polymers.
Glass transition temperature: Temperature where the onset of local or segmental mobility begins.
Heat deflection temperature: Temperature at which a simple loaded beam undergoes a definite deflection.
Index of refraction: Ratio of the velocity of light in a vacuum to that in a transparent material or mixture.

Infrared spectroscopy: Technique used for the characterization of polymers based on their molecular vibration and vibration–rotation spectra.

ISO: International Standards Organization.

Loss factor: Power factor multiplied by the dielectric constant.

Moh's scale: Hardness scale ranging from 1 for talc to 10 for diamond.

Nuclear magnetic resonance spectroscopy (NMR): Based on the absorption of magnetic energy by nuclei that have an uneven number of protons or neutrons; this absorption is dependent on the particular chemical structure and environment of the molecule.

Oxygen index (OI): Test for the minimum oxygen concentration in a mixture of oxygen and nitrogen that will support a candle-like flame of a burning polymer.

Phase contrast microscopy: Measurement of differences in indices of reflection.

Power factor: Electrical energy required to rotate the dipoles in a polymer while in an electrostatic field.

Raman spectroscopy: Based on the interaction of vibrational modes of molecules with relative high-energy radiation where absorption is based on the chemical moieties that undergo changes in polarization.

Scanning probe microscopy (SPM): Group of surface detection techniques that includes atomic force microscopy and scanning tunneling microscopy; measures surface depth differences.

Secondary ion mass spectroscopy (SIMS): Mass spectroscopy that looks at the ion fragments generated from the bombardment of surfaces with an ion beam.

Thermogravimetric analysis (TG, TGA): Measurement of the change in weight when a polymer is heated.

EXERCISES

1. Why are electrical tests for polymers important?
2. Which is the better insulator: a polymer with a (a) low K factor or (b) high K factor?
3. Why are the specific heats of polymers higher than those of metals?
4. Which IR technique might give you good surface information?
5. Why is the use of the term flameproof plastics incorrect?
6. Which plastic should be more resistant when immersed in 25% sulfuric acid at room temperature: (a) HDPE, (b) PMMA, or (c) PVac?
7. Why is it important to know the nature and structure of the surface of materials?
8. Why is it important to know things such as the values of thermal expansion of materials?
9. How is NMR important in characterizing polymers?
10. What is the UV region of the spectrum?
11. Which would absorb in the UV region: (a) polystyrene, (b) heva rubber, or (c) PVC?
12. What technique(s) would you use to determine crystallinity in a polymer?
13. What thermal instrumental technique would you use to determine T_g?

FURTHER READINGS

Alfrey, T. (1948): *Mechanical Behavior of Polymers*, Interscience, New York.

Ando, I. and Askakura, T. (1998): *Solid State NMR of Polymers*, Elsevier, New York.

Armer, M. (2009): *Raman Spectroscopy for Soft Matter Applications*, Wiley, Hoboken, NJ.

Batteas, J. (2005): *Applications of Scanned Probe Microscopy to Polymers*, American Chemical Society, Washington, DC.

Brandolini, A. and Haney, D. (2000): *NMR Spectra of Plastics*, Dekker, New York.

Carraher, C. E. (1977): Resistivity measurements, *J. Chem. Ed.*, 54: 576.

Carswell, T. S. and Nason, H. K. (1944): Classification of polymers, *Mod. Plast.*, 21: 121.

Chalmers, J. and Meier, R. (2008): *Molecular Characterization and Analysis of Polymers*, Elsevier, New York.

Cohen, S. and Lightbody, M. (1999): *Atomic Force Microscopy/Scanning Tunneling Microscopy*, Kluwer, New York.

Craver, C. and Carraher, C. E., (2000): *Applied Polymer Science*, Elsevier, New York.

Everall, N., Chalmers, J., and Griffiths, P. (2007): *Vibrational Spectroscopy of Polymers: Principles and Practice*, Wiley, Hoboken, NJ.

Friebolin, H. (1998): *Basic One- and Two-Dimensional NMR Spectroscopy*, Wiley, New York.

Gregoriou, V. (2004): *Polymer Spectroscopy*, Wiley, Hoboken, NJ.

Grellmann, W. and Seidler, S. (2007): *Polymer Testing,* Hanser-Gardner, Cincinnati, OH.

Groenewoud, W. (2001): *Characterization of Polymer by Thermal Analysis*, Elsevier, New York.

Grulkc, E. (2009): *Devolatilization and Purification of Polymers*, CRC Press, Boca Raton, FL.

Gupta, R. (2000): *Polymer and Composite Rheology*, Dekker, New York.

Hatada, K. (2004): *NMR Spectroscopy of Polymers*, Springer, New York.

Hatakeyama, T. and Quinn, F. (1999): *Thermal Analysis*, Wiley, New York.

Koenig, J. (1999): *Spectroscopy of Polymers*, Elsevier, New York.

Kosmulski, M. (2001): *Chemical Properties of Material Surfaces*, Dekker, New York.

Lee, T. (1998): *A Beginners Guide to Mass Spectral Interpretations*, Wiley, New York.

Li, L.(2008): *Principles and Practice of Polymer Mass Spectrometry*, Wiley, Hoboken, NJ.

Maev, R. (2008): *Acoustic Microscopy: fundamentals and Applications*, Wiley, Hoboken, NJ.

Marcomber, R. (1997): *A Complete Introduction to Modern NMR*, Wiley, New York.

Mark, J. (2004): *Physical Properties of Polymers*, Cambridge University Press, MA.

Menard, K. (2008): *Dynamical Mechanical Analysis: A Practical Introduction*, CRC Press, Boca Raton, FL.

Menczel, J. and Prime, R. (2008): *Thermal Analysis of Polymers: Fundamentals and Application*, Wiley, Hoboken, NJ.

Meyer, V. (1999): *Practical High-Performance Liquid Chromatography*, 3rd ed., Wiley, New York.

Michler, M. (2008): *Electron Microscopy of Polymers*, Springer, New York.

Mirau, P. (2004): *A Practical Guide to Understanding the NMR of Polymers*, Wiley, Hoboken, NJ.

Oliver, R. W. A. (1998): *HPLC of Macromolecules*, 2nd ed., Oxford, University Press, NY.

Petty, M. C., Bryce, M., and Bloor, D. (1995): *An Introduction to Molecular Electronics*, Oxford University Press, New York.

Pham, Q. (2002): *Proton and Carbon NMR Spectra of Polymers*, Wiley, New York.

Rainer, M. (2008): *Mass Spectrometric Identification of Biomolecules—Employing Synthetic Polymers*, VDM Verlag, New York.

Riande, E. (2004): *Electrical Properties of Polymers*, Dekker, New York.

Roe, R. (2000): *Methods of X-Ray and Neutron Scattering in Polymer Science*, Oxford University Press, New York.

Sawyer, L., Grubb, D., and Meyers, G. (2008): *Polymer Microscopy*, Springer, New York.

Scheirs, J. (2000): *Practical Polymer Analysis*, Wiley, New York.

Schmida, M., Antonietti, M., Coelfen, H., Koehler, W., and Schaefer, R. (2000): *New Developments in Polymer Analysis*, Springer-Verlag, New York.

Seymour, R. B. and Carraher, C. E. (1984): *Structure–Property Relationships in Polymers*, Plenum, New York.

Shah, V. (1998): *Handbook of Plastics Testing Technology*, Wiley, New York.

Stuart, B. (2002): *Polymer Analysis*, Wiley, New York.

Solymar, L. and Walsh, D.(1998): *Electrical Properties of Materials*, Oxford University Press, New York.

Tegenfeldt, J. (2008): *Solid-State NMR*, Wiley, Hoboken, NJ.

Turi, E. (1997): *Thermal Characterization of Polymeric Materials*, 2nd ed., Academic Press, Orlando, FL.

Urban, M. (1996): *Attenuated Total Reflectance Spectroscopy of Polymers*, Oxford University Press, New York.

Wallace, G, Spinks, G., Kane-Maguire, L., and Teasdale, P. (2008): *Conductive Electroactive Polymers: Intelligent Polymer Systems*, 3rd ed., CRC Press, Boca Raton, FL.

Zaikiv, G. (2007): *Polymer and Biopolymer Analysis and Characterization*, Nova, Hauppauge, NY.

Zerbi, G. (1999): *Modern Polymer Spectroscopy*, Wiley, New York.

13 Rheology and Physical Tests

Throughout this chapter, please remember that the main reasons for studying rheology (stress–strain relationships) are (1) to determine the suitability of materials to serve specific applications and (2) to relate the results to polymer structures and polymer forms. Understanding structure–property relationships allow a better understanding of the observed results on a molecular level resulting in a more knowledgeable approach to the design of materials. Look for these ideas as you study this chapter.

Polymers are viscoelastic materials meaning they can act as liquids, the "visco" portion, and as solids, the "elastic" portion. Descriptions of the viscoelastic properties of materials generally fall within the area called rheology. Determination of the viscoelastic behavior of materials generally occurs through stress–strain and related measurements. Whether a material behaves as a "viscous" or "elastic" material depends on the temperature, the particular polymer and its prior treatment, structure of the polymer, and the particular measurement or conditions applied to the material. The particular property demonstrated by a material under given conditions allows polymers to act as solid or viscous liquids, as plastics, elastomers, or fibers, etc. This chapter deals with the viscoelastic properties of polymers.

13.1 RHEOLOGY

The branch of science related to the study of deformation and flow of materials was given the name *rheology* by Bingham, whom some consider the father of modern rheology. The prefix *rheo* is derived from the Greek term *rheos*, meaning current or flow. The study of rheology includes two vastly different branches of mechanics called fluid mechanics and solid mechanics. The polymer scientist is usually concerned with viscoelastic materials that act as both solids and liquids.

The elastic component is dominant in solids; hence, their mechanical properties may be described by *Hooke's law* (Equation 13.1), which states that the applied stress (s) is proportional to the resultant strain (γ) but is independent of the rate of this strain ($d\gamma/dt$):

$$s = E\gamma \tag{13.1}$$

Stress is equal to the force per unit area, and strain or elongation is the extension per unit length. For an isotopic solid, that is, one having the same properties regardless of direction, the strain is defined by Poisson's ratio, $V = \gamma_w/\gamma_l$, which is the change in thickness (γ_w; lateral contraction) compared to the change in length (γ_l).

When there is no volume change, as when an elastomer is stretched, Poisson's ratio is 0.5. This value decreases as the T_g of the polymer increases and approaches 0.3 for rigid solids such as PVC and ebonite. For simplicity, the polymers dealt with here will be considered isotropic viscoelastic solids with a Poisson's ratio of 0.5, and only deformations in tension and shear will be considered. Thus, a shear modulus (G) will usually be used in place of Young's modulus of elasticity (E; Equation 13.2) where E is about 2.6 G at temperatures below T_g. For comparison, the moduli (G) for steel, HDPE, and hevea rubber (NR) are 86, 0.087, and 0.0006 dyne/m^2, respectively.

$$ds = G\, d\gamma \quad \text{and} \quad s = G\gamma \tag{13.2}$$

The viscous component is dominant in liquids; hence, their flow properties may be described by *Newton's law* (Equation 13.3) where η is viscosity, which states that the applied stress s is proportional to the rate of strain dγ/dt, but is independent of the strain γ or applied velocity gradient:

$$s = \eta\left(\frac{d\gamma}{dt}\right) \tag{13.3}$$

Both Hooke's and Newton's laws are valid for small changes in strain or the rate of strain, and both are useful in studying the effect of stress on viscoelastic materials. The initial elongation of a stressed polymer below T_g is reversible elongation due to a stretching of covalent bonds and a distortion of the bond angles. Some of the very early stages of elongation by the disentanglement of chains may also be reversible. However, the rate of flow, which is related to a slower disentanglement and a slippage of polymer chains past one another, is irreversible and increases (and η decreases) as the temperature increases in accordance with the following form of the Arrhenius equation (Equation 13.4) in which E is the activation energy for viscous flow:

$$\eta = Ae^{E/RT} \tag{13.4}$$

It is convenient to use a simple weightless Hookean, or ideal, elastic spring with a modulus G and a simple Newtonian (fluid) dashpot or shock absorber having a liquid with a viscosity of η as models to demonstrate the deformation of an elastic solid and an ideal liquid, respectively. The stress–strain curves for these models are shown in Figure 13.1.

In general terms, the Hookean spring represents bond flexing while the Newtonian dashpot represents chain and local segmental movement. It is customary to attempt to relate stress–strain behavior to combinations of dashpots and springs as indicators of the relative importance of bond flexing and segmental movement.

Again, in general terms, below their T_g, polymers can be modeled as having a behavior where the spring portion is more important. Above their T_g where segmental mobility occurs, the dashpot portion is more important.

The relative importance of these two modeling parts, the spring and the dashpot, is also dependent on the rate at which an experiment is carried out. Rapid interaction, such as striking with a hammer, with a polymer is more apt to result in a behavior where bond flexibility is more important, while slow interactions are more apt to allow segmental mobility to occur.

Since polymers are viscoelastic solids, combinations of these models are used to demonstrate the deformation resulting from the application of stress to an isotropic solid polymer. Maxwell

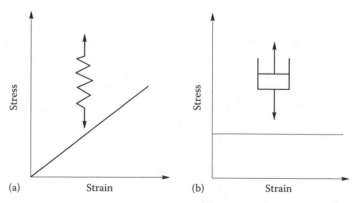

FIGURE 13.1 Stress–strain plots for a Hookean spring (a) where E (Equation 13.1) is the slope, and a Newtonian dashpot (b) where s is a constant (Equation 13.3).

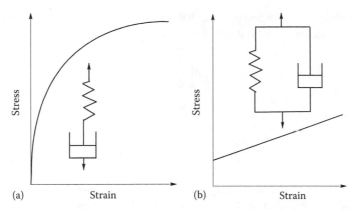

FIGURE 13.2 Stress–strain plot for stress relaxation for the Maxwell model (a) and the Voigt–Kelvin model (b).

joined the two models in series to explain the mechanical properties of pitch and tar (Figure 13.2a). He assumed that the contributions of both the spring and the dashpot to strain were additive and that the application of stress would cause an instantaneous elongation of the spring, followed by a slow response of the piston in the dashpot. Thus, the relaxation time (τ), when the stress and the elongation have reached equilibrium, is equal to η/G.

In the **Maxwell model** for viscoelastic deformation, it is assumed that the total strain is equal to the elastic strain plus the viscous strain. This is expressed in the two following differential equations from Equations 13.2 and 13.3.

$$\frac{d\gamma}{dt} = \frac{s}{\eta} + \frac{ds}{dt}\left(\frac{1}{G}\right) \tag{13.5}$$

The rate of strain, $d\gamma/dt$, is equal to zero under conditions of constant stress(s), that is,

$$\frac{s}{\eta} + \frac{ds}{dt}\left(\frac{1}{G}\right) = 0 \tag{13.6}$$

Then, assuming that $s = s_0$ at zero time, we get

$$s = s_0\, e^{-tG/n} \tag{13.7}$$

Further, since the relaxation time $\tau = \eta/G$, then

$$s = s_0\, e^{-t/T} \tag{13.8}$$

Thus, according to Equation 13.8 for the Maxwell model or element, under conditions of constant strain, the stress will decrease exponentially with time and at the relaxation time, $t = \tau$, s will equal $1/e$, or 0.37 of its original value, s_0.

The parallel dashpot-spring model was developed by Voigt and Kelvin and is known as the **Voigt–Kelvin model** (Figure 13.2b). In this model or element, the applied stress is shared between the spring and the dashpot, and thus the elastic response is retarded by the viscous resistance of the liquid in the dashpot. In this model, the vertical movement of the spring is essentially equal to that of the piston in the dashpot. Thus, if G is much larger than η, the retardation time (η/G) or τ is small, and τ is large if η is much larger than G.

In the Voigt–Kelvin model for viscoelastic deformation, it is assumed that the total stress is equal to the sum of the viscous and elastic stress, $s = s_v + s_0$, so that

$$s = \eta\left(\frac{d\gamma}{dt}\right) + G\gamma \tag{13.9}$$

On integration one obtains

$$\gamma = \frac{s}{G}(1 - e^{-tG/n}) = \frac{s}{G}(1 - e^{-t/T}) \tag{13.10}$$

The retardation time τ is the time for the strain to decrease to $1-(1/e)$ or $1-(1/2.7) = 0.63$ of the original value. The viscoelastic flow of polymers is explained by approximate combinations of the dashpot and the spring. The plots of the real data are compared with those predicted by various models. The relative importance of the various components of the model that fits the experimental data, dashpot and spring combinations, indicate the importance that the types of chain movement represented by the dashpot and the spring have for that particular polymer under the particular experimental conditions

13.1.1 RHEOLOGY AND PHYSICAL TESTS

While polymer melts and non-cross-linked elastomers flow readily when stress is applied, structural plastics must resist irreversible deformation and behave as elastic solids when relatively small stresses are applied. These plastics are called ideal or Bingham plastics with their behavior described mathematically by

$$s - s_0 = \eta\left(\frac{d\gamma}{dt}\right) \tag{13.11}$$

Figure 13.3 contains the stress–strain plots for a number of important flow behaviors. A Bingham plastic exhibits Newtonian flow above the stress yield or stress value (s_0). Liquids that undergo a decrease in viscosity with time are called ***thixotropic*** or false-bodied. Those that undergo an

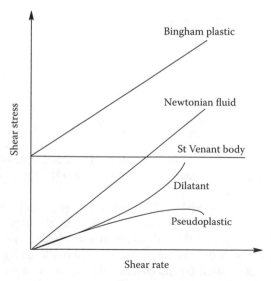

FIGURE 13.3 Various types of polymer flows.

increase in viscosity with time are called *rheopectic* (shear thickening). The term *creep* is used to describe the slow slippage of polymer chains over a long period. The Herschel–Buckley equation (Equation 13.12) is a general equation that reduces to the Bingham equation when $\eta = 1$ and to the Newtonian equation when $\eta = 1$ and $s_0 = 0$ and where ϕ is related to viscosity:

$$(s - s_0)\eta = \phi\left(\frac{d\gamma}{dt}\right) \tag{13.12}$$

Eyring explained liquid flow using a random hole-filling model in which the holes (vacancies or free volume) account for about 15% of the total volume at room temperature. The size of these holes is similar to that for small molecules. The number of holes increases as the temperature increases, and thus flow or hole filling is temperature dependent. Small molecules jump into the holes, leaving empty holes when their energy exceeds the activation energy. The activation energy for jumping or moving into a hole is smaller, per individual unit, for linear molecules that fill the holes by successive correlated jumps of chain segments along the polymer chain. The jump frequency (ϕ) is governed by a segmental factor with both values related to the molecular structure and the temperature.

For convenience and simplicity, polymers have generally been considered to be isotropic, in which the principle force is shear stress. While such assumptions are acceptable for polymers at low shear rates, they fail to account for stresses perpendicular to the plane of the shear stress, which are encountered at high shear rates. For example, an extrudate such as the formation of a pipe or filament expands when it emerges from the die in what is called the Barus or Weissenberg effect, or die swell. This behavior is not explained by simple flow theories.

Viscoelastic behavior can be divided into five subclassifications (Figure 13.4). From 1 to 2, Figure 13.4, the material behaves as a viscous glass or Hookean elastic or glass, where chain segmental motion is quite restricted and involves mainly only bond bending and bond angle deformation. The material behaves as a glass, such as window glass.

At 2, the material is undergoing a glassy transition into the rubbery region, 3–4, which is often referred to as the viscoelastic region where polymer deformation is reversible but time dependent and associated with both side-chain rotation and main-chain rotation. In the rubbery region, local

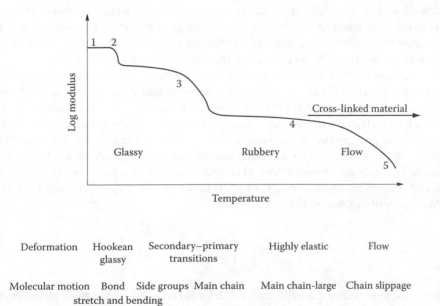

FIGURE 13.4 Characteristic conformational changes as temperature is changed.

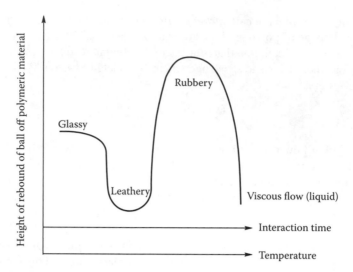

FIGURE 13.5   Regions of material response as a function of interaction (reaction) time and temperature for a typical polymeric material.

segmental mobility occurs but total chain flow is restricted by a physical and/or chemical network. At 5, rubbery flow or viscous flow occurs where irreversible bulk deformation and slippage of chains past one another occurs. Each of these viscoelastic regions is time dependent. Thus, given a short interaction time, window glass acts as a Hookean glass or like a solid, yet the observation of glass over many years would permit the visual observation of flow, with the window glass giving a viscous flow response, thus acting as a fluid. In fact, most polymers give a response, as noted in Figure 13.5, for a ball dropped on to the polymeric material at different rates onto the polymeric material that is heated to different temperatures. Commercial Silly Putty or Nutty Putty easily illustrates three of these regions. When struck rapidly, it shatters as a solid (glass region), when dropped at a moderate rate, it acts as a rubber (rubbery region), and when allowed to reside in its container for some time, it flows to occupy the container contour, acting as a liquid. The exact response (Figure 13.5) varies between materials. The leathery region deserves special comment. This region corresponds to the T_g of a material where the impact energy is absorbed resulting in some segmental mobility rather than simply being released resulting in a rebound. The extent of energy absorption is dependent on a number of factors including the proportion of amorphous area.

There have been many attempts to predict long-range behavior from short-range behavior. This is a dangerous practice if the end prediction is material failure because, for most polymeric materials, failure is catastrophic and not "well behaved." However, for other properties, application of the Boltzman time–temperature superposition master curves can be used to some advantage, particularly pertaining to predicting the behavior as depicted in Figure 13.5. This transposition for amorphous polymers is accomplished using a shift factor (a_T) calculated relative to a reference temperature, T_r, which may be equal to T_g. The relationship of the shift factor to the reference temperature and some other temperature, T, which is between T_g and $T_g + 50\,\text{K}$, may be approximated by the Arrhenius-like Equation 13.13.

$$\log a_T = -\left(\frac{b}{2.3\,T\,T_g}\right)(T - T_g) \qquad (13.13)$$

According to the more widely used Williams, Landel, and Ferry (WLF) equations, all linear, amorphous polymers have similar viscoelastic properties at T_g. At specific temperatures above T_g, such as $T_g + 25\,\text{K}$, where the constants C_1 and C_2 are related to holes or free volume, the following relationship holds.

$$\log a_{\mathrm{T}} = \frac{-C_1(T - T_g)}{[C_2 + (T - T_g)]} \tag{13.14}$$

13.1.2 Response Time

We can get a first approximation of the physical nature of a material from its response time. For a Maxwell element, the relaxation time is the time required for the stress in a stress–strain experiment to decay to $1/e$ or 0.37 of its initial value. A material with a low relaxation time flows easily so it shows relatively rapid stress decay. Thus, whether a viscoelastic material behaves as a solid or fluid is indicated by its response time and the experimental time scale or observation time. This observation was first made by Marcus Reiner who defined the ratio of the material response time to the experimental time scale as the **Deborah Number, D_n**. Presumably, the name was derived by Reiner from the Biblical quote in Judges 5, Song of Deborah, where it says "The mountains flowed before the Lord."

$$D_n = \frac{\text{Response time}}{\text{Experimental time scale}} \tag{13.15}$$

A high Deborah Number designates the solid behavior of a viscoelastic material, while a low Deborah Number corresponds to a viscoelastic material behaving as a fluid. Thus, window glass has a high relaxation time at room temperature. The application of a stress to produce a little strain and looking for it to return to its approximate prestressed state will take a long time as we count our observation time in hours, days, and weeks. Thus, it would have a relatively high Deborah Number under this observation time scale and be acting as a solid. Yet, if were to have as our observation scale millions of years, the return to the prestressed state would be rapid with the glass acting as a viscous liquid and having a low Deborah Number. Again, this represents only a first approximation measure of the behavior of a material.

13.2 TYPICAL STRESS–STRAIN BEHAVIOR

Most physical tests involve nondestructive evaluations. For our purposes, three types of mechanical stress measures (described in Figure 13.6) will be considered. The ratio of stress to strain is called **Young's modulus**. This ratio is also called the modulus of elasticity and tensile modulus. It is calculated by dividing the stress by the strain.

$$\text{Young's modulus} = \frac{\text{stress (Pa)}}{\text{strain (mm/mm)}} \tag{13.16}$$

Large values of Young's modulus indicate that the material is rigid and resistant to elongation and stretching. Many synthetic organic polymers have Young's modulus values in the general range of about 10^4–10^{10} Pa. For comparison, fused quarts have a Young's modulus of about 10^6; cast iron, tungsten, and copper have values of about 10^7; and diamond has a value of about 10^8.

Carswell and Nason assigned five classifications to polymers (Figure 13.7). It must be remembered that the ultimate strength of each of these polymers is the total area under the curve before breaking. The soft and weak class, such as polyisobutylene, is characterized by a low modulus of elasticity, low yield (stress) point, and moderate time-dependent elongation. The Poisson ratio, that is, the ratio of contraction to elongation, for soft and weak polymers is about 0.5, which is similar to that found for many liquids.

In contrast, the Poisson ratio for the hard and brittle class of polymers, such as polystyrene, approaches 0.3. These polymers are characterized by a high modulus of elasticity, a poorly defined

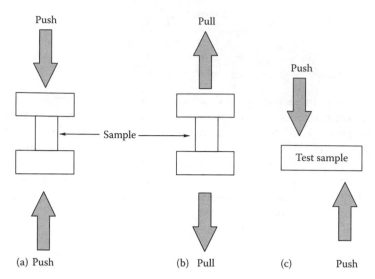

FIGURE 13.6 Major types of stress tests: compressive (a); pulling stress or tensile strength (b); and shear stress (c).

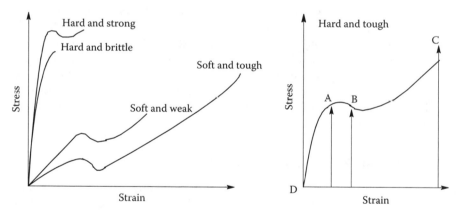

FIGURE 13.7 Typical stress–strain curves for plastics where (right) A is the elongation at yield point, B is the yield stress, C is the elongation at break, and the area under the curve is the ultimate strength.

yield point, and little elongation before failure. However, soft and tough polymers, such as plasticized PVC, have a low modulus of elasticity, high elongation, a Poisson ratio of about 0.5–0.6, and a well-defined yield point. Soft and tough polymers stretch after the yield point with the area under the entire curve, toughness or ultimate strength, greater than that of the hard and brittle polymers.

Rigid PVC is representative of hard and soft polymers. These polymers have a high modulus of elasticity and high yield strength. The curve for hard and tough polymers, such as ABS copolymers, shows moderate elongation prior to the yield point followed by nonrecoverable elongation.

In general, the behavior of all classes of polymer behavior is Hookean prior to the yield point. The reversible recoverable elongation prior to the yield point, called the elastic range, is primarily the result of bending and stretching of covalent bonds in the polymer backbone. This useful portion of the stress–strain curve may also include some recoverable uncoiling of polymer chains. Irreversible slippage of polymer chains is the predominant mechanism after the yield point.

Since these properties are time dependent, the soft and weak polymers may resemble the hard and strong polymers if the stress is rapidly applied, and vice versa. These properties are also temperature dependent. Hence, the properties of soft and tough polymers may resemble hard and brittle when temperature is decreased. The effects of temperature and the mechanisms of elongation are summarized in Figure 13.4.

As noted in Figure 13.6, the major modes for applying stress are axial (compression or tension), flexural (bending or shear), and torsional (twisting; not shown). Superimposed on these modes can be any number of cyclic arrangements. Several common fluctuating stress–time modes are typically employed. In the regular and sinusoidal time-dependent mode, the stress is applied, as both compressional and torsional, in a regular manner with respect to both time and amount. In a second cyclic arrangement stress, again both compressional and torsional are applied in an uneven manner with respect to amount and time so that the maxima and minima are asymmetrical relative to the zero stress level.

Associated with such cyclic application of stress is the term *fatigue* to describe the failure that occurs after repeated applications of stress. The fatigue values are almost always less than measurements, such as tensile strength, obtained under a static load. Thus, it is important that both static and cyclic measurements be made.

Data are often plotted as stress versus the logarithm of the number of cycles, N, to failure for each sample. Generally, tests are taken using about two-thirds of the static parameter value. There is some limiting stress level below which fatigue does not occur (over some reasonably extended time). This value is called the *fatigue* or *endurance limit*. The *fatigue strength* is defined as the stress at which failure will occur for some specified number of cycles. *Fatigue life* is the number of cycles that are needed to cause failure.

Polymeric materials are dissimilar to metals in a number of respects. For instance, the range of strain–strain behavior is more complex and varied for polymers. Highly elastic polymers can exhibit a modulus as low as 7 MPa or as high as 100 GPa for stiff polymeric fibers. The modulus values for metals are higher (50–400 G Pa) but not as varied. Tensile strengths for polymers are on the order of 100 MPa for polymers and for some metal alloys 4000 MPa. By comparison, polymers are more deformable with elastic polymers having elongations in excess of 500% while even "moldable" metals having elongations not exceeding 100%. Further, stress–strain behavior for polymers is more temperature dependent in comparison to metals. For instance, Figure 13.8 contains the stress–strain

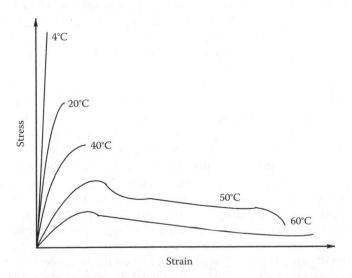

FIGURE 13.8 Influence of temperature on the stress–strain behavior of a sample of poly(methyl methacrylate). (Modeled after Carswell, T.S. and Nason, H.K., Effects of environmental conditions on the mechanical properties of organic plastics, *Symposium on Plastics*, American Society for Testing Materials, Philadelphia, PA, 1944.)

behavior for a poly(methyl methacrylate), PMMA, material. Below the T_g the material behaves as a stiff, brittle solid, similar to the behavior given for hard and brittle polymers (Figure 13.8). As the temperature increases, the PMMA gains enough thermal energy to allow for some segmental mobility. The T_g varies greatly for PMMA depending upon the tacticity, with isotactic PMMA having a T_g of about 45°C and for syndiotactic PMMA the T_g is about 130°C. The material in Figure 13.8 is largely isotactic with a T_g of about 50°C–60°C. By 50°C, the material behaves as a soft and tough material in Figure 13.8.

13.3 STRESS–STRAIN RELATIONSHIPS

Mechanical testing involves a complex of measurements including creep and shear strength, impact strengths, and so on. Stress–strain testing is typically carried out using holders where one member is movable contained within a load frame. Studies typically vary with either the stress or strain fixed and the result response measured. In a variable stress experiment a sample of given geometry is connected to the grips. Stress, load, is applied, generally by the movement of the grip heads either toward one another (compression) or away from one another (elongation). This causes deformation, strain, of the sample. The deformation is recorded as is the force/area necessary to achieve this deformation. The applied load versus deformation is recorded and translated into stress–strain curves such as those given in Figures 13.3, 13.7, and 13.8.

For many materials, between their T_g and T_m, recovery is not complete. This incomplete recovery is called ***creep***. Creep is time dependent and is described as follows:

$$\text{Compliance} = \text{strain} - \text{stress} \qquad (13.17)$$

where the compliance and strain are time dependent.

Creep behavior is similar to viscous flow. The behavior shown by Equation 13.17 shows that compliance and strain are linearly related to strain and inversely related to stress. This linear behavior is typical for most amorphous polymers for small strains over short periods of time. Further, the overall effect of a number of such imposed stresses are additive. Non-creep-related recovery occurs when the applied stress is relieved. Thus, amorphous polymers and polymers containing substantial amounts of amorphous regions act as both elastic solids and liquids above their T_g.

Phase changes, phase transitions, refer to changes of state. Most low molecular weight materials exist as solids, liquids, and gasses. The transitions that separate these states are melting (or freezing) and boiling (or condensing). By contrast, polymers do not vaporize "intact" to a gas, nor do they boil. The state of a polymer, that is, its physical response character, depends on the time allotted for interaction and the temperature as well as the molecular organization (crystalline, amorphous-mix). The term ***relaxation*** refers to the time required for response to a change in pressure, temperature, or other applied parameters. The term ***dispersion*** refers to the absorption or emission of energy at a transition. In practice, the terms relaxation and dispersion are often used interchangeably.

Stress–strain plots for many polymers, that are amorphous or contain a mixture of amorphous/crystalline regions, can be given for various conditions resulting in the general behavior shown in Figure 13.4. As noted before, the viscoelastic behavior of polymers is both temperature and time dependent. One measure of the tendency to deform and rebound is to look at the stress relaxation of materials. Stress relaxation measurements are carried out by rapidly straining a material by tension to a predetermined but low strain level and measuring the stress needed to maintain this strain as a function of time. The stress needed to maintain the strain decreases with time due to a molecular relaxation processes with the various bond flexing and chain and segmental movements influencing this time. Figure 13.9 (similar to Figure 13.4) contains an idealized plot of the logarithm of the relaxation modulus as a function of temperature for a typical vinyl (largely)

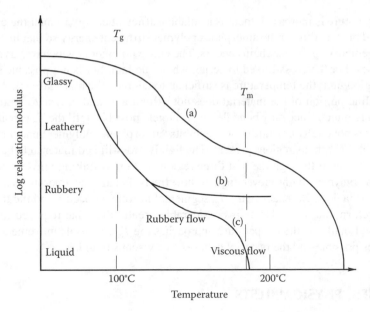

FIGURE 13.9 Logarithm of the relaxation modulus as a function of temperature for three polymer samples. Sample (a) is (largely) crystalline vinyl polymer; sample (b) is an amorphous vinyl polymer that contains light cross-linking; and sample (c) is an amorphous vinyl polymer. The T_g for the amorphous polymer is about 100°C and the T_m for the crystalline polymer is about 180°C.

crystalline polymer, curve a; lightly cross-linked vinyl amorphous polymer, curve b; and an amorphous vinyl polymer, line c. The T_g for the polymer is about 100°C and the T_m is about 180°C for this illustration.

Four general behavior patterns emerge and correspond to the same regions given in Figure 13.5.

The first region is called the glassy region. The Young's modulus for glassy polymers is somewhat constant over a wide range of vinyl polymers having a value of about 3×10^9 Pa. Here the molecular motions are restricted to vibrational and short-range rotational movements, often the former sometimes referred to as bond flexing. Two diverse approaches are often used to describe molecular motion in this range. The cohesive energy density approach measures the energy theoretically required to move a detached segment into the vapor state per unit volume. Using polystyrene as an example, a value of about 3×10^9 Pa is calculated for the modulus. For largely hydrocarbon polymers, the cohesive energy density is typically similar to that of polystyrene and similar theoretical modulus values are found. The second approach is based on the carbon–carbon bonding forces as seen using vibrational frequencies. Again, values about 3×10^9 Pa are found. Thus, in the glassy or solid region, below the T_g, only bond flexing and short-range rotational movement occurs, and the response time is fast for all three samples.

After the precipitous drop that occurs near the T_g, the modulus again is about constant, often on the order of 2×10^6 Pa, until melting is achieved. Over this rubbery range, moderate segmental chain mobility is possible when stress is applied, but wholesale mobility does not occur. Thus, somewhat elastic behavior occurs where segmental mobility allows the material to be stretched often to beyond 100% with an approximate return to the prestretched structure after the release of the stress. The rubbery region is particularly "interaction-time" sensitive. For instance, if applied stress is maintained chain slippage will occur for non-cross-linked materials limiting return to the original molecular conformation after the release of the stress. For the same polymer, the width of this plateau is somewhat molecular weight dependent such that the longer the polymer chains the broader the plateau.

As the temperature is increased, there is available sufficient energy to melt the crystalline polymer, the T_m, and prior to this for the amorphous polymer sufficient energy so that in both cases ready wholesale movement of polymer chains occurs. The entire polymer now behaves as a viscous liquid such as molasses. For the cross-linked material, wholesale mobility is not possible so it remains in the rubbery region until the temperature is sufficient to degrade the material.

The crystalline portion of the material does not exhibit a T_g and only a T_m so the leathery and rubbery areas are muted since only bond flexing is largely possible until the T_m is reached. The crystalline polymer is not totally crystalline so it exhibits "amorphous" polymer behavior corresponding to the amount of amorphous regions present. The lightly cross-linked material behaves similarly to the amorphous material through the first three regions but cross-linking prevents ready wholesale movement of the polymer chains preventing it to flow freely. Finally, the amorphous material, curve c, shows all four behaviors. Here, above the T_g, segmental movement occurs and the time required for the chains to return to a relaxed state is considerably greater than that required for simple return of bond flexing. Finally, as the temperature approaches the T_m wholesale movement of the polymer chains becomes possible and the material behaves as a viscous liquid.

13.4 SPECIFIC PHYSICAL TESTS

13.4.1 Tensile Strength

Tensile strength can be determined by applying force to the material until it breaks as shown in Equation 13.18 (including typical units).

$$\text{Tensile strength (Pa)} = \frac{\text{force required to break sample (N)}}{\text{cross-sectional area (m}^2)} \tag{13.18}$$

Tensile strength, which is a measure of the ability of a polymer to withstand pulling stresses, is usually measured using a dumbbell-shaped specimen (ASTM D-638-72; Figure 13.10). These test specimens are conditioned under standard conditions of humidity and temperature before testing.

The *elastic modulus* (also called tensile modulus or modulus of elasticity) is the ratio of the applied stress to the strain it produces within the region where the relationship between the stress and the strain is linear. The *ultimate tensile strength* is equal to the force required to cause failure divided by the minimum cross-sectional area of the test sample.

When rubbery elasticity is required for sample performance, a high ultimate elongation is desirable. When rigidity is required, it is desirable that the material exhibit a lower ultimate elongation. Some elongation is desirable since it allows the material to absorb rapid impact and shock. The total area under the stress–strain curve is indicative of overall toughness.

Pulling stress is the deformation of a test sample caused by the application of specific loads. It is specifically the change in the length of the test sample divided by the original length. Recoverable strain or elongation is called *elastic strain*. Here, stressed molecules return to their original relative locations after the release of the applied force. Elongation may also be a consequence of wholesale movement of chains past one another, that is, creep or plastic strain. Many samples undergo both reversible and irreversible strains.

FIGURE 13.10   Typical "dumbbell" or "dog-bone" shaped sample used for stress–strain measurements.

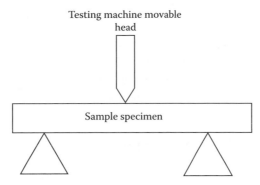

FIGURE 13.11 Typical flexural strength test.

Flexural strength, or cross-breaking strength, is a measure of the bending strength or stiffness or a test bar specimen used as a simple beam in accordance with ASTM D-790. The specimen is placed on supports (Figure 13.11). A load is applied to its center at a specified rate with the loading at failure called the flexural strength. However, because many materials do not break even after being greatly deformed, by agreement, the modulus at 5% strain is used as the flexural strength for these samples.

13.4.2 TENSILE STRENGTH OF INORGANIC AND METALLIC FIBERS AND WHISKERS

The tensile strength of materials is dependent on the treatment and form of the material. Thus, the tensile strength of isotropic bulk nylon-66 is less than that of anisotropic oriented nylon-66 fiber. Inorganics and metals also form fibers and whiskers with varying tensile strengths (Table 13.1). Fibers are generally less crystalline and larger than whiskers.

Many of these inorganic fibers and whiskers are polymeric including many of the oxides (including the so-called ceramic fibers), carbon and graphite materials, and silicon carbide. Carbon and graphite materials are similar but differ in the starting materials and the percentage of carbon. Carbon fibers derived from polyacrylonitrile are about 95% carbon, while graphite fibers are formed at higher temperatures giving a material with 99% carbon.

These specialty fibers and whiskers exhibit some of the highest tensile strengths recorded (Tables 13.1 and 13.2); they are employed in applications where lightweight and high strength are required. Organics offer weight advantages, typically being less dense than most inorganic and metallic fibers.

TABLE 13.1
Tensile Strengths of Inorganic and Metallic Materials as a Function of Form

Material	Form	Tensile Strength (MPa)
Graphite	Bulk	1,000
	Fiber	2,800
	Whisker	15,000
Glass	Bulk	1,000
	Fiber	4,000
Steel	Bulk	2,000
	Fiber	400
	Whisker	10,000

TABLE 13.2
Ultimate Tensile Strength of Representative Organic, Inorganic, and Metallic Fibers

Material	Tensile Strength (MPa)	Tensile Strength/Density
Aluminum silica	4100	1060
Aramid	280	200
Beryllium carbide	1000	400
Beryllium oxide	500	170
Boron–tungsten boride	3450	1500
Carbon	2800	1800
Graphite	2800	1800
UHMW polyethylene	380	3800
Poly(ethylene terephthate)	690	500
Quartz	900	400
Steel	4000	500
Titanium	1900	400
Tungsten	4300	220

They are used in dental fillings, the aircraft industry, the production of lightweight fishing poles, automotive antennas, lightweight strong bicycles, turbine blades, heat-resistant reentry vessels, golf club shafts, etc. Many of them are also used as reinforcing agents in composites.

13.4.3 Compressive Strength

Compressive strength is defined as the pressure required to crush a material defined as

$$\text{Compressive strength (Pa)} = \frac{\text{force (N)}}{\text{cross-sectional area (m}^2)} \tag{13.19}$$

Compressive strength, or the ability of a specimen to resist a crushing force, is measured by crushing a cylindrical specimen (ASTM-D-695).

13.4.4 Impact Strength

Impact strength is a measure of the energy needed to break a sample—it is not a measure of the stress needed to break a sample. The term *toughness* is typically used in describing the impact strength of a material but does not have a universally accepted definition, whereas it is often described as the area under stress–strain curves.

13.4.5 Hardness

The term *hardness* is a relative term. Hardness is the resistance to local deformation that is often measured as the ease or difficulty for a material to be scratched, indented, marred, cut, drilled, or abraded. It involves a number of interrelated properties such as yield strength and elastic modulus. Because polymers present such a range of behavior, they are viscoelastic materials, the test conditions must be carefully described. For instance, elastomeric materials can be easily deformed, but this deformation may be elastic with the indentation disappearing once the force is removed. While many polymeric materials deform in a truly elastic manner returning to the initial state once the load is removed, the range of total elasticity is often small resulting in limited plastic or permanent deformation. Thus, care must be taken in measuring and in drawing conclusions from results of hardness measurements.

13.4.6 FAILURE

The failure of materials can be associated with a number of parameters. Two major causes of failure are creep and fracture. The tensile strength is the nominal stress at the failure of a material. Toughness is related to ductility. For a material to be tough, it often takes a material having a good balance of stiffness and give.

Calculations have been made to determine the theoretical upper limits with respect to the strength of polymers. Real materials show behaviors near to those predicted by the theoretical calculations during the initial stress–strain determination, but vary greatly near the failure of the material. It is believed that the major reasons for the actual tensile strength at failure being smaller than calculated are related to imperfections, the nonhomogeneity of the polymeric structure. These flaws, molecular irregularities, act as the weak link in the polymer's behavior. These irregularities can be dislocations, voids, physical cracks, and energy concentrations. Even with these imperfections, they have high strength-to-mass ratios (Table 13.2) with the highest tensile strength/density value for UHMW polyethylene.

The fracture strengths of polymers are generally lower than those of metals and ceramics. The mode of failure for thermosets is generally referred to as the materials being brittle. Cracks, related to bond breakage, occur at points of excess stress. These cracks create weak spots and may lead to fracture if the applied stress, appropriate to create bond breakage, continues.

The fracture of thermoplastics can occur by either a ductile or brittle fashion, or some combination. The rapid application of stress (short reaction time), lower temperatures, and the application of the stress by a pointed or sharpened object, all increase the chances that failure will occur by a brittle mechanism as expected. For instance, below the T_g thermoplastics behave as solids and above T_g polymer fracture increasingly moves to a ductile mode. Above the T_g they display some plastic deformation before fracture. For some thermoplastics, crazing occurs prior to fracture. These crazes are a form of defect and are associated with some yielding or void formation either through the migration or the elimination of some part of the polymer mix or because of chain migration through chain realignment or other mechanism. Fibrillar connections, from aligned chains, often form between void sides. If sufficient stress is continued, these fibrillar bridges are broken and the microvoids enlarge, eventually leading to crack formation. Crazes can begin from scratches, flaws, molecular inhomogeneities, and molecular-level voids formed from chain migration (such as through crystallization) or small molecule migration from a site as well as the application of stress.

13.5   SUMMARY

1. Rheology is the study of deformation and flow of materials. Polymers are viscoelastic materials meaning they can act as liquids and as solids.
2. Many of the rheological properties of materials are determined using stress–strain measurements.
3. Models are used to describe the behavior of materials. The fluid part or the liquid part of the behavior is described in terms of a Newtonian dashpot or a shock absorber while the elastic part or the solid part of the behavior is described in terms of a Hookean or ideal elastic spring. The Hookean spring represents bond flexing while the Newtonian dashpot represents chain and local segmental movement. In elastomers, the spring also represents chain elongation.

4. In general, the spring-like behavior is more important for polymers below their T_g whereas the dashpot behavior is more important for polymers above their T_g.
5. Mathematical relationships have been developed to relate models composed to varying degrees of these two aspects of behavior to real materials.
6. Stress–strain behavior is closely dependent upon the particular conditions under which the experiment is carried out.
7. There are a number of specific tests that are used to measure various physical properties of materials. Many of these tests are described in detail by the various governing bodies that specify testing procedures and conditions.

GLOSSARY

Barcol impressor: Instrument used to measure the resistance of a polymer to penetration or indentation.

Biaxially stretching: Stretching in two directions perpendicular to each other.

Bingham, E. C.: Father of rheology.

Bingham plastic: Plastic that does not flow until the external stress exceeds a critical threshold value.

Coefficient of expansion: Change in dimensions per degree of temperature.

Compression strength: Resistance to crushing forces.

Creep: Permanent flow of polymer chains.

Dashpot: Model for the Newtonian fluids consisting of a piston and a cylinder containing a viscous liquid.

Dilatant: Shear thickening agent, describes a system where the shear rate does not increase as rapidly as the applied stress.

Elastic range: Area under a stress–strain curve to the yield point.

Flexural strength: Resistance to bending.

Hooke's law: Stress is proportional to strain.

Impact strength: Measure of toughness.

Isotropic: Having similar properties in all directions.

Maxwell element or model: Model in which an ideal spring and a dashpot are connected in series; used to study the stress relaxation of polymers.

Modulus: Stress per unit strain; measure of the stiffness of a polymer.

Newtonian fluid: Fluid whose viscosity is proportional to the applied viscosity gradient.

Newton's law: Stress is proportional to flow.

Poisson's ratio: Ratio of the percentage change in length of a sample under tension to its percentage change in width.

Pseudoplastic: Shear thinning agent, system where the shear rate increases faster than the applied stress.

Relaxation time: Time for stress of a polymer under constant strain to decrease to $1/e$ or 0.37 of its original value.

Retardation time: Time for the stress in a deformed polymer to decrease to 63% of the original value.

Rheology: Science of flow.

Rheopectic: Liquid whose viscosity increases with time.

Shear: Stress caused by planes sliding by each other.

Shear strength: Resistance to shearing forces.

Stress (s): Force per unit area.

Stress relaxation: Relaxation of a stressed sample with time after the load is removed.

Tensile strength: Resistance to pulling stresses.

Thixotropic: Liquid whose viscosity decreases with time.

Velocity gradient: Flow rate.

Viscoelastic: Having the properties of a liquid and a solid.

Voigt–Kelvin model or element: Model consisting of an ideal spring and a dashpot in parallel in which the elastic response is retarded by viscous resistance of the fluid in the dashpot.

Williams, Landel, and Ferry (WLF) equation: Used for predicting viscoelastic properties at temperatures above T_g when these properties are known for one specific temperature.

Yield point: Point on a stress–strain curve below which there is reversible recovery.

EXERCISES

1. What is the difference between morphology and rheology?
2. Which of the following is viscoelastic: (a) steel, (b) polystyrene, (c) diamond, or (d) neoprene?
3. Define G in Hooke's law.
4. Which would be isotropic: (a) a nylon filament, (b) an extruded pipe, or (c) ebonite?
5. What are the major types of stress tests?
6. Describe in molecular terms what a dashpot represents.
7. Which will have the higher relaxation time: (a) Unvulcanized rubber or (b) ebonite?
8. Describe in molecular terms what a spring represents.
9. Define what compliance is.
10. What does viscoelastic mean?
11. Below the glass transition temperature, how should a material behave?
12. At what temperature would the properties of polystyrene resemble those of hevea rubber at 35 K above its T_g?
13. Define Young's modulus. Should it be high or low for a fiber?
14. Define the proportionality constant in Newton's law.
15. What term is used to describe the decrease of stress at constant length with time?
16. In which element or model for a viscoelastic body, (a) Maxwell or (b) Voigt–Kelvin, will the elastic response be retarded by viscous resistance?
17. What is the most important standards organization in the United States?
18. For most polymeric materials, is failure abrupt or continuous?
19. Is it safe to predict long-range properties from short-time period results for plastics?
20. Describe the importance of relating models to stress–strain results.
21. If a sample of polypropylene measuring 5 cm elongates to 12 cm, what is the percentage of elongation?
22. If the tensile strength is 705 kg/cm^2 and the elongation is 0.026 cm, what is the tensile modulus?
23. Define creep.
24. Briefly describe a Voigt–Kelvin model.
25. What changes occur in a polymer under stress before the yield point?
26. What changes occur in a stress polymer after the yield point?
27. How can you estimate the relative toughness of polymer samples from stress–strain curves?
28. What effect will a decrease in testing temperature have on tensile strength?
29. What effect will an increase in the time of testing have on tensile strength?
30. Describe a Maxwell model.

FURTHER READINGS

Alfrey, T. (1948): *Mechanical Behavior of High Polymers*, Interscience, New York.

Boyd, R. and Smith, G. (2007): *Relaxations in Polymers*, Cambridge University Press, New York.

Brinson, H. and Brinson, L. (2007): *Polymer Engineering Science and Viscoelasticity: An Introduction*, Springer, New York.

Brostow, W., D'Souza, N., Menesses, V., and Hess, M. (2000): *Polymer Characterization*, Wiley, New York.

Carswell, T. S. and Nason, H. K., Effects of environmental conditions on the mechanical properties of organic plastics, Symposium on Plastics, American Society for Testing Materials, Philadelphia, PA, 1944.

Greenhalgh, E., Hiley, M., and Meeks, C. (2009): *Failure Analysis and Fractography of Polymer Compositions*, CRC Press, Boca Raton, FL.

Han, C. (2007): *Polymer Rheology*, Oxford University Press, New York.

Hu, J. (2008): *Design Automation*, CRC Press, Boca Raton, FL.

Koppmans, R. (2009): *Polymer Melt Fracture*, CRC Press, Boca Raton, FL.

Moore, D., Pavan, A., and Williams, J. (2001): *Fracture Mechanics Testing Methods for Polymers, Adhesives, and Composites*, Elsevier, New York.

Nielsen, L. E. (1974): *Mechanical Properties of Polymers*, Dekker, New York.

Riande, E. (1999): *Polymer Viscoelasticity*, Dekker, New York.

Scheirs, J. (2000): *Practical Polymer Analysis*, Wiley, New York.

Shaw, M. and MacKnight, W. (2008): *Introduction to Polymer Viscoelasticity*, Wiley, Hoboken, NJ.

Ward, I. (2004): *An Introduction to the Mechanical Properties of Solid Polymers*, Wiley, Hoboken, NJ.

Wineman, A. and Rajagopal, K. (2000): *Mechanical Response of Polymers*, Cambridge University Press, Cambridge, NY.

Xu, Z. (2009): *Surface Engineering of Polymer Membranes*, Springer, New York.

14 Additives

Most polymeric materials are not wholly a single polymer, but they contain chemicals added to them to modify some physical, biological and/or chemical behavior. These added chemicals, or additives, are generally added to modify properties, assist in processing, and introduce new properties to a material. Coloring agents, colorants, are added giving a product or component with a particular color often for identification or ecstatic purposes. Antibacterial agents are added to protect the material from certain microbial attack. Composites contain a fiber and continuous polymer phase resulting in a material that has a greater flexibility and strength than either of the two components. Some of these additives are polymeric while others are not. They may be added as gases, liquids, or solids. Often some combination of additives is present. A typical tire tread recipe has a processing aid, accelerator, activator, antioxidant, antiozonate, softener, finishing aid, retarder, vulcanizing agent, and accelerator as additives. A typical water-based paint has titanium dioxide as the white pigment, China clay as an extender, a fungicide, a defoaming agent, a coalescing liquid, a surfactant-dispersing agent, and calcium carbonate as another extender. In total, additives are essential materials that allow the polymeric portion(s) to perform as needed. Some of these additives are present in minute amounts and others are major amounts of the overall composition. Some additives are expensive and others are added to give simply give bulk and thus are inexpensive. While the identity, amount, and action of many additives were developed on less than a scientific basis, today there is adequate scientific knowledge such that there exists a rational for the use of essentially all additives.

Typical additives include

Antiblocking agents	Lubricants
Antifoaming agents	Mold release agents
Antifogging agents	Odorants or fragrances
Antimicrobial agents	Plasticizers
Antioxidants	Preservatives
Antistatic agents	Reinforcements
Blowing agents	Slip agents
Coloring agents	Stabilizers, including
Coupling agents	Radiation (UV, VIS)
Curing agents	Heat
Fillers	Viscosity modifiers
Flame Retardants	Flow enhancers
Foaming agent	Thickening agents
Impact modifiers	Antistat agents
Low-profile materials	

In this chapter, we will look at a variety of additives.

14.1 FILLERS

According to the American Society for Testing and Materials standard ASTM D-883, a filler is a relatively inert material added to a plastic to modify its strength, permanence, working properties, or other qualities or to lower costs, while a reinforced plastic is one with some strength properties greatly superior to those of the base resin resulting from the presence of high-strength fillers embedded in the composition. The word extender is sometimes used for fillers. Moreover, the notion that fillers simply "fill" without adding some needed property is not always appropriate. Some fillers are more expensive than the polymer resins and do contribute positively to the overall properties.

The behavior of many types of fillers can be treated roughly as though they were spheres. Current theories describing the action of these spherical-acting fillers in polymers are based on the Einstein equation (Equation 14.1). Einstein showed that the viscosity of a viscous Newtonian fluid (η_0) was increased when small, rigid, noninteracting spheres were suspended in a liquid. According to the Einstein equation, the viscosity of the mixture (η) is related to the fractional volume (c) occupied by the spheres, and that was independent of the size of the spheres or the polarity of the liquid.

$$\eta = \eta_0(1 + 2.5c) \quad \text{and} \quad \frac{\eta_{sp}}{c} = 2.5 \tag{14.1}$$

Providing that c is less than 0.1, good agreement with the Einstein equation is found when glass spheres are suspended in ethylene glycol. The Einstein equation has been modified by including a hydrodynamics or crowding factor (β). The modified Mooney equation (Equation 14.2) resembles the Einstein equation when $\beta = 0$.

$$\eta = \eta_0\left(\frac{2.5c}{(1 - \beta c)}\right) \tag{14.2}$$

Many other empirical modifications of the Einstein equation have been made to predict actual viscosities. Since the modulus (M) is related to viscosity, these empirical equations, such as the Einstein–Guth–Gold (EGG) equation (Equation 14.3), have been used to predict changes in modulus when spherical fillers are added.

$$M = M_0(1 + 2.5c - 14.1c^2) \tag{14.3}$$

Since carbon black and amorphous silica tend to form clusters of spheres (grasping effect), an additional modification of the Einstein equation was made to account for the nonspherical shape or aspect ratio ($1/D$). This factor (f) is equal to the ratio of the length (l) to the diameter (D) of the nonspherical particles.

$$\eta = \eta_0(1 + 0.67fc + 1.62f^2c^2) \tag{14.4}$$

Both natural and synthetic fillers are used. Examples of such fillers are given in Table 14.1.

Among the naturally occurring filler materials are cellulosics such as wood flour, α-cellulose, shell flour, and starch, and proteinaceous fillers such as soybean residues. Approximately 40,000 ton of cellulosic fillers are used annually by the U.S. polymer industry. Wood flour, which is produced by the attrition grinding of wood wastes, is used as a filler for phenolic resins, urea resins, polyolefins, and PVC. Shell flour, which lacks the fibrous structure of wood flour, has been used as a replacement for wood flour for some applications.

α-Cellulose, which is more fibrous than wood flour, is used as a filler for urea and melamine plastics. Melamine dishware is a laminated structure consisting of molded resin-impregnated paper. Presumably, the formaldehyde in these thermosetting resins reacts with the hydroxyl groups in cellulose producing a more compatible composite. Starch and soybean fillers have been used to make biodegradable composites and other materials.

Carbon black, which was produced by the smoke impingement process by the Chinese over 1000 years ago, is now the most widely used filler for polymers. Much of the 1.5 million ton produced annually is used for the reinforcement of elastomers. The most widely used carbon black is furnace carbon black.

TABLE 14.1
Types of Fillers for Polymers

I. Organic materials

A. Cellulosic products
  1. Wood products
    a. Kraft paper
    b. Chips
    c. Course flour
    d. Ground flour
  2. Comminuted cellulose products
    a. Chopped paper
    b. Diced resin board
    c. Crepe paper
    d. Pulp preforms
  3. Fibers
    a. α-Cellulose
    b. Pulp preforms
    c. Cotton flock
    d. Textile byproducts
    e. Jute
    f. Jute
    g. Sisal
    h. Rayon
B. Lignin-type products-processed lignin
C. Synthetic fibers
  1. Polyamides (nylons)
  2. Polyesters
  3. Polyacrylonitrile
D. Carbon
  1. Carbon black
    a. Channel black
    b. Furnace black
  2. Ground petroleum coke
  3. Graphite filaments
  4. Graphite whiskers

II. Inorganic materials

A. Silica products
  1. Minerals
    a. Sand
    b. Quarts
    c. Tripoli
  2. Synthetic materials
    a. Wet-processed silica
    b Pyrogenic silica
    c. Silica aerogel

B. Silicates
  1. Minerals
    a. Asbestos
    b. Karolinite (China clay)
    c. Mica
    d. Talc
    e. Wollastonite
  2. Synthetic products
    a. Calcium silicate
    b. Aluminum silicate
C. Glass
  1. Glass flakes
  2. Solid and hollow glass spheres
  3. Milled glass fibers
  4. Fibrous glass
    a. Filament
    b. Rovings and woven rovings
    c. Yarn
    d. Mat
    e. Fabric
D. Metals
E. Boron fibers
F. Metallic oxides
  1. Ground material
    a. Zinc oxide
    b. Alumina
    c. Magnesia
    d. Titania
  2. Whiskers (including nonoxide)
    a. Aluminum oxide
    b. Beryllium oxide
    c. Zirconium oxide
    d. Aluminum nitride
    e. Boron carbide
    f. Silicon carbide and nitride
    g. Tungsten carbide
    h. Beryllium carbide
G. Calcium carbonate
  1. Calk
  2. Limestone
  3. Precipitated calcium carbonate
H. Polyfluorocarbons
I. Barium ferrite and sulfate

Conductive composites are obtained when powered metal fillers, metal flakes, or metal-plated fillers are added to resins. These composites have been used to produce forming tools for the aircraft industry and to overcome electromagnetic interference in office machines.

Calcium carbonate is available as ground natural limestone and as a synthetic chalk. It is widely used in paints, plastics, and elastomers. The volume relationship of calcium carbonate to resin or the pigment volume required to fill voids in the resin composite is called the pigment-volume concentration (PIVC).

14.2   REINFORCEMENTS

According to the ASTM definition, fillers are relatively inert while reinforcements improve the properties of the materials to which they are added. Actually, a few fillers are used that do not improve properties, but reinforcing fibers produce dramatic improvements to the physical properties of the material to which it is added, generally to form composites.

The transverse modulus (M_T) and many other properties of a long fiber–resin composite may be estimated from the law of mixtures. The longitudinal modulus (M_L) may be estimated from the Kelly Tyson equation (Equation 14.5) where the longitudinal modulus is proportional to the sum of the fiber modulus (M_F) and the resin matrix modulus (M_M). Each modulus is based on a fractional volume (c). The constant k is equal to 1 for parallel continuous filaments and decreases for more randomly arranged shorter filaments.

$$M_L = kM_F c_F + M_M c_M \qquad (14.5)$$

Since the contribution of the resin matrix is small in a strong composite, the second term in the Kelly Tyson equation can be disregarded. Thus, the longitudinal modulus is dependent on the reinforcement modulus, which is independent of the diameter of the reinforcing fiber.

As noted before, for the most part, the resulting materials for the use of reinforcements are composites. Composites are materials that contain strong fibers embedded in a continuous phase. The fibers are called "reinforcement" and the continuous phase is called the matrix. While the continuous phase can be a metallic alloy or an inorganic material, the continuous phase is typically an organic polymer that is termed a "resin." Because of the use of new fibers and technology, most of the composites discussed here are referred to as "space-age" and "advanced materials" composites. Composites can be fabricated into almost any shape and after hardening, they can be machined, painted, etc. as desired.

While there is a lot of science and "space-age technology" involved in the construction of composites, many composites have been initially formulated through a combination of this science and "trial-and-error" giving "recipes" that contain the nature and form of the fiber and matrix materials, amounts, additives, and processing conditions.

Composites have high tensile strengths (approximately thousands of MPa), high Young's modulus (approximately hundreds of GPa), and good resistance to weathering exceeding the bulk properties of most metals. The resinous matrix, by itself, is typically not particularly strong relative to the composite. Further, the overall strength of a single fiber is low. In combination, the matrix–fiber composite becomes strong. The resin acts as a transfer agent, transferring and distributing applied stresses to the fibers. Generally, the fibers should have aspect ratios (ratio of length to diameter) exceeding 100, often much larger. Most fibers are thin (less than 20 μm thick, about a tenth the thickness of a human hair). Fibers should have a high tensile strength and most have a high stiffness, that is, low strain for high stress or little elongation as high forces are applied.

There exists a relationship between the "ideal" length of a fiber and the amount of adhesion between the matrix and the fiber. For instance, assume that only the tip, one end, of a fiber is placed in a resin (Figure 14.1—left). The fiber is pulled. The adhesion is insufficient to "hold" the fiber and it is pulled from the resin (Figure 14.1—top right). The experiment is repeated until the fiber is broken (outside the matrix) rather than being pulled (without breaking) from the resin (Figure 14.1—right bottom). Somewhere between the two extremes, there is a length where there exists a balance between the strength of the fiber and the adhesion between the fiber and the matrix. Most modern composites utilize fiber/matrix combinations that exploit this "balance."

Fiber failure is usually of the "catastrophic" type where the failure is sudden. This is typical in polymeric materials where the material is broken at the "weak-link" and where the strength-related property is related to a combination of individual links (chains).

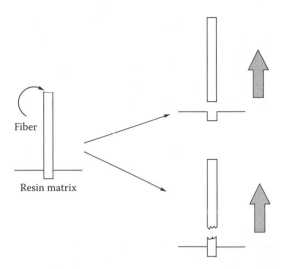

FIGURE 14.1 Tensile loading experiments performed on single fibers embedded in a matrix. The illustration on the left is a preapplication of the tensile loading and those on the right are post-application of the tensile loading.

14.2.1 FIBERS AND RESINS

About 98% of the fibers employed in composites are glass (Section 10.5), carbon (graphite, carbon fibers, etc.; Sections 10.11 through 10.15), and aromatic nylons (often referred to as aramids; Section 5.7). Asbestos, major fiber choice years ago, holds less than 1% of the market because of medical concerns linked to it.

Glass fibers are described using several terms including fiberglass, glass fibers, and fibrous glass. Glass fibers are mainly composed of silicon dioxide glass (Section 10.5). The glass fibers are "pulled" from the glass melt forming fibers that range from 2 to 25 microns in diameter. The pulling action helps to orientate the overall three-dimensional structure giving a material with greater strength and stiffness along the axis of the pull. The ability to pull fibers from molten glass was known for centuries, but fibrous glass was not produced commercially until the 1930s.

As with many polymers, the limits of strength are due to the presence of voids. For glass fibers, these voids generally occur on the surface, thus care is taken to protect these surfaces through surface treatments with methacrylatochromic chloride, vinyl trichlorosilanes, and other silanes. These surface agents chemically react with the fiber surface acting to repel and protect the surface from harmful agents such as moisture.

A number of kinds of carbon-intense fibers are used, the most common being carbon and graphite fibers and carbon black. As in the case of fibrous glass, surface voids are present. Carbon-intense fibers are often surface treated with agents such as low molecular weight epoxy resins. Such surface treatments also aim at increasing the fiber–matrix adhesion.

Two general varieties of aromatic nylons are often employed. A less stiff variety is employed when some flexibility is important, while a stiffer variety is used for applications where greater strength is required. While good adhesion with the resin is often desired, poor adhesion is sometimes an advantage such as in the construction of body armor where "delamination" is a useful mode for absorbing an impact.

As we understand materials better, we are able to utilize them for additional applications. It is known that elongational flow through orifices can result in the stretching and reorientating of polymer chains giving a stronger fiber in the direction of pull. Some polymers become entangled and flow gives additional orientation. Finally, polymer solutions may be stable at rest, but under high rates of extrusion they may be removed from solution, forming a gel phase. These observations have allowed the production of a number of new polyolefin fibers including ultrahigh-modulus polyethylene fibers that are low density but relatively high tensile strength with an elongation at break over two times greater than glass and aromatic nylon fibers.

Both thermoset and thermoplastic resin systems are employed in the construction of composites. The most common thermoset resins are polyamides, unsaturated polyesters, epoxies, phenol-formaldehydes, and amino-formaldehydes. A wide variety of thermoplastic resins has been developed.

14.2.2 APPLICATIONS

Many of the applications for composite materials involve their (relative) lightweight, resistance to weathering and chemicals, and their ability to be easily fabricated and machined. Bulk applications employ relatively inexpensive composites. Combinations of rigorous specifications, low volume, specific machining and fabrication specifications, comparative price to alternative materials, and solutions allow more expensive specialized composites to be developed and utilized.

Applications for composite materials are increasing. Following is a brief description of some of these applications. One of the largest and oldest applications of composites is the construction of water-going vessels from rowboats, sailboats, racing boats, and motor craft to large seagoing ships. The use of fresh water resistant composites and salt water resistant composites allowed the boating industry to grow and today includes a range from individually operated backyard construction to the use of large boatyards producing craft on an assembly line. Most of these crafts are composed of fiberglass and fiberglass/carbon combination composites.

Compositions are also important in the construction of objects to both propel material into and material to exist in outer space. Because of the large amount of fuel required to propel spacecraft into outer space, weight reduction, offered by composites, is essential. The polymeric nature of composites also makes it an ideal material to resist degradation caused by the vacuum of outer space.

Many biomaterials are composites. Bone and skin are relatively light when compared to metals. Composite structures can approach the densities of bone and skin and offer the necessary inertness and strength to act as body-part substitutes.

Power-assisted arms have been made by placing hot-form strips of closed-cell PE foam over the cast of an arm. Grooves are cut into these strips prior to application and carbon/resin added to the grooves. The resulting product is strong, light, and the cushioned PE strips soften the attachment site of the arm to the living body. Artificial legs can be fashioned in glass/polyester and filled with polyurethane foam adding strength to the thin-shelled glass/polyester shell. Artificial legs are also made from carbon/epoxy composite materials. Some of these materials contain a strong interior core with a soft, flexible "skin."

Carbon/epoxy "plates" are now used in bone surgery replacing the titanium plates that had been previously employed. Usually a layer of connective tissue forms about the composite plate.

The rejection of composite materials typically does not occur, but as is the case of all biomaterials, compatibility is a major criteria. Often, the lack of biocompatibility has been found to be the result of impurities (often additives) found in the materials. The removal of these impurities allows the use of the purified materials.

Carbon and carbon/glass composites are used to make "advanced-material" fishing rods, bicycle frames, golf clubs, baseball bats, racquets, skis and ski poles, basketball backboards, and the like. These come in one color—black—because the carbon fibers are black. Even so, they can be coated with about any color desired.

Composites are employed in a number of automotive applications. These include racing car bodies as well as "regular" automobiles. Most automobiles have the lower exterior panels composed of rubbery, plastic blends, and/or composite materials. Other parts of the automobiles such as drive shafts and leaf springs in private cars and heavy trucks, antennas, and bumpers are made from composite materials.

Industrial storage vessels, pipes, reaction vessels, and pumps are now made from composite materials. They offer the needed resistance to corrosion, acids and bases, oils and gases, salt solutions, and the necessary strength and ease of fabrication to allow their continued adoption as a major industrial "building" material.

The Gulf War spotlighted the use of composite materials in the new-age aircraft. The bodies of both the Stealth fighter and the Stealth bomber are mainly carbon composites. The versatility is apparent when one realizes that the Gossamer Albatross, the first plane to cross the English Channel with human power, was largely made up of composite materials including a carbon/epoxy and aromatic nylon composite body, propeller containing a carbon composite core, and the like.

The growth of composite materials in the aerospace industry is generally due to their outstanding strength and resistance to weathering and friction and their lightweight allowing fuel reduction savings. Its growth in the commercial aircraft is seen in the increased use of fiberglass composite material in the succeeding families of Boeing aircraft from about 20 sq yards for the 707, to 200 sq yards for the 727, to 300 sq yards for the 737, and over 1000 for the 747. This amount is increased in the Boeing 767 and includes structural applications of other space-age composites. Thus, the Boeing 767 uses carbon/aromatic nylon/epoxy landing gear doors and wing-to-body fairings.

Until the late 1960s, almost all tactical aircraft were largely titanium. While titanium is relatively light, it is costly and has demanding production requirements so that its use was limited to moderate temperature aircraft applications. Today, most tactical aircraft have a sizable component that is polymeric, mainly composite. The Boeing F/A 18E/F and Lockheed F/A-22 have about 25%, by weight, composite material. It is projected that future military aircraft will have more than 35% composite materials.

Composites have displaced more conventional materials because they are lighter with greater strength and stiffness allowing them to carry a greater payload further. Composites are also relatively insensitive to flaws. The fatigue testing of composites shows that they have a high resistance to cracking and fracture propagation in comparison to metals. They are stable and not subject to corrosion. However, in the design process, particular care must be taken with respect to the metal–composite interface because the galvanic action of some metals will corrode when in contact with certain composites such as the carbon graphite/resin laminates.

The Stealth Bomber, more accurately known as the Northrop Grumman B-2 Spirit Stealth Bomber, is cited as the largest composite structure produced with over 30% of the weight being carbon graphite epoxy composites. The Stealth Bomber was originally slated to hunt Russian mobile missiles that were built in the 1980s. It was deployed in 1993 and has a wingspan of over 52 m, a 21 m length, and is over 5 m high. The reported cost of a single B-2 is varied ranging from about $2.4 billion to a low of $500 million. In any case, it is expensive and only about 21 B-2s have been produced with no plans to build more. The technology associated with the B-2 has been continuously upgraded allowing it to perform tasks with even greater accuracy and stealthiness.

The word stealth comes from the Old English word "stelan" meaning, "to steal." There are a number of stealth aircraft in operation today with others coming on line. To be "stealthy," a plane should

- Be difficult to see with the eye,
- Make little or no noise (achieved by muffling the engines),
- Display little heat from engines and other moving parts,
- Minimize a production of contrails and other signs, and
- Absorb and scatter radar beams.

The most common mode of aircraft detection is radar. Essentially, radar is the detection of radio waves that have been "thrown out" and which bounce off objects returning to the site of origin. Today's radar, if properly used, can help to identify the location, speed, and identity of the aircraft. The radar cross section, RCS, of an aircraft is how much echo the plane sends from radar. Birds have an RCS of about 0.01 m². The Stealth Bomber has a RCS of 0.75 m². The Stealth Bomber, and many stealth aircraft, gain their "stealthiness" from mainly both the shape of the aircraft and the presence of radar absorbing material, RAM. The RAM is made to absorb and eliminate radio waves rather than reflecting them. Most of the RAM materials are polymeric.

The Stealth Bomber has a covering of radar-absorbing, RAM, skin. The basic elements of the composite covering consist of graphite carbon fibers embedded in an epoxy resin. The carbon fibers dissipate some of the absorbed radiation by ohmic heating. The material also contains ferrite particles that are important because they are good "bad" radar absorbers. The good absorbency of radar signals is dependent on a low dielectric constant. In principle, the thicker the coating, the greater the absorption of signals. Thick coatings increase cost and weight and decrease the aerodynamics of the aircraft. (In truth, the flying wing is aerodynamically unstable and is flown with a quadruple-redundant fly-by-wire, FBW, system.) The thickness of the layers needed to provide the necessary radar signature can be reduced using materials with intermediate absorption behavior, the ferrite particles, to smooth the interference between the aircraft and the surrounding air since extreme differences cause strong radar reflections.

The graphite resin layers are laid similar to that employed in making the hull of a fiberglass boat hull. This layering, over a frame or mold, is accomplished using computer-guided assistance. Thin layers of gold foil are also applied to the canopy to give radar reflecting characteristics similar to the rest of the aircraft.

The Stealth Bomber is shaped like a wafer-thin Batman boomerang with smooth, knife-like curved surfaces that lack right angles. The reflective nature of the design can be understood by considering looking straight-on at a mirror. One's image will be reflected back at oneself. Tilt the mirror 45° and it will reflect the image upward. The bomber design is a large system of triangles with flat surfaces angled to deflect radar waves away from the radar emitter. It scrambles the radar all about but not back to the receiver. This shape is also intended to offer a minimum of disruption to the air, that is, the shape brings about a minimum amount of air resistance resulting in a minimization of radar difference between the aircraft and its surroundings.

The use of composites allows a few seams thus reducing radar reflections. For years, seams were believed necessary for large aircraft. For instance, the wings of the large bombers actually "flap" in the wind contributing to the ability of the aircraft to remain in the air for long durations. Without seams and some flexibility these large metal-based aircraft would literally snap apart. Composite assemblies have a minimum of flexibility. To discourage reflections in unwanted directions, the electrical continuity is maintained over the entire surface of the bomber. Thus, all moving and major body parts are tight fitting. The composite nature of the outer skin allows this to be closely achieved.

The engines are buried deep within the interior of the aircraft thereby reducing noise and "heat signature." The heat signature is further reduced by mixing the jet's exhaust with cooled air and exiting it through slit-like vents.

Stealth aircraft are often painted a medium gray or bluish gray that matches the sky so that is does not stand out. Coatings are composed of RAM that penetrates into cracks reducing the number of minor repairs and the use of special tape for repair. Every screw, panel, seam, and gap is covered with a special tape or the RAM coating adding to the bomber's stealthiness.

Because most traditional polymers have low dielectric constants, they can be used as RAMs and many have been used in various applications in military and civil applications. While the B-2 largely depends on a graphite carbon impregnated composite for its integrity and stealthiness, many other polymers are used as RAMs. Many of these polymers are various elastomeric materials. Neoprene is widely used in naval applications as a RAM because of its good weather resistance. Nitrile rubber is used for fuel and oil resistance and fluoro-elastomers are used where wide temperature ranges are encountered. Often, thin flexible sheets or layers of the elastomer are adhered to a metal substrate. While the polysiloxanes offer good RAM behavior, they are difficult to adhere to many metal structures and thus, are not as widely employed in comparison to neoprene and nitrile rubbers. Products are also made using various molding methods. To improve weather resistance, the absorber is often painted with another RAM, an epoxy or urethane-based coating.

14.3 NANOCOMPOSITES

Nature has employed nanomaterials since the beginning of time. Much of the inorganic part of our soil is a nanomaterial with the ability to filter out particles often on a molecular or nano level. The driving force toward many of the nanomaterials is that they can offer new properties or enhanced properties unobtainable with so-called traditional bulk materials. Along with lightweight, high strength to weight features, and small size, new properties may emerge because of the very high surface area to mass where surface atomic and molecular interactions become critical. The nano world is often defined for materials where some dimension is approximately 1–100 nm. In a real way, single linear polymers are nanomaterials since the diameter of the single chain is within this range. The carbon–carbon bond length is approximately 0.15 nm or the average zigzag bond length is about 0.13 nm. While some short to moderate-length vinyl polymers have contour lengths less than 100 nm, higher molecular weight polymers have contour lengths that far exceed this. Even so, individual polymer chains fall within the realm of nanomaterials when they act independently. Since the cumulative attractive forces between chains are large, polymer chains generally act in concert with other polymer chains leading to properties that are dependent on the bulk material. Chain folding, inexact coupling, and chain branching are some of the reasons that bulk properties fall short of theoretical properties, but with the ability to work with chains individually, strength and related properties are approaching theoretical values. Much of the nanorevolution with materials involves how to synthesize and treat materials on an individual basis as well as visualizing the uses for these materials.

The ultimate strength and properties of many materials are dependent upon the intimate contact between the various members. Thus, for ceramics, nano-sized particles allow a more homogeneous structure resulting in stronger ceramic materials.

Nanocomposites have also been with us since almost the beginning of time. Our bones are examples of nanocomposites. The reinforcement material is plate-like crystals of hydroxyapatite, $Ca_{10}(PO_4)_6(OH)_2$, with a continuous phase of collagen fibers. The shell of a mollusk is microlaminated containing as the reinforcement aragonite (a crystalline form of calcium carbonate) and the matrix is a rubbery material. Allowing nature to be a source of ideas is a continuing theme in synthetic polymer science, including the modification of natural polymers. Much of the renewed interest in nanocomposite materials is the direct result of the availability of new nano-building blocks.

Within a composite material, much of the ultimate strength comes from the intimate contact the fiber has with the matrix material. Nano fibers allow more contact between the fiber (on a weight basis) and the matrix resulting is a stronger composite because of an increased fiber surface–matrix interface.

A number of inorganic/organic nanocomposites have been made. These include nano inorganics including nano fibers from silicon nitride, titanium (IV) oxide, zirconia, alumina, titanium carbide, and tungsten carbide. It also includes the use of special clays (layered silicates) mixed with nylons to form nanocomposites. The clay layers are separated giving platelets about 1 nm thick. These nylon–clay microcomposites are used to make the air intake cover of some of the Toyota automobiles. These individual clay platelets have also been used to form nearly single layer polymer chain sheets similar to lignin. The interaction with the silicate surface encourages the polymer chains to take different arrangements. To be effective the hydrophilic silicate surface is generally modified to accommodate the more hydrophobic common monomers and polymers.

While carbon fiber (thickness approximately 1000 nm) composites offer very strong materials, carbon nanotubes make even stronger composites. These carbon nanotubes have aspect ratios of over 1000 (ratio of length to diameter). Further, because some carbon nanotubes are electrically conductive, composites containing them can be made to be conductive. A number of carbon nanotube-matrixes have been made including using a number of engineering resins such as polyesters, nylons, polycarbonates, and poly(phenylene ether).

Individual polymer chains can be more flexible than groups of chains (bulk) even when the polymer is generally considered rigid. This is presumably because single chains have less torsional strain imparted by near neighbors and various chain entanglements and associations are not present. Compared with carbon fibers, carbon nanotubes imbedded within a polymer matrix can withstand much greater deformations before they break. Further, nanomaterials are generally more efficient in transferring applied load from the nanomaterial to the matrix. These factors contribute to the greater strength of carbon nanotube composites.

As noted before, adhesion between the reinforcing agent and the matrix is important. Some matrix materials do not adhere well with certain fibers. This is partially overcome through the introduction of defects or functional groups onto the nanomaterials that act as hooks to anchor them to the matrix material.

Research with tires continues to be active with the end goals of reducing tire weight and increasing tire lifetime. For example, truck tires are now capable of running 750,000 miles with the carcass married to 4 tire threads. The truck tire liners weigh about 9 lb. A decrease in the weight of 50% would translate into a significant increase in the mileage of 3–5 mpg. Solutions should be inexpensive employing readily available and abundant materials.

Nanocomposites are being used in tires, in particular tire inner liners. Here, less permeable inner layers are achieved by the introduction of "clad" layers that allow the use of a thinner inner liner, resulting in an overall lighter tire. Tire inner layers are typically derived from butyl rubber, often halogenated butyl rubber.

Exfoliate means to open, such as to open the pages of a book. New sheet-like nanocomposites are being investigated in the production of tire inner layers. These approaches focus on the use of two different materials, silicates, and graphite. Most silicates and graphite exist in layered groupings.

Layered clay silicates, generally from the intermediate-grained montmorillonite kaolin clay, are often used as fillers in plastics and in the production of pottery and other ceramic items. These silicates consist of the silicate sheets held together mostly by the sodium cation with lesser amounts of other metal ions such as iron, copper, nickel, and the like. There are several approaches to open these silicate layers.

The layers of kaolin clay open up when treated with water. The water acts to remove the cations that hold together the clay matrix. The resulting sheets are negatively charged, since the positively charged sodium ion is removed. These negatively charged sheets repel one another resulting in the formation of nano-sheets as shown in Figure 14.2.

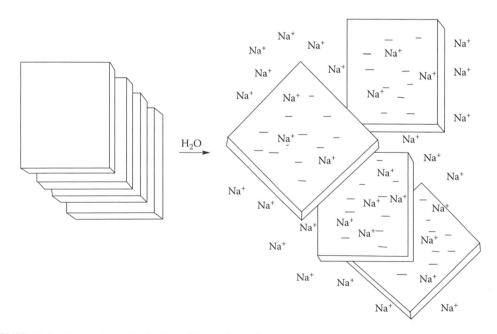

FIGURE 14.2   Formation of individual silicate sheets from exposure to water.

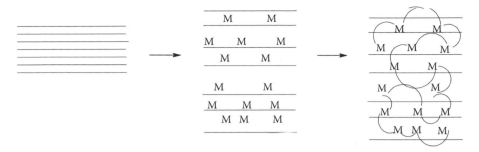

FIGURE 14.3 Silicate sheets impregnated with caprolactam monomer, M, which are eventually polymerized.

Toyota works with karolin clay that has about 10 nm between the sheets. The clay is treated with caprolactam monomer that impregnates the clay sheets increasing the distance between sheets to 30–50 nm. The caprolactam monomer impregnating the silicate sheets is then polymerized giving nylon–silicate sheets. This sequence is pictured in Figure 14.3.

Another approach utilizes highly oriented pyrolytic graphite that can be separated into individual graphite sheets as the inner layer material. In each case, the nanoclay or graphite material is present in about 2% by weight in the tire inner liner.

As the inner liner is processed, the sheets are made to align forming a clad that reduces the ability for air to flow from the tire.

14.4 PLASTICIZERS

Flexibilizing of polymers can be achieved through internal and external plasticizations. Internal plasticization can be produced through copolymerization giving a more flexible polymer backbone or by grafting another polymer onto a given polymer backbone. Thus, poly(vinylchloride-co-vinyl acetate) is internally plasticized because of the increased flexibility brought about by the change in the structure of the polymer chain. The presence of bulky groups on the polymer chain increases the segmental motion and the placement of such groups through grafting acts as an internal plasticizer. Internal plasticization achieves its end goal at least in part through discouraging association between polymer chains. However, grafted linear groups with more than 10 carbon atoms can reduce the flexibility because of side-chain crystallization when the groups are regularly spaced.

External plasticization is achieved through the incorporation of a plasticizing agent into a polymer through mixing and/or heating. The remainder of this section focuses on external plasticization.

Plasticizers should be relatively nonvolatile, nonmobile, inert, inexpensive, nontoxic, and compatible with the system to be plasticized. They can be divided based on their solvating power and compatibility. Primary plasticizers are used as either the sole plasticizer or the major plasticizer with the effect of being compatible with some solvating nature. Secondary plasticizers are materials that are generally blended with a primary plasticizer to improve some performance such as flame resistance, mildew resistance, or to reduce cost. The division between primary and secondary plasticizers is at times arbitrary. Here we will deal with primary plasticizers.

According to the ASTM D-883 definition, a plasticizer is a material incorporated into a plastic to increase its workability, flexibility, or dispensability. The addition of a plasticizer may lower the melt viscosity, elastic modulus, and T_g.

Waldo Semon patented the use of tricresyl phosphate as a plasticizer for PVC in 1933. This was later replaced by the less toxic di-2-ethylhexyl phthalate (DOP), which is now the most widely used plasticizer. The worldwide production of plasticizer is approximately 3.2 million tons annually. Volume-wise, about 90% of the plasticizers are used with PVC and PVC-containing systems.

The effect of plasticizers has been explained by the lubricity, gel, and free volume theories. The lubricity theory states that the plasticizer acts as an internal lubricant and permits the polymers to slip past one another. The gel theory, which is applicable to amorphous polymers, assumes that a polymer, such as PVC, has many intermolecular attractions that are weakened by the presence of a plasticizer. In free-volume theories, it is assumed that the addition of a plasticizer increases the free volume of a polymer and that the free volume is identical for polymers at T_g. There may be some truth in most of these theories. It is believed that a good plasticizer solubilizes segments allowing them some degree of mobility creating free volume through the Brownian movement. In turn, this lowers the temperature where segmental mobility can occur making the material more flexible.

Most plasticizers are classified as to being general purpose, performance, or specialty plasticizers. General-purpose plasticizers are those that offer good performance inexpensively. Most plasticizers are of this grouping. Performance plasticizers offer added performance over general-purpose plasticizers generally with added cost. Performance plasticizers include fast solvating materials such as butyl benzyl phthalate and dihexyl phthalate; low temperature plasticizers such as di-*n*-undecyl phthalate and di-2-ethylhexyl adipate; and so-called permanent plasticizers such as tri-2-ethylhexyl trimellitate, triisononyl trimellitate, and diisodecyl phthalate. Specialty plasticizers include materials that provide important properties such as reduced migration, improved stress–strain behavior, flame resistance, and increased stabilization.

The three main chemical groups of plasticizers are phthalate esters, trimellitate esters, and adipate esters. In all the three cases, performance is varied through the introduction of different alcohols into the final plasticizer product. There is a balance between compatibility and migration. Generally, the larger the ester grouping the less the migration up to a point where compatibility becomes a problem and where compatibility now becomes the limiting factor.

Compatibility is the ability to mix without forming different phases. A good approach to measure the likelihood of compatibility is solubility parameters (Section 3.2). Plasticizers with the solubility parameters and the type of bonding similar to those of the polymer are more apt to be compatible than when the solubility parameters are different. The solubility parameter for PVC (9.66) is near that of good plasticizers for PVC such as DOP (8.85). Table 14.2 contains solubility parameters for typical plasticizers.

The development of plasticizers has been plagued with toxicity problems. Thus, the use of highly toxic polychlorinated biphenyls (PCBs) has been discontinued. Phthalic acid esters, such as DOP, may be extracted from blood stored in plasticized PVC blood bags and tubing. These aromatic esters

TABLE 14.2
Solubility Parameters of Typical Plasticizers

Plasticizer	Solubility Parameter (*H*)
Paraffinic oils	7.5
Dioctyl phthalate	7.9
Dibutoxyethyl phthalate	8.0
Tricresyl phosphate	8.4
Dioctyl sebacate	8.6
Triphenyl phosphate	8.6
Dihexyl phthalate	8.9
Hydrogenated terphenyl	9.0
Dibutyl sebacate	9.2
Dibutyl phthalate	9.3
Dipropyl phthalate	9.7
Diethyl phthalate	10.0
Glycerol	16.5

are also distilled from PVC upholstery in closed automobiles in hot weather. These problems have been solved by using oligomeric polyesters as nonmigrating plasticizers instead of DOP, in appropriate situations. Recently, some limited tests have indicated a relationship of prenatal exposure to phthalates and reproductive abnormalities of male babies. The chemical industry is monitoring these results taking appropriate steps to limit or eliminate the exposure to dangerous materials.

Experimental investigations and assessments show that the average person takes in about 2 g a year of external plasticizers. Most of this is from traces of DOA migrating from food packaging. The so-called no observed effect level (NOEL) for DOA in rodents is about 40 mg per kg of body weight per day. Extrapolation for a person equates to 1000 g of plasticizer for a "safety factor" of about 500. Even so, increased efforts to evaluate the safety of plasticizers continue.

Plasticizers extend the lower temperature range for use of materials since they discourage polymer chain associative behavior and encourage segmental flexibility increasing the rotational freedom effectively decreasing the material's typical T_g.

When present in small amounts, plasticizers can act as antiplasticizers increasing the hardness and decreasing the elongation of the material partly due to their ability to fill voids. Inefficient plasticizers require larger amounts of plasticizers to overcome the initial antiplasticization. However, good plasticizers such as DOP change from being antiplasticizers to plasticizers when less than 10% of the plasticizer is added to PVC.

External plasticizers are not permanent. Plasticizer molecules associate with one another eventually creating "preferred" migration routes to the material's surface where the plasticizer is rubbed or washed away. The preferential association of plasticizers also leaves some sites less flexible and creates variations in the material's stress–strain and expansion–contraction behaviors.

14.5 ANTIOXIDANTS

Antioxidants retard oxidative degradation. Heat, mechanical shear, and UV radiation can be responsible for the formation of free radicals, which in turn, can act to shorten polymer chains and increase cross-linking, both leading to a deterioration in material properties. Free radical production often begins a chain reaction. Primary antioxidants donate active hydrogen atoms to free radical sites thereby quenching or stopping the chain reaction. Secondary antioxidants or synergists act to decompose free radicals to more stable products.

Polymers such as polypropylene (PP) are not usable outdoors without appropriate stabilizers because of the presence of readily removable hydrogen atoms on the tertiary carbon atoms. PP and many other polymers (R-H) are attacked during processing or outdoor use in the absence of stabilizers because of chain degradation reactions. These degradation reactions are retarded by the presence of small amounts of antioxidants. Naturally occurring antioxidants are present in many plants and trees such as the hevea rubber trees. The first synthetic antioxidants were synthesized independently by Caldwell, and by Winkelman and Gray by the condensation of aromatic amines with aliphatic aldehydes.

Many naturally occurring antioxidants are derivatives of phenol and hindered phenols, such as di-*tert*-butyl paracresol (14.6). It has the ability to act as a chain transfer agent forming a stable free radical that does not initiate chain radical degradation. However, the phenoxy free radical may react with other free radicals producing quinone derivatives.

$$(14.6)$$

Since carbon black has many stable free radicals, it may be added to polymers such as polyolefins to retard a free radical by attracting and absorbing other free radicals. It is customary to add small amounts of other antioxidants to enhance the stabilization by a "synergistic effect" whereby many antioxidant combinations are more stable than using only one antioxidant.

14.6 HEAT STABILIZERS

Heat stabilizers are added to materials to impart protection against heat-induced decomposition. Such stabilizers are needed to protect a material when it is subjected to a thermal-intense process (such as melt extrusion) or when the material is employed under conditions where increased heat stability is needed.

In addition to the free radical chain degradation described for polyolefins, other types of degradation occur including dehydrohalogenation that occurs for PVC. When heated, PVC may lose hydrogen chloride forming chromophoric conjugated polyene structures eventually yielding fused aromatic ring systems.

This type of degradation is accelerated in the presence of iron salts, oxygen, and hydrogen chloride. Toxic lead, barium salt, and cadmium salt act as scavengers for hydrogen chloride and may be used for some applications such as wire coating. Mixtures of magnesium and calcium sterarates are less toxic and effective. Among the most widely used heat stabilizers for PVC are the organotin products. Many of these products are based on organotin polymers discovered by Carraher.

$$R + HCl \qquad (14.7)$$

14.7 ULTRAVIOLET STABILIZERS

Ultraviolet stabilizers act to quench UV radiation. While much of the sun's high-energy radiation is absorbed by the atmosphere, some radiation in the 280–400 nm (UV) range reaches the Earth's surface. Since the energy of this radiation is approximately 70–100 kcal/mol (280–400 kJ/mol) UV radiation, is sufficient to break many chemical bonds, and thus cause polymer chains to break. This breakage generally does one of two things—(1) It can increase the amount of cross-linking through the subsequent formation of bonds after the initial bond breakage; or (2) Bond breakage can result in a decreased chain length. Either of these results in a decreased overall strength of the polymeric material, and often the yellowing of many organic polymers. Cross-linking also results in the embrittlement of these polymers.

PE, PVC, PS, PET, and PP are degraded at wavelengths of 300, 310, 319, 325, and 370 nm, respectively. The bond energy required to cleave the tertiary carbon hydrogen bond in PP is only 320 kJ/mol that corresponds to a wavelength of 318 nm.

Since the effect of UV radiation on synthetic polymers is similar to its effect on the human skin, it is not surprising that UV stabilizers such as phenyl salicyclate have been used for commercial polymers and in suntanning lotions for many years. Phenyl salicylate rearranges in the presence of UV radiation forming 2,2′-dihydroxybenzophenone. The latter and other 2-hydroxybenozophenones act as energy transfer agents, that is, they absorb energy-forming chelates that release energy at longer wavelengths by the formation of quinone derivatives.

$$\text{Light} \qquad (14.8)$$

Over 100,000 ton of UV stabilizers are used by the U.S. polymer industry annually.

14.8   FLAME RETARDANTS

Flame retardants inhibit or resist the ignition and spread of fire. There are several general types of flame retardants that have been added to synthetic polymers. These include synthetic materials generally halocarbons and phosphorus-containing materials. Another class of flame retardant is minerals such as aluminum hydroxide, boron compounds, magnesium hydroxide, and antimony trioxide. Many of these, including most of the halocarbons such as PCBs, present health problems.

Flame retardants resist fire by several mechanisms. Some minerals, such as magnesium hydroxide and aluminum hydroxide decompose when exposed to fire through endothermic means removing heat from the fire. Inert fillers such as calcium carbonate and talc dilute the portion of material that can burn. Thus, a home made of concrete blocks is less apt to burn completely than a home that is made of wood since concrete blocks reduce the amount of flammable material.

While some polymers such as PVC are not readily ignited, most organic polymers, like hydrocarbons, will burn. Some polymers will support combustion, such as polyolefins, SBR, wood, and paper, when lit with a match or some other source of flame. The major products for much of this combustion are carbon dioxide (or carbon monoxide if insufficient oxygen is present) and water.

Since many polymers are used as shelter and clothing and in household furnishing, it is essential that they have good flame resistance. Combustion is a chain reaction that may be initiated and propagated by free radicals. Since halides and phosphorus radicals couple with free radicals produced in combustion terminating the reaction, many flame retardants are halide (halogen) or phosphorus-containing compounds. These may be additives; external retardants sprayed onto the material such as antimony oxide and organic bromides; or internal retardants such as tetrabromophthalic anhydride that are introduced during the polymerization process so that they are part of the polymer chain.

Fuel, oxygen, and high temperature are essential for the combustion process. Thus, polyfluorocarbons, phosphazenes, and some composites are flame resistant because they are not good fuels. Fillers such as alumina trihydrate (ATH) release water when heated and hence reduce the temperature of the combustion process. Compounds such as sodium carbonate, which releases carbon dioxide when heated, shields the reactants from oxygen. Char, formed in some combustion processes also shield the reactants from a ready source of oxygen and retard the outward diffusion of volatile combustible products. Aromatic polymers, such as PS, tend to char and some phosphorus and boron compounds catalyze char formation aiding in controlling the combustion process.

Synergistic flame retardants such as a mixture of antimony trioxide and an organic bromo compound are more effective than single flame retardants. Thus, mixtures are often employed to protect materials and people.

Since combustion is subject to many variables, tests for flame retardancy may not correctly predict flame resistance under unusual conditions. Thus, a disclaimer stating that flame retardancy tests do not predict performance in an actual fire must accompany all flame-retardant products. Flame retardants, like many organic compounds, may be toxic or may produce toxic gases when burned. Hence, care must be exercised with using fabrics or other polymers treated with flame retardants.

14.9   COLORANTS

Color is a subjective phenomenon whose aesthetic value has been recognized for centuries. Since it is dependent on the light source, the object, and the observer, color is not subject to direct measurement, though instruments can measure a color for reproducibility. Colorants in polymers may be soluble dyes or comminuted pigments.

Some polymeric objects, such as rubber tires, are black because of the presence of high proportions of carbon black filler. Many other products, including some paints, are white because of the presence of titanium dioxide (titanium (IV) oxide), the most widely used inorganic pigment. Over 50,000 tons of colorants are used annually by the polymer industry.

Pigments are classified as organic or inorganic. Organic pigments are brighter, less dense, and smaller than the more widely used, more opaque inorganic colorants. Iron oxides or ochers are available as yellow, red, black, brown, and tan.

Carbon black is the most widely used organic pigment, but phthalocyanine blues and greens are available in many shades and are also widely used.

14.10 CURING AGENTS

The use of curing agents began with the serendipitous discovery of vulcanization of hevea rubber with sulfur by Charles Goodyear in 1838. The conversion of an A- or B- stage phenolic novolac resin with hexamethylenetetramine in the early 1900s was another relatively early use of curing (cross-linking) agents. Organic accelerators, or catalysts, for the sulfur vulcanization of rubber were discovered by Oenslager in 1912. While these accelerators are not completely innocuous, they are less toxic than aniline, used prior to the discovery of accelerators. Other widely used accelerators are thiocarbanilide and 2-mercaptobenzothiazole (Captax).

Captax (Structure 14.9) is used to the extent of 1% with hevea rubber and accounts for the major part of the over 30,000 tons of accelerators used annually in the United States. Other widely used accelerators include 2-mercaptobenzothiazole sulfonamide (Santocure; Structure 14.10), used for the vulcanization of SBR; dithiocarbamates and thiuram disulfides. Thiuram disulfide (Structure 14.11) is a member of a group called ultra-accelerators that allow the curing of rubber at moderate temperatures and may be used in the absence of sulfur.

2-Mercaptobenzo thiazole

(14.9)

2-Mercaptobenzothiazole sulfenamide

(14.10)

Tetramethyl thiuram disulfide

(14.11)

Initiators such as benzoyl peroxide are used not only for the initiation of chain reaction polymerization, but also for the curing of polyesters and ethylene–propylene copolymers, and for the grafting of styrene on elastomeric polymer chains.

Unsaturated polymers such as alkyd resins can be cured or "dried" in the presence of oxygen, a heavy metal, and an organic acid called a drier. The most common organic acids are linoleic, abietic, naphthenic, octoic, and tall oil fatty acids.

14.11 ANTISTATIC AGENTS—ANTISTATS

Antistatic agents (antistats) dissipate static electrical charges. Insulating materials, including most organic plastics, fibers, films, and elastomers, can build up electrical charge. Because these largely organic materials are insulators, they are not able to dissipate the charge. Such charge build up is particularly noticeable in cold, dry climates and led to dust attraction and sparking.

Antistatic agents can be either internal or external. External antistats are applied to the surface by wiping, spraying, etc. These surface treatments act to prevent static charge build up. Internal antistats are added during the processing and become an integral part of the bulk composition of the material. Because surface treatments are often worn away through washing, waxing, and handling, the external antistats must be replenished. Internal antistats are added to allow the antistats to come ("bleed") to the surface over a long time, giving the material long-term protection. Many antistatic agents are long-chain aliphatic amines and amides, esters of phosphoric acid, poly(ethylene glycol) esters, PEGs, and quaternary ammonium salts such as docosyltrimethylammonium chloride (14.12) or cocamidopropyl betaine (14.13).

Docosyltrimethylammonium chloride (14.12)

Cocamidopropyl betaine (14.13)

Antistatic agents often have both a hydrophobic (above the hydrocarbon chain; 14.12 and 14.13) and a hydrophilic (salts, quaternary amine, amide; 14.12 and 14.13) portion. The hydrophobic portion is attracted to the hydrophobic part of most vinyl polymers while the hydrophilic portion interacts with the moisture in the air.

14.12 CHEMICAL BLOWING AGENTS

Chemical blowing agents, CBAs, are employed to create lighter weight material through the formation of a foam. Physical CBAs are volatile liquids and gases that expand, volatilize during processing through the control of pressure and temperature.

Cellular polymers not only provide insulation and resiliency, but are usually stronger on a weight basis than solid polymers. Fluid polymers may be foamed by the addition of low-boiling liquids such as pentane or fluorocarbons by blowing with compressed nitrogen gas, by mechanical heating, and by the addition of foaming agents. While some carbon dioxide is produced when polyurethanes are produced in the presence of moisture, auxiliary propellants are also added to the prepolymer mixture to give the desired amount of foaming. The most widely used foaming agents are nitrogen-producing compounds such as azobisformamide. Other foaming agents that decompose at various temperatures are also available.

14.13 COMPATIBILIZERS

Compatibilizers are compounds that provide miscibility or compatibility to materials that are otherwise immiscible or only partially miscible yielding a homogeneous product that does not separate into its components. Typically compatibilizers act to reduce the interfacial tension and are concentrated at phase boundaries. Reactive compatibilizers chemically react with the materials they are to make compatible. Nonreactive compatibilizers perform their task by physically making the various component materials compatible.

14.14 IMPACT MODIFIERS

Impact modifiers improve the resistance of materials to stress. Most impact modifiers are elastomers such as ABS, BS, methacrylate–butadiene–styrene, acrylic, ethylene–vinyl acetate, and chlorinated polyethylene.

14.15 PROCESSING AIDS

Processing aids are added to improve the processing characteristics of a material. They may increase the rheological and mechanical properties of a melted material. Acrylate copolymers are often utilized as processing aids.

14.16 LUBRICANTS

Lubricants are added to improve the flow characteristics of a material during its processing. They operate by reducing the melt viscosity or by decreasing adhesion between the metallic surfaces of the processing equipment and the material being processed. Internal lubricants reduce molecular friction consequently decreasing the material's melt viscosity allowing it to flow more easily. External lubricants act by increasing the flow of the material by decreasing the friction of the melted material as it comes in contact with surrounding surfaces. In reality, lubricants such as waxes, amides, esters, acids, and metallic stearates act as both external and internal lubricants.

14.17 MICROORGANISM INHIBITORS

While most synthetic polymers are not directly attacked by microorganisms such as fungi, yeast and bacteria, they often allow growth on their surfaces. Further, naturally occurring polymeric materials such as cellulosics, starch, protein, and vegetable oil-based coatings are often subject to microbiologic deterioration. Finally, some synthetics that contain linkages that can be "recognized" by enzymes contained with the microorganism (such as amide and ester linkages) may also be susceptible to attack.

One major antimicrobial grouping once was organotin-containing compounds. These monomeric organotin-containing compounds are now outlawed because of the high "leaching" rates of these materials affecting the surrounding areas. Even so, polymeric versions are acceptable and can be considered "nonleaching" or slowly leaching.

Organic fungistatic and bacteriostatic additives are currently employed, but in all cases, the formation of resistant strains and the toxicity of the bioactive additive must be considered.

14.18 SUMMARY

1. Most polymers contain materials added to modify some chemical and/or physical properties that allow them to better fulfill their intended use. These added materials are called additives.
2. Fillers are relatively inert materials that usually add bulk but when well chosen, they can enhance physical and chemical properties. Many natural and synthetic materials are used as fillers today. These include polysaccharides (cellulosics), lignin, carbon-based materials, glass, and other inorganic materials.

3. The most important reinforced materials are composites. Composites contain strong fibers embedded in a continuous phase. They form the basis of many of the advanced and space-age products. They are important because they offer strength without weight and good resistance to weathering. Typical fibers are fibrous glass, carbon-based aromatic nylons, and polyolefins. Typical resins are polyimides, polyesters, epoxys, phenol-formaldehyde, and many synthetic polymers. Applications include biomedical, boating, aerospace and outer space, sports, automotive, and industry.

4. Nanocomposites employ nanofibers that allow for much greater fiber–resin surface contact per mass of fiber and consequently they generally offer greater strength in comparison to similar nonnanocomposites.

5. Plasticizers allow polymer chains to move past one another allowing wholesale flexibility. They can enhance flexibility above and below the glass transition temperature of a polymer. Plasticization can occur through the addition of an external chemical agent or may be incorporated within the polymer itself.

6. Antioxidants retard oxidative degradation.

7. Heat stabilizers allow some protection to heat-induced decomposition.

8. Ultraviolet stabilizers act to quench UV radiation. UV radiation is strong enough to break chemical bonds, and thus cause polymer chains to break either increasing the amount of cross-linking through the subsequent formation of bonds or can result in a decreased chain length. Either of these will result in decreased overall strength of the polymeric material.

9. Flame retardants impart to the polymers some ability to resist ready combustion. Since fuel, oxygen, and high temperature are essential for the combustion of polymers, the removal of any of these prerequisites retards combustion. Flame retardants act through a variety of mechanisms including char formation, combination with free radical species that promote further combustion, through the release of water, and so on.

10. A variety of inorganic and organic compounds are used to color polymers.

11. The rate of cross-linking of polymers, such as natural rubber, is increased using accelerators, often misnamed as catalysts.

12. Antistats reduce the electrostatic charge on the surface of polymers.

13. Gas-producing additives are essential for the formation of cellular products such as foam cushions.

14. Biocides are used to prevent or retard attack on polymers by microorganisms.

15. Lubricants serve as processing aids that discourage the sticking of polymers to metal surfaces during processing.

16. Other important additives are coloring agents, curing agents, antistatic agents, chemical blowing agents, and microorganism inhibitors.

17. Each of these additives perform a critical role in allowing polymers to be processed and utilized giving the variety of useful products we have today.

GLOSSARY

Accelerator: Catalyst for the vulcanization of rubber.
Acicular: Needle-shaped.
Antioxidant: Additive that retards polymer degradation by oxidative modes.
Antiplasticization: Hardening and stiffening effect observed when small amounts of a plasticizer are added to a polymer.
Antistat: Additive that reduces static charges on polymers.
Aspect ratio: Ratio of length to diameter.
Biocide: Additive that retards attack by microorganisms.
Blocking: Sticking of sheets of film to one another.
Bulk molding compound (BMC): Resin-impregnated bundles of fibers.
Bound rubber: Rubber adsorbed on carbon black and that is insoluble in benzene.

Carbon black: Finely divided carbon made by the incomplete combustion of hydrocarbons.

Cellular polymers: Foams.

Chemical blowing agent: Volatile liquids and gases that expand and/or volatilize during the processing of a material creating pockets leading to lighter weight materials.

Comminuted: Finely divided.

Composites: Materials that often contain strong fibers embedded in a continuous phase called a matrix or resin.

Colorant: Color causing material; usually a dye or a pigment.

Compatibilizers: Chemicals that provide miscibility or compatibility to materials that are otherwise immiscible or only partially miscible, giving a more homogeneous material.

Continuous phase: Resin in a composite.

Coupling agents: Products that improve the interfacial bond between the filler and the resin.

Curing agent: Additive that causes cross-linking.

Diatomaceous earth: Siliceous skeletons of diatoms.

Discontinuous phase: Discrete filler additive, such as fibers, in a composite.

Drier: Catalyst that aids the reaction of polymers with oxygen.

Drying: Cross-linking of an unsaturated polymer chain.

Energy transfer agent: Molecule that absorbs high energy and reradiates it in the form of lower energy.

Extender: Term sometimes applied to an inexpensive filler.

Fiberglass: Trade name for fibrous glass.

Fibrous filler: Fiber with an aspect ratio of at least 150:1.

Fibrous glass: Filaments made from molten glass.

Filament winding: Process in which resin-impregnated continuous filaments are wound on a mandrel and the composite is cured.

Filler: Usually a relatively inert material used as the discontinuous additive; generally inexpensive.

Flame retardant: Additive that increases the flame resistance of a polymer.

Foaming agent: Gas producer.

Free volume: Volume not occupied by polymer chains.

Fuller's earth: Diatomaceous earth.

Gel theory: Theory that assumes that in the presence of a pseudo-three-dimensional structure intermolecular attractions are weakened by the presence of a plasticizer.

Graphite fibers: Fibers made by the pyrolysis of polyacrylonitrile fibers.

Heat stabilizers: Additives that retard the decomposition of polymers at elevated temperatures.

Impact modifiers: Materials that improve the resistance of materials to stress.

Lamellar: Sheetlike.

Laminate: Composite consisting of layers adhered by a resin.

Low-profile resins: Finely divided incompatible resins that made a rough surface smooth.

Lubricants: Materials that improve the flow characteristics of materials, generally during processing.

Lubricity theory: Theory that plasticization occurs because of increased polymer chain slippage.

Mica: Naturally occurring lamellar silicate.

Microballoons: Hollow glass spheres.

Mold release agent: Lubricant that prevents polymers from sticking to mold cavities.

Novaculite: Finely ground quartzite rock.

Plasticizer: Material that increases the flexibility of a material.

Plastisol: Suspension of finely divided polymer in a liquid plasticizer with the plasticizer penetrating and plasticizing the polymer when heated.

Promotor: Coupling agent.

Pultrusion: Process in which bundles of resin-impregnated filaments are passed through an orifice and cured.

Reinforced plastic: Composite whose additional strength is dependent on a fibrous additive.

Roving: Bundle of untwisted strands.

Side-chain crystallization: Stiffening effect noted when long, regularly spaced pendent groups are present on a polymer chain.

Sheet molding compound (SMC): Resin-impregnated mat.

Strand: Bundle of filaments.

Synergistic effect: Enhanced effect beyond that expected by simply an additive effect.

Syntactic foam: Composite of resin and hollow spheres.

Talc: Naturally occurring hydrated magnesium silicate.

Ultra-accelerator: Catalyst that cures rubber at low temperatures.

Ultraviolet stabilizer: Additive that retards degradation caused by UV radiation.

Vulcanization: Cross-linking with heat and sulfur.

Whiskers: Single crystals used as reinforcement; extremely strong.

Wood flour: Attrition-ground, slightly fibrous wood particles.

EXERCISES

1. Name three unfilled polymers.
2. What is the continuous phase in wood?
3. What filler is used in Bakelite?
4. Name three laminated plastics.
5. How would you change a glass sphere from an extender to a reinforcing filler?
6. If one stirs a 5 mL volume of glass beads in 1 L of glycerol, which will have the higher viscosity, small or large beads?
7. When used in equal volume, which will have the higher viscosity: (a) a suspension of loosely packed spheres or (b) a suspension of tightly packed spheres?
8. Why is the segmental mobility of a polymer reduced by the presence of a filler?
9. What effect does a filler have on the T_g?
10. Which would yield the stronger composite: (a) peanut shell flour, or (b) wood flour?
11. What is the advantage and disadvantage, if any, of α-cellulose over wood flour?
12. What filler is used in decorative laminates such as Formica table tops?
13. What are the potential advantages of natural fillers?
14. What is a disadvantage to using chipped glass as a filler in comparison to glass spheres?
15. What is the advantage of using BMC and SMC over hand lay-up techniques such as those used to make some boat hulls?
16. Which would be stronger: (a) a chair made from PP or (b) one of equal weight made from cellular PP?
17. What advantage would a barium ferrite-filled PVC strip have over an iron magnet?
18. Providing that the volumes of the fibers are similar, which would give the stronger composite: fibers with (a) small or (b) large cross sections?
19. How would you make an abrasive foam from polyurethane?
20. Why is a good interfacial bond between the filler surface and the resin important?
21. Cellulose nitrate explodes when softened by heat. What would you add to permit processing at lower temperatures?
22. PVC was produced in the 1830s but no used until the 1930s. Why?
23. What was the source of "fog" on the inside of the windshield until the 1930s?
24. Can you propose a mechanism for antiplasticization?
25. What naturally occurring fiber is more resistant to microbiological attack than nylon?
26. The T_g decreases progressively as the size of the alkoxy group increases from methyl to decyl in polyalkyl methacrylates, but then increases. Explain.
27. PP is now used in indoor/outdoor carpets. However, the first PP products deteriorated rapidly when subjected to sunlight because of the presence of tertiary hydrogen atoms present. Explain.

28. Which is more resistant to attack by microorganisms: (a) PVC or (b) plasticized PVC?
29. Lead stearate is an effective thermal stabilizer for PVC, yet its use in PVC pipe is not permitted. Why?
30. When PVC sheeting fails when exposed long times to sunlight it goes through a series of color changes before becoming black. Explain.
31. What is the advantage of using epoxidized soybean oil as a stabilizer for PVC?
32. Why do PVC films deteriorate more rapidly when used outdoors in comparison to indoors?
33. Which of the following is more resistant to UV light: (a) PP, (b) PE, or (c) PVC?
34. Sometimes molded plastic has opaque material called "bloom" on the surface. Explain.
35. Why are the presence of static charges on polymers surfaces not desired?
36. Was carbon black always used as a reinforcing filler for tires?
37. Problem—Why does a tire filled with oxygen lose pressure faster than a tire filled with nitrogen gas?

FURTHER READINGS

Bart, J. (2005), *Additives in Polymers*, Wiley, Hoboken, NJ.
Bolgar, M., Groger, J., Hubball, J., and Meronek, S. (2007): *Handbook for the Chemical Analysis of Plastics and Additives*, CRC Press, Boca Raton, FL.
Bukharov, S., Nuqumanova, G., and Kochner, A. (2009): *Polyfunctional Stabilizers of Polymers*, Nova, Hauppauge, NY.
Craver, C. and Carraher, C. E. (2000): *Applied Polymer Science*, Elsevier, New York.
Datta, S. and Lohse, D. (1996): *Polymeric Compatibilizers*, Hanser Gardner, Cincinnati, OH.
Lutz, J. and Grossman, R. (2000): *Polymer Modifiers and Additives*, Dekker, New York.
Marklund Sundkvist, A. (2008): *Organophosphorus Flame Retardants and Plasticizers*, VDM Verlag, New York.
Merhari, L. (2009): *Hybrid Nanocomposites for Nanotechnology: Electronic, Optical, Magnetic, and Biomedical Applications*, Springer, New York.
Mikitaev, A., Kozlov, G., and Zaikov, G. (2009). *Polymer Nanocomposites: Variety of Structural Forms and Applications*, Nova, Hauppauge, NY.
Okdmsn, K. and Sain, M. (2008): *Wood–Polymer Composites*, CRC Press, Boca Raton, FL.
Pethrick, R., Zaikov, G., and Horak, D. (2007): *Polymers and Composites: Synthesis, Properties, and Applications*, Nova, Hauppauge, NY.
Powell, G. M. and Brister, J. E. (1900): Plastigels, U.S. Patent 2,427,507.
Wilkie, C. and Morgan, A. (2009): *Fire Retardancy of Polymeric Material*, 2nd ed., CRC Press, Boca Raton, FL.

15 Synthesis of Reactants and Intermediates for Polymers

Many of the reactants used for the production of polymers are standard organic chemicals. However, because of the high purity requirements and large amounts needed, special conditions have been developed that allow large amounts of high purity reactants to be made in high yield. The first section of this chapter deals with the availability of the general feedstocks. The remaining sections deal with the synthesis of particular polymer reactants.

There is an industrial turn toward green materials. Much of this is found in developing processes that allow the creation of monomers from natural resources. You will see the move toward the creation of synthetic polymers from natural materials as you view Sections 15.2 and 15.3. The emphasis on green chemistry is also found when reactions produce "environmentally friendly" by-products.

15.1 MONOMER SYNTHESIS FROM BASIC FEEDSTOCKS

Most of the monomers widely employed for both vinyl and condensation polymers are derived indirectly from simple feedstock molecules. This synthesis of monomers is a lesson in inventiveness. The application of the saying "necessity is the mother of invention" has led to the sequence of chemical reactions where little is wasted and by-products from one reaction are employed as integral materials in another. Following is a brief look at some of these pathways traced from basic feedstock materials. It must be remembered that often many years of effort were involved in discovering the conditions of pressure, temperature, catalysts, etc., that must be present as one goes from the starting materials to the end products.

Fossil fuels refer to materials formed from the decomposition of once living matter. Because these once living materials contain sulfur and heavy metals such as iron and cobalt, they must be removed either prior or subsequent to use.

The major fossil fuels are coal and petroleum. Marine organisms were typically deposited in muds and underwater bodies, where anaerobic decay occurred. The major decomposition products are hydrocarbons, carbon dioxide, water, and ammonium. These deposits form much of the basis for our petroleum resources. Many of these deposits are located so that the evaporation of the more volatile products such as water and ammonia occurred, giving petroleum resources with little nitrogen- or oxygen-containing products. By comparison, coal is formed from plant material that has decayed to graphite carbon and methane.

Only about 5% of the fossil fuels consumed today are used as feedstocks for the production of today's synthetic carbon-based products. This includes the products produced by the chemical and drug industries with a major portion acting as the feedstocks for plastics, elastomers, coatings, fibers, etc.

The major petroleum resources contain linear, saturated hydrocarbons (alkanes), cyclic alkanes, and aromatics. For the most part, this material is considered to have low free energy content.

Raw or crude petroleum materials are separated into groups of compounds with similar boiling points by a process called fractionation. Table 15.1 contains a brief listing of typical fraction-separated materials. Accompanying or subsequent to this fractionation occurs a process called "cracking" whereby the hydrocarbon molecules are heated over catalysts that allow the hydro-carbon molecules to break up and then reform into structures that contain more branching that

TABLE 15.1
Typical Straight Chain Hydrocarbon Fractions Obtained from Distillation of Petroleum Resources

Boiling Range (°C)	Average Number of Carbon Atoms	Name	Uses
<30	1–4	Gas	Heating
30–180	5–10	Gasoline	Automotive fuel
180–230	11,12	Kerosene	Jet fuel, heating
230–300	13–17	Light gas oil	Diesel fuel, heating
300–400	18–25	Heavy gas oil	Heating

allow for appropriate combustion in our automobiles and trucks. Under other conditions, the cracking allows the formation of other desired feedstock molecules including methane, ethane, ethylene, propylene, benzene, etc., that eventually become our plastics, fibers, elastomers, sealants, coatings, composites, etc.

In 1925, Phillips Petroleum Company was only one of dozens of small oil companies in Oklahoma. The only distinction was the large amount of natural gasoline (or naphtha), the lightest liquid fraction (Table 15.1), found in its crude oil. As was customary, Phillips, and most of the other oil companies of the time, employed a distillation process to isolate butane and propane. Even so, they were sued for the use of this distillation process probably because they were small, had no real research capacity of their own, and no real legal defense team. Frank Phillips elected to fight supposedly including an argument that the ancient Egyptians had used a similar process to create an alcoholic equivalent to an Egyptian alcoholic drink. Phillips won the suit but became convinced that if the company was to remain successful it would need to have a research effort.

During the early dust bowl years, in 1935, they established the oil industry's first research team in Bartlesville, Oklahoma. George Oberfell, hired by Phillips to fight the lawsuit, planned the initial research efforts that involved three main initiatives. First, develop technology to use light hydrocarbons in new ways as motor fuels. Second, develop markets for butane and propane. Finally, find new uses for the light hydrocarbons outside the fuel market. All the three objectives were achieved.

Frederick Frey and Walter Shultze were instrumental early researchers. Frey was among the first to dehydrogenate paraffins catalytically to olefins and then the olefins to diolefins that serve as feedstocks to the production of many of today's polymers. In competition with Bakelite, he discovered the preparation of polysulfone polymers made from the reaction of sulfur dioxide and olefins creating a hard Bakelite-like material. Frey and Schultz also developed a process that allowed the production of 1,3-butadiene from butane that allowed the synthesis of synthetic rubber.

Probably Frey's most important invention involved the use of hydrogen fluoride to convert light olefins, produced as by-products of a catalytic cracker, into high-octane motor and aviation fuels. This process is still widely used. It came at a critical time for the United States' World War II efforts allowing fuel production for the Allied forces. This fuel allowed aircraft faster liftoffs, more power, and higher efficiency.

The major one-carbon feedstock is methane and it serves as the feedstock to a number of important monomers including hexamethylenetetramine and melamine, used in the synthesis of a number of cross-linked thermosets as well as vinyl acetate, ethylene, ethylene glycol, and methyl methacrylate (Table 15.1).

Formaldehyde, produced in the methane stream, serves as the basis for the formaldehyde-intensive resins, namely phenol–formaldehyde, urea–formaldehyde, and melamine–formaldehyde resins, as noted above. Formaldehyde is also involved in the synthesis of ethylene glycol, one of the two comonomers used in the production of PET. Formaldehyde also serves as the basic feedstock for the synthesis of polyacetals.

Another important use for methane is its conversion to synthesis gas (or syn-gas), a mixture of hydrogen gas and carbon monoxide as shown in Figure 15.1. Synthesis gas can also be derived from coal. When this occurs, it is called water gas. Interestingly, the reaction of methane giving carbon monoxide and hydrogen can be reversed so that methane can be produced from coal through this route.

One major two-carbon feedstock is ethylene. From Figure 15.2, you can see that a number of the monomers are directly synthesized from ethylene. Again, while the "react" arrow goes directly from ethylene to the product, as noted above, it often took years to develop an economical procedure to obtain the product in essentially 100% yield. Here, depending on the reaction conditions, a wide variety of intermediates and products are formed that allow the synthesis of a number of polymers including poly(acrylic acid), poly(vinyl chloride), polystyrene, poly(vinyl acetate), polyesters (in particular PET), and poly(methyl acrylate). Of course, ethylene is itself part of the polymer feedstock pool being the feedstock for all of the polyethylenes.

FIGURE 15.1 Monomer synthesis chemical flow diagram based on methane feedstock.

FIGURE 15.2 Monomer synthesis chemical flow diagram based on ethylene feedstock.

Another two-carbon feedstock is acetylene. Acetylene is typically obtained from coal by converting coke calcium carbide and then treating the calcium carbide with water. As shown in Figure 15.3, a number of important monomers can be made from acetylene. Even so, because of the abundance of other feedstocks from petroleum reserves, only some of the routes shown in Figure 15.3 are widely used.

Propylene is the basic three-carbon building block (Figure 15.4). Again, its polymerization gives polypropylene. The ingeniousness of some of the synthetic routes is shown in the conversion of benzene, through the reaction with propylene, to cumene and the consequent oxidation forming phenol and acetone that is subsequently converted to bisphenol A, a basic building block for certain polyesters. Phenol is involved in the synthesis of the phenol–formaldehyde resins, adipic acid, and 1,6-hexamethylenediamine. Acetone, in turn, is also involved in numerous important synthetic steps either as a reactant or as a solvent. It is involved in the synthesis of methyl methacrylate and isoprene.

The major four-carbon feedstock molecules are 1,3-butadiene and isobutylene, both involved in the synthesis of many monomers and intermediates. Butadiene is copolymerized with styrene to form SBR rubber and with acrylonitrile to form ABS rubbers.

Benzene forms the basis for a number of monomers (Figure 15.5) including those that retain their aromatic character like styrene, and those that do not, like adipic acid.

In summary, monomer synthesis from basic, readily available inexpensive feedstocks based on fossil fuels is both an art and a science developed over the past half-century or so. It represents a delicate balance and interrelationship between feedstocks and so-called by-products from one reaction that are critical reactants in another reaction. Monomer and polymer synthesis continues to undergo change and improvement as the natural environment and societal and worker health continue to be dominant factors.

Many of these and other monomers are now being produced by various microbes. Some of these efforts are described in Sections 16.15 and 16.16.

FIGURE 15.3 Monomer synthesis chemical flow diagram based on acetylene feedstock.

FIGURE 15.4 Monomer synthesis chemical flow diagram based on propylene feedstock.

FIGURE 15.5 Monomer synthesis chemical flow diagram based on benzene feedstock.

15.2 REACTANTS FOR STEP-REACTION POLYMERIZATION

Adipic acid (1,4-butanedicarboxylic acid) is used for the production of nylon-66 and may be produced from the oxidation of cyclohexane as shown in Equation 15.1. Cyclohexane is obtained by the Raney nickel-catalytic hydrogenation of benzene. Both the cyclohexanol and cyclohexanone are oxidized to adipic acid by heating with nitric acid.

Benzene   Cyclohexane   Cyclohexanol Cyclohexanone   Adipic acid

(15.1)

A new method that emphasizes green chemistry principles involves oxidizing cyclohexene with hydrogen peroxide through a tungsten catalyst and a phase transfer catalyst producing adipic acid and water as the by-product.

Adipic acid can also be made from tetrahydrofurane (THF), obtained from furfural, a naturally derived material (Equation 15.2). It is carbonylated in the presence of nickel carbonyl–nickel iodide catalyst. Furfural is a chemurgic product obtained by the steam-acid digestion of corncobs, oat hulls, bagasse, or rice hulls.

THF   Adipic acid

(15.2)

Adiponitrile may be produced from the hydrodimerization of acrylonitrile or from 1,3-butadiene via 1,4-dicyanobutene-2. Adiponitrile is then hydrogenated forming 1,6-hexanediamine (Equation 15.3).

$$(15.3)$$

1,6-Hexanediamine can also be made by the liquid phase catalytic hydrogenation of adiponitrile or adipamide, which is made from adipic acid (Equation 15.4).

$$(15.4)$$

Sebacic acid (1,8-octane dicarboxylic acid), which is used to make nylon-610, has been produced from 1,3-butadiene and by the dry distillation of castor oil (ricinolein). The cleavage of ricinoleic acid, from the caster plant, gives 2-octanol and the salt of sebacic acid (Equation 15.5).

$$(15.5)$$

Phthalic acid (1,2-benzene dicarboxylic acid), terephthalic acid (1,4-benzene dicarboxylic acid; Equation 15.6), and isophthalic acid (1,3-benzene dicarboxylic acid; Equation 15.7) are made by the selective oxidation of the corresponding xylenes. The oxidation of p-xylene by oxygen from the air is generally done using acetic acid as a solvent in the presence of a catalyst such as cobalt–manganese. The yield is close to 100%.

$$(15.6)$$

p-Xylene   Terephthalic acid

$$(15.7)$$

m-Xylene   Isophthalic acid

Terephthalic acid may also be produced from the oxidation of naphthalene and by the hydrolysis of terephthalonitrile.

   Malic anhydride (2,5-furandione) is made as a by-product in the production of phthalic anhydride, and by the vapor phase oxidation of butylene or crotonaldehyde (Equation 15.8). It is also obtained by the dehydration of maleic acid and by the oxidation of benzene. Maleic anhydride is used for the production of unsaturated polyester resin. This reactant, like many reactants, is fairly toxic and should be treated as such.

$$(15.8)$$

2-Pyrrolidone is a lactone used for the production of nylon-4. This reactant may be produced by the reduction ammonation of maleic anhydride. ε-Caprolactaom, used in the production of nylon-6, may be produced using several approaches. The oxime may be produced by the catalytic hydrogenation of nitrobenzene, the photolytic nitrosylation of cyclohexane (Equation 15.9), or the reaction of cyclohexanone and hydroxylamine (Equation 15.10) or by the Beckman rearrangement of cyclohexanone oxime (Equation 15.11). Nearly one half of the production of caprolactam is derived from phenol.

$$(15.9)$$

$$(15.10)$$

$$(15.11)$$

Ethylene oxide, used for the production of ethylene glycol and poly(ethylene oxide) (PEO), is obtained by the catalytic oxidation of ethylene (Equation 15.12). Ethylene glycol, used in the production of PET, is produced by the hydrolysis of ethylene oxide.

$$(15.12)$$

Glycerol, used for the production of alkyds, is produced by the catalytic hydroxylation or the hypochlorination of allyl alcohol. Allyl alcohol is produced by the reduction of acrolein, which is in turn, obtained by the oxidation of propylene (Equation 15.13).

$$(15.13)$$

Until recently, glycerol was obtained mainly from epichlorohydrin. However, because glycerol forms the backbone of fats, it is produced as the fats are degraded giving the fatty acid and glycerol. It is also a by-product of the production of biodiesel production through the transesterification of vegetable oils. Thus, glycerol is almost solely obtained today from these sources.

Pentaerythritol, used in the production of alkyds, is produced by a crossed Cannizzaro reaction of the aldol condensation product of formaldehyde and acetaldehyde (Equation 15.14). The by-product formate salt is a major source of formic acid.

| Formaldehyde | Acetaldehyde | | Trimethylol acetaldehyde | | Pentaerythritol salt | Formate |

$$(15.14)$$

2,4-Toluene diisocyanate (TDI), used for the production of polyurethanes and polyureas, is obtained by the phosgenation of 2,4-toluenediamine (Equation 15.15). Phosgene is obtained by the reaction of chlorine and carbon monoxide.

(15.15)

Formaldehyde is employed as a basic unit for many industrial adhesives such as the phenolic plastics formed from the reaction of phenol and formaldehyde. It also serves as one of the reactants in the formation of amino plastics in the production of urea–formaldehyde, UF, resins. Formaldehyde can self condense forming the cyclic trimer trioxane and the polymer paraformaldehyde. It is industrially produced from the catalytic oxidation of methanol. In turn, hexamethylenetetramine is produced by the condensation of ammonia and 30% aqueous formaldehyde (formalin; Equation 15.16).

(15.16)

While some phenol is produced by the nucleophilic substitution of chlorine in chlorobenzene by the hydroxyl group (Equation 15.17), most of it is produced by the acidic decomposition of cumene hydroperoxide (Equation 15.18) that also gives acetone along with the phenol. Some of the new processes for synthesizing phenol are the dehydrogenation of cyclohexanol, the decarboxylation of benzoic acid, and the hydrogen peroxide hydroxylation of benzene.

(15.17)

(15.18)

Urea is highly water soluble and offers an efficient avenue for the human body to expel excess nitrogen. The individual atoms come from water, carbon dioxide, aspartate, and ammonia and are involved in the urea cycle metabolic pathway. Urea, which is used for the production of urea-formaldehyde resins, is made by the *in situ* decomposition of ammonium carbamate, which is made by the condensation of ammonia and carbon dioxide at 200°C and 200 atm (Equation 15.19).

(15.19)

Melamine (cyanuramide), used in the production of melamine–formaldehyde resins, is obtained by heating dicyanodiamide (Equation 15.20), which is obtained by heating cyanamide. Today, most melamine is produced by heating urea (Equation 15.21)

$$(15.20)$$

$$(15.21)$$

Bisphenol A [(bis-4-hydroxphenol)dimethylmethane], used for the production of epoxy resins and polycarbonates, is obtained by the acidic condensation of phenol and acetone (Equation 15.22). Here, the carbonium ion produced by the protonation of acetone attacks the phenol molecule at the para position producing a quinoidal oxonium ion that loses water and rearranges to a *p*-isopropylphenol carbonium ion. The water attacks another phenol molecule, also in the para position, giving another quinoidal structure that rearranges to bisphenol A. It has been found that bisphenol A may be involved in one of the endocrine systems. The consequences are still being determined.

$$(15.22)$$

Epichlorohydrin (chloropropylene oxide) is used for the production of epoxy resins. It is produced by the dehydrochlorination of 2,3-dichloro-1-propanol (Equation 15.23). The hydrin is produced by the chlorohydrination of allyl chloride.

$$(15.23)$$

Methyltrichlorosilane is produced by the Grignard reaction of silicon tetrachloride and methylmagnesium chloride (Equation 15.24). Dimethyldichlorosilane, used in the synthesis of polydimethylsiolxane, is obtained by the reaction of methylmagnesium chloride and methyltrichlorosilane (Equation 15.25).

$$H_3CMgCl + SiCl_4 \rightarrow H_2CSiCl_3 + MgCl_2 \tag{15.24}$$

$$H_3CMgCl + H_3CSiCl_3 \rightarrow (H_3C)_2SiCl_2 + MgCl_2 \tag{15.25}$$

15.3 SYNTHESIS OF VINYL MONOMERS

Styrene is generally produced by the catalytic vapor phase dehydrogenation of ethylbenzene (Equation 15.26). Ethylbenzene is made by the Friedel–Crafts condensation of ethylene and benzene. Ethylbenzene, in the vapor phase, is passed over a solid catalyst bed. Most catalysts are based on iron(III) oxide containing potassium oxide or potassium carbonate. Steam serves several roles acting as an energy source and it removes coke that forms on the iron oxide catalyst. Typically, several reactors are used in series with each "pass" of the vapor ethylbenzene resulting in increased styrene production. The main by-products are benzene and toluene. Styrene is also produced by the palladium acetate-catalyzed condensation of ethylene and benzene and by the dehydration of methylphenylcarbinol obtained by the propylation of ethylbenzene. Because of the toxicity of styrene, its concentration in the atmosphere must be severely limited.

$$(15.26)$$

A process to produce styrene monomer and propylene oxide simultaneously was introduced in 1969 and it is also employed to produce styrene industrially (Equation 15.27).

$$(15.27)$$

Vinyl chloride, formerly obtained from acetylene, is now produced by the transcatalytic process where the chlorination of ethylene, the oxychlorination of the by-product hydrogen chloride, and the dehydrochlorination occur in a single reactor (Equation 15.28).

$$(15.28)$$

Vinyl chloride is also produced by the direct chlorination of ethylene and the reaction of acetylene and hydrogen chloride (Equation 15.29). The hydrogen chloride generated in the chlorination of ethylene can be employed in the reaction with acetylene allowing a useful coupling of these two reactions (Equation 15.30). Today, most vinyl chloride is produced from the reaction of ethylene and chlorine forming ethylene dichloride. The ethylene dichloride is heated under pressure resulting in its decomposition to vinyl chloride and hydrogen chloride (Equation 15.29).

$$H_2C = CH_2 + Cl—Cl \longrightarrow Cl\diagup\diagdown\diagup^{Cl} \longrightarrow \underset{CH_2}{\overset{Cl}{\parallel}} + H—Cl \qquad (15.29)$$

$$HC \equiv CH + H—Cl \longrightarrow \underset{CH_2}{\overset{Cl}{\parallel}} \qquad (15.30)$$

Vinylidene chloride, or 1,1-dichloroethene, is produced by the pyrolysis of 1,1,2-trichloroethane at 400°C in the presence of lime or base (Equation 15.31). Since both vinylidene chloride and vinyl chloride are carcinogenic, their concentrations must be kept low. It was widely used to form polymer, mainly poly(vinylidene chloride), which formed the basis for a cling wrap called Saran warp. Research suggested that Saran wrap may pose a health hazard by leaching when microwaved. Because of this potent risk, most cling wrap formulations have changed to polythene.

$$Cl\diagup\diagdown\underset{Cl}{\overset{Cl}{|}} \longrightarrow H_2C{=}\underset{Cl}{\overset{Cl}{\diagdown}} + H{-}Cl \qquad (15.31)$$

Vinyl acetate, was produced by the catalytic acetylation of acetylene, but this monomer is now produced by the catalytic oxidative condensation of acetic acid and ethylene (Equation 15.32). Other vinyl esters can be produced by the transesterification of vinyl acetate with higher boiling carboxylic acids.

$$H_2C = CH_2 + H_3C{-}\overset{O}{\underset{OH}{\diagup\diagdown}} \longrightarrow H_2C{=}\diagdown\diagup_{O}{-}\overset{CH_3}{\underset{O}{\diagup\diagdown}} \qquad (15.32)$$

Acrylonitrile (vinyl cyanide) is produced by the Sohio process involving the ammoxidation of propylene (Equation 15.33). Again, since this monomer is carcinogenic, care must be taken to minimize exposure to it.

$$\underset{H_2C}{\diagup}{\diagdown}{-}CH_3 + NH_3 + O_2 \longrightarrow \underset{CH_2}{\overset{N}{\parallel}} + H_2O \qquad (15.33)$$

Tetrafluoroethylene is produced from the thermal dehydrochlorination of chlorodifluoromethane (Equation 15.35) which, in turn, is produced from chloroform and HF (Equation 15.34).

$$H—F + \underset{Cl}{\overset{H}{\underset{Cl}{\diagup|\diagdown}}}_{Cl} \longrightarrow H—Cl + \underset{Cl}{\overset{H}{\underset{F}{\diagup|}}}_{F} \qquad (15.34)$$

$$\underset{Cl}{\overset{H}{\underset{F}{\diagup|}}}_{F} \longrightarrow H—Cl + \underset{F}{\overset{F}{\diagdown}}{=}\underset{F}{\overset{F}{\diagup}} \qquad (15.35)$$

Trifluoromonochloroethylene is obtained from the zinc metal dechlorination of trichlorotrifluoro-ethane (Equation 15.36). The latter is produced by the fluorination of hexachloroethane.

$$\text{(15.36)}$$

Vinylidene fluoride is produced by the thermal dehydrochlorination of 1-monochloro-1,1,-difluoroethane (Equation 15.37).

$$\text{(15.37)}$$

Vinyl fluoride may be obtained by the catalytic hydrofluorination of acetylene (Equation 15.38).

$$\text{(15.38)}$$

Vinyl ethyl ether is obtained by the ethanolysis of acetylene in the presence of potassium ethoxide (Equation 15.39).

$$\text{(15.39)}$$

1,3-Butadiene, used for the production of elastomers, is produced by the catalytic thermal cracking of butane and as a by-product of other cracking reactions (Equation 15.40).

$$\text{(15.40)}$$

While butadiene is produced as a by-product of the steam cracking process used in the production of ethylene and other olefins in the United States, Europe, and Japan, it is produced in other parts of the world from ethanol, a green material. In the single-step process, ethanol is converted to buta-diene, hydrogen, and water through passing the ethanol over metal oxide catalysts. In a two-step process, ethanol is oxidized to acetaldehyde, which then reacts with additional ethanol producing butadiene.

The isoprene monomer is not readily available from direct cracking processes. Several routes are employed for its synthesis. One route begins with the extraction of isoamylene fractions from catalytically cracked gasoline streams. Isoprene is produced by subsequent catalytic dehydrogena-tion (Equation 15.41).

$$(15.41)$$

The dimerization of propylene is also used to produce isoprene. Several steps are involved in this process. Initially, the dimerization of propylene to 2-methyl-1-pentene occurs. Then isomerization to 2-methyl-2-pentene is effected. Finally, the 2-methyl-2-pentene is pyrolized to isoprene and methane. Another multistep synthesis starts with acetylene and acetone. Perhaps the most attractive route involves formaldehyde and isobutylene (Equation 15.42).

$$(15.42)$$

Chloroprene, used for the production of neoprene rubber, is obtained by the dehydrochlorination of dichlrobutene (Equation 15.43). The latter is produced by the chlorination of 1,3-butadiene, which, in turn, is synthesized from acetylene.

$$(15.43)$$

Acrylic acid can be prepared by the catalytic oxidative carbonylation of ethylene or by heating formaldehyde and acetic acid in the presence of KOH (Equation 15.44).

$$H_2C = CH_2 + CO + O_2 \longrightarrow$$

$$(15.44)$$

Methyl acrylate may be obtained by the addition of methanol to the reactants in the previous synthesis (15.44) for acrylic acid or by the methanolysis of acrylonitrile (Equation 15.45).

$$(15.45)$$

Methyl methacrylate may be prepared by the catalytic oxidative carbonylation of propylene in the presence of methanol (Equation 15.46).

$$
\text{(structure)} \quad + \quad H_3C\text{——}OH + CO + O_2 \quad \longrightarrow \quad \text{(structure)} \tag{15.46}
$$

Most production employs acetone and hydrogen cyanide as the beginning materials. The intermediate cyanohydrin is converted through the reaction with sulfuric acid to give a sulfate ester of the methacrylamide. This sulfate ester is subsequently hydrolyzed giving ammonium bisulfate and methyl methacrylate.

A newer process employs ethylene, carbon monoxide, and methanol passed over a catalyst to produce methyl propionate. The methyl propionate is combined with formaldehyde producing (green chemistry) water and methyl methacrylate.

15.4   SUMMARY

1. Feedstocks for the synthesis of monomers of basic polymeric materials must be readily available and inexpensive because they are utilized in polymer synthesis in large quantities, allowing the polymeric materials to be inexpensive. Basic feedstocks are petrochemical and coal based.
2. Monomer synthesis is both an art and a science developed by major and ongoing research efforts allowing the inexpensive and safe availability of the starting materials upon which the polymer industry is based. Commercial monomer synthesis is based on both the availability of inexpensive materials and on an "interconsecutiveness" between products and synthetic by-products that are essential to the synthesis of other essential materials.
3. The precise conditions of synthesis are continually being refined. They are based on "high" science.
4. There is an increased emphasis on green chemistry in the production of monomers.

GLOSSARY

Cannizzaro reaction: An internal oxidation-reduction reaction of aldehydes.
Carbamide: Urea.
Carcinogenic: Cancer causing.
Chemurgic compound: Compound made from a plant source.
Friedel–Crafts condensation: Condensation that takes place in the presence of a Lewis acid such as aluminum chloride.
Grignard reagent: RMgX.
Raney nickel: A porous nickel catalyst produced from a nickel-aluminum alloy.

EXERCISES

1. Why are there so many methods for the preparation of adipic acid?
2. Write equations for the industrial synthesis of the following:

   a. Adipic acid   b. Hexamethylenediamine   c. Sebacic acid
   d. Terephthalic acid   e. Maleic anhydride   f. ε-Caprolactam
   g. Ethylene glycol   h. Glycerol   i. Pentaerythritol
   j. TDI   k. Hexamethylenetetramine   l. Phenol
   m. Urea   n. Melamine   o. Bisphenol A
   p. Epichlorohydrin   q. Methyltrichlorosilane   r. Styrene
   s. Vinyl chloride   t. Vinyl acetate   u. Acrylonitrile
   v. Vinyl ethyl ether   w. Methyl methacrylate

3. Name a reactant or monomer produced by the following:
   a. Grignard reaction   b. Friedel–Crafts reaction
   c. Beckman rearrangement   d. A chemurgic process
   e. A crossed Cannizzaro reaction
4. Name three monomers whose synthesis illustrates green chemistry.

FURTHER READINGS

Belgacem, M. and Gandini, A. (2008): *Monomers, Polymers, and Composites from Renewable Resources*, Elsevier, New York.

Boundy, R. H., Boyer, R. F., and Stroesser, S. M. (1965): *Styrene*, Hafner, New York.

Leonard, E. C. (1970): *Vinyl and Diene Monomers*, Wiley, New York.

Pethrick, R., Zaikov, G., and Pielichowski, J. (2008): *Progress in Monomers, Oligomers, Polymers, Composites and Nanocomposites*, Nova, Hauppauge, NY.

Yokum, R. H. and Nyquist, E. B. (1974): *Functional Monomers, Their Preparation, Polymerization and Application*, Dekker, New York.

16 Polymer Technology

Today, nearly 10,000 U.S. companies are active in the general area of synthetic polymers. Following is a brief description of these companies divided according to their function.

Manufacturers: There are over 200 major manufacturers of general-purpose polymers and numerous other manufacturers of specialty polymers.

Processors: Some companies manufacture their own polymeric materials for subsequent processing, but the majority purchases the necessary polymeric materials from other companies. Processors may specialize in the use of selected polymers, such as nylons and polycarbonates, or focus on particular techniques of processing, such as coatings, films, sheets, laminates, and bulk molded and reinforced plastics.

Fabricators and finishers: The majority of companies are involved in the fabrication and finishing of polymers, that is, the production of the end products for industrial and general public consumption. Fabrication can be divided into three broad areas: machining, forming, and fashioning. Machining includes grinding, sawing, screwing, and other techniques. Forming includes molding and other methods of shaping and joining by welding, gluing, screwing, and other techniques. Fashioning includes cutting, sewing, sheeting, and sealing. Fabrication sequences vary with the polymeric material and desired end product.

While much classic polymer technology was developed without the benefit of science, modern polymer technology and polymer science are closely associated. The technology of fibers, elastomers, coatings, composites, drug delivery, and plastics is discussed in this chapter.

Chemistry is moving center stage in many areas of medicine, biology, engineering, environmental science, and physics. While solid-state physics is traditionally based on silicon, polymers offer a much wider vista of opportunities for application and fine-tuning those applications. Some areas are based on single crystals that may be small in our sight, but are large when compared to individual molecules. Even single silicon wafers with a minimum pattern dimension of 200 nm are on the order of ten times the size of individual molecules. Eventually, electronic, photonic, and stress–strain behavior individuality can be placed into single giant chains creating chains that behave as entire assemblies behave today.

16.1 POLYMER PROCESSING

Polymer processing can be defined as the process whereby raw materials are converted into products of desired shape and properties. Thermoplastic resins are generally supplied as pellets, marbles, or chips of varying sizes and they may contain some or all of the desired additives. When heated above their T_g, thermoplastic materials soften and flow as viscous liquids that can be shaped using a variety of techniques and then cooled to "lock" in the micro and gross structures.

Thermosetting feedstocks are normally supplied as meltable and/or flowable prepolymer, oligomers, or lightly or non-cross-linked polymers that are subsequently cross-linked forming the thermoset article.

The processing operation can be divided into three general steps: preshaping, shaping, and postshaping. In preshaping, the intention is to produce a material that can be shaped by the application of heat and/or pressure. Important considerations include

- Handling of solids and liquids including mixing, low compaction, and packing
- Softening through the application of heat and/or pressure
- Addition and mixing/dispersion of added materials
- Movement of the resin to the shaping apparatus through the application of heat and/or pressure and other flow aiding processes
- Removal and recycling of unwanted solvent, unreacted monomer(s), by-products, and waste (flash)

The shaping step may include any single or combination of the following:

- Die forming (including sheet and film formation, tube and pipe formation, fiber formation, coating, and extrusion)
- Molding and casting
- Secondary shaping (such as film and blow molding, thermoforming)
- Surface treatments (coating and calendering)

Post-shaping processes include welding, bonding, fastening, decorating, cutting, milling, drilling, dying, and gluing.

Polymer processing operations can be divided into five broad categories:

- Spinning (generally for fibers)
- Calendering
- Coating
- Molding
- Injection

Table 16.1 lists some of the major shapes produced by each of these processing groups.

Essentially all of the various processing types utilize computer-assisted design (CAD) and computer-assisted manufacture (CAM). CAD allows the design of a part and incorporates operating conditions to predict the behavior of the pieces prior to real operation. CAD also transfers particular designs and design specifications to other CAMs, that allow the actual construction of the part or total apparatus. CAM systems operate most modern processing systems many allowing feedback to influence machine operation.

Processing and performance are interrelated to one another and to additional factors. Jaffe relates these major groups of factors in an interactive diamond shown in Figure 16.1. Understanding these factors and their interrelationships becomes increasingly important as the specific performance requirements become more specific. Performance is related to the chemical and physical structure and to the particular processing performed on the material during its lifetime. The physical structure is a reflection of both the chemical structure and the total history of the synthesis and subsequent exposure of the material to additional forces. These "additional" forces are included under the broad idea of processing and include any influence that contributes to the secondary (and greater) structure–stress–strain, light, chemical, etc. The portion of the diamond relating processing to physical structure encompasses the study of structure–property relationships. A single material may be processed using only a single process somewhat unique to that material (such as liquid crystals) or by a variety of processes (such as polyethylene) where the particular technique is dictated by such factors as end use and cost.

16.2 SECONDARY STRUCTURES—MESOPHASES

The primary and secondary structures greatly influence possible processing scenarios. Here, the secondary structure is generally the same as the physical structure and the primary structure is generally the same as the chemical structure. The end properties and uses are governed by intrinsic

TABLE 16.1
Major Forms of Polymer Processing Groupings

Process	Typical Form of Product
Calendering	Films, sheets
Coating	Film
Injection	Solid
Reaction injected	
Reciprocating screw	
Two-stage	
Molding	
Blow	Hollow
Displacement	
Extrusion	
Injection/transfer	
Stretch	
Cold solid	
Compression	Solid, hollow
Rotational	Solid, hollow
Thermoforming	Hollow
Transfer	Solid
Spinning	Fibers
Dry	
Gel	
Melt	
Reaction	
Wet	

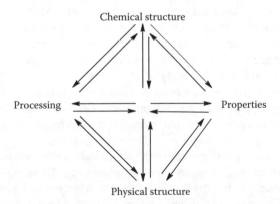

FIGURE 16.1 Relationships that influence the important interrelationships that exist for polymeric materials with respect to processing and end-product properties.

properties that in turn are related to the primary and secondary structures—the chemical and physical structures.

The term "meso" will be used to describe local chain organizations that occur within the nano- and micro-scale regions (Figure 16.2). While the terms mesophase and meso regions have been

FIGURE 16.2   Relationship between organizational level and size.

employed in describing order within liquid crystals, the definition will be broaden to refer to include other ordered regions within a material's physical or secondary structures. We generally describe polymer secondary structure in simple terms of ordered and/or disordered or crystalline and/or amorphous. The ordered regions can be further described in terms of meso regions or mesophases according to their permanency and ability to influence changes within and about these regions.

Jaffe describes four mesophase classifications. *Permanent mesogens* are materials whose microstructures are highly fixed such as in liquid crystalline polymers, LCs. LCs are characterized by highly ordered structures in the quiescent state. They exhibit relatively low viscosities in uniaxiel flows and can be easily realigned through the application of processing forces such as extrusion. To be processed, such polymers must be in the mesogenic state below their decomposition temperature. This can be achieved using a specific solvent or the introduction of special comonomer units that allow them to melt (or soften), but that are introduced in such a manner as to preserve their LC character. As solids, such materials exhibit high molecular orientation, high tensile moduli (near to theoretical), poor compression (i.e., little unoccupied volume), poor shear behavior, and high tensile strengths (on the order of 4 GPa). Such materials are anisotropic conductors and generally offer good liquid and gas barrier properties. Properties are controlled by the inherent chemical structure, molecular orientation, defect occurrences, and stress transfer mechanisms. Defects often act as the "weak links" in a chain limiting mechanical properties so that defect detection, and elimination/curtailment are important and can be dependent on the processing conditions. Only certain processing techniques are suitable.

Accessible mesogens are formed from polymers that are thermotropic (i.e., polymers that have a phase organization that is temperature dependent) but have an accessible isotrophic phase below their decomposition temperature. Such polymers can be processed either when the material is in its mesogen or ordered state using LC-type processing forming strong well-ordered products, or at temperatures where the ordered mesogen structure is absent. In temperature-assisted systems, the material is rapidly cooled, quenched, preventing the mesogen structure formation producing a metastable isotropic glass or rubber. The metastable material can be processed employing less energy and force followed by a simple annealing and slower cooling that allows the formation of the ordered mesogenic structures along with the appearance of associated properties. Examples of *assembled mesogens* include groups of polymer coils and polymers with side chains that can form such mesogens. In the former case, tertiary-mesophase structures can be formed when the bundles of coiled chains come together.

Transient mesogens are regions present in flexible, random coil polymers often caused by the application of external forces including simple flowing/shearing. These regions occur through local segmental chain movements that happen within the chain network at points of minimum chain entropy such as sites of entanglements. They are fibrillar-like and appear to be the nucleating phase and key to the row and shish-kebab-like structures in oriented polymer crystals.

These latter groups include many of the so-called crystallites and crystalline regions of common polymers.

Understanding the factors that govern the formation of mesogens will assist in determining the processing conditions for the production of materials with specified amounts, sizes, and distribution of such crystalline microstructures. Mesophases can be local or permeate the entire structure. They can be large or small, and present in a random or more ordered arrangement.

16.3 FIBERS

16.3.1 POLYMER PROCESSING—SPINNING AND FIBER PRODUCTION

16.3.1.1 Introduction

Most polymeric materials are controlled by the Federal Trade Commission, FTC, with respect to the relationship between the name and content including fibers. While the FTC controls industry in the United States, the international standards are generally determined by the International Organization for Standardization, ISO. Table 16.2 contains a brief listing of some of the ISO and FTC names for some of the most utilized fibers.

Fiber production continues to increase for most general groupings.

The dimensions of a filament or yarn are expressed in terms of a unit called the "*tex*" which is a measure of the fineness or linear density. One tex is 1 g per 1000 m or 10^{-6} kg/m. The tex has replaced denier as a measure of the density of the fiber. One denier is 1 g per 9000 m, so 1 denier = 0.1111 tex.

While some natural polymers produced "natural" fibers, fibers from synthetic and regenerated natural polymers are generally produced using one of the spinning processing techniques. Three spinning processes are generally employed in the large-scale commercial production of fibers. The first produces fiber from the melted polymer-melt spinning. The other two techniques form fibers from concentrated polymer solutions—dry spinning and wet spinning. Figure 16.3 illustrates the essentials of these three spinning techniques. Table 16.3 is a listing of the most common polymers made into fibers by these three processes.

16.3.1.2 Melt Spinning

Melt spinning was developed in the 1930s. In melt spinning, the polymer is melted or extruded, clarified by filtration, and pumped through a die having one or more small holes. The die is called a spinneret. The number, shape, and size of the hole can vary considerably. The number of holes ranges from several holes to several thousand holes. For continuous filament formation, the number of holes is in the order of 10–100. The exit hole is usually circular giving round fibers. Other shaped holes are also employed that produce fibers with varying shapes.

The extruded fiber is then often uniaxially stretched by take-up rollers rotating at different speeds. The fiber stretching encourages the polymer chains to align on a molecular level producing increased strength in the direction of the pull.

TABLE 16.2
Generic Names for Synthetic Fibers according to the ISO and FTC

ISO	FTC
Acetate	Acetate
Acrylic	Acrylic
Aramid	Aramid
Chlorofiber	Vinyon/saran
Cupro/viscose/modal/deacetylated acetate	Cupra/rayon
Elastane	Spandex
Glass	Glass
Modacrylic	Modacrylic
Nylon/polyamide	Nylon/polyamide
Polyester	Polyester
Polyethylene/polypropylene–polyolefin olefin	Vinylal/vinal

FIGURE 16.3 Fiber production using the three major spinning techniques.

TABLE 16.3
(Preferred) Spinning Processes

		Wet Spinning	
Melt Spinning	**Dry Spinning**	**Coagulation**	**Regeneration**
Nylon	Acetate	Acrylic	Viscose
Polyester	Triacetate	Aramid	Cupro
Polyethylene	Acrylic	Elastane/spandex	
Polypropylene	Elastine/spandex	Poly(vinyl chloride)	
Poly(vinylene chloride)	Poly(vinyl chloride)		
	Aramid		
	Modacrylic		
	Vinyal		

To produce the melted polymer, the polymer chips, rods, marbles, or sheets are heated forming a melted pool of material. In order to minimize oxidation, the melted polymer is blanketed by an inert gas such as nitrogen or argon. The fluidicity (inverse of viscosity) of the melt increases with increased temperature as does the cost to provide the necessary energy and tendency for unwanted reactions to occur. Thus, the polymer melt is generally assisted to and through the spinneret by means of an extruder that may also be used to supply some or all of the heating.

Many nylon and polyester assemblies are configured so that there is a continuous progression from the melt formation of the polymer, and without hardening, the melted polymer is melt spun into fibers.

Monofilament is produced at a lower spinning speed, in comparison with chopped filament, because of the problem of heat buildup within the monofilament. The monofilament is generally cooled by passing it through cold water or by winding it on to a cold quench roll.

16.3.1.3 Dry Spinning

Polymer concentrations in the order of 20%–40% are employed in dry and wet spinning. In the dry spinning process, the solution is filtered and then forced through a spinneret into a spinning cabinet through which heated air is passed to dry the filament. For economical reasons, the gas is usually air, but inert gasses such as nitrogen and superheated water are sometimes used.

Volatile solvents are used to assist in the drying process. Water has been used for some systems, such as poly(vinyl alcohol), where the polymer is water soluble. Solvent removal and recycling is important. Spinning is usually carried out using either low (about 1%–2%) or high (10%–50%) solvent in the filament. The amount of solvent influences the drawing process. In high solvent cases, the filaments are plasticized allowing greater extension of the filament and greater alignment of the polymer chains to occur at lower temperatures and lower stresses. The extra solvent is removed just prior to, during, or subsequent to stretching. Just prior to extrusion, the polymer solution is heated to just above its boiling point, increasing the likelihood for the ready removal of the solvent.

Dry spinning produced fibers have lower void concentrations in comparison to melt spun fibers because the presence of solvent molecules causes voids that are often "remembered" by the polymer. This is reflected by greater densities and lower dyeability for the dry spun fibers.

Fibrous glass is the most important inorganic fiber. It is produced by melt spinning in both a continuous filament and a staple form. The molten glass is fed directly from the furnace, or melted from rods or marbles, to the spinneret. As the fibrous glass emerges, it is attenuated, quenched, lubricated, and wound forming a yarn or continuous filament. The temperature for spinning is on the order of 1200°C–1500°C. This temperature is important since it controls the output, and in conjunction with the removal speed, helps control the properties of the resultant fiber including thickness and density.

16.3.1.4 Wet Spinning

Wet spinning is similar to dry spinning except that fiber formation results from the coagulation of the polymer solution as it is introduced into a nonsolvent bath. Since the coagulation process is relatively long, the linear velocity of wet spinning is less than for either melt or dry spinning. Wet spinning allows the placement of holes in the spinneret face to be closer together allowing productivity to be increased. Even so, it remains the slowest of the traditional spinning processes. The equipment used for wet spinning is similar to that used in dry spinning though it is not necessary to heat the polymer solution to a high temperature. The spinnerets are immersed in tanks containing the nonsolvent. Wet spinning is the most complex of the three spinning processes typically including washing, stretching, drying, crimping, finish application, and controlled relaxation to form tow material. Spinning of natural-derived materials generally include additional steps including ageing or ripening to achieve the desired viscosity and chain length.

Fibers made from wet spinning generally have high void contents in comparison to all of the other processes giving them increased dyeability. The surface is rougher with longitudinal serrations and from a round die hole, it has an approximately circular to bean-shaped diameter.

Hollow fibers for gas and liquid separations are prepared through passing air through the material just prior to entrance into the nonsolvent bath.

16.4 ELASTOMERS

Prior to World War II, hevea rubber accounted for over 99% of all elastomers used, but synthetic elastomers account for more than 70% of all rubber used today. Natural rubber and many synthetic elastomers are available in latex form. The latex may be used, as such, for adhering carpet fibers or for dipped articles, such as gloves, but most of the latex is coagulated and the dried coagulant used for the production of tires and mechanical goods.

Over 5.5 billion pounds of synthetic rubber is produced annually in the United States. The principle elastomer is the copolymer of butadiene (75%) and styrene (25%) (SBR) produced at an annual rate of over 1 million ton by the emulsion polymerization of butadiene and styrene. The copolymer of butadiene and acrylonitrile (Buna-H, NBR) is also produced by the emulsion process at an annual rate of about 200 million pounds. Likewise, neoprene is produced by the emulsion polymerization of chloroprene at an annual rate of over 125,000 ton. Butyl rubber is produced by the low-temperature cationic copolymerization of isobutylene (90%) and isoprene (10%) at an annual rate of about 150,000 ton. Polybutadiene, polyisoprene, and ethylene–propylene copolymer rubber (EPDM) are produced by the anionic polymerization of about 600,000, 100,000, and 350,000 ton, respectively. Many other elastomers are also produced.

16.4.1 ELASTOMER PROCESSING

The processing of elastomeric (rubbery) material is quite varied depending on the end use, form of the material (i.e., dry or in solution), and material processed. Latex forms of rubber can be properly mixed with additives using simple (or more complex) stirring and agitation. The mixing/agitation should be such as to not cause a separation or breakdown of the latex or foam formation. Straight, coagulant, and heat-assisted dipping processes are commonly used to produce a variety of tubes, gloves, etc. Latexes are also used to make thread for the garment industry, and adhesives for shoes, carpets, and tape formation.

In the following, we will look at the processing involving bulk rubber. The manufacture of rubber products from this material can be divided into the following four steps:

1. Mastication
2. Incorporation or compounding
3. Shaping
4. Vulcanization

The shaping and vulcanization steps are combined in a number of processes such as transfer and injection molding or may be separated as in the extrusion and subsequent vulcanization sequence. An outline of these steps is given in Figure 16.4.

We will look at the processing of natural dry rubber first since its processing is similar to other elastomers and because of its historical importance. Natural rubber is a dispersion of rubber particles in water. Unvulcanized raw rubber obtained by coagulation and drying has large chains with chain lengths on the order of 10^5 carbons.

Mastication is intended to bring the material to the necessary consistency to accept the compounding ingredients. Mastication results in a lowering of chain size to an average chain length on the order of 10^4 carbons. Two basic types of internal mixers are in use. The Banbury has rotors

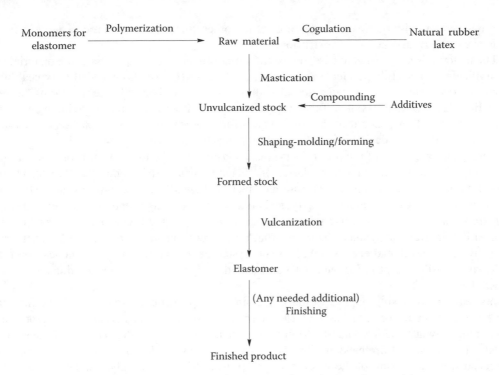

FIGURE 16.4   Outline of steps involved in the processing to form elastomeric materials.

rotating at different speeds creating a "kneading" action such as that employed in handling bread dough. A shearing action between the rotors and the walls of the mixer is also achieved. The Shaw Intermix employs rotors that turn at the same speed and closely intermesh causing an intra-compound friction for mixing; thus, closely resembling a mill's mixing action.

The next step is the incorporation of various additives—*compounding*. Typical additives include filler, processing aids, activators, processing aids, age resistors, sulfur, antioxidants and anti-ozone compounds, extenders, plasticizers, blowing agents, pigments, and accelerators.

An important aspect in the compounding is the amount of crystallization of the rubber. If the rubber is in a highly crystalline state, it will mix poorly if at all. Thus, partially crystallized rubber must be heated before it will yield to mixing.

Stabilizers are materials that help the rubber to withstand oxidative ageing and ozone attack. They act by intercepting the active free radicals breaking the free-radical-associated degradation process. Amines and phenols are generally employed. Reinforcing fillers, of which carbon black is the most important, are added to improve the mechanical properties such as hardness, abrasion resistance, modulus, and tear resistance. It is believed that the rubber adheres to the carbon surface. Carbon black also helps in retarding UV degradation and increases the electrical conductivity, reducing triboelectric charging and acting as an antistatic material.

Natural rubber can be compounded without fillers to give a vulcanized material with high elongation (to about 800%) and high tensile strength (about 28 MPa).

The internal mixers fragment the large rubber molecules by high shear forces. Depending on the particular assembly and ingredients, the created free radicals can combine to give larger molecules or may form smaller chains. Breakdown is often assisted by the use of chemical peptizers such as thiophenols, mixtures of salts of saturated fatty acids and aromatic disulfides. The fatty acids generally act mainly as dispersing agents and processing aids.

The viscous prerubber is now *shaped* by addition to a mold of the desired shape. Addition can be achieved by simply pouring the material into the mold but usually the material is added to the mold employing the usual molding addition (extrusion, compression, and transfer) techniques.

The material can also be treated using most of the other "thermoplastic" processing techniques such as calendering, coating, and extrusion.

The material is now heated to cure, set, or *vulcanize* (all terms are appropriate) the material into the (typically) finished shape. Between 1% and 5% of sulfur (by weight) is added in typical black rubber mixes giving a vulcanized material with an average of about 500 carbon atoms between cross-links. Larger amounts of sulfur will give a tougher material eventually giving a somewhat brittle, but quite strong, ebonite as the amount of sulfur is increased to about 40%. Sometimes additional finishing may be desirable including painting, machining, grinding, and cutting.

These steps are typical for most of the synthetic elastomers. The use of sulfur for vulcanization is common for the production of most elastomers. Magnesium and zinc oxides are often used for the cross-linking of polychloroprene, CR. Saturated materials such as ethylene–propylene (EPM) and fluoroelastomers are cross-linked using typical organic cross-linking agents such as peroxides.

Carbon black is widely used as a reinforcing agent for most synthetic elastomers. Carbon black is especially important for synthetic elastomers such as styrene–butadiene rubber (SBR), nitrile rubber (NBR), and polybutadiene (BR) that do not crystallize at high strains. Thus, noncarbon filled SBR has a tensile strength of about 2 MPa and with the addition of carbon black this increases to about 20 MPa.

The earlier processing applies to the processing of typical bulk carbon backbone-intensive elastomers. Other important classes of elastomers are also available. Polyurethanes represent a broad range of elastomeric materials. Most polyurethanes are either hydroxyl or isocyanate terminated. Three groups of urethane elastomers are commercially produced. Millabile elastomers are produced from the curing of the isocyanate group using trifunctional glycols. These elastomers are made from high polymers made by the chain extension of the polyurethane through the reaction of the terminal isocyanate groups with a polyether or polyester. Low molecular weight isocyanate terminated polyurethanes are cured through a combination of chain extension by reaction with a hydroxyl-terminated polyether or polyester and trifunctional glycols giving cast elastomers. Thermoplastic elastomers are block copolymers formed from the reaction of isocyanate-terminated polyurethanes with hydroxyl-terminated polyethers or polyesters. These are generally processed as thermoplastic materials as are the thermoplastic elastomers. Many of these materials have little or no chemical cross-linking. The elastomeric behavior is due to the presence of physical hard domains that act as cross-links. Thus, SBR consists of soft butadiene blocks sandwiched between polystyrene hard blocks. These hard blocks also act as a well dispersed fine-particle reinforcing material increasing the tensile strength and modulus. The effectiveness of these hard blocks greatly decreases above the T_g (about 100°C) of polystyrene.

Polysiloxanes (silicons) form another group of important elastomers. Again, processing typically does not involve either carbon black or sulfur.

16.5   FILMS AND SHEETS

Films, such as regenerated cellulose (cellophane), are produced by precipitating a polymeric solution after it has passed through a slit die. Other films, such as cellulose acetate, are cast from a solution of the polymers, but most films are produced by the extrusion process. Some relatively thick films and coextruded films are extruded through a flat slit die, but most thermoplastic films, such as PE film, are produced by air blowing of a warm extruded tube as it emerges from a circular die. Films and sheets are also produced employing calendering. Calendering is also used to apply coatings to textiles or other supporting materials.

The most widely used films are LDPE, cellophane, PET, PVC, cellulose acetate, polyfluorocarbons, nylons, PP, PS, and LLDPE. The strength of many films is improved by biaxial orientation, stretching. Most of the thermoplastics used as films may also be extruded as relatively thick sheets. These sheets may also be produced by pressing a stack of film at elevated temperature (laminating) or by the calendering process.

Wire is coated by being passed through a plastic extruder, but most materials are coated from solutions, emulsions, or hot powders. The classic brushing process has been replaced by roll coating, spraying, and hot powder coating. The application of polymers from water dispersions to large objects, such as automobile frames, has been improved by the electrodeposition of the polymer onto the metal surface.

Printing inks are highly filled solutions of resins. The classic printing inks were drying oil-based systems but the trend in this almost billion-dollar business is toward solvent-free inks.

16.5.1 CALENDERING

Calendering is simply the squeezing or extruding of a material between pairs of corotating, parallel rollers to form film and sheets. It can also be used to mix and impregnate such as in the case of embedding fiber into a slightly melted matrix material to form impregnated composite tapes. It can also be used to combine sheets of material such as sheets of impregnated paper and fiber woven and nonwoven mats to form laminar composite materials. It is also used in processing certain rubber material and textiles. Calendering is also employed in conjunction with other processing techniques such as extrusion in the formation of films from extruded material. It is also used to coat, seal, laminate, sandwich, finish, and emboss.

Because of the variation in flexibility, the terms film and sheet vary with materials. For PVC, films have a thickness of 6 mil (0.15 mm) and less while sheets are thicker than this. While PVC is relatively rigid with a tensil modulus greater than about 690 MPa (105 psi), thin films are easily folded. Films are generally shipped as rolls with the PVC rolled about a central rigid core. Sheets are generally shipped as flat-layered sheets.

Films and thin sheets are typically drawn to impart additional unidimensional strength. For films, both unidirectional and bidirectional drawings are used.

16.6 POLYMERIC FOAMS

Prior to 1920, the only flexible foam available was the natural sponge, but chemically foamed rubber and mechanically foamed rubber latex were introduced prior to World War II. These foams may consist of discrete unit cells (unicellular, closed cell), or they may be composed of interconnecting cells (multicellular, open cells) depending on the viscosity of the system at the time the blowing agent is introduced. Over 1.5 million ton of foamed plastic is produced annually in the United States.

Unicellular foams are used for insulation, buoyancy, and flotation applications, while multicellular foams are used for upholstery, carpet backing, and laminated textiles. Expanded polystyrene, which is produced by the extrusion of PS beads containing a volatile liquid, is used to produce low-density moldings such as foamed drinking cups and insulation boards. Foamed products are also produced from PVC, LDPE, urea resins, ABS, and PU. PU foams are versatile materials that range from hard (rigid) to soft (flexible). These foams are produced by the reaction of a polyols and a diisocyanate.

16.7 REINFORCED PLASTICS (COMPOSITES) AND LAMINATES

16.7.1 COMPOSITES

Theoretical and material considerations for composites are dealt with in Sections 14.2 and 14.3. Here we will focus more on processing considerations.

Composites are generally composed of two phases, one called the ***continuous*** or ***matrix phase*** that surround the second ***discontinuous*** or ***dispersed phase***. There are a variety of polymer-intense composites that can be classified as shown in Figure 16.5. Many of these composite groups are used

FIGURE 16.5 Classification of polymer-intense composites.

in combination with other materials including different types of composites and like types of composites except differing in orientation.

Many naturally occurring materials such as wood are reinforced composites consisting of a resinous continuous phase and a discontinuous fibrous reinforcing phase.

The modern ski is a good example of the use of composites to make a product with unique properties (Figure 16.6). The top and sides are composed of ABS polymer that has a low T_g allowing it to remain flexible even at low temperatures. It is employed for cosmetic and containment purposes. Polyurethane forms the core and a damping layer that acts as a filler and to improve chatter resistance. The base is a carbon-impregnated matrix composite that is hard, strong, and with good abrasion resistance. There are numerous layers of fibrous glass that are a mixture of bidirectional layers to provide torsional stiffness, unidirectional layers that provide longitudinal stiffness with bidirectional layers of fiberglass acting as outer layers to the polyurethane layers composing a torsional box. The only major noncomposite material is the hardened steel edge that

FIGURE 16.6 Cut-away illustration of a modern ski.

FIGURE 16.7 Use of graphite (solid) and graphite/Kevlar (TM) (dotted) composites in the exterior of the Boeing 767 passenger jet. Sites include wing tips, fixed trailing edge panels, inboard and outboard spoilers, and inboard and outboard ailerons for the large wings and for the tail wings the fin tip, rudder, elevators, stabilizer tips and fin fixed trailing edge panels.

assists in turning by cutting into the ice. These materials work together to give a light, flexible, shock absorbing, tough ski.

Composites are also used extensively where light but very strong materials are needed such as in the construction of the new Boeing 767 where composites play a critical role in the construction of the exterior (Figure 16.7).

Here we will briefly look at each of the main groupings of composites.

16.7.2 Particle-Reinforced Composites—Large-Particle Composites

Some materials to which fillers have been added can be considered as composites. These include a number of the so-called cements including concrete (Section 10.2). As long as the added particles are relatively small, of roughly the same size, and evenly distributed throughout the mixture there can be a reinforcing effect. The major materials in Portland cement concrete are the Portland cement, a fine aggregate (sand), course aggregate (gravel and small rocks), and water. The aggregate particles act as inexpensive fillers. The water is also inexpensive. The relatively expensive material is the Portland cement. Good strength is gained by having a mixture of these materials such that there is a dense packing of the aggregates and good interfacial contact, both achieved by having a mixture of aggregate sizes—thus the use of large gravel and small sand. The sand helps fill the voids between the various larger gravel particles. Mixing and contact are achieved with the correct amount of water. Enough water must be present to allow a wetting of the surfaces to occur along with providing some of the reactants for the setting up of the cement. Too much water creates large voids and weakens the concrete.

16.7.3 Fiber-Reinforced Composites

Mathematically the critical fiber length necessary for effective strengthening and stiffening can be described as follows:

Critical fiber length = [Ultimate or tensile strength times fiber diameter/2] times the fiber–matrix bond strength OR the shear yield strength of the matrix—which ever is smaller.

Fibers where the fiber length is greater than this critical fiber length are called *continuous fibers* while those that are less than this critical length are called *discontinuous* or short *fibers*.

Little transference of stress and thus little reinforcement is achieved for short fibers. Thus, fibers whose lengths exceed the critical fiber length are used.

Fibers can be divided according to their diameters. Whiskers are very thin single crystals that have large length to diameter ratios. They have a high degree of crystalline perfection and are essentially flaw free. They are some of the strongest materials known. Whisker materials include graphite, silicon carbide, aluminum oxide, and silicon nitride. Fine wires of tungsten, steel, and molybdenum are also used but here, even though they are fine relatively to other metal wires, they have large diameters. The most used fibers are "organic fibers" that are either crystalline or amorphous or semicrystalline with small diameters.

16.7.4 PROCESSING OF FIBER-REINFORCED COMPOSITES

These exists a wide variety of particular operations but briefly they can be described in terms of filament winding, preimpregnation of the fiber with the partially cured resin, and pultrusion. Pultrusion is used to produce rods, tubes, beams, etc. with continuous fibers that have a constant cross-sectional shape. The fiber (as a continuous fiber bundle, weave, or tow) is impregnated with a thermosetting resin and pulled through a die that shapes and establishes the fiber to resin ratio. This stock is then pulled, though a curing die that can machine or cut producing the final shape such as filled and hollow tubes and sheets.

The term used for continuous fiber reinforcement *preimpregnation* with a polymer resin that is only partially cured is "prepreg." Prepreg material is generally delivered to the customer in the form of a tape. The customer then molds and forms the tape material into the desired shape, finally curing the material without having to add any additional resin. Preparation of the prepreg can be carried out using a calendering process. Briefly, fibers from many spools are sandwiched and pressed between sheets of heated resin with the resin heated to allow impregnation but not so high as to be very fluid.

Thus, the fiber is impregnated in the partially cured resin. Depending upon the assembly the fiber is usually unidirectional, but can be made so that the fibers are bidirectional or some other combinations. The process of fitting the prepreg into, generally onto, the mold is called "lay-up." Generally, a number of layers of prepreg are used. The lay-up may be done by hand, called hand lay-up, or done automatically, or some combination of automatic and hand lay-up. As expected, hand lay-up is more costly but is needed where one-of-a-kind products are produced.

In *filament winding*, the fiber is wound to form a desired pattern, usually but not necessarily hollow and cylindrical. The fiber is passed through the resin and then spun onto a mandrel. After the desired number of layers of fiber is added, it is cured. Prepregs can be filament wound. With the advent of new machinery, complex shapes and designs of the filament can be readily wound.

16.7.5 STRUCTURAL COMPOSITES

Structural composites can be combinations of homogeneous and composite materials. Laminar composites are composed of two-dimensional sheets that generally have a preferred high-strength direction. The layers are stacked so that the preferred high-strength directions are different, generally at right angles to one another. The composition is held together by a resin. This resin can be applied simply as an adhesive to the various surfaces of the individual sheets or the sheet can be soaked in the resin prior to laying the sheets together. In either case, the bonding is usually of a physical type. Plywood is an example of a laminar composite. Laminar fibrous glass sheets are included as part of the modern ski construction. These fibrous glass sheets are fiber-reinforced composites used together as laminar composites.

Laminar materials are produced by a variety of techniques. Co-extrusion blow molding produces a number of common food containers that consist of multi-layers such as layers consisting of polypropylene/adhesive/poly(vinyl alcohol)/adhesive/adhesive/ polypropylene.

FIGURE 16.8 Assembly employed for the fabrication of laminates (left) and reinforced plastics (right).

Sandwich composites are combinations where a central core(s) is surrounded generally by stronger outer layers. Sandwich composites are present in the modern ski and as high temperature stable materials used in the space program. Some cores are very light acting something like filler with respect to high strength, with the strength provided by the outer panels. Simple corrugated cardboard is an example of a honeycomb core sandwich structure except that the outer paper-intense layers are not particularly strong. Even in the case of similar polyethylene and polypropylene corrugated structures, the outer layers are not appreciatively stronger than the inner layers. In these cases the combination acts to give a lightweight somewhat strong combination, but they are not truly composites rather they simply exploit a common construction.

16.7.6 Laminating

Laminating is a simple binding together of different layers of materials. The binding materials are often thermosetting plastics and resins. The materials to be bound together can be paper, cloth, wood, or fibrous glass. These materials are often referred to as the reinforcing materials. Typically, sheets impregnated by a binding material, are stacked between highly polished metal plates, subjected to high pressure and heat in a hydraulic press producing a bonded product that may be subsequently treated, depending on its final use (Figure 16.8, left). The end product may be flat, rod-shaped, tubular, rounded, or some other formed shape.

Reinforced plastics differ from high-pressure laminates in that little or no pressure is employed. For instance, in making formed shapes, impregnated reinforcing material is cut to a desired shape, the various layers are added to a mold, and the molding is completed by heating the mold (Figure 16.8, right). This process is favored over the high-pressure process because of the use of a simpler, lower cost mold and the production of strain-free products.

16.8 MOLDING

Molding is a general technique that can be used with plastics and thermosetting materials when employing mobile prepolymer. Molding is used to produce sheet-like, foamed, hollow, or solid materials from very small to very large objects. Here we will look at various molding processes.

16.8.1 Injection Molding

The most widely employed processing techniques for thermoplastics are extrusion and injection molding. Injection molding is also used to produce some thermoset products utilizing fluid prepolymer.

Injection molding involves forcing, injecting, a molten polymer into a mold where it cools becoming solid. The mold separates allowing the molded material to be released. The mold parts are again joined and the process begins again.

Injection molding allows the rapid, economical production of small to large parts. It provides close tolerances and the same machine can be used to mold many different articles. Parts can be molded combining the polymer with other polymers and with any number of additives. Further, it can be run so that various parts can be easily married as part of an entire or combined-parts production assembly of an article. The ability to easily modify the operating conditions of the injection molding machine is important because of the variety of articles that may be needed, variety of material employed to produce the same (general) article, variety of materials to produce different injection molded articles, and the variability of supposedly the same polymer material from batch to batch.

Injection molding is not new. A patent was issued in 1872 for an injection-molding machine for camphor-plasticized cellulose nitrate, celluloid. Almost all of the machines used today are reciprocating or two-stage screw types. Both types employ a reciprocating Archimedean-like screw similar to that of a screw extruder. A few are of the plunger type.

A traditional injection apparatus consists of a hopper that feeds the molding powder to a heated cylinder where the polymer is melted and forced forward by a reciprocating plunger or screw. The cooled part is ejected when the mold opens and then the cycle is repeated. The molten material passes from the nozzle through a tapered sprue, a channel or runner, and a small gate into the cooled mold cavity. The polymer in the mold is easily broken off at the gate site and the materials in the sprue, runner, and gate are ground and remolded. An illustration of such an injection molding press is given in Figure 16.9. The hopper (a) feeds the molding powder to a heated cylinder (b) where the polymer is melted and forced forward by a reciprocating plunger (c) (or screw). The molten material advances toward a spreader or torpedo into a cool, closed, (here) two-piece mold (d). The cooled part is ejected when the mold opens and then the cycle is repeated. The molten plastic is passed from the nozzle through a tapered sprue, runner, and a small gate into the cooled mold cavity. The plastic in the narrow gate section is easily broken off and excess material remaining within the sprue, runner, and gate ground and remolded.

FIGURE 16.9   Cross section of an injection-molding press. (From Seymour, R., *Modern Plastics Technology*, Reston Publishing Co., Reston, VA, 1975.)

In a reciprocating screw machine, the material is collected in front of the screw that continues to move backward as additional material is melted. The area where the melted material is collected corresponds to the heating chamber or pot in a two-stage system. The material is melted by the internally generated heat caused by the friction of the polymer segments and chains rubbing against one another. The screw is also good at mixing so that additive introduction and mixing of different polymers can be achieved in the same step with overall polymer melting.

As the size of the molded product becomes larger, it is more difficult to control uniformity and to maintain a sufficient clamping force to keep the mold closed during filling. Reaction injection molding, RIM, overcomes these problems by largely carrying out the polymerization reaction in the mold. The most widely used RIM materials are polyurethane and polyurethane reinforced elastomeric materials. Most of the automotive interior panels (such as dashboards) are produced using RIM.

On a molecular level, partially crystalline polymers to amorphous polymers are normally used. As the material is heated, Brownian motion occurs resulting in a more random chain arrangement. When a unidirectional force is applied to a resting polymer melt, the chains tend to move away from the applied force. If the applied force is slow enough to allow the Brownian movement to continue to keep the polymers in a somewhat random conformation then the movement of the polymer melt is proportional to the applied stress; that is, the flow is Newtonian.

As the rate of movement increases, chain alignment occurs along the direction of flow with movement too fast for the Brownian factors to return the system to a somewhat random state and flow is then non-Newtonian. Most systems are operated, at least at the injection stage, under non-Newtonian conditions so that some polymer alignment occurs. If the polymer melt flow rate continues to increase, polymer chains align parallel to the flow plane and eventually reach a point where it again becomes Newtonian. Even so, the polymer chains have been aligned as the flow processes moved through the non-Newtonian flow range.

As the molten polymer is injected into the cold mold, it rapidly solidifies locking in at least some of the "orientated" chain conformations. As the material enters the cold mold, the flow turbulence occurring with the outermost layers is generally sufficient to result in a more randomized, more amorphous outer structure. As the outermost chains cool, they "drag" the next chains effectively aligning them giving a more ordered structure. Finally, the cooling of the inner material is slowed because of the heat uptake of the outer layers allowing the Brownian movement to again somewhat randomize these chains. Thus, the structure of the molded part is varied and can be further varied by controlling the flow rate, cooling rate, and flow and cooling temperatures for a specific injected-produced material.

16.8.2 BLOW MOLDING

Most molded materials, as well as most processed materials, will have a different surface or skin composition compared with the bulk or core materials. Take a look at a common disposable polystyrene foam plate. The surface or skin is smooth. Break it and look at the core and it is different being more cellular. This difference is greater than having simply a difference in appearance. There also exist different fine molecular-level differences. Molecular structures, and associated bulk properties, are controlled in part by the particular processing and processing particulars.

Blow molding has been used for many years in the creation of glass bottles. In about 1872, the blow molding of thermoplastic objects began by the clamping of two sheets of cellulose nitrate between two mold cavities. Steam was injected between the two sheets softening the sheets and pushing the material against the mold cavities. However, it was not until the late 1950s that large-scale use of blow molding began with the introduction of blow molded high-density polyethylene, HDPE, articles.

Figure 16.10 contains a sketch of an extrusion blow-molding scheme. Here a heat-softened hollow plastic tube, or parison, is forced against the walls of the mold by air pressure. The sequence of material introduction into the mold and the subsequent rejection of the material from the mold are

Polymer injected   Air injected

FIGURE 16.10   Steps employed in simple extrusion blow molding of a test tube. From left to right: mold closed, soften material introduced, air or other gas injected forcing the softened thermoplastic against the walls of the mold, and after suitable cooling, the mold is opened giving the molded plastic test tube.

generally rapid and automated. Approximately 1 million ton of thermoplastics are produced by this technique annually.

While there are a wide variety of blow-molding techniques, there are three main blow-molding procedures:

- Injection blow molding that employs injection molded "test-tube" shaped preforms or parisons
- Extrusion blow molding that uses an extruded tube preform or parison
- Stretch blow molding that employs an injection molded, extrusion blow molded preform, or extruded tube preform

The major difference between injection and extrusion blow molding is the way the soft hollow tube (called a preform or parison) is made. In injection blow molding two different molds are used. One mold forms the preform and the other mold is used in the actual blow molding operation to give the final shaped article. In the molding process, the soften material preform, from the preform mold, is introduced into the blowing mold and blow molded to fit the cavity of the second "finished" blow mold. This process is sometimes also called transfer blow molding because the injected preform is transferred from the preform mold to the final blow mold. This allows better control of the product wall thickness and the thickness of the various curved locations.

Injection blow molding is typically used to produce smaller articles, generally with a total volume of 500 mL or less. Because two molds are used, there is little waste material that must be recycled and there is no bottom weld joint. It allows the production of small articles that at times are very difficult to manufacture in any other way.

Extrusion blow molding is the most common process used to produce hollow articles larger than 250 mL up to about 10,000 L. In extrusion blow molding the soften material is extruded continuously or intermittently. The preform is introduced, the mold haves close, and air or other gas is introduced forcing the preform material against the mold surfaces. After cooling, the mold is opened and the formed article rejected. Articles with handles and off-set necks can be manufactured using extrusion blow molding. Unlike injection blow molding, waste that must be cut away and recycled is produced as the two halves of the mold are pressed together.

In continuous extrusion blow molding the preform is continuously produced at the same rate as the article is molded, cooled, and released. To avoid interference with the preform formation, the mold-clamping step must be rapid to capture the preform and move it to the blow mold station. There are various modifications of this that allow essentially continuous operation.

The stretching is best done just above the materials T_g allowing a balance between good alignment because of ease in chain movement, and a decreased tendency to form crystalline areas in the melt allowing ready flow of material.

16.8.3 ROTATIONAL MOLDING

In rotational molding, also known as rotomolding, the mold (or cavity) is filled with material, either as a solid power, or as a liquid. The mold is closed, placed in a heated oven, and then rotated biaxially. The mold is then cooled, opened, and the article recovered. Powders of about 35 mesh (500 μm) are typical though different sizes are also employed. The distribution of particles and additives is determined by mixing and rotation ratio.

Almost any mold design can be incorporated into rotational molding. Tanks used for agricultural, chemical, and recreational vehicle industries are made using rotational molding as are containers used for packaging and material handling, battery cases, portable toilets, vacuum cleaner housings, light globes, and garbage containers. Rotational molding produces little waste and produces a material with uniform wall thickness as well as strong corner sections.

16.8.4 COMPRESSION AND TRANSFER MOLDINGS

While there are a number of molding processes, compression molding and transfer molding are the main techniques for molding articles from thermosetting materials. In compression molding or transfer molding, the material, thermoplastic or prethermoset material, is heated sufficiently to soften or plasticize the material to allow it to enter the mold cavity. The soften material is held against the mold by pressure. For thermoplastics, it is then cooled below the T_g thus locking in the shape. For thermosets, it is held until the cross-linking occurs thereby locking in the article shape.

The most widely employed molding process is compression molding where the material is placed in the bottom half of an open heated mold. The second half of the mold is closed and brings heat and pressure against the material softening the material further and eventually allowing it to cross-link, if it is a thermoset. When completed, the pressure is released and article is removed from the mold. Generally, excess material, or flash, is produced. Figure 16.11 contains a representation of a simple compression molding assembly.

Compression molding is one of the oldest materials handling processes. Ancient Chinese employed compression molding to form articles from paper mache. Rubber articles were made in the early nineteenth century from composites of woody fibers and gum shellac. Baekeland used compression molding to make many of his early phenol–formaldehyde products.

Transfer (or plunger) molding introduces the to be molded material after the mold is closed through a small opening or gate. This process can be used when additional materials, such as a glass globe or other designed object, are placed in the mold prior to closing the mold.

Applied pressure

Sample and mold cavity

FIGURE 16.11 Representation of a compression molding assembly.

In true transfer molding or pot-type molding, the mold is closed and placed in a press. The soften material is introduced into an open port at the top of the mold. The plunger is placed into the pot and the press closed. As the press closes, it pushes against the plunger forcing the molding material into the mold cavity. Excess molding compound is used to ensure that there is sufficient material to fill the mold. After the material is cured and/or cooled, the plunger is removed and the part is also removed from the mold. In plunger molding, the plunger is part of the press rather than part of the mold. Because of this feature, it can be smaller than the pot-type plunger. The clamping action of the press keeps the mold closed. Here there is less material waste compared with the pot-type molding.

In cold molding, the compound is compacted in a mold at around room temperature. The compressing operation is similar to that employed in the production of KBr pellets from powdered KBr. The compound generally contains a lot of fillers and binders. The compacted material is removed from the mold and placed in an oven where it becomes cross-linked. Ceramic materials are often produced using cold molding.

16.8.5 THERMOFORMING

Thermoforming involves heating a sheet or thick film just above its T_g or T_m, stretching it against a rigid mold, cooling, and trimming the formed part. Inexpensive aluminum, wood, epoxy, and steel molds are often employed. This allows the construction of inexpensive molds that allow the production of low-volume articles. All thermoplastic materials that can be formed into sheets can be thermoformed provided that the heating does not exceed the ability of the sheet to support itself.

Thermoforming is employed to convert extruded sheets into smaller items such as packaging containers, plates, trays, bathtubs, pickup truck liners, freezer liners, cabinetry, and cups. The skin packaging that involves a flexible plastic skin drawn tightly over an article on a card backing is made by thermoforming. Thermoforming permits the production of small to large articles including those with thin walls such as drinking cups. Thus, thermoforming is employed to produce articles with a relatively high surface to thickness ratio.

Multilayered materials can be readily formed using thermoforming including food packaging that may involve the inclusion of layers of ethylene–vinyl alcohol copolymers, polystyrene, polyolefins, and/or copolymers of vinylene dichloride and vinyl chloride. Microwavable food trays from (crystallized) poly(ethylene terephthate) are manufactured using thermoforming.

Polystyrene is the most widely used resin material for thermoforming. High impact polystyrene, HIPS, is the most widely used being employed largely in the packaging areas including disposables (foam drinking cups, lids, lunch trays, and food service containers), packaging for medical devices, and food packaging (meat and poultry trays, egg cartons, dairy and delicatessen containers, and barrier packages). ABS is thermoformed to produce more durable articles such as refrigerator inner door liners, recreational vehicle and boat parts, automotive panels, picnic coolers, and luggage exteriors. High-density polyethylene is used in the manufacture of pickup truck liners, golf cart tops, and sleds. Crystallized poly(ethylene terephthalate), PET, is used in the thermoforming of food trays that can be heated in a microwave or regular oven. Coextruded polycarbonate–polyetherimide is used to make frozen-food oven-friendly trays. Polycarbonates and poly(methyl methacrylate) are thermoformed to produce skylights, windows, tub and shower stalls, and outdoor display signs. Poly(vinyl chloride) is used in the production of blister packaging of pharmaceuticals, foods, cosmetics, and hardware.

16.9 CASTING

Casting is employed in making special shapes, sheets, films, tubes, and rods from both thermoplastic and thermoset materials. The essential difference between most molding processes and casting is that no added pressure is employed in casting. In casting, the polymer or prepolymer is heated to a fluid, poured into a mold, cured at a specific temperature, and removed from the mold. Casting of films and sheets can be done on a wheel or belt or by precipitation. In the case of a wheel or belt, the polymer is spread to the desired thickness onto a moving belt as the temperature is increased. The film is dried and then stripped off. "Drying" may occur though solvent evaporation, polymerization, or cross-linking.

16.10 EXTRUSION

Extrusion involves a number of processing operations and is widely used. We will look at extrusion as it is involved in several of these processes. These processing operations are used together or separately. A representative extruder is shown in Figure 16.12. The extruder accepts granulated thermoplastic in a hopper (c), and forces it from the feed throat (d) through a die (f). The die may be circular for the production of a rod or pipe, or flat for the production of a sheet, or it may have any desired profile for the continuous production of almost any uniformly shaped product. The screw (a) advances the polymer through a heated cylinder (barrel) (b) to a breaker plate and protective screen pack (e) before it enters the die (f). The extrusion process may be divided into a feed or transport zone, a compression or transition zone, and a metering zone. Over 1 million ton of extruded pipes are produced annually in the United States.

In extrusion, a fluid material, generally rendered a fluid material through heating, is forced through a shaping device. Since there is a need for quickness and because the preshaped material is quite viscous, extrusion requires high pressure to drive or force the melt through a die. The melts can be extruded as pipers, sheets, films, or into molds.

Along with moving and shaping the molten material, extruders also act to disperse additives and are often the agent for creating heat thus enabling the material to become molten.

16.11 COATINGS

The fundamental purpose for painting is decorating, whereas the purpose for coating is for protection. In truth, we often do not differentiate between the two terms. Government edicts concerning air, water, solid particulates, and worker conditions are having real effects on the coatings industry with the generation of new coating techniques. Paint solvents, in particular, are being looked at in view of increased environmental standards. The volatile organic compound (VOC) regulations under Titles I and VI of the Clean Air Act specify the phasing out of ozone-depleting

FIGURE 16.12 Sketch details of screw and extruded zones. (From Seymour, R., *Modern Plastics Technology*, Reston Publishing Co., Reston, VA, 1975.)

chemicals—namely chlorinated solvents. Baseline solvent emissions are to be decreased. These, and related regulations, affect the emission of all organic volatiles, whether in coatings or other volatile-containing materials.

A major driving force in coatings continues to be a move toward water-based coatings. Another is to eliminate the "odor" of the coating. Most waterborne coatings actually contain about 8%–10% nonaqueous solvent. The odor we get as the coating dries is mainly due to the evaporation of the solvent. Work continues to develop the right balance of properties and materials allowing the latex particles to flow together and coalesce into suitable films without the need of nonaqueous liquids.

Another area of active research is the development of paints that dry under extreme or unusual conditions including under water and on cool substrates. The latter allows the painting season for exterior coating to be extended, particularly in the northern states.

Work continues on making more durable exterior paints. Remember that there is a difference in requirements of exterior and interior paints. For instance, interior paints are generally required to be faster drying and more durable against scraps and punctures since it is the inside of the house that generally experiences such traumatic events. By comparison, exterior paints need to remain flexible and adhered under a wide variety of humidity and temperature conditions. A more durable exterior coating should allow it a longer lifetime because it can better withstand stress caused by the pounding of the rain, sticks, and human afflicted dings and dents. Binders or coatings resins are critical to the performance of coatings. They bind the components together. Since the primary cost of most commercial application of coatings is labor, the market will allow price increases for products that give added positive properties. Table 1.5 contains a listing of the production of paints for the year 2004.

Paint manufactures in the United States sell about 1300 million gallon of coating material annually or about 4 gal for every man, woman, and child. Paint is typically a mixture of a liquid and one or more colorants (pigments). The liquid is called a *vehicle* or binder (adhesive) and may include a solvent or a thinner along with the coating agent. The colored powders are called *pigments* that may be prime or inert. Prime pigments give the paint its color. These pigments may be inorganic, such as titanium dioxide (titanium (IV) oxide; the most widely used pigment by far) for white (but contained in many colored paints as well), iron oxides for browns, yellows, and reds, or organic compounds such as phthalocyanine for greens and blues. Inert pigments such as clay, talc, calcium carbonate, and magnesium silicate make the paint last longer and may contribute to the protective coating as do mica chips in some latex paints that actually form a clad on drying. The paint may also contain special additives and catalysts. Thus, many paints in wet areas contain an agent to fight fungus, rot, and mold.

Vehicles include liquids such as oils (both natural and modified natural) and resins and water. A latex vehicle is made by suspending synthetic resins, such as poly(ethyl acrylate), in water. This suspension is called an emulsion, and paints using such a vehicle are called latex, waterborne, or emulsion paints. When the vehicle comes in contact with air, it dries or evaporates, leaving behind a solid coating. For latexes, the water evaporates, leaving behind a film of the resin.

Paints are specially formulated for specific purposes and locations. Following is a brief description of the most popular paint types.

Oil paints: Oil paints consist of a suspension of pigments in a drying oil, such as linseed oil. The film is formed by a reaction involving atmospheric oxygen that polymerizes and cross-links the drying oil. Catalysts may be added to promote the cross-linking reaction. Oil paints, once dried, are no longer soluble, although they can be removed through polymer degradation using the appropriate paint stripper.

Oil varnishes: Varnish coatings consist of a polymer, either natural or synthetic, dissolved in a drying oil together with appropriate additives as catalysts to promote cross-linking with oxygen. When dried, they produce a clear, tough film. The drying oil is generally incorporated, along with the dissolved polymer, into the coating.

Enamels: Classical enamel is an oil varnish with a pigment added. The added polymer is typically selected to provide a harder, glossier coating than the oil varnish mixture. Today, there are latex enamels that are similar to the oil enamels except no natural oil is present.

Lacquers: Lacquers consist of polymer solutions to which pigments have been added. The film is formed through the simple evaporation of the solvent leaving the polymer film as the coating. These coatings are formed without subsequent cross-linking, thus the surface exhibits poor resistance to some organic solvents.

Latex paints: Latex paints today account for more than one-half of the commercial paint sold. They are characterized by quick drying (generally several minutes to several hours), little order, and easy cleanup (with water). Latex paints are polymer latexes to which pigments have been added. The film is formed by the coalescence of the polymer particles on the evaporation of the water. The polymer itself is not water soluble, though these paints are called waterborne coatings.

The composition of latex paints is variable depending on the end use, manufacturer, and intended cost. Ethyl acrylate (16.1) and butyl acrylate (16.2) are frequently employed as the synthetic resin monomer. Most synthetic resins for coatings are copolymers of the acrylics. Vinyl acetate or other monomers are sometimes incorporated.

Ethyl acrylate derived unit (16.1)   Butyl acrylate derived unit (16.2)

The T_g of the copolymer must be below the application temperature to allow diffusion from the latex as it is drying forming the protective film.

16.11.1 Processing

Many different process are used to apply a thin layer of liquid-melted polymer or polymer solution or dispersion including rollers, spraying, calendering, and brushing. Here we will look at the industrial application of coatings onto film and sheet-like materials. Figure 16.14 contains examples of general coating processes employed to achieve this. The moving sheet is called a web. In *roll coating*, Figure 16.13, left side, the lower roller picks up the coatings material that then transfers it to the second roller and finally to one surface of the mat. Spacing of the rollers, viscosity of the polymer solution in the dip tank, and roller speed and size control the thickness of the applied coating.

In *blade coating*, as shown in the upper right hand side of Figure 16.13, a flexible blade helps control the coating thickness. The blade may be located after the bath or within the bath as part of the roller system. The blade is flexible and adjustable so that the amount of coating material can be controlled by the application of force on the blade and/or distance the blade is from the web or the roller. Both of these processes are forms of calendering. The third main coating process involves the direct application of the coating material forming a so-called *fluid curtain*, Figure 16.13, lower right side. This process, *curtain coating*, is usually used in conjunction with a curing process. The coating thickness is controlled by the web speed, the polymer concentration, and the rate of application from the extruder.

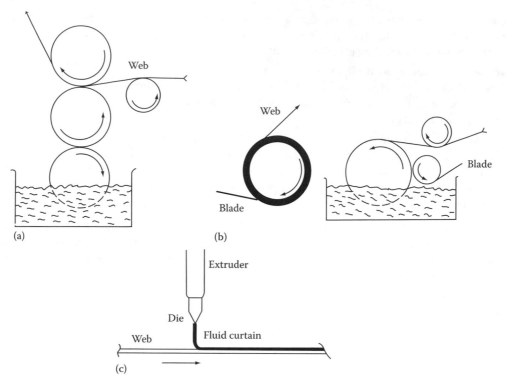

FIGURE 16.13 Three principle industrial coatings processes for films and thin sheets. (a) Roll coating, left; (b) blade coating, upper right; and (c) curtain coating, lower right.

Coatings can be divided according to formulation, drying mechanism or other system of categorizing. A coating is normally a mixture of various components. For instance, the label on a latex emulsion-type paint might have as major components poly(ethyl acrylate), titanium dioxide as the white pigment, and water. It could also have china clay and calcium carbonate as extenders, carboxymethylcellulose as a colloid thickener, a defoaming agent, a plasticizer, a surfactant dispersing aid, additional coloring agents, and an added fungicide.

The hiding power is a measure of the ability of the coating to achieve a specified degree of "hiding" or obliteration. Industrially, it is often tested by comparing the reflectance of the coated surface overpainting a black surface (i.e., the tested paint applied over a black surface) with white panels. The ability to cover or hide is related to the scattering of incident light hitting the surface and returning to the observer or the light meter. As the film surface increases, the ability of light to penetrate the surface coating and be scattered from the (black for tests) under-coating lessens. For a simple white latex paint, no absorption occurs and we can consider the scattering occurring at the interfaces of the transparent polymer matrix and the dispersed pigment particles. The scattering coefficient of polymers can be obtained from reflectance measurements. While the refractive indices for most polymers do not widely vary (generally about 1.5), the scattering coefficients can vary widely. For good scattering, the refractive index of the polymer should differ from that of the pigment. For instance, while calcium carbonate, with a refractive index of about 1.6, is often used as a pigment in paints, it has a much lower hiding power than titanium dioxide (rutile titanium (IV) oxide), with a refractive index of about 2.8.

Scattering efficiency increases as the pigment surface area becomes larger, thus smaller particles aid in increasing the scattering to a lower limit determined by the wavelength of light, thereafter reduced size produces a rapid loss in scattering efficiency. For good scattering, and good hiding power, the particles should be dispersed in a homogeneous manner so that dispersing agents are commonly used.

16.12 ADHESIVES

In contrast to coatings, which must adhere to one surface only, adhesives are used to join two surfaces together. Resinous adhesives were used by the Egyptians at least 6000 years ago for bonding ceramic vessels. Other adhesives, such as casein from milk, starch, sugar, and glues from animals and fish, were first used at least 3500 years ago.

Adhesion occurs generally through one or more of the following mechanisms. Mechanical adhesion with interlocking occurs when the adhesive mixture flows about and into two rough substrate faces. This can be likened to a hook and eye, where the stiff plastic hooks get caught in the fuzz-like maze of more flexible fibers. Chemical adhesion is the bonding of primary chemical groups. Specific or secondary adhesion occurs when hydrogen bonding or polar (dipolar) bonding occurs. Viscosity adhesion occurs when movement is restricted simply due to the viscous nature of the adhesive material.

Adhesives can be divided according to the type of delivery of the adhesive or by type of polymer employed in the adhesive. Following are short summaries of adhesives divided according to these two factors.

Solvent-based adhesives: Solvent-based adhesion occurs through the action of the adhesive on the substrate. Solidification occurs on the evaporation of the solvent. Bonding is assisted if the solvent partially interacts, or, in the case of model airplane glues and PVC piping glues, actually dissolves some of the plastic (the adherent). Thus, model airplane glues and PVC glues often contain volatile solvents that dissolve the plastic, forming what is called a solvent weld. Some of the PVC glues actually contain dissolved PVC to assist in forming a good weld. A major difference between home-use PVC adhesive and industrial PVC adhesive is added color for the commercial adhesive, so building inspectors can rapidly identify that appropriate solvent welding has been accomplished.

Latex adhesives: Latex adhesives are based on polymer latexes and require that the polymers be near their T_g so that they can flow and provide good surface contact when the water evaporates. It is not surprising that the same polymers that are useful as latex paints are also useful as latex adhesives (such as PMMA). Latex adhesives are widely employed for bonding pile to carpet backings.

Pressure-sensitive adhesives: Pressure-sensitive adhesions are actually viscous polymer melts at room temperature. The polymers must be applied at temperatures above their T_g to permit rapid flow. The adhesive is caused to flow by the application of pressure. When the pressure is removed, the viscosity of the polymer is sufficient to hold and adhere to the surface. Many tapes are of this type where the back is smooth and coated with a nonpolar coating so as not to bond with the sticky surface. The two adhering surfaces can be separated, but only with some difficulty.

Hot-melt adhesives: Hot-melt adhesives are thermoplastics that form good adhesives simply by melting, followed by subsequent cooling after the plastic has filled surface voids. Nylons are frequently employed in this manner. Electric glue guns typically operate on this principle.

Reactive adhesives: Reactive adhesives are either low molecular weight polymers or monomers that solidify by polymerization and/or cross-linking reactions after application. Cyanoacrylates, phenolics, silicon rubbers, and epoxies are examples of this type of adhesives. Plywood is formed from the impregnation of thin sheets of wood with resin, with the impregnation occurring after the resin is placed between the wooden sheets.

Thermosets: A number of thermosets have been used as adhesives. Phenolic resins were used as adhesives by Leo Baekeland in the early 1900s. Phenolic resins are still used to bind together thin sheets of wood to make plywood. Urea resins have been used since 1930 as binders for wood chips in the manufacture of particleboard. Unsaturated polyester resins are used for body repair and polyurethanes are used to bond polyester cord to rubber in tires, bond vinyl film to particleboard, and to function as industrial sealants. Epoxy resins are used in the construction of automobiles and aircraft and as a component of plastic cement.

Elastomers: Solutions of natural rubber have been used for laminating textiles for over a century. The Macintosh raincoat, invented in 1825, consisted of two sheets of cotton adhered by an inner layer of natural rubber. Styrene–butadiene rubber (SBR) is used as an adhesive in carpet backing and packaging. Neoprene (polychloroprene) may be blended with a terpene or phenolic resin and used as a contact adhesive for shoes and furniture.

Pressure-sensitive tape: Pressure-sensitive tape, consisting of a coating of a solution of a blend of natural rubber and an ester of glycerol and abietic acid (rosin) on cellophane, was developed over half a century ago. More recently, natural rubber latex and synthetic rubber have been used in place of the natural rubber solution. The requirement for pressure-sensitive adhesives is that the elastomers have a T_g below room temperature. Today, there are many other formulations used in the production of pressure-sensitive tapes.

Contact adhesives: Contact adhesives are usually applied to both surfaces, which are then pressed together. Liquid copolymers of butadiene and acrylonitrile with carboxyl end groups are used as contact adhesives in the automotive industry.

Thermoplastics: A number of thermoplastics have been used as adhesives. Polyamides and copolymers of ethylene and vinyl acetate (EVA) are used as melt adhesives. Copolymers of methyl methacrylate and other monomers are used as adhesives in the textile industry. Poly(vinyl acetate) is often used in school glues.

Anaerobic adhesives: Anaerobic adhesives consist of mixtures of dimethacrylates and hydroperoxides (initiators) that polymerize in the absence of oxygen. They are used for anchoring bolts.

Cyanoacrylates: One of the most interesting and strongly bonded adhesives is cyanoacrylate (Super Glue; Krazy Glue). These monomers, such as butyl-α-cyanoacrylate (16.3), polymerize spontaneously in the presence of moist air, producing excellent adhesives. These adhesives, which have both the cyano and ester polar groups, are used for household adhesive problems as well as in surgery, mechanical assemblies, and as a "fast fix" for athletic cuts such as in boxing.

(16.3)

Many seemingly simple adhesive applications are actually complex. The labels on commercial dry cell batteries can contain over a dozen layers each present for a specific purpose. While price is a major consideration, the ease of application is another. Thus, while many naturally derived adhesives are less expensive, synthetic materials may be chosen because of ready application and consistency of the end product.

Important factors involved with pressure-sensitive adhesion are a balance between allowing molecular interaction between the adhesive and adherent (often referred to as "wetting') and the dynamic modulus of the adhesive mixture. This also involves a balance between "pull-off-rate" and "wetting rate." Mechanical adhesion with interlocking and diffusion factors is less important than for permanent adhesion. Pressure-sensitive adhesives, such as present in "pull-off" tabs such as Post-it Notes, contain components similar to those present in more permanent Scotch Tape, except that particles of emulsified glass polymer are added to reduce the contact area between the adhesive and the substrate. Some polymers, such as PE, might appear to be decent adhesive materials, but even in its melt, it is not exceptionally tacky. This is believed a result of the high degree of chain entanglement. Since the dynamic modulus increases with increasing chain entanglement, PE is not "tacky" (does not easily contact and wet a substrate), and is not a useful pressure-sensitive adhesive.

16.13 CONDUCTIVE POLYMERIC MATERIALS

Conductance behavior is dependent on the material and what is conducted. For instance, polymeric materials are considered poor conductors of sound, heat, electricity, and applied forces in comparison to metals. Typical polymers have the ability to transfer and "mute" these factors. For instance, as force is applied, a polymer network transfers the forces between neighboring parts of the polymer chain and between neighboring chains. Because the polymer matrix is seldom as closely packed as a metal, the various polymer units are able to absorb (mute; absorption through simple translation or movement of polymer atoms, vibrational and rotational changes) as well as transfer (share) this energy. Similar explanations can be given for the relatively poor conductance of other physical forces.

Even so, polymers can be designed that compete with metals and other nonpolymer materials in the area of conductance. The following text describes some of these. Covered elsewhere is described other materials that act similar to metals in the conductance of specific factors. For instance, force transference of ceramics is similar in some ways to that of metals because of a number of factors including the inability of the tightly packed ceramics to mute applied forces and their ability to directly "pass along" the results of such applied forces or stresses.

16.13.1 PHOTOCONDUCTIVE AND PHOTONIC POLYMERS

Some polymeric materials become conductive when illuminated with light. For instance, poly(N-vinylcarbazole), 16.4, is an insulator in the dark, but when exposed to UV radiation it becomes conductive. Addition of electron acceptors and sensitizing dyes allows the photoconductive response to be extended into the visible and NIR regions. In general, such photoconductivity is dependent on the material's ability to create free-charge carriers, electron holes, through absorption of light, and to "move" these carriers when a current is applied.

(16.4)

Poly(N-vinylcarbazole)

Related to this photoconductivity are materials whose response to applied light varies according to the intensity of the applied light. This kind of behavior is referred to as nonlinear behavior. In general, polymers with whole chain delocalization or large area delocalization where electrons are optically excited may exhibit such nonlinear optical behavior.

A photoresponsive sunglass whose color or tint varies with the intensity of the sunlight is an example of nonlinear optical material. Some of the so-called smart windows are also composed of polymeric materials whose tint varies according to the incident light. Currently much material is stored using electronic means but optical storage is becoming commonplace with the use of CD-ROM and WORM devices. Such storage has the advantages of rapid retrieval and increased knowledge density (i.e., more information stored in a smaller space).

Since the discovery of doped polyacetylene becoming electronically conductive, a range of polymer-intense semiconductor devices has been studied including normal transistors and field-effect transistors, FETs, and photodiodes and light-emitting diodes, LEDs. Like conductive polymers, these materials obtain their properties because of their electronic nature, specifically the presence of conjugated pi-bonding systems.

In electrochemical light emitting cells, the semiconductive polymer can be surrounded asymmetrically with a hole-injecting material on one side and a low work function electron injecting metal (such as magnesium, calcium, or aluminum) on the other side. The emission of light results from a radiative charge carrier recombining in the polymer as electrons from one side and holes from the other recombine.

Poly(p-phenylene vinylene), PPV; 16.5, was the first reported (1990) polymer to exhibit electroluminescence. PPV is employed as a semiconductor. A layer of PPV is sandwiched between a hole-injecting electrode on one side and an electron-injecting metal on the other side. PPV has an energy gap of about 2.5 eV and thus produces a yellow–green luminescence. Today, other materials are available that give a variety of colors.

$$ (16.5) $$

Poly(p-phenylene vinylene)

A number of poly(arylene vinylene), PPV, derivatives have been prepared. Attachment of electron-donating substituents, such as dimethoxy groups (16.6), acts to stabilize the doped cationic form and thus lower the ionization potential. These polymers exhibit both solvatochromism (color change as solvent is changed) and thermochromism (color is temperature dependent).

$$ (16.6) $$

Poly(2,5-dimethoxy-p-phenylene vinylene)

16.13.2 ELECTRICALLY CONDUCTIVE POLYMERS

The search for flexible, noncorrosive inexpensive conductive materials has recently focused on polymeric materials. This search has increased to include, for some applications, nano-sized fibrils, and tubes. The conductivity for general materials is noted in Figure 16.14. As seen, most polymers are nonconductive and in fact, are employed in the electronics industry as insulators.

The Nobel Prize in Chemistry for 2000 was given to Alan MacDiarmid, Alan Heeger, and Hideki Shirakawa for the discovery and development of electrically conductive polymers. In 1975, MacDiarmid and Heeger began studying the metallic properties of inorganic poly(sulfur nitride) but shifted their efforts to polyacetylene after visiting with Shirakawa. While the synthesis of polyacetylene was known for years, Shirakawa and coworkers, using a Ziegler–Natta catalyst, prepared it as a silvery film in 1974. However, in spite of its metallic appearance, it was not a conductor. In 1977, using techniques MacDiarmid and Heeger developed for poly(sulfur nitride) Shirakawa, MacDiarmid, and Heeger were able to increase the conductivity of *trans*-polyacetylene

FIGURE 16.14   Electrical conductivity for various materials.

samples, after doping, to a conductivity of about 10^3 S/m, or 1 S/cm. They found that oxidation with chlorine, bromine, or iodine vapor made polyacetylene film 10^9 times more conductive than the nontreated film. This treatment with a halogen was called "doping" by analogy with the doping employed with semiconductors. Other oxidizing "doping" agents have been used including arsenic pentafluoride. Reducing agents such as metallic sodium have also been successfully used. This chemical doping transforms the polyacetylene from an insulator or a semiconductor to a conductor.

A critical structural feature for conductive polymers is the presence of conjugated double bonds. For polyacetylene, every backbone carbon is connected by a stronger localized sigma-bond. They are also connected by a network of alternating less localized and weaker pi-bonds. While conjugation is present, it is localized enough to prevent ready delocalization of the pi-bond electrons. The dopants cause the introduction of sites with increased electrons or deficiency of electrons. When there is a deficiency of electrons, or hole created, electrons flow to fill this hole with the newly created hole causing other electrons to flow to fill the new hole, etc. allowing charge to migrate within and between the polyacetylene chains.

In general, the Huckel theory predicts that pi-electrons can be delocalized over an entire chain containing adjacent pi-bonds with the activation energy for delocalization decreasing as the chain length increases. Doping provides a ready mechanism for delocalization to occur.

Ordinary polyacetylene is composed of small fibers (fibrils) that are randomly oriented. Conductivity (Sections 12.7 and 12.8) is decreased because of the contacts between the various random fibrils. Two approaches have been taken to align the polyacetylene fibrils. The first approach is to employ a liquid crystal solvent for the acetylene polymerization and to form the polymer under external perturbation. The second approach is to mechanically stretch the polyacetylene material causing the fibrils to align. The conductivity of polyacetylene is about 100 greater in the direction of the "stretch" in comparison to that perpendicular to the stretch direction. Thus, the conductivity is isotropic. By comparison, the conductivity of metals such as copper and silver is anisotropic. Of interest is the nonconductivity of diamond, which has only ordered sigma-bonds and thus no "movable" electrons and the isotropic behavior of graphite. Graphite, similar to polyacetylene, has a series of alternating pi-bonds (Equation 10.6) where the conductivity in the plane of the graphite rings is about 10^6 times that at right angles to this plane.

Polyacetylene has been produced by several methods, many utilizing the Zeigler–Natta polymerization systems. Both *cis* and *trans* isomers exist (16.7 and 16.8). The *cis*-polyacetylene is copper-colored with films having a conductivity of about 10^{-8} S/m. By comparison, the *trans*-polyacetylene is silver-colored with films having a much greater conductivity on the order of 10^{-3} S/m. The *cis* isomer is changed into the thermodynamically more stable *trans* isomer by heating. As noted earlier, conductivity is greatly increased when the *trans*-polyacetylene is doped (to about 10^2–10^4 S/cm). Conductivity is dependent on the micro- or fine structure of the fibrils, doping agent, extent, and technique, and aging of the sample.

$$(16.7)$$

$$(16.8)$$

Polyacetylene was initially produced using Ziegler–Natta systems producing what have become known as Shirakawa polyacetylene. These materials are not easily processable and are mainly fibrillar. Recently other approaches have been taken. In the Durham route, the metathesis polymerization of 7,8-bis(trifluoromethyl)tricyclo[4.2.2.0]deca-3,7,9-triene gives a high molecular weight soluble precursor polymer that is thermally converted to polyacetylene (16.9). The precursor polymer is soluble in common organic liquids and is easily purified by reprecipitation. The end product can be aligned giving a more compact material with bulk densities on the order of 1.05–1.1 g/cc.

$$(16.9)$$

Grubbs and others have used the ring-opening metathesis polymerization to produce thick films of polyacetylene and polyacetylene derivatives (16.10).

$$(16.10)$$

Polyacetylene has good inert atmospheric thermal stability but oxidizes easily in the presence of air. The doped samples are even more susceptible to air. Polyacetylene films have a lustrous, silvery appearance and some flexibility. Other polymers have been found to be conductive. These include poly(p-phenylene) prepared by the Freidel–Crafts polymerization of benzene, polythiophene and derivatives, poly(phenylene vinylene), polypyrrole, and polyaniline. The first polymers commercialized as conductive polymers were polypyrrole and polythiophene because of their greater stability to air and the ability to directly produce these polymers in a doped form. While their conductivities (often on the order of 10^4 S/m) are lower than that of polyacetylene, this is sufficient for many applications.

Doped polyaniline (16.11) is employed as a conductor and as an electromagnetic shielding for electronic circuits. Poly(ethylenedioxythiophene), PEDOT, doped with polystyrenesulfonic acid is used as an antistatic coating material to prevent electrical discharge exposure on photographic emulsions and is also used as a hole injecting electrode material in polymer light-emitting devices. Organic soluble substituted polythiophenes with good conductivities have been prepared. Poly(3-hexylthiophene) has a room temperature conductivity of about 100 S/cm; poly(3-metnylthiophene) has a conductivity of 500 S/cm; and a poly(3-alkylether)thiphene) with a conductivity of about 1000 S/cm reported. The unsubstituted polythiophene has a conductivity in the range of 50–100 S/cm. The fact that all of these substituted polythiophenes have similar conductivities indicates that there is little twisting of the backbone conjugation as alkyl substituents are added.

Polypyrrole (16.12) is being used as microwave-absorbing "stealth" screen coatings and in sensing devices. Poly(phenylene vinylidene) derivatives are being used in the production of electroluminescent displays. Polythiophene (16.13) derivatives are being used in field-effect transistors.

Following are the structures of some of the more common conjugated polymers, along with poly(acetylene), which can be made conductive through doping. As noted before, doping causes and electrical imbalance that allows electrons to flow when an electrical potential is applied. The band gap is the energy needed to promote an election from the valence band to the empty energy or conductive band. Metals have zero band gaps while insulators such as polyethylene have large band gaps meaning that a lot of energy is needed to promote an electron to an empty band. Semiconductors have small band gaps where valence electrons can be moved into the conductance band through the application of relatively small potential energies.

Polyalanine (16.11)   Polypyrrole (16.12)   Polythiophene (16.13)

The optical behavior can be used to help in the understanding of these conductive materials. Pi-conjugation often occurs in the visible region so that most, if not all, of these conductive polymers are colored. Color changes are an important probe in assisting and characterizing the effects of various doping agents. Changes in the spectra allow various mechanisms to be studied. Because many of the polythiophene derivatives are organic soluble, spectral changes can be more easily studied. The electronic absorption spectra of a variety of polythiophene derivatives indicate that the band edge for conductivity begins about 2 eV.

Water-soluble derivatives of polythiophene have been made allowing counterions bound to the polymer backbone to "self-dope" with the protons (such as lithium and sodium ions) injecting electrons into the pi-system. Thus, combinations of sodium salts and so-called proton salts (such as prepared from poly-3-(2-ethanesulfonate) thiophene) have been prepared that are both water soluble and conducting.

Another area of activity involves the synthesis of material with small band gaps that would allow activation to occur at room temperature without doping. Polyisothianaphthene has been produced with a band gap of about 1 eV. More recently polymers with alternating donor and acceptor units with band gaps of about 0.5 eV and less have been developed.

While the amount of electricity that can be conducted by polymer films and "wires" is limited, on a weight basis the conductivity is comparable to that of copper. These polymeric conductors are lighter, some more flexible, and they can be "laid down" in "wires" that approach being one atom thick. They are being used as cathodes and solid electrolytes in batteries, and potential uses include in fuel cells, "smart" windows, nonlinear optical materials, light-emitting diodes, conductive coatings, sensors, electronic displays, and in electromagnetic shielding.

There is a large potential for conducting polymers as corrosion inhibiting coatings. For instance, the corrosion protection ability of polyaniline is pH dependent. At lower pH's polyaniline-coated steel corrodes about 100 more slowly than noncoated steel. By comparison, at a pH of about 7 the corrosion protection time is only twice for polyaniline-coated steel. Another area of application involves creation of solid-state rechargeable batteries and electrochromic cells. Polyheterocycles have been cycled thousands of times with retention of over 50% of the electrochromic activity for some materials after 10,000 cycles. Infrared polarizers based on polyaniline have been shown to be as good as metal wire polarizers.

They will also find uses in nonlinear optical devices such as in optoelectronics, that is, for signal processing and optical communication. Some of the new conducting polymers offer such benefits as flexibility, high damage threshold, ultra-fast response in the subpicosecond range, and good mechanical strength. Polyheterocyclic conducting polymers have shown a wide variation in color as they are electronically converted between oxidized and reduced forms. The instability of some of the polymers is being used to monitor moisture, radiation, mechanical, and chemical destruction.

16.14 DRUG DESIGN AND ACTIVITY

Most drugs are smaller molecules. However, there are some drugs that are polymeric offering advantages over smaller molecules. Polymer-containing drug formulations are common. Here, the drug is contained within some polymeric matrix that assists in controlling the release either through the control of drug diffusion through the chains or by erosion, degradation or solubility, of the polymer network.

Along with polymers being simply depositories of the drugs, polymers can perform active roles in drug therapy. Polymeric drugs can act in two divergent ways—as a control release agent and as a drug itself. The polymeric drug may act in a control release manner. Thus, a polymer that contains a known therapeutic portion can degrade releasing this portion over some time. Thus, a polymer that contains the drug L-dopa (16.14) used to treat Parkinson's disease degrades over time releasing L-dopa over a period of time.

(16.14)

Polymer containing L-dopa

The polymer may act as a drug itself or the polymer drug may also act in both roles.

Following is a brief discussion of the advantages of polymeric drugs that may apply to either mode of control release or/and as a drug itself. Because of their size, polymers with chain lengths of about 100 units and greater typically are unable to easily move through biological membranes, thus movement is restricted. This can result in limiting negative side effects, such as damage to the kidneys, because the polymer can reside only certain body sites. In addition, this limited mobility can assist in directing the polymer drug. This mobility also allows the use of lower and more level drug concentrations reducing renal, kidney, etc. damage. In some cases, multiple attachments by the same polymer chain can be beneficial. Thus, many cancer drugs act to "tie up" the DNA chains inhibiting cell replication. Here, multiple attachments to the DNA, possible for polymer drugs that contain numerous binding units within each single chain, might be beneficial in decreasing the cell's effectiveness in overcoming single DNA "chemical knots" (chemical bonding together of the two DNA strands).

The polymeric nature may inhibit premature drug deactivation. Thus, cisplatin (16.15), the most widely used anticancer drug, is converted into numerous inactive, but more toxic, platinum-containing compounds before it arrives at the targeted cancer cells. Placement of the active platinum-containing platinum moiety into a polymer (16.16) decreases this tendency to hydrolyze into these unwanted cisplatin compounds because of the greater hydrophobic character of the polymeric drug.

Cisplatin (16.15)   Polymer containing a cisplatin-like unit (16.16)

Polymers may also evade the microorganism's defense. Recently, it was found that many cells become drug resistant after repeated treatment. Cells have groups of molecules that protect them from outside invasion. The small molecule drugs are considered "outside" invaders. Thus, drugs introduced into their environment are considered as "outside" invaders and the cell manufactures greater numbers of these "house-keepers" to rid the cell of the invaders. These "house-keeping" agents are often not very specific and remove other similar drugs that intrude their cell. These "house-keeping" proteins may not be as effective at eliminating polymer drugs.

Polymer drugs are also known to remain within human hosts longer than smaller molecules. Depending on the illness and treatment, this preferential retention may be simply due to physical retention through entanglement with biological outer-layer materials or through some specific additional interaction. Tumor-associated cells are frequently hyper-permeable to plasma proteins and other macromolecules. These "leaky" vasculatures and limited lymphatic drainage, typical of tumor and missing in normal tissue, result in the accumulation of polymers. Thus, such polymers reside in the interstitial space of these cells. This activity results in enhanced permeability and retention, EPR, of large chains.

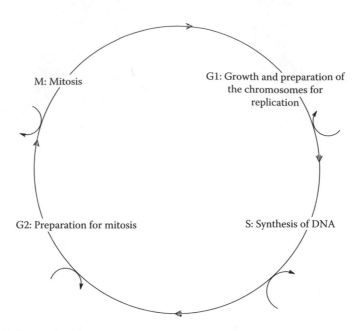

FIGURE 16.15 Cell growth cycle.

Drug design, today, typically aims at certain specific biological activities. Thus, for cancer treatment the focus is on controlling cell growth (Figure 16.15). Cell growth can be considered as occurring in four dependent steps. Drugs are designed to control cell growth at any one of these steps directly or indirectly. The indirect intervention of cell growth has as a target any one of many steps that influence the cell growth cycle. Essentially all chemotherapies are based on the continual growth of cancer cells compared with healthy cells generally being in some rest state.

Drugs aiming to control the same problem may target different sites. For instance, in the treatment of breast cancer, tamoxifen acts to control cell growth by blocking estrogen receptors on the cancer cell while arimidex acts to reduce the body's estrogen production.

16.15 SYNTHETIC BIOMEDICAL POLYMERS

Synthetic polymers have been studied for some time for their use in the general field of medicine. Initial results were average. More recently, because of a better understanding of the importance of surface, avoidance of contaminants, and control of molecular weight and molecular weight distribution, the use of synthetic polymers in the biomedical sciences is increasing. Polymers are viewed as important biomedical materials for a number of reasons, some of which appear contradictory—but only because different uses require different properties. Some of the important properties are the ability to tailor-make structures, surface control, strength, flexibility, rigidity, inertness/reactivity, lightweight, ease of fabrication, ability to achieve a high degree of purity, lack of and because of their water solubility/compatability, bioerodablity, and the ability of some of them to withstand long-term exposure to the human body—a truly hostile environment. Fighting against some of the biomaterials are their limited (by volume) use—thus, researchers are often limited to using commercial materials made for other applications, but as the use of these important materials increases, manufactures will become more interested in tailor-making materials if for no other reason than the high cost per unit volume.

Long-term inertness without loss of strength, flexibility, or other necessary physical property is needed for use in artificial organs, prostheses, skeletal joints, etc. Bioerodability is needed when the polymer is used as a carrier such as in the controlled release of drugs, the removal of unwanted materials, or where the purpose of materials is short lived such as in their use as sutures and as frames for natural growth.

Material surface is critical. While the nature of the material is important, the surface of the material is critical. The human body wants to wrap around or connect to bodies within its domain. In some cases, the desired situation is little or no buildup on the polymer. Here, surface "slickness" is needed. Siloxanes and fluroinated materials such as PTFE are generally slick materials but other materials can be made slick through surface treatments that presents to the body few voids and irregularities at the atomic level. In other cases, bodily buildup is desired and surfaces and materials that assist this growth are desired. Surface hydrophobicity/hydrophilicity, presence/absence of ionic groups, chemical and physical (solid or gel) surface are all important considerations as one designs a material for a specific application.

The ability to function long term is an ongoing problem. In general, polyurethanes degrade after about 1.5 years, nylons lose much of their mechanical strength after about 3 years, and polyacrylonitrile loses about 25% strength after 2 years. On the long side, PTFE loses less than 10% strength after about a year and siloxanes retain most of their properties after 1.5 years.

Following is a brief look at some of the varied uses, real and projected, for synthetic polymers. These are intended to be illustrative only. The field of biomedical materials is rapidly growing and it is extensive.

There has been a lot of effort to construct artificial hearts for human use. These hearts are largely composed of polymeric materials. Even with the problems associated with organ transplant, the use of artificial hearts may never be wide spread. This is because of several trends. First, many biological solutions to biological problems reside with the use of biological materials to solve biological problems. With the advent of increased effectiveness of cloning and related advances suitable biological replacements may be grown from a person's cells; thus, ensuring compatability. Second, related to the first, the regeneration of essential parts of vital organs is becoming more practical. Third, because of nanotechnology and related electrical and optical advances, surgery to locate and repair imperfections is improving. Forth, our autoimmune system is divided into two main systems—one guards against "small" invading organisms such as viruses, bacteria and pollen. The second group acts to reject whole organisms such as the heart through the rejection of foreign body tissue. Purine nucleoside phosphorylase, PNP, is a human enzyme that serves at least two major functions. First, PNP acts to degrade unleashed molecules including foreign nucleic acids. PNP is necessary to our immune system as it fights disease. Some anticancer drugs are synthetic nucleosides and nucleotides such as polyIpolyC, that are employed to directly attack selected cancerous tumors. PNP degrades such nucleic acid-related materials before they reach the tumor. Neutralization of PNP just prior to administration of the synthetic nucleic acid-related material would allow lower dosages to be used. Second, PNP is an integral part of the body's T-cell immunity system that rejects foreign tissue. Effort is currently underway to effectively inhibit only the PNP action allowing the first autoimmune system to work while allowing organ transplantation without rejection. Recently, through the use of zero gravity conditions, crystals of PNP have been growth of sufficient size to allow structural determination. With this structure, efforts are underway to detect sites of activity and drugs that would allow these sited to be neutralized when needed.

Even so, synthetic polymers have been important in replacing parts of our essential organs. Thus, silicon balls are used in the construction of mechanical heart valves. Many of these fail after some time and they are being replaced by a flap valve made from pyrolytic carbon or polyoxymethylene.

Aneurisms can be repaired through the reinforcement of the artery wall with a tube of woven PTFE or polyester. The replacement of sections of the artery can be done using a tube of porous PTFE. One remaining problem is the difference in elasticity between the woven and porous materials and the arteries themselves.

Carbon fiber composites are replacing screws for bone fracture repair and joint replacements. These fiber composites are equally as strong and are chemically inert. By comparison, the metals they replace are often alloys that may contain metals that the patient may be allergic to.

Polymers are also used as sutures. Fighters and other athletics have used poly(α-cyanoacrylates), super glues, to quickly stop blood flow in surface cuts. Today, super glue is also used for, in place of, or along with more traditional polymeric suture threads for selected surface wounds, internal surgery, and in retinal and corneal surgery. The α-cyanoacrylates undergo anionic polymerization in the presence of water forming polycyanoacrylates (16.17).

Poly(α-cyanoacrylate)   (16.17)

Siloxanes are the most extensively used synthetic biomaterial. They are used for a number of reasons including flexibility, chemical and biological inertness, low capacity to bring about blood clotting, overall low degree of biological toxicity, and good stability within biological environments.

Artificial skin had been made from a bilayer fabricated from a cross-linked mixture of bovine hide, collagen, and chondroitin-b-sulfate derived from shark cartilage with a thin top layer of siloxane. The siloxane layer acts as a moisture and oxygen permeable support and to protect the lower layer from the "outer world" allowing skin formation to occur in conjunction with the lower layer. Poly(amino acid) films have also been used as an "artificial" skin. Research continues in search of a skin that can be effectively used to cover extensive wounds and for burn patients.

Controlled release of drugs using polymer intensive materials is becoming more commonplace. The release "pack" can be attached externally such as many of the "nicotine patches" that deliver controlled amounts of nicotine transdermally. The release "pack" can also be introduced beneath the skin or within the body as is the case with many diabetes treatment assemblies.

Controlled release of drugs can be envisioned as occurring via three major routes. One approach utilizes diffusion-controlled release through membranes or matrices. Here, the rate of release is controlled by the permeability of the membrane or matrix. In the second approach, the drug is captured within a matrix that undergoes degradation, usually through aqueous-assisted solubilization or degradation (including hydrolysis). Here, the rate of drug release is dependent on the break up of the typically polymeric matrix. For the second approach, a number of polymers have been used including poly(glycolic acid) and polyanhydrides. The third approach involves simple degradation of a drug-containing polymer where the drug moiety is present as part of the polymer backbone or as side chains. Degradation of the polymer results in the release of the drug in some fashion.

Implant materials can be divided into two general categories dependent on the time requirement. Those that are present for the release of a drug or to hold a broken bond in place until sufficient healing occurs are termed short-term implant materials. The second group includes materials that are to function over a longer time such as for the life of the patient. In the first case, degradation is generally required while for the longer term material inertness and long-term stability are typically required. There are times when this is not true. For instance, some of the newer biomaterials act as scaffolds that promote tissue growth by providing a three-dimensional framework with properties that encourage favorable cell growth. This material may be designed to be either short term or long term. One approach to designing scaffolding material involves placing certain amino acid-containing units on the polymeric scaffold that encourage cell growth.

Another aspect related to control the release of drugs concerns the type of structures that currently appear to be working. Not unexpectedly, because of compatibility and degradation purposes, most of the effort on the control release formulations includes polymers that have both a hydrophobic and hydrophilic portion with the material necessarily containing atoms in addition to carbon. Another concern is that the products of degradation are not toxic or do not go on to form toxic materials. It has also been found that amorphous materials appear to be better since they are more flexible and permit more ready entrance of potential degradative compounds.

Another area of activity involves the synthesis of supermolecular layers that are connected through cross-linking giving essentially one molecule thick micelles. Depending on the particular template and solvents employed, these monolayers can be designed to have almost any combination of hydrophilic and hydrophobic sites. Again, specific control of release rates, degradation times, and routes, biocompatibility, or incompatibility is possible. Many of these micelle-based delivery systems are based on a poly(ethylene oxide)-*b*-poly(propylene oxide)-*b*-poly(ethylene oxide) triblock or on a polypeptide and poly(ethylene oxide) combination. Drug delivery has also been achieved using conducting electroactive polymers formed through the controlled ionic transport of counterions (dopants) in and out of membranes.

Hydrogels that shepherd drugs though the stomach and into the more alkaline intestine have been used. Hydrogels are cross-linked, hydrophilic polymer networks that allow the smaller drugs access to their interior and that can be designed to inflate, swell at the desired site, to deliver the drug. These hydrogels have largely been formed from materials with a poly(acrylic acid) backbone.

Along with the organic materials, a number of inorganic materials are also being considered as degradable implant material. A porous form of β-tricalcium phosphate had been used for the repair of bone voids and related defects in the spine, pelvis, arms, and legs.

Moreover, included in the general grouping of biomaterials are the new electronics that are being developed including nanotechnology. These electronic biomaterials will need to be encapsulated sufficiently to protect them from the body and to protect the body from hostile actions against the presence of the foreign object or they will need to be made with a coatings material that will act as an encapsulating material to the electronics device. Special care will need to be taken with respect to adhesive materials that connect the electrical device to the particular site for activity and/or interactions between the electromagnetic messages and surrounding tissue. Again, adhesive, encapsulating, and shielding materials will be largely polymeric.

A number of siloxane-containing controlled release packs have been devised and are being used. Glaucoma, motion sickness, and diabetics have been treated using drugs dispersed in a silicon matrix. This kind of pack needs to be placed near the site of intended activity for greatest effectiveness.

Elastomeric siloxanes have also been used in encapsulating drugs, implant devices, and in maxillofacial applications to replace facial portions lost through surgery or trauma. Transcutaneous nerve simulators are made from "conductive" siloxanes. These are employed in the treatment of chronic severe pain through the application of low-level voltage to the nerves disrupting the transmission of pain impulses to the brain. Siloxanes are also used in extracorporeal blood oxygenation employed in the heart–lung assist machine that is routinely used in open-heart surgery. The "heart" of the apparatus is the membrane that must allow ready transport of oxygen and carbon dioxide yet retain moisture and the blood cells. The siloxane membranes can be made using a polyester or PTFE screen in an organic dispersion of silicon rubber. When dried, thin films are obtained that are used in the heart–lung assist device.

Silioxane-containing devices have been also used as contact lenses (Section 11.3), tracheostomy vents, tracheal stents, antireflux cuffs, extracorporeal dialysis, ureteral stents, tibial cups, synovial fluids, toe joints, testes penile prosthesis, gluteal pads, hip implants, pacemakers, intra-aortic balloon pumps, heart valves, eustachian tubes, wrist joints, ear frames, finger joints, and in construction of brain membranes. Almost all of the siloxane polymers are based on various polydimethylsiloxanes.

The kidney removes waste material from our blood. Because of partial or total kidney failure, many persons are on hematolysis. The first hemodialysis units were large and by today's standards not very effective and the semipermeable tubes, made mainly of cellophane, had very limited lifetimes. Initially, dialysis treatment was expected to be for only terminal patients but as the life expectancy of dialysis patients increased as did the demand for dialysis, smaller, more efficient dialysis machines have emerged. At the heart of these advances is the filtering material. Today bundles of micro-hollow fibers are used in the construction of hemodialysis cells. The fibers are "heparinized" to discourage blood clotting. The fibers are mainly made of polyacrylonitrile. Polycarbonate, cellulose acetate, and rayon fibers are also being used.

16.16 DENTAL MATERIALS

Polymers are used extensively in dentistry. Somewhat permanent dental materials exist in a hostile environment, the mouth, and often must resist fracture and wear under extreme forces. They must perform without fatiguing. They must also be compatible with the biological environment.

The original modern day filling or amalgams contain various metals and alloys. More recently, composite resins often referred to as plastic or white fillings are being employed. They often consist of a mixture of powered glass and polymer resin. They are strong, compatible, resistant, and cosmetically approach the appearance of natural teeth. Even crowns are generally metallic with a ceramic coating to approach the appearance of natural teeth.

Dental restorative composites generally consist of di- and tri-functional monomer systems that provide the cross-linking necessary to form a strong matrix once polymerized. Reinforcing fillers such as silanized quarts, ceramics; hybrid fillers such as prepolymerized resins on fumed or pyrogenic silica; and ceramics act as the dispersed phase of the composite. The diluent or continuous phase is the mono- and difunctional monomer that decreases the viscosity of di- and tri-functional monomers that finally act to form the stable composite matrix. Dental sealants are similar except the filler material is either not present or present in lesser concentration. Many of the restorative composites are based on modified dimethacrylates, the most common being 2,2-bis[p-(2'-2-hydroxy-3'-methacryloxypropoxy)phenyl]-propane, bis-GMA (16.18), and triethyleneglycol dimethacrylate, TEGDMA (16.19).

(16.18)

bis-GMA

(16.19)

TEGDMA

These two are used either separately or as a mixture. Because of the presence of the somewhat flexible ethylene oxide and related units and the use of appropriate fillers, these materials give composite fillings with lower polymerization shrinkage, enhanced mechanical properties, lower solubility and water adsorption, better thermal expansion characteristics, good biocompatibility, with aesthetic properties closely matching those of the tooth itself.

Some other fillings employ urethanedimethacrylate (1,6-bis(methacryloxy-2-ethoxy carbonylamino)-2,4,4-trimethylhexane), UDMA in place of bis-GMA. This is an active area of research with new monomer systems being introduced in an ongoing manner.

Cavity varnishes are used to seal the exposed dentinal tubules and protect the pulp from the irritation of chemicals in the filling materials. They are generally largely natural rubber or a synthetic polymeric resin such as 2-hydroxyethyl methacrylate, HEMA.

Fillings should be viewed as nonpermanent with a lifetime often of a decade. Many materials are employed in fillings including metals, alloys, amalgams, composites, and glass ionomer cements. Glass ionomer cements are a mixture of glass and an ionomer referred to as a polyalkenoic acid, which is simply an ionomer with carboxylic salt groups as the chelating moieties. The ionometric polymer chelates both the tooth and glass binding them together into the filling. Today's glass ionomer cements are especially designed for dental application. For instance, many contain releasable fluoride that prevents carious lesions with the fluoride content recharged using fluoride-containing toothpaste. Newer formulations containing light-cured resins do not release fluoride as readily.

Almost all denture bases are made of methacrylic (acrylic) resins that give good fit and a natural appearance. A compression molding process is used where the monomer–polymer dough or slurry contain poly(methyl methacrylate) or poly(methyl acrylate). Often there is a change in the contour of the soft tissue and a liner is fitted onto the denture base. Silicon reliners are often used for this purpose.

Plastic acrylic denture teeth are made by injection or transfer molding. Acrylic teeth have a higher strength than porcelain teeth and break less readily. However, they cold flow, have a greater water absorption, and they have a higher wear rate than porcelain teeth.

Many of the dental polymer cements are glass-ionomer combinations made from ionomer-forming polymers that contain acid groups such as poly(acrylic acid), poly(itaconic acid) (16.20), poly(maleic acid) (16.21), and poly(2-butene-1,2,3-tricarboxylic acid) (16.22). These polyalkenoate cements are set up through reaction with an alumino-silicate-fluoride glass with the polyacids hydrolyzing the glass network releasing the aluminum and calcium ions and forming a silaceous hydrogel. The acid groups chelate with the released metal cations forming a cross-linked matrix.

Poly(itaconic acid)
(16.20)

Poly(maleic acid)
(16.21)

Poly(2-butene-1,2,3-tricarboxylic acid)
(16.22)

16.17 EMERGING POLYMERS

A number of small-scale polymeric materials will continue to enter the marketplace on a regular basis. These include biomaterials and electronics materials where the cost per pound is high and the poundage is low, generally well less than a 100 tons a year. These are materials that fulfill specific needs.

The number of new larger scale giant molecules that enter the marketplace will be small. It has been estimated that it takes about $1 billion to introduce and establish a new material. It is a daunting task with no guarantee of success. In the past, new giant molecules could be introduced that offered improvements in a number of areas and thus would attract a market share in a number of application areas. Today, there are already a wide range of materials for most large-scale application areas that compete for that particular market share so that it is difficult for any material to significantly break into any market area. A new material needs a "flagship" property that a particular market needs.

Dupont and Shell have developed a new polyester, poly(trimethylene terephthalate) (PTT; 16.23) that is structurally similar to poly(ethylene terephthalate), PET, except that 1,3-propanediol, PDO is used in place of ethylene glycol. The extra carbon in Sorona allows the fiber to be more easily colored giving a textile material that is softer to the touch with greater stretch. Further, it offers good wear and stain resistance for carpet use. The ready availability of the monomer PDO is a major consideration with efforts underway to create PDO from the fermentation of sugar through the use of biocatalysts for this conversion. Sorona and Lycra blends have already been successfully marketed. Sorona is also targeted for use as a resin and film.

(16.23)

Poly(trimethylene terephthalate), PTT

DSM introduced in 1990 nylon 4,6 called Stanyl (16.24), based on the reaction between adipic acid and 1,4-diaminobutane. Stanyl can withstand temperature of about 310°C allowing it to create a niche between conventional nylons and high-performance materials. It was not able to break into the film market and has only now begun to be accepted for tire cord applications. About 22 million pounds of Stanyl was produced in 2001.

(16.24)

In 2007, DSM announced the development of PA4T nylon (16.25). It is intended to address new needs for miniaturization and the convergence of electronic devices such as cell phones and computers. It offers good dimensional stability, compatibility with lead free soldering, high melting point, good mechanical strength at elevated temperatures, high stiffness, and good processability. Specific applications include CPU sockets, high temperature bobbins, notebook computer memory module connectors, and memory card connectors.

(16.25)

PA4T

The development of PA4T nylon illustrates the use of polymers in the intermediate synthesis scale, larger than specialty chemical use and smaller than bulk chemical use. The development of polymers that fits into this category will greatly increase as special properties are needed and the price per weight is high.

In 1997, Dow introduced syndiotactic polystyrene under the trade name Questra. The technology for the production of Questra is based on relatively new technology and science involving soluble stereoregulating catalysts that produce polystyrene that has a fixed and repeating geometry as each new styrene monomer unit is added to the growing polystyrene chain. Targeted areas include medical, automotive, and electronic applications.

Several other produces have been developed based on the relatively new soluble stereoregulating catalysis systems. Index, an ethylene–styrene interpolymer, was introduced in 1998 and is intended to compete with block copolymers such as styrene–butadiene–styrene, flexible PVC, polyurethanes and polyolefins. It is being used as a modifier for polystyrene and polyethylene. Dow is also developing soundproofing and packaging foam applications for Index. Hoechst Celanese (now Ticona) developed Topas, a cyclo-olefin copolymer, in the 1980s and in 2000 began commercial production of it. Topas has high moisture-barrier properties and is being considered for use in blister packaging for pharmaceuticals. It is also being used in resealable packages where it provides stiffness to the sealing strip. Other uses include in toner resin applications and it is being blended with linear low-density polyethylene and low-density polyethylene providing stiffness and improving sealing properties.

A number of new materials are looking toward being involved in the upcoming move toward blue-light CDs. For any of these to become important materials in this area they will need to improve on the present polycarbonate-based materials.

GE introduced in 2000 a new polyester carbonate based on resorcinol arylates it called W-4. It is now marketed as Sollx. Sollx does not need to be painted, it offers good weather, chip, scratch, and chemical resistance and is being used as the fenders for the new Segway Human Transporter. It is also aimed at automotive uses including as body panels. Sollx is coextruded into two layers, one clear and one colored, to simulate automotive paint. It is then thermoformed and molded into the finished product.

A number of new materials have been developed because of the health fears associated with the monomer Bisphenol A, which is the comonomer for most polycarbonates. The replacement should possess similar properties to polycarbonates and also be available in large quantity and inexpensive.

One material that has become available is a copolymer polyester developed by Eastman. The precise structure for this material, Tritan, is proprietary, but is believed to be based on the diol tetramethylcyclobutandiol, 16.26. The ring system contributes the necessary stiffness and the methylenes supply the flexibility.

$$\text{(16.26)}$$

Several new ventures are based on using natural, renewable materials as the starting materials instead of petrochemicals. These products are known as "green" products since they are made from renewable resources and they can be composted. Along with the microbial production of 1,3-propanediol by Shell and Dupont to produce nylon 4,6, Cargill Dow is making PLA beginning with corn-derived dextrose. The polylactide, PLA (16.27), is made from corn-derived dextrose, which is fermented making lactic acid. The lactic acid is converted into lactide, a ring compound that is polymerized through ring opening.

$$\text{(16.27)}$$

Lactic acid   Lactide   Polylactide, PLA

PLA looks and processes like polystyrene. It has the stiffness and tensile strength of PET, and offers good odor barrier and resists fats, oils, and greases. PLA is being considered for use in fibers and in packaging. As a film, PLA has good deadfold properties, that is, it has the ability to be folded and to stay folded. It is being used as a fiber for apparel and carpeting applications. It is being sold as a bridge between synthetic and natural fibers in that it processes like synthetic fibers but has the touch, comfort, and moisture-management of natural fibers.

Many of the so-called new materials are actually "old materials" arranged in different settings. Following is a brief description of three of these materials. Trgris, developed by Milliken, is a polypropylene, PP, thermoplastic composite. It is made from coextruded PP tape yarn that has a highly drawn core creating strength additional strength for the PP yarn. The pulling encourages better alignment of the PP chains. The tape yarn is then woven into a fabric, layered, and finally heated with nonyarn PP, with pressure applied forming rigid sheets that are used for panel applications or premold materials. Thus, the yarn acts as the discontinuous phase with the nonoriented PP used as the continuous phase forming a PP–PP composite. In some fashion, Trgris is similar to plywood in that various layers are stacked, generally with the sheets placed at different angles to one another. The entire PP composite is strong and it is lighter than most composites. It also has good impact resistance since the entire composite is more intimately mixed allowing more efficient transfer of impact energies. Because it is one single material, it is easy to recycle lending to it being a more environmentally friendly material. It is being used in automotive applications including at the racetrack. Because of its outstanding impact resistance, it is being used as the race car splitter and to reinforce other parts of the race cars. If it does fail, unlike other composite materials that often shatter leaving unwanted debris on the track, Trgris materials retain their entirety. Further, because of the layering, fractures do not propagate throughout the entire Trgris part. Other uses are in transportation as liners and floors; in water sports, in the construction of small boats and watercraft; in construction of blast resistance materials; and it is being used in the construction of playground and outdoor objects.

A second material recently introduced as Impaxx is an energy absorbing foam developed by Dow Automotive. It reduces weight, space, cost, and improves safety. It is composed of foamed polystyrene, PS, so is similar to disposable coffee cups. Impaxx is a closed foam material. On impact, the closed foam cells compress and buckle absorbing energy and when excessive impact energy is applied the cells rupture again absorbing energy. Impaxx is used in over 2 million automobiles including the Honda Pilot, Chevy Malibu, and Ford Crown Victoria as well as a number of racing cars. It is used as headliners and pillar trims, steering column and lower instrument panels, under carpet, front and rear doors, and in bumper systems. Again, it is advertised as being green friendly since it is all PS and can be readily and easily recycled.

CarbonX is made by Chapman Thermal Products and is in competition with Nomex and similar materials as a flame resistant material. CarbonX is made by a controlled heating of polyacrylonitrile forming a fully oxidized fused polycyclic composition similar to that described in Section 10.12. The material can be heated to 500°C, 1000°C, or higher giving a material that is then stable to that temperature since material that is unstable has already been driven off. The oxidized polyacrylonitrile, OPAN, is forced through small holes resulting in the formation of tiny fiber-like materials that are not particularly flexible. Thus, the OPAN is often woven with other fibers including steel fibers, giving the material additional strength and flexibility.

Even if some burning occurs, the OPAN fabric expands removing oxygen. CarbonX materials are replacing some of the fabrics previously made by Nomex. Nomex and CarbonX garments are made for use by fire fighters, steel mill workers, military, racetrack drivers and other personnel, and police and include undergarments, socks, suits, head gear, shoelaces, and escape blankets. CarbonX comes in your choice of colors as long as you want black. Other colors are achieved through blending colored fibers in with the CarbonX fibers.

16.18 GREEN MATERIALS

There is an increasing emphasis in so-called "green materials." There are other terms often associated with the term green materials. These other terms include renewable resource materials and natural materials. The emphasis is the replacement of nongreen materials by these green materials. Each of these terms has slightly different meanings. The term "renewable materials" is generally employed for rapidly renewable materials such that are replenished within a short time such as a year. The term natural materials emphasizes materials that are derived from nature. Along with these descriptions, a green material also encompasses the energy requirements, processing procedures, and recycling ability of the material.

Natural materials include materials that are "naturally" found about us. This includes oil, coal, and natural gas. These materials are not considered green materials because they are not readily replenished. Other natural materials include silicon dioxide-intense materials such as mica that is utilized in some coatings producing clad-like coverings and diatomaceous earth that is being employed in the construction of some tires to retain tire pressure. It also includes many readily renewable materials such as cellulose, chitosan, vegetable oils, and lignin. These renewable materials are readily available in the billions of tons yearly, renewable on a regular yearly cycle, and are greatly underused materials. Carraher and others have worked with chitosan to produce anticancer drugs and lignin to produce materials with structural integrity as alternative industrial materials. Sperling and others have employed naturally derived oils producing various rubbers.

Here we will describe other green materials. We emphasize polymers that are commercially available and that are really derived from natural sources, not simply said to be possibly derived from natural sources. Nylon-66 was initially advertised to be derivable from various natural sources, and it is, but today it is manufactured from monomers that are part of the stream of petroleum-derived feedstocks. Thus, nylon-66 is not included in this discussion as a green material though it may become part of the green material revolution as natural, biomass-derived sources for the reactants become available on a large scale.

The movement from a petroleum-based material to a green material is not straightforward and involves many considerations. It is a journey that we need to be careful of but a journey we need to begin. We also need to be aware that the large bulk of petroleum-based use is not as materials but rather as fuel.

There are some concerns that need to be considered in the production of products from green materials. A major emphasis in the production of materials from synthetic polymers is that they are inert, not offering unwanted biological activity and decomposition. The idea that green materials should be naturally recyclable runs against this concept. Another concern involves the actual net energy and resources necessary to produce the monomers/polymer/product. An analysis needs to be made concerning the various components such as energy and resource (including water) requirements before a green material is accepted as a replacement. Another consideration is the ready availability of the particular green material. A material may be green but if it is not readily available or can be seen as readily available in sufficient amount then it may not be a strong candidate to be employed on a large-scale commercial basis. In addition, the behavior or properties of the green material should at least approach the materials that may be replaced.

As our bioengineering revolution matures, it should be possible to employ various biomasses as resources that can be converted into classical monomers such as ethylene, propylene, styrene, terephthalic acid, etc. from reactions caused by genetically designed microbes. In fact, it may be possible for combinations of microbes to produce finished polymers that may have better tacticity and chain orientation producing polymers with superior physical properties.

The latest major emphasis on a major green material is polylactide, PLA, discussed in the previous section. The use of rayon and other material derived from cellulose has also been discussed.

There are many vegetable oils that are currently available on an industrial scale. These include palm, soybean, cottonseed, castor, and rapeseed oils. While these have been employed in the production of many commercial materials such as coatings, pharmaceuticals, plasticizers, and building materials the particular fatty acids are being used as polymeric materials. Three of the five most common fatty acid substituents, oleic acid, linoleic acid, and linolenic acid, are unsaturated with these sites of unsaturation available for cross-linking. These are pictured in Figure 16.16. These fatty acids have been directly polymerized through the double bond(s).

Castor oil produces ricinoleic acid on its hydrolysis. Ricinoleic acid has a single olefinic site as well as both an acid and alcohol group (Figure 16.16). Thus, it can be polymerized by reaction through its double bond or by condensation, as shown by Carraher and coworkers, with a Lewis acid. Polyesters have also been produced through the formation of the lactone followed by ring opening polymerization, ROP.

One area of activity involves the production of polyurethanes employing fatty acids as well as the oils themselves. Often the oil is functionalized with hydroxyl groups through reaction at the unsaturated sites. The reaction is analogous to that employed to produce a number of soft–hard block polyurethanes.

Carbon dioxide is readily available and renewable. It is employed in the production of polycarbonates but can also be utilized in the production of various polymers through reaction with different heterocycles such as aziridines, episulfides, and epoxides. The production of polypropylene carbonate, PPC (16.28) has been known for about 40 years but it has yet to be commercially produced.

$$O=C=O \ + \ \underset{CH_3}{\triangle} \longrightarrow R-\underset{CH_3}{\overset{O}{\underset{|}{C}}}\text{...} \qquad (16.28)$$

FIGURE 16.16 Vegetable oil derived fatty acids.

There are a number of biodegradable polyesters including poly(glycolic acid) (PGA; 16.29), PLA, poly-3-hydroxybutyrate; PHB; 16.30), and polycaprolactone (PCL; 16.31). PGA, PLA, and PCL are synthesized from the acid-catalyzed ring opening polymerization, ROP, of the internal ester. PHB is made from microorganisms such as *Alcaligenes eutrophus* and *Bacillus megaterium* from natural materials such as starch or glucose. PHB has properties similar to those of polypropylene. It is stiff and brittle, has a relatively high melting point of about 180°C, and unlike PP it is biodegradable. However, PP is much tougher. PHB is commercially available from Metabolix. It is an example of a larger grouping of synthetic polyesters called PHAs.

PGA (16.29) PHB (16.30) PCL (16.31)

This theme of employing bioengineers, namely bacteria, to produce polymers is illustrated in the synthesis of a group of over 150 linear polyesters called polyhydroxyalkanoates, PHAs; 16.32. Over 150 species of both gram-negative and gram-positive bacteria have been found to produce a variety of PHA. For instance, *Alcaligenes eutrophus* is placed in a medium that allows it to rapidly reproduce.

After a sufficient population is obtained, the conditions are changed resulting in the production of a range of PHBs with general structures as shown later. *Alcaligenes eutrophus* typically produces so-called short-chain PHAs where $n = 3–5$. By comparison, *Pseudomonas oleovorans* produce longer-chained PHAs where $n = 6–14$.

$$\text{(16.32)}$$

PHA general structure

PHB has $n = 1$ and R = methyl. PHB is biocompatible and is a metabolite normally present in blood. Thus, a wide range of biomaterial uses can be envisioned for it. The copolymer of PHB and PHV (where R = ethyl and $n =1$), poly(-3-hydroxybutyrate-co-3-hydroxyvalerate) is considered as a packaging material. Compared to PHB the copolymer is tougher, less rigid, and easier to process.

The general properties can be predicted knowing the values of R and n. As expected, the smaller the value of n the stiffer the backbone and the larger the R the greater the flexibility because the ability of the chains to coalesce is less because of greater steric requirements. Values of X generally range from 100 to over 30,000.

PHAs are viewed by microorganisms as an attainable energy source so are biodegradable. The end products under aerobic conditions are carbon dioxide and water while methane is produced in anaerobic environments. They can also degrade under simple physical conditions with the ester moiety attacked by water, acid, and base.

Both nylon-9 and nylon-11 (16.33) can be synthesized from biomaterials. Nylon-11 is commercially available and synthesized from monomers produced from vegetable oils, particularly from the castor plant.

$$\text{(16.33)}$$

Nylon-11

Dupont has developed several polymers derived from monomers created from renewable resources. Sorona, 3GT, is a thermoplastic polyester fiber based on the reaction between 1,3-propanediol, PDO (16.34), where the PDO is created from corn, and terephthalic acid creating the poly(trimethylene terephthalate), 3GT (16.35). The corn is converted to sugar and the sugar is fermented in the presence of certain bacteria that convert it to PDO.

PDO (16.34)   Terephthalic acid   Poly(trimethylene terephthalate), 3GT (16.35)

Hytrel is a thermoplastic polyester and elastomer, TPC-ET, containing 20%–60% polyol, Cerenol, derived from corn. Cerenol is synthesized from the self-reaction of PDO with itself and includes a variety of polyols of different chain lengths with molecular weights from about 500 to 2000.

Below, 16.36, shows the formation of a trimethylene ether glycol from reaction with three molecules of PDO. This diol is then reacted in the usual fashion giving polyesters with the expected hard (terephthalic derived moiety) and soft (polyol derived moiety) segments.

(16.36)

There are a variety of blends that are using various biomaterials including celluloses and proteins that are also being investigated and employed.

Thus, there exist a number of possible green polymeric materials that are possible for commercialization. More will be developed.

16.19   SUMMARY

1. The processing of polymeric materials is well advanced in science, engineering, and technology and continues to advance. The properties of the end product are dependent on the chemical structure, processing, and physical structure. The properties depend on the meso-structure of the particular polymer.

2. Fibers are processed using mainly one of three processes: melt spinning where the polymer is melted and forced through spinnerets, dry spinning where the dissolved polymer is forced through a spinneret and the solvent removed, and wet spinning where the dissolved polymer is forced through a spinneret into a nonsolvent bath. Fibers are usually stretched as they are formed allowing the chains to further align giving added strength in the direction of the pull.

3. Elastomers are processed using four basic steps: mastication, incorporation or compounding, shaping, and vulcanization.

4. Film and sheets are made from precipitation and/or regeneration from a polymer solution or melt. Calendering is also used where the polymer is passed between a series of counter-rotating rollers.

5. Unicellular (closed cell, discrete unit cells) and multicellular (open cells, interconnecting cells) foams are made through the chemical and physical introduction of gasses.

6. A wide variety of composites are made. Composites have two phases, the continuous or matrix phase that surrounds the discontinuous or dispersed phase, usually a fiber material. Composites are strong with many uses.

7. Molding employs a mobile prepolymer that may be thermoset or using a thermoplastic polymer. The polymer can be injection molded (often for solid objects), blow molded (for hollow objects such as bottles), rotational molded, compression molded, transfer molded, and thermoformed. Casting is closely related to molding except pressure is typically not used.

8. In extrusion, a fluid material is forced through a shaping device. The melts are extruded as pipes, sheets, films, or into molds.

9. Coatings and paints are used for decoration and protection. There are many different kinds of coatings with the majority being latex based where the polymer is present in a suspension that forms a film when the water evaporates. Industrially, coatings are applied using roll coating, blade coating, and curtain coating processes.

10. The difference between coatings and adhesives is that coatings must adhere on only a single surface, but adhesives must bind together two surfaces. There is a wide variety of adhesives that make use of the various polymer properties. Adhesion can occur through physical locking together, through chemical (primary and secondary bond formation) joining, or through simply highly viscous material holding together materials.

11. Polymers are major materials in the nanotechnology revolution including as conductive (photo and electronic) materials. Delocalization of electrons throughout a polymer chain or matrix is important for electronic conductance. This is often accomplished through doping that encourages flow of electrons.
12. Polymers are also major materials in the biomedical areas as materials and in the delivery of drugs.
13. For new polymeric materials to enter the marketplace, many things must be in place including a "flagship" property that meets a particular market need, ready availability, and money. It takes about $1 billion to introduce and establish a new polymer into the marketplace.
14. There is an increased awareness and much activity involving the development of green materials and green practices. This awareness involves public awareness, increased petroleum cost, recognition that the petroleum reserves are limited, and increased ability to employ microorganisms to synthesize needed materials. This awareness also involves realizing that we need to take care of our environment with us being responsible caretakers of the world in which we reside.

GLOSSARY

Abaca: Hemp-like fiber.
Acrilan: Polyacrylonitrile-based fibers.
Acrylic fibers: Polyacrylonitrile-based fibers.
Adhesive: Material that binds, holding together two surfaces.
Biaxial orientation: Process where a material, normally a film, is stretched in two directions at right angles to each other.
Buna-N: Acrylonitrile- butadiene copolymer.
Calender: Machine for making polymeric sheets using counterrotating rolls.
Casting: Production of film by evaporation of a polymeric solution.
Cellophane: Regenerated cellulosic film.
Charge: Amount of polymer used in each molding or processing cycle.
Coextruded film: Film produced by the simultaneous extrusion of two or more polymers.
Dacron: Trade name for PET fiber.
Draw: Depth of mold cavity.
Drying oils: Liquids employed in coatings that will be cured, cross-linked.
Dry spinning: Process for obtaining fiber by forcing a solution of a polymer through holes in a spinneret and evaporating the solvent from the extruded material (extrudate).
Elastomer: Rubber.
Electrodeposition: Use of an electric charge to deposit polymer film or aqueous dispersions onto a metal substrate.
Engineering material: Material that can be machined, cut, drilled, sawed, etc.; must have sufficient dimensional stability to allow these actions to occur.
Extrusion: Fabrication process in which a heat-softened polymer is continually forced by a screw through a die.
Fibrillation: Process for producing fiber by heating and pulling twisted filmstrips.
Filament: Continuous thread.
Filament winding: Process in which filaments are dipped in a prepolymer, wound on a mandrel, and cured.
Gate: Thin sections of runner at the entrance of a mold cavity.
Green materials: Includes materials that do not have a negative impact on the environment; biomass-derived materials
Hemp: Fiber from plants of the nettle family.

Hevea rubber: Natural rubber; *Hevea brasiliensis*.
Hycar: Trade name for Buna-N elastomer.
Jute: Plant fiber used for making burlap.
Kodel: Trade name for PET fiber.
Lacquers: Polymer solutions to which pigments have been added.
Lamination: Plying up of sheets.
Latex: Stable dispersion of a polymer in water.
Mechanical goods: Generally industrial rubber products like belts.
Melt spinning: Process of obtaining fibers by forcing molten polymer through holes in a spinneret and cooling the filaments.
Molding powder or compound: Premix of resin and other additives used as a molding resin.
Multicellular: Open celled.
Neoprene: Trade name for polychloroprene.
Nonwoven textiles: Sheet produced by binding fibers with a heated thermoplastic.
Oil paints: Suspension of pigments in a drying oil.
Oil varnish: Polymer dissolved in a drying oil.
Parison: Short plastic tube that is heated and expanded by air in the blow-molding process.
Perlon: Trade name for some polyurethane fibers.
Photoconductive: Material that is conductive when exposed to light.
Photoresponsive: Material whose properties change when exposed to light.
Pigment: Coloring material; colorant.
Polyacetylene: Polymer whose conductivity increases when doped to be a conductor.
Printing ink: Highly pigmented coatings used in printing.
Pultrusion: Process in which filaments are dipped in a prepolymer, passed through a die, and cured.
Rotational molding: Polymer added to a warm, rotating mold; centrifugal force distributes the polymer evenly.
Rovings: Multiple untwisted strands of filaments.
Runner: Channel between the spruce and the mold cavity.
Screen pack: Metal screen that prevents foreign material form reaching the die in an extruder.
Specific strength: Strength based on mass rather than area.
Spinneret: Metal plate between the nozzle and the runner.
Sprue: Tapered orifice between nozzle and runner. Term also used for plastic material in the sprue.
Structural foams: Polymeric foamed article with a dense surface.
Styrofoam: Trade name for foamed polystyrene.
Technology: Applied science.
Tenacity: Fiber strength.
Thermoforming: Shaping of a hot thermoplastic sheet.
Transfer molding: Process in which a preheated briquette or preform is forced through an orifice into a heated mold cavity.
Vehicle: Liquid in a coating.
Wet spinning: Obtaining fibers by precipitation of polymeric solutions.

EXERCISES

1. Which is more important: (a) polymer science or (b) polymer technology?
2. Name three important natural fibers.
3. Name an important regenerated fiber.
4. Why is secondary cellulose acetate more widely used than the tertiary cellulose acetate?
5. What is the difference between rayon and cellophane?
6. Name three important synthetic fibers.

7. Name an elastomer produced by (a) cationic, (b) anionic, (c) free-radical, and (d) step-reaction polymerization techniques.
8. How is LDPE film produced?
9. Why is there a trend toward the use of less solvent in polymeric coatings?
10. What is meant by trade sales?
11. What are general steps needed before a new drug comes to market.
12. How would you produce a unicellular foam?
13. Which foam is preferable for upholstery: (a) unicellular or (b) multicellular?
14. Why do non-flame-retardant foams burn readily?
15. Why has reinforced plastic been used as an automobile body?
16. Why is graphite-reinforced epoxy resin a good candidate for parts in future automobiles?
17. Why are molded thermoplastics used more than molded thermosets?
18. Which of the following might you expect would increase their conductivity when doped: (a) PS, (b) PPO, (c) nylon-66, (d), aramids?
19. Why are structural foams used for complex furniture?
20. What are some of the advantages of a blow-molded PET bottle?
21. Why would an article be thermoformed instead of molded?
22. What is the limit to the length of an extrudate such as PVC pipe?
23. Name three popular laminates.
24. Why are the terms painting and coating often used interchangeably?
25. Why have latex waterborne coatings been popular with the general public?
26. Differentiate between oil paints, oil varnishes, latexes, enamels, and lacquers.
27. Briefly discuss the popular mechanisms related to adhesion and the general types of adhesives.
28. Why is there an industrial move toward so-called green material?
29. Why is there such a broad definition as to what green materials mean?
30. Name three polymeric green materials.

FURTHER READINGS

Aldissi, M. (1992): *Intrinsically Conducting Polymers*, Kluwer, Dordrecht, the Netherlands.
Bachmann, K. J. (1995): *Materials Science of Microelectronics*, VCH, New York.
Baird, D. and Collias, D. (1998): *Polymer Processing Principles and Design*, Wiley, New York.
Barford, W. (2009): *Electronic and Optical Properties of Conjugated Polymers*, Oxford University Press, Ithaca, NY.
Belcher, S. (1999): *Practical Extrusion Blow Molding*, Dekker, New York.
Bhattacharya, S., Kamal, M., and Gupta, R. (2007): *Polymeric Nanocomposites: Theory and Practice*, Hanser-Gardner, Cincinnati, OH.
Bieleman, J. (2000): *Additives for Coatings*, Wiley, New York.
Boyd, R. (2008): *Nonlinear Optics*, Elsevier, New York.
Bundy, K. (2007): *Fundamentals and Biomaterials: Science and Applications*, Springer, New York.
Cahn, R., Haasen, P., and Kramer, E. (1996): *Processing of Polymers*, VCH, New York.
Carraher, S., Carraher, C. ISO process and practices in a series of articles appearing in *Polymer News* 19:373; 20:147; 20:278; 21:21; 21:167: 22:16.
Celina, M. and Assink, R. (2007): *Polymer Durability and Radiation Effects*, Oxford University Press, Ithaca, NY.
Chanda, M. and Roy, S. (2008): *Plastics Fundamentals, Properties, and Testing*, CRC Press, Boca Raton, FL.
Dadmun, M. (2008): *Modification of Interfaces with Polymers*, Wiley, Hoboken, NJ.
Ericksen, J. (2008): *Orienting Polymers*, Springer, New York.
Freund, M. (2007): *Self-Doped Conducting Polymers*, Wiley, Hoboken, NJ.
Fried, J. R. (1995): *Polymer Science and Technology*, Prentice Hall, Upper Saddle River, NJ.
Galaev, I. and Mattiasson, B. (2007): *Smart Polymers: Applications in Biotechnology and Biomedicine*, CRC Press, Boca Raton, FL.
Grulke, E. (1993): *Introduction to Polymer Processing*, Prentice Hall, Englewood Cliffs, NJ.
Hearle, J. and Morton, W. (2008): *Physical Properties of Textile Fibres*, CRC Press, Boca Raton, FL.
Inzelt, G. (2008): *Conductin Polymers*, Springer, New York.

Jenkins, M. (2007): *Biomedical Polymers*, CRC Press, Boca Raton, FL.

Jimenez, A. and Zaikov, G. (2008): *Recent Advances in Research on Biodegradable Polymers and Sustainable Composites*, Nova, Hauppauge, NY.

Johnson, P. (2001): *Rubber Processing*, Hanser, Gardner, Cincinnati, OH.

Kestleman, V (2001): *Adhesion of Polymers*, McGraw-Hill, New York.

King, E. (2008): *Shape-Memory Polymers: Industrial Applications of Smart Materials*, Wiley, Hoboken, NJ.

Marder, S. and Lee, K. (2008): *Photoresponsive Polymers*, Springer, New York.

Martin, J., Dickie, R., Ryntz, R., and Chin, J. (2008): *Service Life Prediction of Polymeric Materials: Global Perspectives*, Springer, New York.

O'Brian, K. T. (1999): *Applications of Computer Modeling for Extrusion and Other Continuous Polymer Processes*, Oxford University Press, Oxford, U.K.

Olmsted, B. (2001): *Practical Injection Molding*, Dekker, New York.

Piringer, O. and Baner, A. (2008): *Plastic Packaging Materials: For Food and Pharmaceuticals*, Wiley, Hoboken, NJ.

Reyne, M. (2008): *Plastic Forming Processes*, Wiley, Hoboken, NJ.

Rockett, A. (2008): *The Materials Science of Semiconductors*, Springer, New York.

Rudin, A. (1998): *Elements of Polymer Science and Technology*, 2nd ed., Academic Press, Orlando, FL.

Schnabel, W. (2007): *Polymers and Light: Fundamentals and Technical Applications*, Wiley, Hoboken, NJ.

Seymour, R. (1975): *Modern Plastics Technology*, Reston Publishing Co., Reston, VA.

Shibaev, V. and Lam, L. (1993): *Liquid Crystalline and Mesomorphic Polymers*, Springer-Verlag, New York.

Sinha, R. (2002): *Outlines of Polymer Technology: Manufacture of Polymers and Processing Polymers*, Prentice Hall, Englewood Cliffs, NJ.

Sudarshan, T. and Dahorte, N. (1999): *High-Temperature Coatings*, Dekker, New York.

Vollath, D. (2008): *Nanomaterials: An Introduction to Synthesis, Properties, and Applications*, Wiley, Hoboken, NJ.

Wallace, G. and Spinks, G. (1996): *Conductive Electroactive Polymers,* Technomic, Lancaster, PA.

Wicks, Z., Jones, F., and Papas, S. P. (1999): *Organic Coatings Science and Technology*, Wiley, New York.

Williams, P. (2007): *Handbook of Industrial Water Soluble Polymers*, Wiley, Hoboken, NJ.

Appendix A: Structures of Common Polymers

Acrylonitrile–butadiene–styrene (ABS)

Butyl rubber

Ethylene–methacrylic acid copolymer (ionomer)

Ethylene–propylene elastomer

Melamine–formaldehyde resin (MF)

Nitrile rubber (NRB)

Phenol–formaldehyde resin (PF)

Polyacetaldehyde

Polyacrolein

483

Polyacrylamide

Poly(acrylic anhydride)

Polyacrylonitrile

Poly(β-alanine); nylon-3

Polyallene

Polybutadiene, butadiene
rubber (BR)

1,2-Polybutadiene

Poly(butylene terephthalate (PBT))

Poly(n-butyraldehyde)

Polychloroprene

Poly(2,3-
dimethylbutadiene)

Poly(3,5-dimethyl-1,4-
phenylene sulfonate)

Polydimethylsiloxane

Polyethylene (PE)

Poly(ethylene glycol)
(PEG)

Poly(ethylene terephthalamide)

Poly(ethylene terephthalate)

(PET) Poly(glycolic ester)

Poly(hexamethylene adipamide) (nylon-66)

Poly(hexamethylene sebacamide) (nylon-610)

Poly(hexamethylene thioether)

Poly(hexamethylene urea)

Polyisobutylene (PIB)

Polyisoprene

Poly-3,4-isoprene

Poly-*trans*-1,4-isoprene

Poly(methyl acrylate)

Poly(methyl methacrylate)

Poly(methyl vinyl ketone)

Polyoxymethylene Polyacetal

Poly(1,4-phenylene adipate)

Poly(phenylene oxide) (PPO) Poly(2,6-dimethyl-*p*-phenylene ether)

Poly(phenylene sulfide) (PPS)

Polypropylene (PP)

Poly(propylene glycol) (PPG)

Polystyrene

Polytetrafluoroethylene (PTFE)

Poly(vinyl butyral)

Poly(vinyl chloride) (PVC)

Poly(vinyl acetate) (PVAc)

Poly(vinyl alcohol) (PVA)

Poly(vinyl formal) (PVF)

Poly(vinylidene chloride)

Poly(vinyl pyridine)

Poly(vinyl pyrrolidone)

Urea–formaldehyde resin (UF)

Styrene–acrylonitrile (SAN)

Styrene–butadiene rubber (SBR)

Appendix B: Syllabus

Three interrelated questions can be addressed when considering the construction of a course. These questions are (1) topics to be covered, (2) order in which these topics should be covered, and (3) proportion of time to be spent on each topic. Just as in any other area of science and engineering, there exists a healthy variety of topics, the extent of coverage of each topic, and the order of covering the material. There is no "right" answer, and this text is developed so that chapters can be dealt with in essentially any order with the most important topics dealt with near the beginning of most chapters and (possibly) less important material covered later in the chapter.

Polymers are expanding at a rapid rate with too much fundamental material to be handed in a single introductory course, yet the basic elements can be included in such a course and are included in this book. Some topics that are today considered to be fundamental were not known a decade ago. Each of the fundamental topics is placed into perspective in the current book, building upon the core courses of chemistry—organic, physical, inorganic, and analytical chemistry.

One assumption agreed upon by most academic and industrial polymer scientists and engineers and associated education committees is that there should be both a core of material common to introductory courses and a portion that reflects individual interests and training of teachers, student bodies, and local preferences and circumstances. Thus, not every topic needs to be covered to present a meaningful introductory polymer course. Some years ago, PolyEd, the education arm of Polymer Chemistry and Polymeric Materials: Science and Engineering, developed, with the help of polymer scientists and engineers, a listing of basic topics and preferred coverage. The results are given in Table B.1.

Basically, the committee proposed that all lecture courses include portions of the first seven topics with the level and extent of coverage guided by factors such as available class time, additional topics covered as per interest of instructor, student interests, class composition, etc. It must be emphasized that the "optional topics" listed should not be considered limiting and that additional topics can be introduced.

TABLE B.1
Preferred Topics in Introductory Polymers

Topics	Amount of Course Time (%)	Chapters
Introduction	5	1
Morphology	10	2
Stereochemistry		
Molecular interactions		
Crystallinity/amorphousity		
Molecular weights	10	3
General types		
Solubility		
Testing and characterization	10	12, 13
Structure/property relationships		
Physical tests		
Spectral identification		

(continued)

TABLE B.1 (continued)
Preferred Topics in Introductory Polymers

Topics	Amount of Course Time (%)	Chapters
Stepwise polymerizations and condensation polymers	10	5
Chain-reaction polymerizations and addition polymers	10	6, 7
Ionic and free-radical kinetics of polymerization		
Polymers produced by chain-reaction polymerization		
Copolymerization	10	8
Kinetics		
Types of copolymers		
Blends		
Principal copolymers		
Optional topics	35	
Natural and biomedical polymers		4
Organometallic polymers		9
Inorganic polymers		10
Reactions of polymers		11
Rheology (flow properties, viscoelasticity)		13
Additives		14
Synthesis of polymer reactants		15
Polymer technology		16

Appendix C: American Chemical Society Committee on Professional Training Requirements

The American Chemical Society Committee on Professional Training in their latest set of requirements (2008) state that "students should be exposed to the principles of macromolecules across foundational areas." This is ideally fulfilled through integration of polymer topics within the foundational courses. It can be enhanced with a course in polymers using this book. This book is written so that a polymer course based on the materials found in it act toward fulfilling the in-depth course requirements.

Appendix D: Comments on Health

Most polymers are nontoxic for normal and intended uses. (Some biopolymers, such as snake venom, should not be dealt with except under strictly controlled conditions.) Most of the additives employed are also relatively nontoxic. Even so, care should be exercised when dealing with many of the monomers of synthetic polymers, and polymeric materials under extreme conditions, such as in commercial and domestic fires.

FIRE

Fire hazards involve not only burning but also the ingestion of volatiles produced by the fire, which causes most deaths. Carbon monoxide, the major cause of death, causes unconsciousness in less than 3 min due to its preferential attack on hemoglobin.

Interestingly, one important observation with regard to burning, in general, is the production of colored smoke. Some materials produce a lot of dark-colored smoke when burnt. Some of this colored smoke may be due to the production of aromatic systems including fused ring systems, which may contain harmful chemicals including respiratory toxins and cancer-causing agents.

MEASURES OF TOXICITY

Toxicity involves the effects of various materials on living objects, including bacteria, plants, mice, fish, and humans. Tests to determine the toxicity of materials are typically performed in a number of ways, including inhalation, simple skin contact, and injection under the skin.

Although mainly concerned with the effects of various agents on humans, most standard tests are carried out on animals; often used are suitable test animals that are believed to be good models for transferring the results from these tests to humans. Table D.1 contains the description of some of the toxicity measures found in today's literature.

While commercially available synthetic polymers are relatively nontoxic, monomers vary greatly in toxicity. This points out the need for the removal of monomers and other potentially toxic chemicals from the polymers. Table D.2 contains the time–weight average (TWA) for some monomers, as cited by the United States Occupational Standards. For comparison, entries for some well-known toxic materials have been added.

CUMULATIVE EFFECTS

While exposure of the general public to toxins is to be avoided, exposure of people that deal with commercial chemicals on a daily basis is even more important. Such people must take special care to avoid exposure since the toxicities of many of these chemicals accumulate in our bodies, slowly building to levels that may be unhealthy.

Most of the toxic, environmentally unwanted chemicals used a decade ago have been eliminated from the common workplace. These include halogenated hydrocarbons, such as carbon tetrachloride and aromatic hydrocarbons like benzene and toluene. Further, chemicals that are known to be potentially toxic, such as some monomers, are being eliminated from polymeric materials, within the limits of detection.

Industrial recognition of customer and employee safety is a major factor and is included in international programs such as ISO 9000 and ISO 14000. Further, a number of national and state agencies and associations deal with the aspects of environmental and personal safety issues on an ongoing basis.

TABLE D.1
Description of Toxic Measures

Toxic dose low (TDLo)	The lowest dose introduced by any route other than inhalation over any period of time that produces any toxic effect in humans or carcinogenic, teratogenic, mutagenic, or neoplastic effects in humans and animals
Toxic concentration low (TCLo)	Any concentration in air that produces any toxic effect in humans or carcinogenic, teratogenic, mutagenic, or neoplastic effects in humans or animals
Lethal dose low (LDLo)	The lowest dose introduced by any route other than inhalation over a time that can cause death of humans, or the lowest single dose that can cause death of animals
Lethal dose fifty (LD_{50})	A calculated dose expected to cause the death of 50% of a tested population from exposure by any route other than inhalation
Lethal concentration low (LCLo)	The lowest concentration in air that can cause death of humans or animals when exposed for 24 h or less
Lethal concentration fifty (LC_{50})	A calculated concentration of a substance in air that can cause the death of 50% of a test population from exposure for 24 h or less
Emergency exposure guideline level (EEGL)	Exposure limits for a very short exposure
Workplace environmental exposure level (WEEL)	Exposure limits for healthy workers exposed repeatedly without adverse health effects
Occupational exposure limits (OEL)	Worker exposure guide
Permissible exposure limits (PREL)	Worker exposure limits for no ill effects
No observable adverse effect level (NOAEL)	Safe usage level
Lowest observable adverse effect level (LOAEL)	Safe usage level

TABLE D.2
TWA Values for Selected Monomers and Additional Recognized Toxins

Chemical	TWA (ppm)	Chemical	TWA (ppm)
Acetic anhydride	5	Ethylene oxide	50
Acrylonitrile	20	Formaldehyde	3
Benzene	10	Hydrazine	1
1,3-Butadiene	1000	Hydrogen cyanide	10
Carbon monoxide	50	Phenol	5
Chloroprene	25	Styrene	100
1,2-Ethylenediamine	10	Vinyl chloride	500

ENVIRONMENT

Today, industry and business recognize that taking care of the environment is part of doing business. Along with various agencies, industry and business are finding that good environmental practices constitute good business. Advances continue with respect to lowering potentially harmful emissions into water and air shared by all of us. Chemical industries are taking the lead in this cleanup process. A combination of watchful vigilance and trust is needed to continue this effort.

Appendix E: ISO 9000 and 14000

The International Organization for Standardization (ISO) has members from about 100 countries working to develop common global standards.

The ISO 9000 series encompasses the product development sequence from strategic planning to customer service. Currently, it is a series of five quality system standards with two of the standards focusing on guidance and three on contractual standards.

ISO 9000 certification is often obtained to promote a company's perceived quality level, for supplier control, and to promote certain management practices, such as total quality management (TQM). It acts as a global standardizing tool in its broadest sense with respect to business and industry, including banking, volunteer organizations, and most aspects of the chemical (including the polymer) industry.

ISO 9000 requires what is called a "third-party" assessment but involves the development of "first-" and "second-party" strategies. "First party" refers to the supplier company that requests ISO 9000 certification. "Second party" refers to the customer whose "needs" have been met by the "first party" through the use of quality management procedures achieved through ISO 9000 compliance. "Third party" refers to an outside reviewer who certifies that the "first party" has satisfied ISO 9000 requirements.

While ISO 9000 is a management tool, it affects the way "industry does business" and deals with quality control issues such as how machinery and parts manufactured by a company are monitored for quality. It focuses on satisfying the "customer," the "purchaser" of raw materials, manufactured parts, or assembled items, and includes the eventual "end-customer"—the general public. It is an attempt to assure quality goods.

ISO 14000 is a series of standards intended to assist in managing the impact of manufactured materials including finished products and original "feedstocks." It addresses the need to have one internationally accepted environmental management system that involves "cradle to grave" responsibility for manufactured materials emphasizing the impact of products, operations, and services on the environment.

Appendix F: Symbols

A	Arrhenius constant
ABS	copolymer of acrylonitrile, butadiene, and styrene
AIBN	2,2′-azobisisobutyronitrile
ANSI	American National Standards Institute
ASTM	American Society for Testing Materials
ATR	attenuated total reflectance spectroscopy
AU	polyurethane
BPO	benzoyl peroxide
BSI	British Standards Institute
bp	boiling point
CA	cellulose acetate
CAB	cellulose acetate butyrate
CED	cohesive energy density
CMC	carboxymethylcellulose
CN	cellulose nitrate
CR	neoprene
C_p	specific heat
C_s	chain transfer constant
DNA	deoxyribonucleic acid
DP	degree of polymerization
DRS	dynamic reflectance spectroscopy
DS	degree of substitution
DSC	differential scanning calorimetry
DTA	differential thermal analysis
E_a	activation energy or energy of activation
E	Young's modulus
EGG	Einstein–Guth–Gold equation
EP	epoxy resin
EPM	poly(ethylene-co-propylene)
EPDM	poly(ethylene-co-propylene) cross-linked
EPR	electron paramagnetic resonance spectroscopy
ESR	electron spin resonance spectroscopy
ET	Thiokol
ETA	electrothermal analysis
f	aspect ratio
G	Gibbs free energy; modulus; molar attraction constant
GPC	gel permeation chromatography
GRS	poly(butadiene-co-styrene)
HDPE	high-density polyethylene
HIPS	high-impact polystyrene
IIR	butyl rubber
IR	infrared
ISO	International Standards Organization
IUPAC	International Union of Pure and Applied Chemistry
K	constant in Mark–Houwink equation; Kelvin

LDPE	low-density polyethylene
LLDPE	linear, low-density polyethylene
MF	melamine–formaldehyde resin
MWD	molecular weight distribution
$\bar{M}_n$	number-average molecular weight
$\bar{M}_v$	viscosity-average molecular weight
$\bar{M}_w$	weight-average molecular weight
$\bar{M}_z$	z-average molecular weight
NBR	poly(butadiene-*co*-acrylonitrile)
NBS	National Bureau of Standards
NMR	nuclear magnetic resonance spectroscopy
NR	natural rubber
OI	oxygen index
PA	polyamide
PBI	polybenzimidazole
PBT	poly(butylene terephthalate)
PC	polycarbonate
PCB	polychlorinated biphenyls
PE	Polyethylene
PEEK	poly(ether ether ketone)
PET	poly(ethylene terephthalate)
PF	phenol–formaldehyde resin
PGC	pyrolysis gas chromatography
PMMA	poly(methyl methacrylate)
PMR	proton magnetic resonance spectroscopy
PNF	poly(phosphonitrilic fluorides)
PolyEd	Polymer Education Committee
POM	polyoxymethylene, polyformaldehyde, acetals
PP	Polypropylene
PPO	poly(phenylene oxide)
PPS	poly(phenylene sulfide)
PS	Polystyrene
PTFE	polytetrafluoroethylene, Teflon
PU	polyurethane
PVA	poly(vinyl alcohol)
PVAc	poly(vinyl acetate)
PVB	poly(vinyl butyral)
PVC	poly(vinyl chloride)
PVDC	poly(vinylidene chloride)
RIM	reaction injection molding
RNA	ribonucleic acid
S	entropy; radius of gyration
SAM	poly(styrene-*co*-acrylonitrile)
SBR	poly(butadiene-*co*-styrene) elastomer
SEM	scanning electron microscopy
SI	silicon
SMA	poly(styrene-*co*-maleic anhydride)
SMC	sheet-molding compound
SN	sulfur nitride

SPE   Society of Plastics Engineering
SPI   Society of the Plastics Industry
SR   synthetic rubber
TDI   toluenediisocyanate
TGA, TG   thermal gravimetric analysis
TMMV   threshold molecular weight value
TPE   thermoplastic elastomer
TPX   poly-4-methylpentene
T_c ceiling temperature; cloud point temperature
T_g glass transition temperature
T_m melting point
UF   urea–formaldehyde resin
UHMWPE   ultrahigh molecular weight polyethylene
UV   ultraviolet
WLF   Williams–Landel–Ferry equation
WS   polyurethane

Appendix G: Web Information

The information concerning polymers on the World Wide Web is rapidly expanding. This is a valuable source of information, giving both applied and fundamental information on a wide range of polymer-related topics. As you search, please be aware that there exist specific pooled information sites on many topics including those considered "hot" topics such as the human genome. There are also Web clusters that deal with special topics such as nanomaterials and electrically conductive materials. Have fun "surfing the Web." It is an important source of information on polymers. Wikipedia is also a great place to start with for general information about polymers and polymeric materials.

Sites that you should consider visiting and that are not obvious because of their names are as follows:

www.polyed.org: This is a general site for PolyEd, the joint polymer education committee of the American Chemical Society divisions of Polymer Chemistry and Polymeric Materials: Science and Engineering. It contains many connections to other important Web sites dealing with polymers.

www.uwsp.edu/chemistry/ipec: This is a general site for the Intersociety Polymer Educational Council, which is a joint society venture that focuses on K-12 science education employing polymers as the connective vehicle. The Polymer Ambassadors have their own site at www.polymerambassadors.org.

www.pslc.ws/macrog/index.htm: You can also simply type in "macrogalleria" and you will be taken to a fun and educational Web site that focuses on the relationship between everyday items and their polymeric nature, and the fundamentals that underlie their use in these materials.

Index